D1503350

Synthetic Membrane Processes
FUNDAMENTALS AND WATER APPLICATIONS

WATER POLLUTION

A Series of Monographs

D. T. O'Laoghaire and D. M. Himmelblau. *Optimal Expansion of a Water Resources System*. 1974

C. W. Hart, Jr., and Samuel L. H. Fuller (eds.). *Pollution Ecology of Freshwater Invertebrates*. 1974

H. Shuval (ed.). *Water Renovation and Reuse*. 1977

C. W. Hart, Jr., and Samuel L. H. Fuller (eds.). *Pollution Ecology of Estuarine Invertebrates*. 1979

H. Shuval (ed.). *Water Quality Management under Conditions of Scarcity: Israel as a Case Study*. 1980

S. Mandel and Z. L. Shiftan. *Groundwater Resources: Investigation and Development*. 1981

Georges Belfort (ed.). *Synthetic Membrane Processes: Fundamentals and Water Applications*. 1984

Synthetic Membrane Processes

FUNDAMENTALS
AND WATER APPLICATIONS

Edited by

GEORGES BELFORT

Department of Chemical Engineering and Environmental Engineering
Rensselaer Polytechnic Institute
Troy, New York

1984

ACADEMIC PRESS, INC.

(Harcourt Brace Jovanovich, Publishers)

Orlando San Diego San Francisco New York London
Toronto Montreal Sydney Tokyo São Paulo

ACADEMIC PRESS, INC.
Orlando, Florida 32887

United Kingdom Edition published by
ACADEMIC PRESS, INC. (LONDON) LTD.
24/28 Oval Road, London NW1 7DX

Library of Congress Cataloging in Publication Data

Main entry under title:

Synthetic membrane processes.

(Water pollution)
Includes bibliographical references and index.
1. Saline water conversion--Reverse osmosis process. 2. Saline water conversion--Electrodialysis process. 3. Membranes (Technology) I. Belfort, Georges. II. Series.
TD480.4.S94 1983 628.1'64 83-2654
ISBN 0-12-085480-5

PRINTED IN THE UNITED STATES OF AMERICA

84 85 86 87 9 8 7 6 5 4 3 2 1

TO: DAVID, GABRIEL, JONATHAN, AND MARLENE

Contents

Contributors

Numbers in parentheses indicate the pages on which the authors' contributions begin.

Nathan Arad (479), Engineering Division, Mekorot Water Company, Ltd., Tel Aviv, Israel

David G. Argo (377), Orange County Water District, Fountain Valley, California 92708

Georges Belfort (1, 131, 221), Department of Chemical and Environmental Engineering, Rensselaer Polytechnic Institute, Troy, New York

C. E. Boesen (101), Instituttet for Kemiindustri, The Technical University of Denmark, DK-2800 Lynby, Denmark

Pinhas Glueckstern (479), Mekorot Water Company, 8, Tel Aviv, Israel

G. Jonsson (101), Instituttet for Kemiindustri, The Technical University of Denmark, DK-2800 Lynby, Denmark

C. Kleinstreuer (131), Department of Chemical and Environmental Engineering, Rensselaer Polytechnic Institute, Troy, New York

E. Korngold (191), Division of Membranes and Ion Exchangers, Applied Research Institute, Research and Development Authority, Ben-Gurion University of the Negev, Beer-Sheva 84110, Israel

W. A. P. Luck (21), Fachbereich Physikalische Chemie, Philipps Universität Marburg, Federal Republic of Germany

I. Nusbaum (377), San Diego, California 92115

Alfred N. Rogers[1] (437, 509), Research and Engineering, Bechtel Group, Inc., San Francisco, California 94119

David C. Sammon (73), Chemistry Division, AERE Harwell, Didcot, Oxon, United Kingdom

Eberhard Staude (281), Institut für Technische Chemie, Fachbereich Chemie, Universität Essen, D-4300 Essen, Federal Republic of Germany

H. Strathmann (343), Membrane and Biotechnology, Fraunhofer-Institut für Grenzflächen und Bioverfahrenstechnik (IGB), 7000 Stuttgart 80, Federal Republic of Germany

[1]Present address: Engineering Consultant, Pleasanton, CA 94566.

Preface

The purpose of this book is to present a coherent summary of some of the latest theoretical developments in membrane and fluid transport and to review water and wastewater hyperfiltration, ultrafiltration, and electrodialysis. The economics of these processes is also covered. The topic is generally approached from the standpoint of chemical engineering. The contributors are all highly regarded in their respective fields. I have attempted to impose some uniformity, thereby, it is hoped, making the book easier to read.

The book is aimed at those in the water and wastewater field but should also be generally useful for teaching and for anyone interested in adapting membrane technology to new separation or concentration applications. In the latter regard the description of procedures for maintaining reasonable fluxes with a balanced pretreatment, cleaning, and fluid management program could be invaluable. Biotechnology is an example of a field to which membrane technology is expected to make a major contribution; much of the experience in membrane technology is easily translated to biotechnology; especially in regard to the operation and choice of an appropriate module for a particular separation.

Several omissions have purposely been made. For example, in keeping with the engineering approach, many of the modern developments in the chemical and morphological structure of membranes have been left out. For such topics the reader is referred to recent reviews in the membrane literature.

My deep appreciation goes to Sam Spiegler for his support and encouragement. I also thank all the authors for eventually bending to my pleas to submit, update, or shorten their chapters. Without their contributions this text would not have become a reality during the long hard climb to publication. It has also been a pleasure to deal with the staff of Academic Press.

1

Membrane Methods in Water and Wastewater Treatment: An Overview

GEORGES BELFORT
Rensselaer Polytechnic Institute
Troy, New York

I. Introduction

A. NEW WATER SOURCES AND POLLUTION REDUCTION

Several semiarid regions in the world, including areas in the Middle East, South Africa, and the Southwest United States, are actively searching for supplementary sources of water to help fulfill future demands. This quest includes new unconventional sources of water such as the renovation and

ISBN 0-12-085480-5

reuse of wastewater (Shuval, 1977) and desalination of brackish and seawater (Spiegler and Laird, 1980).

New technological developments necessary for producing greater quantities of water and protecting the quality of various waters have begun to appear, particularly since the 1960s. Several new unit processes for water and wastewater treatment have been developed. These include the membrane processes as a group, which can be divided into pressure-driven [reverse osmosis (RO) and ultrafiltration (UF)] and electrically driven [electrodialysis (ED) and transport-depletion] processes. The membrane separation processes are thought to be especially useful in water renovation because they allow separation of dissolved materials from one another or from a solvent, with no phase change.

B. WHY MEMBRANE PROCESSES?

Several questions arise regarding membrane processes. Why and how were these processes developed in the first place? Why do we think they will play an important and unique role in the future with respect to water and wastewater treatment? What are the advantages and disadvantages of these processes within the spectrum of available and new unit processes? Finally, what is the state of the art both fundamentally and in application of these membrane processes with respect to water and wastewater treatment? The first and second questions will be discussed below, while the answers to the last two questions are covered in detail in the chapters that follow.

From the outset, it should be made clear that widescale acceptance and usage of membrane processes for water and wastewater treatment is a recent development. Although several large commercial applications exist or are in the planning stage, much of the data discussed in this text are from small commercial applications and experimental pilot plants and small-scale research studies.

Reverse osmosis and electrodialysis (and, later, transport depletion) have been developed during the past 30 years for the purpose of removing salt from brackish and sea waters with a total dissolved solids concentration from about 1,000 to 35,000 ppm. Most or all of the dissolved solids in the brackish and sea feedwaters are inorganic (ionic) in nature, with low concentrations of dissolved organic species present. In the mid-sixties, various research laboratories involved in desalination research and development realized that these same processes could also be used in both municipal and industrial wastewater treatment as single elements in the train of unit processes for recycling and/or treatment of water prior to disposal or reuse. The application of membrane processes to wastewater treatment is about 20 years old. A classification of the membrane processes according to their type of driving potential and their general behavior with respect to feedwaters is presented in Table I. Figure 1 presents the removal range of particle sizes for various separation processes. Note that electrodialysis and reverse osmosis cover essentially the same particle removal size range whereas UF covers a particle size range of more than three orders of magnitude. The spectrum of substances to be removed from municipal, industrial, and wastewater streams can vary across the whole range of particle size, as shown in Fig. 1.

It was Reid's (1959) proposal to the U.S. Office of Saline Water in the late 1950s that provided the impetus to develop a "new" desalting process by reversing the osmotic flow through a permselective membrane. Immediately thereafter at UCLA, Loeb and Sourirajan (1962) developed a practical reverse osmosis or hyperfiltration (HF) asymmetric membrane with reasonably high water flux and excellent salt rejection. This was the major technological breakthrough that established HF as a viable, economically attractive process having many potential applications.

Ultrafiltration using asymmetric membranes is a modern outgrowth of HF in that the ability to tailor-make these membranes for specific

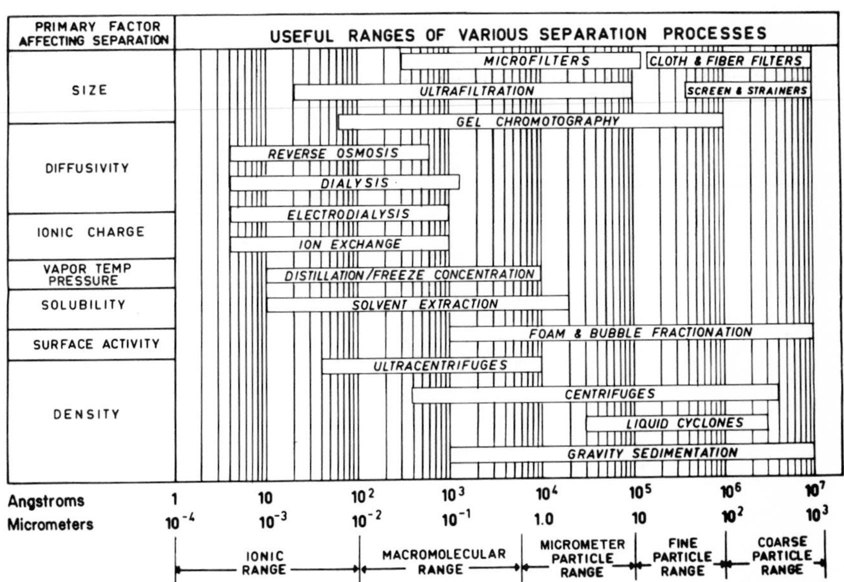

Fig. 1 Useful ranges of various separation processes. (Courtesy Dorr-Oliver, Inc., Stamford, Conn.)

TABLE I

Classification of Membrane Processes[a]

Process	Driving potential	Constituents removed from feedwater	Constituents remaining in product (other than water)	Possible size ranges of permeable species (Å)
Hyperfiltration (reverse osmosis)	Pressure (as high as 40 atm)	Water without dissolved and nondissolved inorganic and organic constituents	Little salt (owing to membrane leakage) BO_3^-, NO_3^-, urea, low MW organics[b]	4–300
Ultrafiltration	Pressure (usually below 10 atm)	Water without dissolved and nondissolved organic constituents	All the salt and low molecular weight organics	$20–10^5$
Electrodialysis	Electrical	Dissolved inorganic ions	Little salt, all the organics (dissolved and nondissolved) including viruses, bacteria, etc.	4–300
Transport depletion	Electrical	Dissolved inorganic ions	More than a little salt, all the organics (dissolved and nondissolved) including viruses, bacteria, etc.	10–1000

[a] The feed is assumed to contain a range of dissolved and nondissolved inorganic and organic constituents.
[b] With the Loeb–Sourirajan asymmetric cellulose acetate membrane.

applications [with different molecular weight (MW) cutoffs] resulted from major advances in HF membrane research. This modification of UF is thus a more recent process than either HF or ED.

Ultrafiltration, like RO, is a pressure-driven membrane process using permselective membranes. The distinguishing features of the two processes are that RO is a high-pressure (200–1500 psi) process with membranes capable of rejecting salt molecules, whereas UF is a low-pressure (5–100 psi) process with membranes only capable of retaining molecules with a MW of about 500 or higher.* Notice, however, from Fig. 1 that their particle size separation ranges overlap. This is because both RO and UF membranes can be "tailor-made" for rejection of larger or smaller particles. The upper MW cutoff for UF is about 300,000 to 500,000. Above this MW range another membrane process called microfiltration (MF), which allows the passage of solvent and most solute molecules but impedes the passage of large colloids and small particulate matter, is operative. The reason UF is a low-pressure process as compared with RO is because the large molecules being retained exert very little osmotic pressure. By their very nature, UF membranes are "looser" and able to pass much higher fluxes of product water than RO membranes. Thus UF membranes will not retain salt molecules or small low MW organic molecules.

Unlike the pressure-driven processes, ED employs an electric field to remove charged ionic species from the feed or dialysate stream. Anion and cation exchange membranes allow anions and cations, respectively, to pass selectively from the dilute dialysate to the concentrate brine solution. By stacking many cell pairs of membranes and streams between the electrodes and manifolding the different streams, ED was converted into a practical commercial large-scale process. Early developments in the 1950s at TNO in the Netherlands and a large-scale commercial mine-drainage reclamation application in South Africa (Wilson, 1960) essentially established the viability of the process. It should be noted that electrically driven processes have not proved as attractive as HF for treating effluents with substantial amounts of microbiological contaminents and dissolved organic compounds. Early work on ED treatment of municipal secondary effluents indicated that for adequate performance virtually all the dissolved organics had to be removed from the feed prior to treatment (Smith and Eisenman, 1964, 1967). Thus the chief function of the ED process is the removal of inorganic ions, which leave bacteria, viruses, and neutral organics in the dilute stream. This could become a serious problem when recycling for potable use.

* "Retain" and "reject" are synonymous terms, although they are used for UF and RO, respectively.

Although the transport-depletion process is very similar to the ED process, it is marked by two important differences: (1) the troublesome anion-exchange membranes used in ED are replaced by near-neutral membranes; (2) although conventional ED is a well established process with existing plants operating on a brackish feedwater capacity of more than one million gallons per day (mgd), transport depletion is still a small laboratory pilot plant curiosity without commercialization. In spite of this, there is some evidence that, where municipal effluents are concerned, transport depletion would perform better than ED (Lacey and Huffman, 1971). The objection with respect to bacteria, viruses, and neutral organics in the dilute stream also holds for the transport depletion (TD) process.

Other interesting membrane separation processes that are in the developmental stage but are not discussed in this text include dialysis, piezodialysis, Donnan dialysis, gas permeation, and pervaporation (Lacey, 1972).

Because multiple recycling results in a buildup of conservative constituents,* one very important question is whether the wastewater will be reused only once or will be recycled many times. Dissolved inorganic ions, refractory organics, viruses, and some bacteria are examples of conservative constituents for normal biological secondary treatment. The removal of these elements may become necessary if, on recycle and buildup, they become detrimental for the intended reuse (Shuval and Gruener, 1973; World Health Organization, 1973). It is with the purpose of removing these conservative constituents that several advanced treatment techniques are being developed. These include activated-carbon adsorption, ion exchange (IE), chemical precipitation, and clarification, and membrane processes.

In this text, we shall concern ourselves with the treatment by HF or RO, UF, and ED of various surface and groundwaters, municipal and industrial wastewaters, and polluted river waters. A general overview of these applications is presented in Section III, and details are presented in Chapters 7–11 of this text.

C. ECONOMICS OF MEMBRANE PROCESSES

In Table II, we compare the approximate removal efficiencies and costs for RO, ED and IE for treating carbon-treated secondary effluent (Garrison and Miele, 1977). These data were obtained from research ex-

* A conservative constituent is not removed from the water during treatment.

TABLE II

Approximate Removal Efficiencies and Costs for Water Reclamation[a]

	Removal efficiency[b] (%)				Cost[c] (¢/1000 gal)		
Process	TDS	Hardness as $CaCO_3$	Total COD	Turbidity	Capital	Operating	Total
Reverse osmosis	91	97	90	92	11.3	38.7	50.0
Electrodialysis	34	52	30	50	5.4	17.5	22.9
Ion exchange	90	99	59	92	5.1	25.2	30.3

[a] After Garrison and Miele, 1977.

[b] Influent to all systems was carbon-treated secondary effluent.

[c] Assumptions were: Engrg. News Record Cost Indexes, 2500; plant scale, 0.4 m^3/s (10 mgd); amortization for twenty years at 7 percent interest; power cost, 2.5 c/kWh.

periments conducted at the Pomona Water Renovation Plant by the Los Angeles County Sanitary District during the late 1960s and early 1970s. Since then the relative cost of RO has been reduced so as to make it competitive with ED and IE. From Table II it can be seen that RO removes greater than 90% of all the quality parameters listed. Normally, ED is able to remove more than 90% of the total dissolved solids TDS, the relatively poor performance here is probably because of the presence of unacceptable amounts of the other three quality parameters (Smith and Eisenman, 1964, 1967). Ion-exchange performance is acceptable except for the removal of total chemical oxygen demand COD. These results confirm our suspicion that the three desalting processes are not entirely comparable. Also, RO is capable of removing organics, viruses, bacteria, and soluble inorganic ions from the product, whereas ED and IE will not usually do this.

Total product-water costs for membrane processes can be roughly divided into: (1) capital costs or fixed charges based on capital recovery of initial investment, interest, and insurance; and (2) operating costs for fuel, power, labor, pretreatment, and membrane cleaning. These two group costs are shown diagrammatically in Fig. 2 as a function of membrane transfer area. The operating costs per unit volume of product will decrease with decreased driving force, which in turn will decrease with increased membrane transfer area. The capital cost per unit product will increase with increased membrane transfer area. The total cost curve in Fig. 2 is the sum of the operating cost curve and the capital cost curve and has a minimum that defines the optimum transfer area.

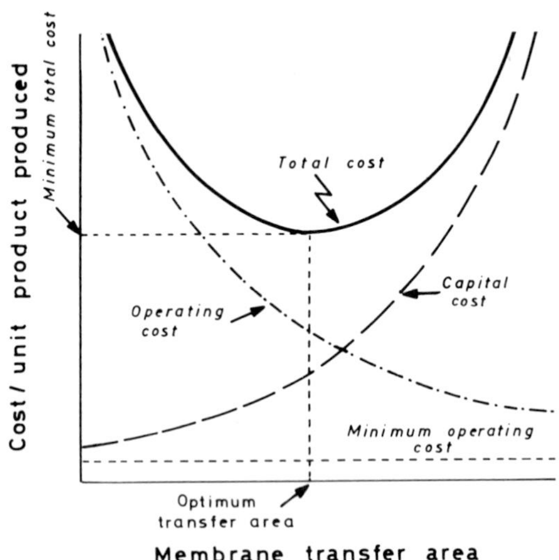

Fig. 2 Cost per unit product of treated wastewater as a function of system membrane-transfer area for a fixed production rate (after Probstein, 1973).

With respect to wastewater treatment technology, the traditional biological treatment processes have relatively lower operating and maintenance costs, but higher capital investment than the "advanced" physicochemical treatment processes, such as the pressure-driven membrane processes discussed here. Hence, municipalities in the United States which have various financing options such as very large grants-in-aid provided to the public sector for construction of sewage-treatment works may be reluctant to burden themselves with the high long-term operating and maintenance costs. Thus it appears in the United States that these advanced processes will find easier acceptance in the industrial sector. Furthermore, pressure-driven membrane processes of a capacity of only several million gallons per day and less are currently operating. For purposes of municipal water supply, these volumes are very small, although the situation will probably change with completion of the Yuma Desalting Plant (~100 mgd). These smaller units are, in the short run, most appropriate for the industrial and small municipal sectors (Channabasappa, 1969). In the long run, as larger plants become available, this picture will change.

Detailed economics of the application of membrane processes to the desalting of brackish and sea waters and the renovation of wastewaters are presented in Chapters 12 and 13.

II. Principles of Membrane Processes

A. DEFINITION

A membrane process, as discussed in the context of this chapter, is defined as one whose purpose is to separate, using selective membranes, one (or more) component(s) from a two (or more) component system using a differential driving potential across the membrane. We begin with a feed solution, nominally called the water or wastewater stream, from which we would like to remove either (1) the unwanted solutes or pollutants such as dissolved organics or inorganics; or (2) relatively clean water, and leave behind a more concentrated solute or polluted water. A differential driving potential across the membrane thickness is needed to attract or push the mobile component through the membrane. The choice of a driving force is a function of the type of membrane used. The driving force across the membrane may be the result of differences in concentration, as in dialysis; electrical potential, as in ED and transport depletion; or hydrostatic pressure, as in RO and UF. Several kinds of driving force may be operable simultaneously in any one process.

B. IDEAL MINIMUM WORK

For desalting processes it is useful to discuss the ideal work needed to separate salt from water or vice versa for any given process. It is known that the equilibrium vapor pressure of a salt solution is less than that of pure water under isothermal conditions. This is because the activity of the water is lower in the solution than in pure water. If two large reservoirs, one containing the solution and the other the pure water, were connected and sealed from the outer environment, work would have to be supplied, for example by a compressor, to prevent the movement of water vapor from the pure water reservoir to the solution reservoir. This energy is the ideal minimum work and has been calculated at approximately 3 kW h/1000 gal of freshwater produced from seawater at standard temperature (Probstein, 1972). In practical terms, the actual energy consumed by an operating membrane process for desalting is usually several times higher than the ideal minimum energy. This is a result of the existence of power losses or inefficiencies and of a finite flow rate and driving potential imposed on the system by economic requirements.

C. NO PHASE CHANGE

One of the most important factors responsible for recent interest in membrane processes is that they are able to separate dissolved species from one another or a solvent without phase-change. The large heat transfer requirements associated with the evaporation and crystallization processes are therefore avoided in membrane processes. Because heat transfer costs are a major part of the operating costs, use of membrane processes is an attractive alternative to vaporization or crystallization processes. In addition, many volatile low MW organics may not be easily separated from water by vaporization due to the proximity of their boiling points.

D. MEMBRANES

The membranes used for the pressure and electrically driven processes are functionally different, but they have similar features. They must separate two fluid-containing compartments without leakage and provide for differential transport rates through the membrane for different molecules, i.e., be permselective. Membranes of this kind may be visualized as consisting of many long-chain organic polymers randomly associated and cross linked. The void spaces between the chains represent the interstitial volume in the membrane through which transferring species pass. Depending on the function and type of membrane, the long-chain polymers will: (1) have long or short lengths; (2) be crystalline, amorphous or cross linked; (3) be homogeneous or heterogeneous; and (4) have neutral or highly-charged functional groups (positive or negative) associated with (or grafted onto) the chains. Specific details describing RO and UF membranes are presented in Chapter 7 and Chapter 9, respectively, and ED membranes are discussed in Chapter 6. The stability and lifetime of HF membranes are discussed in Chapter 3.

E. TRANSPORT EQUATIONS AND COEFFICIENTS

Because we are interested in the relative motion of various components through a membrane, it is convenient to be able to describe this motion quantitatively and thus be able to establish some basis for membrane performance. To attempt this, most researchers have invoked thermodynamics of irreversible processes (de Groot, 1959; de Groot and Mazur, 1962; Katchalsky and Curran, 1965; Haase, 1969). It is not our purpose here to develop this theory for membrane processes. Only the major results useful for our discussion will be presented; for further details the reader is referred to various references. Before proceeding, however, two additional points should be made. The first is that thermodynamics of irreversible processes is a phenomenological description of the relative motion of various components within the membrane, which is itself considered to be a "black box." This implies that the microscopic mechanism of flow (and rejection) will not and cannot be explained by this theory. To the extent that this theory is combined with some "internal" membrane model, such as the solution-diffusion model in RO, a mechanism can be inferred. The second point is that the thermodynamics of irreversible processes have been applied more frequently to the pressure-driven as opposed to the other membrane processes. Details of this approach are presented in Chapter 7.

Based on the thermodynamics of irreversible processes, several approaches have been used to develop the basic transport equation for RO, which relate the fluxes of solvents and solutes (J_i) with their respective driving forces (X_i) (Spiegler and Kedem, 1966; Staverman, 1951). These equations describe a coupling phenomenon that occurs between species when moving through the membrane. In general, processes in which generalized fluxes and forces are proportional to each other, one can write the following linear flux equations:

$$J_i = \sum_{j=1}^{m} L_{ij} X_j, \qquad i = 1,2,\ldots,m. \tag{1}$$

Onsager has shown theoretically, and others have verified experimentally for membrane transport in particular (Katchalsky and Curran, 1965), that the following symmetry exists for the phenomenological coefficients:

$$L_{ij} = L_{ji}, \qquad i,j = 1,2,\ldots,m. \tag{2}$$

Other restrictions on the coefficients are operable and are due to second-law considerations. They include

$$L_{ii} > 0, \qquad i = 1,2,\ldots,m \tag{3}$$

and

$$L_{ii} L_{jj} - L_{ij}^2 > 0, \qquad i,j = 1,2,\ldots,m \quad \text{and} \quad i \neq j. \tag{4}$$

The approach discussed in the following text uses the methods described by Haase (1969) to obtain the generalized equations for the isothermal heterogeneous (discontinuous) membrane system. Thus, we shall merely define the system and present the results obtained using the procedure described below.

Here, we consider two liquid subsystems separated from each other by a semipermeable membrane. Let the two homogeneous subsystems of our heterogeneous system (see Table III) be designated as phase′ and phase″. According to this, we can attach a definite value for the pressure (P' or P''), for composition (molar concentrations C'_k or C''_k), and for the electrical potential (ϕ' or ϕ'') to each phase at constant temperature and at any arbitrary instant.

TABLE III

Heterogeneous (Discontinuous) System Consisting of Two Homogeneous Isotropic Subsystems (Phase′ and Phase″)

Phase′	Phase″
Pressure P'	Pressure P''
Composition variable C'_k	Composition variable C''_k
Electrical potential ϕ	Electrical potential ϕ

After performing a mass and energy balance across the membrane and determining the entropy change resulting from the process, an explicit expression of the dissipation function is derived. Then the fluxes (J_i) and generalized forces (X_i) acting on the membrane system are determined. All these quantities are independent of each other and disappear at equilibrium. Additional details are presented in Katchalsky and Curran (1965), while a detailed step-by-step procedure is presented in Chartier *et al.* (1975).

For this system, ignoring gravitational and magnetic effects, the following types of forces result at constant temperature (Haase, 1969):

$$X_i = \text{grad } \mu_i = v_i \text{ grad } P + (\partial\mu_i/\partial C_j)_{T,P} \text{ grad } C_j + z_i F \text{ grad } \phi, \qquad (5)$$

where v_i is the partial molar volume of species i, μ_i the chemical potential of species i, z_i the valency of species i, F Faraday's constant, and grad refers to the gradient of a function such as μ_i between phase″ and phase′.

Equation (5) or a variant thereof, along with some approximations, will be used in Chapters 4–6 to develop the practical transport equations for the RO and ED processes. The three terms on the right-hand side of Eq. (5) represent the pressure, concentration, and electrical driving forces, respectively. For RO both pressure and concentration forces are usually predominant, while for ED, the electrical and concentration forces are most important. Only pressure forces are usually considered for UF.

Because of the important role of water and aqueous solutions in the three processes discussed in this text, Chapter 2 is included. Its main purpose is to present a fundamental picture of the state of water in solutions and membranes and to possibly relate this to their desalting characteristics. This physico–chemical microscopic approach is supposed to complement the phenomenological description of transport.

F. FLUID MECHANICS

It is well to emphasize the central role played by the movement of bulk fluid over the surface of the membranes in all membrane processes. Mass and viscous boundary layers are either growing or present at steady-state thickness in all these membrane processes. The mass boundary layer is due to the relative motion of various components through the membrane. In RO and UF, for example, because the water is forced through the membranes at a much higher rate than the solute molecules, a buildup of the solute species is observed at the solution-membrane interface. This phenomenon, known as concentration polarization, is schematically portrayed in Fig. 3. The viscous boundary layer is a function of the water removed and the gross fluid mechanics of the system. For fully developed flow, the viscous boundary layer thickness δ is equal to half the channel width or the radius of the tube. The importance of the hydrodynamic condition of the brine stream expresses itself through the shear force that is exerted at the membrane-solution interface. The higher this shear force, the easier it is for the solute molecules concentrated there to

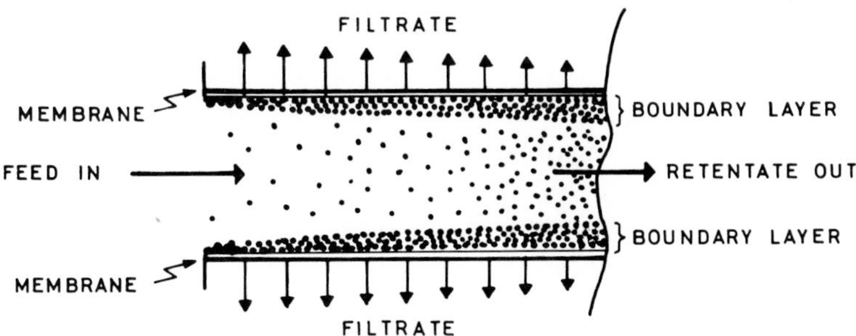

Fig. 3 Development of the mass or polarized boundary layer known as concentration polarization.

diffuse back into the bulk solution. The wall shear also plays a role in scouring the suspended matter near the membrane surface and thus reducing membrane fouling (Copas and Middleman, 1973; Thomas *et al.*, 1973). This author and co-workers have measured directly for the first time the solute concentration profile adjacent to a hyperfiltration membranes in a batch nonflow cell (Mahlab *et al.*, 1979).

In ED, critical concentration polarization usually occurs at the anionic permselective membrane on the dialysate solution side, where the current-carrying ions become depleted. When this occurs, water splitting begins and protons and hydroxyl ions begin to carry the current. This is associated with pH changes and potential scaling or membrane fouling from precipitation. The major consequence of concentration polarization as described previously is a catastrophic decrease in efficiency with increased power consumption.

The role of concentration polarization and fluid-solute dynamics are discussed in Chapters 4 and 5, respectively, for membrane processes.

G. Membrane Fouling

In practice, the single most important critical problem facing the application of membrane processes to wastewater renovation is membrane fouling (Belfort, 1977). Although a few scattered attempts to investigate fouling have been initiated (Belfort and Marx, 1979; Belfort *et al.*, 1976; Grossman and Sonin, 1972; Gutman, 1977; Jackson and Landolt, 1973; Porter, 1972; Sheppard *et al.*, 1972), the mechanism of fouling remains unclear.

The process of fouling is similar to the classic two-step flocculation–attachment process of coagulation in wastewater treatment. In the first step, concentration polarization by convective or electrical transport and Brownian diffusion, the foulant is carried to the membrane-solution interface. The efficiency of the second step, attachment, will depend on the usual force vector interaction (gravitational, London, electrokinetic, and hydrophobic) between the foulant and itself and/or the membrane (Hung and Tien, 1976). What little research has been conducted on membrane fouling to date has concentrated on specific foulants such as ferric oxide flocs (Jackson and Landolt, 1973), artificial latex (Belfort *et al.*, 1976; Belfort and Marx, 1979) and bovine serum albumin (Probstein *et al.*, 1978).

For high-flux membranes with relatively low Reynolds number for feed solution cross flow, the first step could ostensibly dominate the fouling process irrespective of the chemical nature of the foulant–membrane interaction. Thus the call for the development of improved performance membrane, i.e., higher chemical stability with higher permeation fluxes and solute rejections (Channabasappa, 1977) *must* go hand-in-hand with an equivalent reduction in concentration polarization. This is usually accomplished hydrodynamically with increased Reynolds numbers for the feed solution cross flow. The fouling problem is discussed in detail in Chapter 7.

III. Applications of Membrane Processes

It is not the purpose of this introductory chapter to describe in detail the applications of membrane processes. This is presented for each process in Chapters 7–11 of this text. The purpose, however, is to review the current trends in membrane applications including current plans for large-scale plants.

The application of membranes processes falls readily into two areas. The first area and the one developed initially was the desalination of brackish and (later) seawater to potable water suitable for domestic and industrial needs. The latest available inventory of desalting plants of capacity greater than 25,000 gpm is summarized in Table IV (El-Ramly and Congdon, 1981). In absolute terms, membrane processes make up approximately 24% of all plant capacity and about 56% of the number of plants. To put these numbers in temporal perspective, the equivalent percentages obtained from the Second United Nations Desalination Plant Operating Survey (1973), which surveyed plants greater in capacity than 10,000 gpd up to 1968, were 1.6 and 16%, respectively. Thus during the intermediate 13 years both the capacity and number of membrane plants, especially of the RO type, have increased faster than for any other process. For example, in the 1973 survey only one RO plant of 100,000 gpd capacity was included, but in Table IV 929 RO plants with a total capacity of 390.5 mgd are included. Although the minimum capacity of plants for inclusion in the two surveys differed, a phenomenal increase in the number and capacity of membrane plants has occurred world wide. One possible reason for this is the lower overall costs and especially those costs associated with energy needs for the membrane methods when compared with the phase-change processes. Belfort (1974) pointed out this potential advantage for membrane processes directly after the energy crises of 1973. Providing present trends continue, the future of membrane desalting processes looks extremely bright.

TABLE IV

Summary of Desalting Plants by Process[a]

Type of process	Number of plants	Plant capacity (M G D)
Distillation		
Single-stage flash	77	11.3
Multistage flash	415	1,292.6
Thin-film vertical tube	104	75.8
Vertical tube–multistage flash	7	4.7
Thin–film horizontal tube	51	22.7
Submerged tube	127	21.6
Vapor compression	168	22.1
Vertical tube–vapor compression	16	8.3
	965	1,459.1
Membrane		
Electrodialysis	82	35.7
Reverse osmosis	929	390.5
Electrodialysis–reversing	228	36.8
	1,239	463.0
Freezing		
Vacuum freezing–vapor compression	1	—
All types of processes	2,205	1,922.2

[a] After El-Ramly and Congdon (1981).

The second area of application for membrane processes is in wastewater renovation for reuse or disposal to meet effluent quality standards. The unique advantages of ambient temperature operation and relatively low-power requirements suggest the use of membrane techniques for the treatment of large volumes of municipal, industrial, and agricultural wastewaters for multiple reuse. Major limitations for these applications have been membrane stability and high flux-decline rates resulting from membrane fouling and compaction. This has necessitated the development of new membranes able to withstand large pH ranges, unusual temperature conditions, and corrosive chemical conditions (Channabasappa, 1977). To control the degradation in performance because of fouling, the most effective economical methods of pretreating and periodic cleaning have been sought (Belfort, 1977). Others have tried to overcome this problem by choosing as little pretreatment and cleaning as possible associated with unusually high pumping rates, i.e., high Reynolds numbers and wall shear rates (Sachs, 1976).

An example of treating 5 mgd carbon-filtered tertiary effluent by RO for injection into underground reservoirs is currently in operation at Orange County Water District in California. This example of the treatment of municipal effluent by membranes is described in detail in Chapter 10.

Compared with the relatively constant composition of municipal sewage, the wastes from manufacturing plants vary widely in physical characteristics and chemical composition. Precisely because of this diversity, membrane processes are predicted to play an important role in treating these industrial wastes. The reason for this is the ability to match membranes with special characteristics to particular types of industrial wastewaters. In addition to renovating the water for reuse, recycle, or disposal, there also exists the possibility of simultaneously concentrating valuable by-products, such as proteins in whey, heavy metals in plating and metal finishing wastes, silver in photographic processing waters, etc. Membrane processes are also beginning to play a useful role in the new gene-splicing industry for concentration and recovery of various biologically derived products. With the current emphasis on environmental and health protection from potential pathogenic contaminants in water, the economic incentive to recycle water and concentrate by-products should increase.

Although the present use of membrane processes to renovate water economically for agricultural use appears to be prohibitive, special applications do appear possible. Adaptions of reverse osmosis and ED could be used to control the sodium adsorption ratio $SAR = [Na]/(([Ca] + [Mg])/2)^{1/2}$ below 15 and to specifically remove high concentrations of boron from the water. One application of the treatment of agricultural runoff is the largest proposed membrane plant in the world, the Yuma Desalting Plant (Leitz and Ewoldsen, 1978). The motivation in deciding to build this plant was not based on economics but rather determined by geopolitical considerations. A section by Belfort in Chapter 7 is devoted to this application.

IV. Conclusions

It was the express purpose of this chapter to present an overview of membrane methods in water and wastewater treatment and to direct the reader requiring additional details to the relevant chapters herein.

The historical development by HF, UF and ED is presented. Each process is classified according to its driving potential and range of application. Their major advantages and disadvantages are discussed and the differences highlighted. Some performance costs are presented for treating carbon-treated secondary effluent. The major cost parameters are also discussed.

A general presentation of the principles of membrane processes is outlined. This includes a definition of membrane processes, discussions on the ideal minimum work necessary to desalt sea water, and that phase changes with high heat-transfer requirements are not associated with membrane processes. Also, the chemical, physical, and morphological characteristics of membranes are introduced, as is the thermodynamic approach for irreversible processes presented as one way of describing membrane transport. Because of their importance, both fluid mechanics and membrane fouling are discussed separately.

Finally, the application of membrane processes to the desalting of brackish and sea water and to the renovation of municipal, industrial and agricultural wastewaters is discussed.

In conclusion, with large membrane plants, such as the Yuma Desalting Plant (100 mgd), the new seawater (3.2 mgd) and brackish water (31 mgd) plants in Saudi Arabia, and the Orange County municipal effluent plant (5 mgd), it appears that membrane technology has come of age. With reliable operation of these large plants, the traditional conservatism of water planners, designers, and administrators could conceivably be overcome, resulting in the wide application of membrane processes to all aspects of water treatment and renovation.

References

Belfort, G. (1974). *Desalination* **15,** 143–144.

Belfort, G. (1977). *Desalination* **21,** 285–300.

Belfort, G., and Marx, B. (1979). *Desalination* **25,** 13–30.

Belfort, G., Alexandrowicz, G., and Marx, B. (1976). *Desalination* **19,** 127–138.

Channabasappa, K. C. (1969). *Chem. Eng. Prog. Symp. Ser.* **65,** 140–147.

Channabasappa, K. C. (1977). *Desalination* **23,** 495–514.

Charitier, P., Gross, M., and Spiegler, K. S. (1975). "Applications de la Thermodynamique du Non-equilibre." Hermann, Paris.

Copas, A. L., and Middleman, S. (1973). "The Use of Convective Promotion in Ultrafiltration of a Gel-Forming Solute" (Paper No. 53a). Presented at 66th Annual AIChE Meeting, Philadelphia, 1973.

de Groot, S. R. (1959). "Thermodynamics of Irreversible Processes." North-Holland, Amsterdam.

de Groot, S. R., and Mazur, P. (1962). "Non-Equilibrium Thermodynamics." North Holland, Amsterdam.

El-Ramly, N. A., and Congdon, C. F. (1981). Desalting Plants Inventory (Report No. 7). National Water Supply Improvement Assoc., Ipswich, Massachusetts.

Garrison, W. E., and Miele, R. P. (1977). *J. AWWA,* **69,** 364–369.

Grossman, G., and Sonin, A. P. (1972). "Membrane Fouling in Electrodialysis: A Model and Experiments" (Fluid Mechanics Laboratory Report No. 72–2). Department of Mechanical Engineering, MIT, Cambridge.

Gutman, R. G. (1977). *Chem. Eng.* **322,** 510–513, 521–523.

Haase, R. (1969). "Thermodynamics of Irreversible Processes." Addison-Wesley, Reading, Massachusetts.

Hung, Ching-cheh, and Tien, Chi (1976). Effects of Particle Deposition on the Reduction of Water Flux in Reverse Osmosis. *Int. Symp. Fresh Water Sea 5th* **4,** 335–345.

Jackson, J. M., and Landolt, D. (1973). *Desalination* **12,** 361.

Katchalsky, A., and Curran, P. F. (1965). "Non-Equilibrium Thermodynamics in Biophysics." Harvard University Press.

Lacey, R. E. (1972). *Chem. Eng. London* September 4th, 56–74.

Lacey, R. E., and Huffman, E. L. (1971). *Water Pollut. Contr. Res. Ser.* 17040 EUN02/71.

Leitz, F. B., and Ewoldsen, E. I. (1978). *Desalination* **24,** 321–340.

Loeb, S., and Sourirajan, S. (1962). *Adv. Chem. Ser.* **38,** 117.

Mahlab, D., Ben Yosef, N., and Belfort, G. (1979). "Interferometric Measurement of Concentration Polarization Profile for Dissolved Species in Unstirred Batch Hyperfiltration (Reverse Osmosis)." Presented at the 72nd Annual AIChE Meeting, San Francisco, Nov. 25–29.

Porter, M. C. (1972). *Ind. Eng. Chem. Res. Dev.* **11**(3), 234.

Probstein, R. F. (1972). *Trans. ASME* June, 266–313.

Probstein, R. F. (1973). *Desalination Am. Sci.* **61**(3), 280–293.

Probstein, R. F., Shen, J. S., and Leung, W. F. (1978). *Desalination* **24,** 1–16.

Reid, C. E., and Breton, E. J. (1959). *Chem. Eng. Prog. Symp. Ser.* **55**(24), 171.

Sachs, S. B., Zisner, E., Herscouri, G., and Shelef (1976). Hybrid reverse osmosis-ultrafiltration membranes. *Proc. 5th Int. Symp. Fresh Water Sea, Algheso, Italy* May, 16–20.

Sheppard, J. D., Thomas D. G., and Channabasappa, K. C. (1972). *Desalination* **11,** 385–398.

Shuval, H. I. (ed.) (1977). "Water Renovation and Reuse." In Water Pollution Series (K. S. Spiegler and J. Bregman, eds.-in-chief) Vol. 3. Academic Press, New York.

Shuval, H. I., and Gruener, N. (1973). *Environ. Sci. Technol.* **7,** 600–604.

Smith, J. D., and Eisenman, J. L. (1964). *Eng. Bull. Purdue Univ. Eng. Ext. Ser.* **117,** 738–760.

Smith, J. D., and Eisenman, J. L. (1967). Federal Water Pollution Control Administration (Report WP-20-AWTR-18).

Spiegler, K. S., and Kedem, O. (1966). *Desalination* **1,** 311.

Spiegler, K. S., and Laird, A. D. K. (eds.) (1980). "Principles of Desalination." 2nd Edition. Academic Press, New York.

Staverman, A. J. (1951). *Rec. Trav. Chim. Pays-Bas Belg.* **70,** 344.

Thomas, D. G., Gallaher, R. B., and Johnson, J. S., Jr. (1973). Hydrodynamic Flux control for wastewater application of hyperfiltration system (Environmental Protection Technology Series, Office of Research and Development, U.S. Environmental Protection Agency, Washington, D.C. 20460, EPA-R2-73-228).

United Nations (1973). "Second U.N. Desalination Plant Operating Survey." Resources and Transport Division, Center for Economic and Social Information, New York/Geneva.

Wilson, J. R. (1960). "Demineralization by Electrodialysis." Buterworth, London.

World Health Organization (1973). "Reuse of Effluents; Methods of Wastewater Treatment and Health Safeguarding" (Tech. Rep. Ser. 517). World Health Organization.

2

Structure of Water and Aqueous Systems*

W. A. P. LUCK

Fachbereich Physikalische Chemie, Universität Marburg
Marburg, Federal Republic of Germany

* Dedicated to Prof. Dr. Seefelder, President of BASF Ludwigshafen, West Germany on his 60th birthday.

ISBN 0-12-085480-5

List of Symbols

A^-	Anion
$[OH_{free}]$	Concentration of non-hydrogen-bonded OH groups
$[C_0]$	Concentration by weight
$[OH_b]$	Concentration of hydrogen-bonded OH groups
$\bar{C}_s$	Salt concentration mol/liter
CS	Chondroitine sulfate
dO_F/dT	Slope of O_F in Fig. 1
dw	Dry weight
E, A	Absorbance
Extinction, (Ext).	Absorbance
ĒO, ĀO	Ether group
F–A	Force–area diagrams
HN	Hydration number
ΔH_H	Hydrogen-bond energy
ΔH_{hyd}	Hydration energy of ions
IR	Infrared spectroscopy
K^+	Cation
$N(T)$	Average number of bonded H_2O molecules
ΔM_{max}	Maximum weight increase at 100% rh
O_F	Fraction of non-hydrogen-bonded OH groups
OH_F	Non-bonded or "free" OH
OH_b	Hydrogen-bonded OH
OH_xA^-	Hydrated anion with x = hydration number
θ_yK^+	Hydrated cation with y = hydration number
PIOP-,!	*p*-isooctylphenol with *n*-ethylene oxide groups
RS	Reciprocal solubility
rh	Relative humidity
t	Time
T	Temperature
T_c	Critical temperature
T_{str}	Structure temperature of solution (T of pure water)
T_K	Lowest temperature of two-phase formation
V_1	Partial molar volume of water in electrolyte solutions
β	Hydrogen-bond angle ($\beta = 0$ if angle between axis OH and lone pair electrons is zero)
ε	Extinction coefficient
ε_m	Extinction coefficient at band maximum
ν_F	Frequency of the band maximum of non-hydrogen-bonded OH groups
ν_b	Frequency of the band maximum of hydrogen-bonded OH groups
$\Delta\nu$	Frequency shift between: free OH and hydrogen-bonded OH
$\Delta\nu_\beta$	Frequency shift at hydrogen-bond angle β
$\Delta\nu_{1/2}$	Half-width of bands
ρ	Density
σ	Ion reflection coefficient of membranes
θ	Lone-pair electrons

I. Introduction

During the last two decades our knowledge of the structure of liquid water has increased rapidly. Many details, however, are still not clear. For example, the polydimensional partition function of distances and orientations is not known. Given this, it would appear to be difficult to describe complex aqueous solutions such as electrolytes, aqueous systems associated with biological cells or membranes, and the influence of ions on active membrane transport. Chemists have developed techniques, however, that begin with simplified idealized models and optimize them step by step by comparing observed and predicted results. This chapter outlines the salient features of an idealized model of liquid water and attempts to apply it to aqueous solutions, especially to the structure of water in membranes. The consequences of the model on the membrane separation mechanisms will also be discussed. Because of the complex conditions within membranes, only an approximate model of the aqueous system is proposed.

II. Structure of Liquid Water

A. AN APPROXIMATE TWO-STATE MODEL

The anomalous properties of water are based on its high concentration of hydrogen bonds (110 mol/liter at room temperature) and the fact its concentration of OH groups and lone-pair electrons $[\theta]$ are equal. Some of the different properties of alcohols are based on the fact of $[\theta] = 2[\mathrm{OH}]$. An estimate of the hydrogen bond energy ΔH_{H} in ice and in liquid water at room temperature is about two-thirds of the total intermolecular energy (In ice, ΔH_{H} is 8 kcal/mol, and the dispersion energy is about 3.6 kcal/mol.)

Because of the evidence and importance of infrared (IR) fundamental spectroscopy, it is usually not recognized that IR overtone spectroscopy is one of the best methods to study hydrogen bonds quantitatively. Hydrogen bonds shift the frequency, ν, of the OH stretching overtone or combination bands more than the fundamental bands, but hardly change their intensity $\int \varepsilon d\nu$. The intensity of the fundamental bands is increased through hydrogen bonding by a factor of 20–30 or more. In addition, the photometric accuracy of most overtone instruments is much higher than that of equivalent fundamental instruments. Solutions of molecules with OH or NH groups have a distinct sharp band, ν_F, of undisturbed—so-called free—OH or NH groups, free is used here to indicate no hydrogen bonds. However, the molecules are not free of other interactions, such as dispersion forces and broad bands, ν_b, of OH or NH groups with hydrogen bond interaction, with large half-width $\Delta\nu_{1/2}$. The frequency shift ($\nu_b - \nu_F$) = $\Delta\nu$ is proportional to the hydrogen bond interaction energy ΔH_H. These solutions in CCl_4 could be described quantitatively with high precision by the equilibrium equation

$$OH_{free} + \theta_{free} \rightleftharpoons OH_b, \qquad K = [OH_b]/[OH_F][\theta_{free}], \tag{1}$$

where $[\theta_{free}]$ is the concentration of non-hydrogen-bonded lone-pair electrons. The equilibrium constant, K, values determined by IR overtone spectra are regarded as extremely reliable. For example, K values for the hydrogen-bond equilibrium of lactams with cis-amide groups are constant for a concentration variation of 10^3 (Luck, 1965a, 1967b). The concentration of hydrogen-bonded hydroxyl groups ($[OH_b]$) in Eq. (1) can be determined, in principle, by the intensity of hydrogen-bond bands. The accuracy of this method is less than the determination of $[OH_{free}]$ by band-overlapping or by one-quantum excitation of OH vibrations in two neighboring hydrogen-bonded molecules (Schiöberg *et al.*, 1979). Therefore, it is preferred to determine $[OH_b]$ as follows: $[OH_b] = [C_0] - [OH_{free}]$, where $[C_0]$ is the total concentration by weight of OH or NH. When $[OH_{free}] = [O_F][C_0] = [\theta_{free}]$ then $[OH_b] = [C_0] - [OH_{free}] = (1 - [O_F])$. In the case of H_2O then K in Eq. (1) can be expressed as:

$$K = (1 - [O_F])/[O_F]^2[C_0]. \tag{2}$$

A comparison of the concentration dependence of overtone spectra of solutions of CH_3OH or C_2H_5OH with a change in the temperature (T) of pure liquids has demonstrated[3,4] that the hydrogen-bond equilibrium is similar. The band of free OH groups appears to increase with increasing temperature. Experiments up to the critical temperature T_c have shown the extinction coefficient ε_M of the band maximum is smaller in the liquid state but $\varepsilon_M\Delta\nu_{1/2}$(liquid) $\approx \varepsilon_M\Delta\nu_{1/2}$(solution), or by calibrating the data at T_c, $[OH_{free}]$ can also be determined in liquid alcohols or amides (Luck and Ditter, 1967, 1968). The hydrogen bond bands of liquids show a more extended partition by simultaneous one-quantum excitation of two neighboring hydrogen-bonded molecules (Schiöberg *et al.*, 1979), and probably by a broader partition of hydrogen-bond angles, β (Luck, 1965a, 1967b, 1976a). A comparison of four overtone bands of liquid H_2O, D_2O, or HOD with liquid alcohols leads to the conclusion this method can be usefully applied to determine the hydrogen-bond state of liquid water (Luck, 1965d, 1974, 1976b, 1976d, 1978, 1979a; Luck and Ditter, 1969).

In the lower half of Fig. 1 the absorbance of the first HOD overtone for $10 < T < 90$°C is presented. In the upper half of Fig. 1, the differences of spectra at different temperature and $T = 10$°C is shown (Schiöberg *et al.*, 1979). These experiments demonstrate a distinct increase in the frequency region 7300–6800 cm^{-1} with temperature. From solution spectra of HOD/CCl_4, H_2O/CCl_4 or by comparison with alcohol spectra, this can be assigned to the free OH. Secondly, the intensity in the region 6800–6200 cm^{-1}, known as the region of hydrogen bonds, decreases with increasing temperature. The isosbestic point at 6800 cm^{-1} exists up to about 150°C and indicates that there exist two temperature-dependent types of OH groups with different absorption bands (Luck and Ditter, 1969). Above 150°C, the isosbestic point flattens out (Luck and Ditter, 1969). This effect can be described quantitatively in detail up to about 400°C by a third very broad absorption band with a maximum at 6850 cm^{-1}. This band may be caused by hydrogen bonds with unfavored angles around the antiparallel orientations

```
H···O—H
|   |
H—O···H.
```

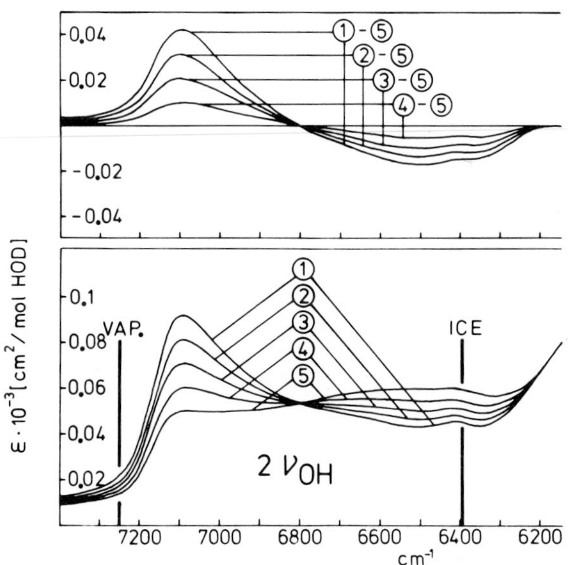

Fig. 1 Below: Overtone band HOD in D_2O: 7260–6800 cm^{-1} region of non-hydrogen-bonded or weak hydrogen-bonded OH absorption, 6800–6300 cm^{-1} region of hydrogen-bonded OH. (1) 90°C; (2) 70°C; (3) 50°C; (4) 30°C; (5) 10°C; Top: Difference spectra: 90°C–10°C, 70°C–10°C, etc.

Up to about 150°C, however, such orientations are nearly temperature-independent (Luck and Ditter, 1969).

Using the existence of the isosbestic point with the two assigned overtone regions described previously and the equilibrium defined in Eq. (1), simple and useful probes as measured by IR overtone spectroscopy can be used to study the hydrogen-bond state of liquid water (Luck, 1965d, 1974, 1976b, 1976d, 1978, 1979a, 1980a, 1980b). Using the model proposed above, K values plotted versus inverse temperature are linear in Fig. 2 up to about 200°C (Luck, 1973). This procedure has similar success for alcohols (Luck, 1973). Above 200°C, the method suggests small aggregates with less than 10 hydrogen-bonded molecules (Luck, 1973). Under these conditions, Eq. (1) is probably not valid. The H_2O hydrogen-bond interaction energy calculated from the slope in Fig. 2 is given by ΔH_H =

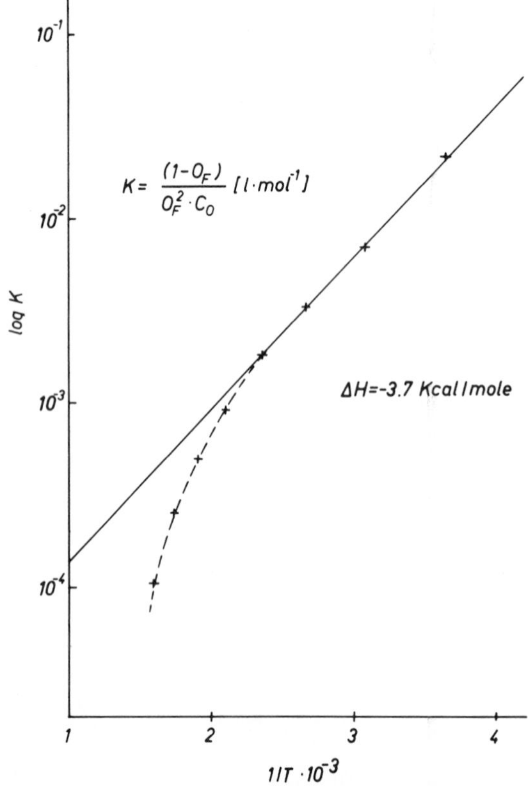

Fig. 2 Equilibrium constant K of the hydrogen bonds of liquid HOD for the equilibrium: $OH_{free} + \theta_{free} \rightleftharpoons OH_{bond}$. The experimental O_F values are from Fig. 3.

−3.7 kcal/mol. A similar value has also been established by another spectroscopic method (Schiöberg *et al.*, 1979). In Fig. 3 the spectroscopically determined values of [OH_{free}] as a function of temperature under saturation conditions at vapor–liquid equilibrium is presented. The importance of this curve is that many abnormal properties of water, such as specific heat, enthalpy, heat of vaporization, surface energy, surface tension, and density, ρ, (including its 4°C maximum) can be calculated quantitatively in the region $0 < T < 400°C$ without any adjustable constants (ρ with constants) (Luck, 1967, 1973, 1980a, 1980b). At the melting point of ice about 10% of hydrogen bonds break. This result agrees with the measured heat of melting of 1.4 kcal/mol (Luck, 1980a, 1980b), but not with earlier theories of water. Because of the cooperativity property of hydrogen bonds in liquids and its associated angle dependence (Luck, 1965a, 1967b, 1976a), an estimate of the average size of hydrogen-bonded clusters (i.e., N = number of water monomers) can be obtained from the data presented in Fig. 3 together with the idealized assumption that all free OH bonds are arranged at fissure planes between hydrogen-bonded aggregates. Such an estimate is presented here: $N(0°C) \approx 400$; $N(50°C) \approx 100$; $N(100°C) \approx 40$; $N(150°C) \approx 20$; $N(200°C) \approx 10$; $N(250°C) \approx 6$; and $N(300°C) \approx 3$.

B. Angle Dependence of Hydrogen Bonds

The hydrogen-bond interaction differs from the dispersion forces by larger intensity and orientation dependence. An energy minimum of the hydrogen bond arises from the angle $\beta = 0$ between the axes of OH and the lone-pair electron orbital. An estimate of the sharpness of this minimum was sought by the matrix spectra of H_2O or CH_3OH in solid Ar or N_2

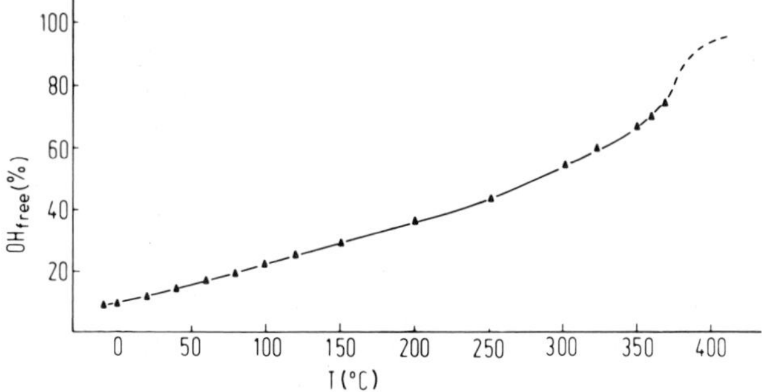

Fig. 3 Spectroscopically determined content of non-hydrogen-bonded OH (O_F) of liquid water in equilibrium with vapor.

(Hallam 1973; Thiel *et al.*, 1957a, 1957b). Four sharp bands of different hydrogen-bonded OH groups are observed. Pimentel *et al.* correlated these bands to cyclic dimers, trimers, tetramers and polymers (Thiel *et al.*, 1957a, 1957b). Figure 4 is the result of this correlation, and shows $\Delta\nu \approx \Delta H_H$ of the hydrogen-bond bands as a function of the hydrogen-bond angle β. Some doubts as to the existence of cyclic dimers have been published (Ayers and Pullin, 1976; Fredin *et al.*, 1975a, 1975b; Tursi and Nixon, 1970), but measurements by Luck *et al.* (Behrens and Luck, 1979; Luck and Schrems, 1979) weaken these doubts and support Pimentel's assumption. Thus, the major result of Fig. 4 is $\Delta\nu \sim \Delta H_H = f(\beta)$. This agrees with the fact of the hydrogen-bond angle, β, in crystalline hydrates generally differs only by about 10° from $\beta = 0$; (Falk and Knop, 1973), also $\beta = 0$ exists in ice I, where six H_2O molecules form ring structures. The result in Fig. 4 also demonstrates the cooperativity of hydrogen bonds with angle dependence. Newer results indicate that the dimer band position may be induced by cooperativity effects of the hydrogen-bond distance too (Luck, 1982). Thus, if a hole defect in an icelike ring is induced, the other five molecules form a ring with unfavorable angles of an average $\beta = 10°$. Therefore, on the basis of Fig. 4, the probability to induce a second defect into this disturbed ring would be higher than in an undisturbed ring.

The cyclic dimer with $\beta \approx 110°$ would have an antiparallel orientation of

H···O
|╳|
O···H

The O···O distance, $r_{O—O}$, would be about 20% shorter than in a linear orientation. Therefore, its dispersion energy, which is proportional to $1/r_{O—O}{}^6$, should be about 1 kcal/mol larger (Luck, 1980a, 1980b). This hydrogen-bond angle dependence is of fundamental importance to biochemistry and aqueous systems.

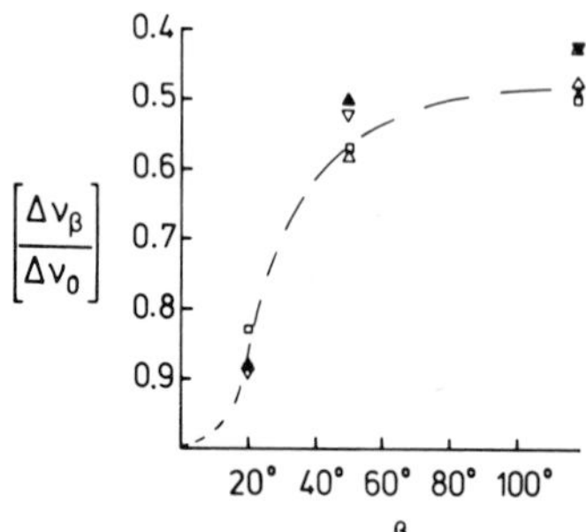

Fig. 4 Angle dependence of $\Delta\nu$ and therefore of hydrogen-bonded energy ΔH_H. ($\Delta\nu_0$ corresponds to $\beta = 0$). △▲ represents H_2O matrix; □ represents D_2O matrix; ▽ represents CH_3OH matrix; x represents R—OH solutions/CCl_4.

C. COMPUTER SIMULATION CALCULATIONS

Stillinger, Rahman, and Ben Naim (Ben-Naim, 1974; Rahman and Stillinger, 1971; Stillinger and Rahman, 1972, 1974) have calculated the water properties with success, using a four-point charge model and a neon like dispersion potential. These calculations agree with the IR spectroscopic results in Section II,A. Care should be taken, however, investing too much confidence to these computer models, because the assumed dispersion energy was too low and the four-charge potential had too small a repulsion term. The authors compensated for the second problem by adding a "switch potential." In the most interesting r_{O-O} region, the switch potential dominates the other two terms. This switch potential has been adjusted to describe water properties, but it lacks physical meaning and appears to be a fitting parameter. The computer and spectroscopic results agree in the following respects: At room temperature the analysis of water leads to an unambiguous division in pairs into hydrogen-bonded and non-hydrogen-bonded (Stillinger and Rahman, 1972) with an oscillation (Geiger, 1979; Geiger *et al.*, 1919) of the bonded OH around $\beta = 0$ and a non-hydrogen-bonded state, whose "entities persist for longer than H_2O molecules' vibrational periods." (Rahman and Stillinger, 1971). Stillinger's and Rahman's conclusion that "the computer simulation would exclude a two state model" has been based on calculated coordination numbers. This assertion is, therefore, relevant only to models with different types of molecules and not to our model, which has different types of OH groups.

Water is thus described as a simplified two-state model of bonded and unbonded (O_F) OH groups. The small content of O_F indicates that monomeric H_2O molecules with two free hydroxyls can be neglected for $T <$ 200°C. Indeed, the associated errors of the IR overtone method are more likely to overestimate the number of O_F bonds. This possible overestimation of O_F seriously questions the validity of earlier water models of Eucken (Eucken, 1946, 1948, 1949) (assumed small aggregates), and of Némethy and Scheraga (Némethy and Scheraga, 1962). These latter two theories require the adjustment of five constants, a method that needs experimental verification.

The hydrogen-bond bands can be separated into two main bands with broad half-width indicating a certain partition of hydrogen-bond angles (Luck and Ditter, 1969). One band can be coordinated to hydrogen-bond angles near $\beta = 0$, and a second to angles around the antiparallel $\begin{matrix} H\cdots O \\ | \quad\ \ | \\ O\cdots H \end{matrix}$ orientation with $\beta = 110°$. This latter orientation may be stabilized by additional dispersion forces, which suggest that the two hydrogen-bond states do not differ much energetically (Luck, 1980a, 1980b). This may favor the possibility of a simplified two-state model, wherein the hydrogen-bond state may correspond to an oscillation around $\beta = 0$, which has been predicted by computer simulation studies (Geiger, 1979; Geiger *et al.*, 1979).

A detailed band analysis should clear up additional details of this angle and distance orientation. X-Ray data (Narten, 1974) and the small density change on increasing the temperature to 200°C may indicate that the angle partition dominates the distance partition.

D. SUMMARY

In contrast to fundamental IR spectroscopy, the overtone spectra represent an extremely useful tool for studying the properties of liquid water. These spectra provide the basis for describing water and its anomalous properties up to about 400°C. A simplified two-state model of free or non-hydrogen-bonded OH groups and hydrogen-bonded OH groups is proposed. The hydrogen-bonded angles are assumed to oscillate around the energy minimum. The hydrogen-bond energy of water can be estimated to be about 3.7 kcal/mol of OH groups. The anomalous properties of water depend on (1) the high concentration of hydrogen bonds (110 mol/liter at room temperature), (2) equal amounts of OH and lone-pair electrons, and (3) the angle dependence of the hydrogen-bond energy. The content of non-hydrogen-bonded OH groups in liquid water is about 12% at room temperature. The association of hydrogen-bonded water molecules may produce aggregates consisting of about 100 molecules at room temperature.

III. Electrolyte Solutions

A. THE HOFMEISTER ION SERIES

Ionic solutions were once thought to be dominated by long-range Coulombic forces, although colloid chemists have known since the 1800s of special properties of different ions. They described this empirically by the Hofmeister, or "lyotropic ion series," (Hofmeister, 1890, 1891; Stauf, 1960), based on its efficiency of colloid flocculation:

$Th^{4+} > Al^{3+} > H^+ > Ba^{2+} > Sr^{2+} > Ca^{2+} > Cs^+ > Rb^+ > K^+ > Na^+ >$

Citrate > Tartrate > SO_4^{2-} > Acetate > Cl^- > NO_3^- > Br^- > I^- > CNS^-.

The reason for the order of this series was unknown, although a connection to ionic size was presumed.

As a first approximation, the effect of adding different ionic species to water was found to change the overtone spectra of electrolyte solutions in a manner similar to the effect observed as a result of temperature variation on the pure water spectra. (Luck, 1965c). An example of this analogy is shown in Fig. 5, in which one overtone band of pure water (full lines) is compared with 1 M KSCN solutions. The result suggests a fictitious positive temperature shift of about 8°C. Bernal and Fowler described electrolyte solutions qualitatively by using a structure temperature T_{str}, which is defined as the temperature of pure water with a hydrogen-bond state similar to the electrolyte solution under consideration.

In Fig. 6 a quantitative definition or ranking is established with the following conclusions:

(1) The ion series measured in this way is similar to the Hofmeister series. Indeed, it may be concluded that the Hofmeister ion series describes a series of water-structure change (Luck, 1964, 1965c, 1974, 1976b, 1976d, 1978, 1979a).

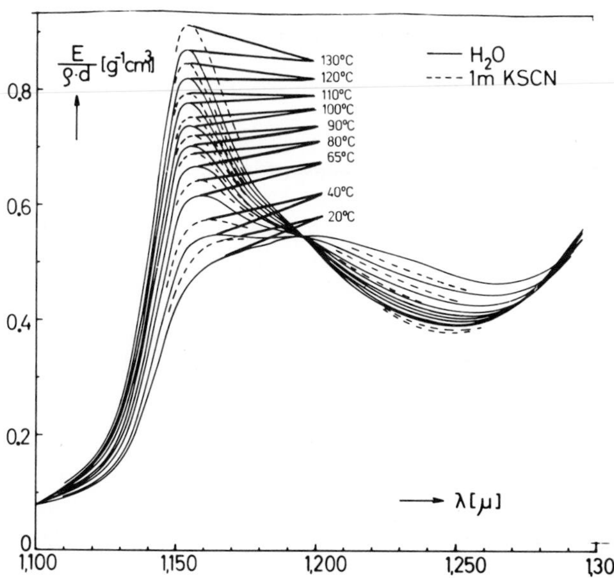

Fig. 5 Full lines: Combination band of liquid H_2O. Dotted lines: Combination band of aqueous solutions of 1 M KSCN.

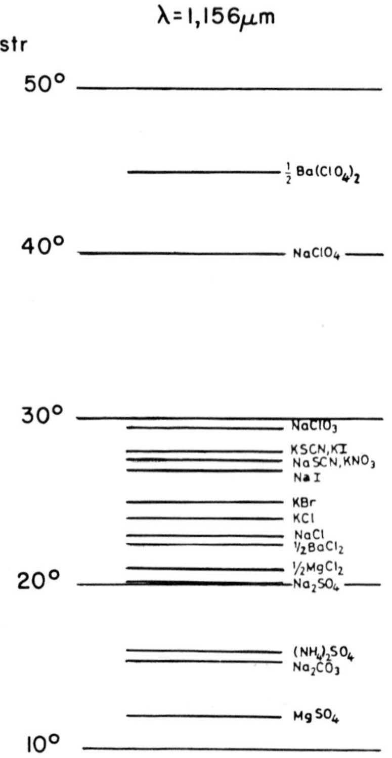

Fig. 6 Structure temperatures T_{str} of different salt solutions in H_2O at the solution temperature $T = 20°C$ determined at 1.156 μm. Salt concentration was 1 mol/liter.

(2) A minimum of about 0.2 mol of ions/liter is needed to recognize changes in water spectra.

(3) There are ions with $T_{str} > T$(solution) called structure-breaking ions and ions with $T_{str} < T$ (solution) called structure-making ions.

(4) The boundary between breaking and making ions depends on the solution temperature (Luck, 1965c) and the ion concentration.

(5) Anions usually exhibit more influence on T_{str} than do cations.

Several additional comments are also possible. A solute concentration greater than 0.2 mol/liter corresponds to about 300 water molecules per ion pair, a number equivalent to the average cluster size.

That $T_{str} > T$(actual solution temperature) implies the solutions of structure-breaking ions have on an average more free or weak hydrogen-bonded OH groups and are more hydrophilic than water at the same temperature. Therefore, some solutes can be salted-in (Hallban *et al.*, 1935). That $T_{str} < T$ implies the solutions have relatively less free OH. The resultant salt-out effect is important for many technical processes such as dying or washing with detergents.

The boundary between the salt-in or salt-out effect also depends on the type of organic solute being dissolved. There is a competition between the interactions: $H_2O \cdots H_2O$, $H_2O \cdots$ions, and $H_2O \cdots$organic solute, depending on the hydrogen-bond acceptor and/or donator strength of the organic molecules.

B. THE STRUCTURE TEMPERATURE APPROACH

1. *Examples of Utility*

1. Turbidity points of Polyethylene oxides. Polyethylene oxides of the type

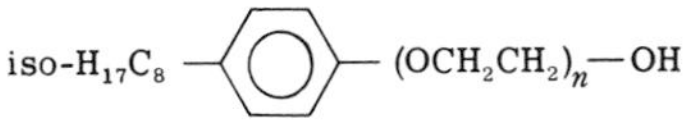

$$\text{iso-}H_{17}C_8-\langle\bigcirc\rangle-(OCH_2CH_2)_n-OH$$

PIOP-n

are water-soluble as a result of hydrogen bonds between H_2O and ether oxygens, with $n = 9$. At $T = 64°C$ a turbidity point appears; for $T_K > 64°C$ there exist two phases, one H_2O and one organic (i.e., PIOP-9) T_K is sensitive to the Hofmeister ion series (as shown in Fig. 7) and is proportional to the spectroscopically determined structure temperature T_{str} (Luck, 1964, 1965c). T_K values vary with salt concentration; salt-in effects an increase in T_K and salt-out effects a decrease.

2. The transition temperature of ribonuclease is also sensitive to the concentration of ions (Hippel and Wrong, 1965; Luck, 1979a). T_{str} measured spectroscopically at various salt concentrations at the measured transition temperature of ribonuclease demonstrates T_{str} is similar to the transition temperature in pure water (Luck, 1979a). This supports the presumption that T_{str} describes the hydrogen-bonded state. In the presence of ions, water has a hydrogen-bond state similar to that in pure water at the transition temperature of ribonuclease. Comparing these two effects, the following can be concluded:

(1) The salt-in effect on PIOP-9 suggests that the existence of two phases appears at a higher temperature.

(2) The salt-in effect on ribonuclease suggests an earlier appearance of a phase with more water interaction.

(3) The melting temperature in D_2O of RNase of *E. coli* ribosomes is

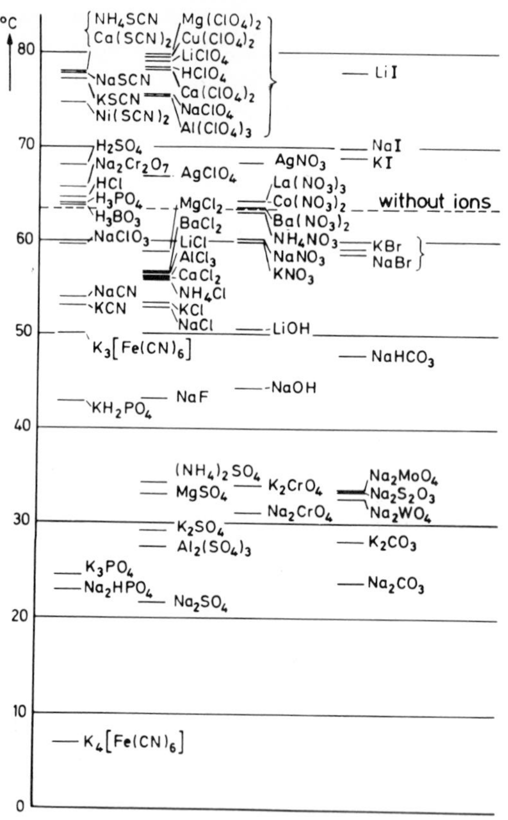

Fig. 7 Turbidity point T_K if iso-C_8H_{17}—⌬—$(OCH_2CH_2)_9OH$ in presence of 0.5 M salts. Example of the Hofmeister ion series and the anion influence on salt effects.

3.8°C higher than in H_2O. This effect could be explained as follows: The structure temperature of D_2O, $T_{str}(D_2O) \approx T(H_2O) + 4°C$ because of its stronger hydrogen bonds. The dye pseudoisocyanine forms an aqueous gel in a concentration ratio of 1 part dye to 2000 parts H_2O. In the gel state a sharp new absorption band of the dye appears (Luck, 1976c). This provides a useful method for measuring the "melting" region of the gel (Luck, 1976c). In D_2O this melting appears 3.8°C higher than in H_2O. A similar example includes the death rate of *E. coli B* during heating at 52°C is smaller in D_2O than in H_2O (Hübner *et al.*, 1970; Luck, 1976b).

(4) Philipp and Joliceur (1973) found a linear relationship between T_{str} determined by a similar method to Luck (1965c) and the heat of ion transfer, ΔH^0, from H_2O to D_2O. Both authors demonstrated the proportionality of T_{str} to ion–water interaction energy. (Buanam-Om *et al.*, 1979; Phillip and Joliceur, 1973).

(5) The nmr relaxation measurements have also been used to study the structure-breaking and structure-making effects of ions (Hertz, 1974) Samoilov (1957) has obtained similar results based on diffusion studies.

Discussion

The structure temperature has been defined as the temperature of pure water exhibiting the same free OH fraction as is found in the ionic electrolyte solution (Luck, 1976c). At different wavelengths, the value of T_{str} may vary slightly. If salts increase the O_F fraction, they will also decrease the intensity of hydrogen-bond bands. (Choppin and Buijs, 1963). There are also special ionic effects depending on the type of hydrogen-bond interactions (Luck and Zukovskij, 1974; Paquette and Joliceur, 1977). T_{str} was defined in the free OH region because the solubility of other solutes seemed to occur at free OH or weakly hydrogen-bonded OH groups.

C. DOMINANT ROLE OF ANIONS?

The dominant influence of anions on T_{str}, as shown in Fig. 6, may depend on the larger influence of anions on OH stretching vibrations and could thus be construed as an artifact of the measurement method. Anions, however, disturb the two-phase formation of PIOP-9 much more than do cations (as shown in Fig. 7). This indicates the ion effect is a changed water structure effect and not an artifact. If ionic adsorption had occurred, preferred cation adsorption on the lone-pair ether electrons would be expected. Anions also play a dominant role in affecting the partial molar volume, V_1, of water. If, in place of the salt concentration, the anion concentration of $MgCl_2$ is plotted against partial molar volume (Fig. 8), then NaCl and $MgCl_2$ have the same influence on the V_1 of water. The cause of this anion effect may be such that, because of the stronger interaction between cations and H_2O, the cations are shielded by stronger hydration spheres than the anions, and the surrounding environment of the anions is more accessible for other solute interactions.

D. THE CAUSE OF THE STRUCTURE BREAKER EFFECT

The structure-maker effect can easily be explained by the orientation of water dipoles in the strong coulombic field of the ions (Debye, 1929). What is the cause for the breaker effect? To answer this question, Raman

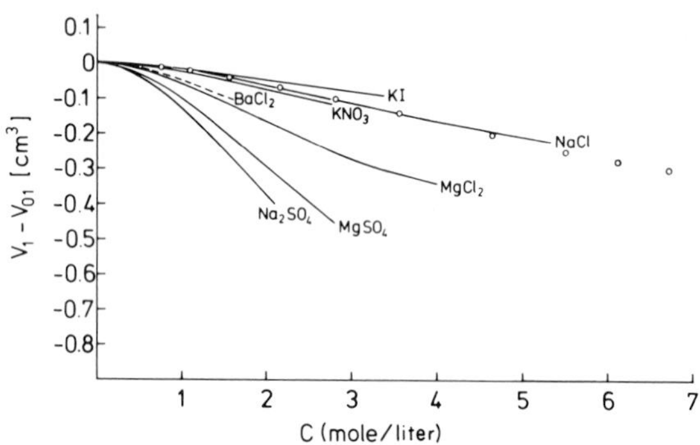

Fig. 8 Differences of partial molar volume V_1 of water in electrolyte solutions minus molar volume of pure water V_{01} at 20°C as function of salt concentration (full lines). Open circles: (V_1-V_{01}) as function of Cl^- concentration in $MgCl_2$ solutions (anions determine V_1).

spectra of water in crystalline hydrates were studied (Buanam-Om *et al.*, 1979). In crystalline hydrates, most of the water is in an optimal position with respect to its hydrogen bonds (Falk and Knop, 1973). Therefore, hydrate spectra should provide information on the water–ion interaction. Such results are shown in Fig. 9, where the shifts of band maxima in hydrates as a function of ionic species is shown. The shifts are between the H_2O vapor and ice positions.

There are many examples showing that $\Delta\nu$ of OH vibrations are proportional to the hydrogen-bond energy ΔH_H (Badger–Bauer rule). In applying this to the hydrate spectra, it could be concluded that most anions have a smaller interaction energy with H_2O than H_2O has with itself. This seems paradoxical because hydration energies ΔH_{hyd} are known to vary from about 100–200 kcal/mol or more. Is the conclusion wrong? "Per mole" means in this case "per ion-pair mole," remembering that the main part of ions have a finite solubility, despite their high values of ΔH_{hyd}. For instance, at room temperature NaCl needs a minimum of nine H_2O molecules per ion pair. Does this mean that nine H_2O molecules are needed to produce ΔH_{hyd} of NaCl? If the answer is yes, ΔH_{hyd} should be divided by 36 (i.e., 4 × 9) to calculate the value per OH or per lone pair of electrons. Indeed, the result $\Delta H_{hyd}/36$ is on the order of 5 kcal/mol, the hydrogen-bond energy per OH in ice.

Lower solubilities of ions could mean that such salts need more water of hydration to produce ΔH_{hyd}. It is from this viewpoint that the order of the Hofmeister series can be explained.

Exceptions to these observations where $\Delta\nu(\text{ion–}H_2O) > \Delta\nu(\text{ice})$ are $AlCl_3$ and CsF. It is known that AlCl can induce hydrolysis, suggesting

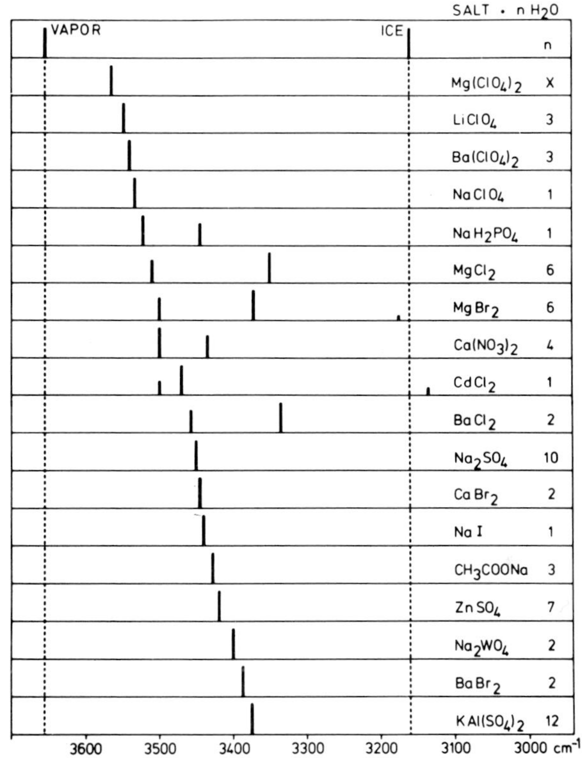

Fig. 9 Frequency maxima of Raman spectra for water of crystalline hydrates with *n* H_2O per ion pair. The range $\Delta\nu$ is between water vapor and ice. Water–anion interaction is smaller than water–water interaction.

strong interactions, whereas CsF has a large solubility. Both may be induced by stronger interactions with water than with normal electrolytes. The largest group of exceptions to the ionic rule $\Delta\nu(\text{ion–H}_2\text{O}) > \Delta\nu(\text{ice})$ are strong acids and bases. They induce broad and large water shifts in the fundamental (Zundel, 1969) or overtone (Luck, 1964) regions of the IR (see Fig. 10). The broad half-width $\Delta\nu_{1/2}$ has been correlated frɔm experience as follows: $\Delta\nu_{1/2} \approx \Delta\nu$ (Buanam-Om *et al.*, 1979).

In agreement with their excellent solubilities, the IR spectra of strong acids and bases indicate a much stronger interaction with water than water–water interactions. This suggests a possible reason for the greater influence of pH as compared with concentrations of neutral salts.

Summarizing the previous section, the Eq. (1) has been expanded to

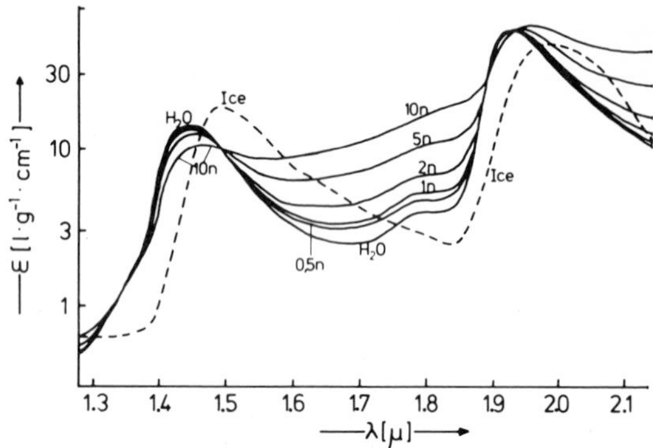

Fig. 10 Overtone and combination band of water and: 0.5, 1, 2, 5, and 10 M aqueous HCl solutions at 20°C and ice (ordinate log ε). Spectra demonstrate strong interaction water . . . H_3O^+.

$$x\ \mathrm{OH_{free}} + \mathrm{A^-} \rightleftharpoons \mathrm{OH}_x\mathrm{A}^-, \qquad K_\mathrm{A} = \frac{[\mathrm{OH}_x\mathrm{A}^-]}{[\mathrm{OH_F}]^x[\mathrm{A}^-]} \tag{3}$$

$$y\ \theta_\mathrm{free} + \mathrm{K^+} \rightleftharpoons \theta_y\mathrm{K}^+, \qquad K_\mathrm{K} = \frac{[\theta_y\mathrm{K}^+]}{[\theta]^y[\mathrm{K}^+]} \tag{4}$$

$$\frac{K}{(K_\mathrm{A})^{1/x}(K_\mathrm{K})^{1/y}} = \frac{[\mathrm{OH_b}]}{[\mathrm{OH}_x\mathrm{A}^-][\theta_y\mathrm{K}^+]}\,[\mathrm{K}^+]^{1/y}[\mathrm{A}^-]^{1/x}. \tag{5}$$

Equation (3) or (4) include serial hydration steps, for example:

$$\mathrm{OH_{free}} + \mathrm{OH}_{x-1}\mathrm{A}^- \rightleftharpoons \mathrm{OH}_x\mathrm{A}^-. \tag{3a}$$

To describe electrolyte solutions on the basis of the model just presented, the values of the different equilibrium constants in the presence of ions K_A and K_K and the corresponding K-values for the serial hydration equations such as (3a), (4a), etc., including the hydration numbers (HN) are necessary. Until this has been accomplished, the simplified T_str approach will prove useful.

E. DETERMINATION OF HYDRATION NUMBER: AN EXAMPLE

The ions ClO_4^- or BF_4^- induce sharp water bands near the bands of free OH or OD (Fig. 11). The frequency shifts $\Delta\nu$ indicate a weak interaction of these structure-breaking ions with water. The sharpness of these ion-induced bands is based on the rule $\Delta\nu_{1/2} \sim \Delta\nu$. These sharp bands allow for a detailed analysis of the ionic hydration by studying the coupling of the vibrations of H_2O and D_2O in asymmetric 1 : 1 complexes

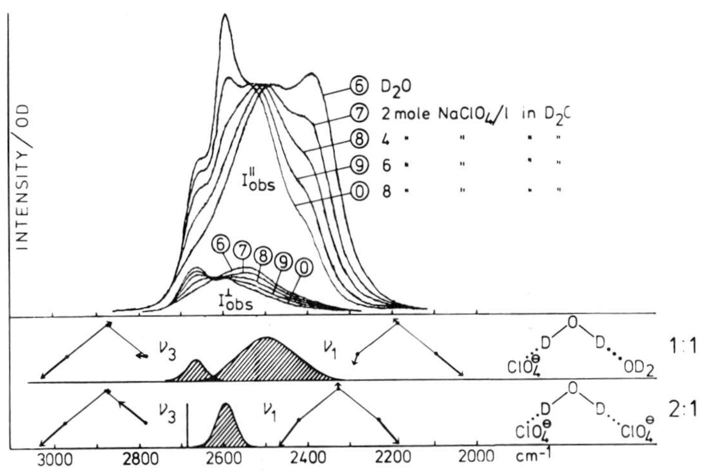

Fig. 11 Top: Raman spectra of D_2O and $NaClO_4/D_2O$ solutions at room temperature. Below: Coordination of bands 1 : 1 and 2 : 1 complexes D_2O–ClO_4^- and the calculated vibration amplitudes.

$$
\begin{array}{ccc} & O & \\ D & & D \\ \vdots & & \vdots \\ ClO_4^- & & D_2O \end{array}
$$

or 2 : 1 complexes (Luck, 1979a; Schiöberg and Luck, 1979).

$$
\begin{array}{ccc} & O & \\ D & & D \\ \vdots & & \vdots \\ ClO_4^- & & ClO_4^- \end{array}
$$

The 1 : 1 complexes induce two Raman active bands, whereas the 2 : 1 complexes induce one such band. A detailed quantitative analysis (Buanam-Om *et al.*, 1979) and comparisons with HOD and IR spectra enable calculation of the concentrations of the different species (Luck and Schiöberg, 1979). The results of the calculations are shown in Fig. 12. The top curve in Fig. 12 describes the reduction of the molar water concentration by the addition of $NaClO_4$. The second curve from the top gives the concentration of undisturbed liquidlike water, and the next two curves below that show, respectively, the concentration of the 1 : 1 and 1 : 2 complexes $H_2O\cdots ClO_4^-$. The lower half of Fig. 12 gives the D_2O hydration numbers per ClO_4 in the 1 : 1 or the 2 : 1 state. The upper curve gives the sum of D_2O hydration numbers in both complexes, and D_2O in the 1 : 1 positions begin with a HN of about 4, decreasing with $NaClO_4$ concentra-

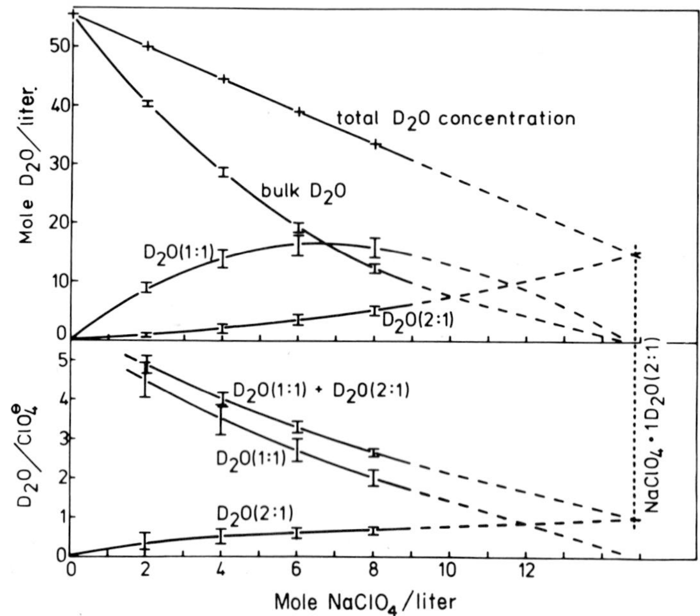

Fig. 12 Top: Total D_2O concentration and concentration of different types of D_2O molecules in $NaClO_4$ solutions. Bottom: Hydration numbers of ClO_4^- in 1 : 1 and 2 : 1 complexes at room temperature.

tion. The HN of the 2 : 1 complexes is about 1, the same value as in the crystalline hydrate.

The following comments are made with reference to Fig. 12:

(1) Coupling between cation and anion hydration would help in understanding the concentration dependence shown in the figure.

(2) The term "2 : 1 complex" could be interpreted as symmetric by involving either two ClO_4^- ions or one ClO_4^- ion between the two OD portions of one D_2O molecule. The asymmetric 1 : 1 complex with its predominance at low concentrations, indicates that the 2 : 1 complex means two ClO_4^- ions in positions with OD axes "dotting" to the ion center. This location of anions is also predicted by computer simulation calculations (Palinkas *et al.*, 1977).

The interesting decrease of water concentration, as affected by ions, also depends on their position in the Hofmeister ion series, as shown in Fig. 13. In 1 or 0.5 M Na_2SO_4 at $120 < T < 250°C$, the water content per unit volume is hardly influenced by the presence of these ions (Buanam-

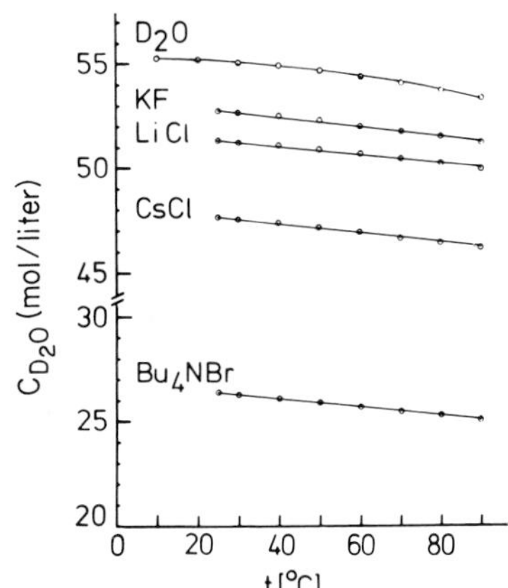

Fig. 13 The reduction of the water concentration by ions (concentration of water to salt is 15 : 1) at different temperatures.

Om *et al.*, 1979). This may mean the electrostriction effect of these ions is of the same order as the volume of ions (Zerahn, 1976). The decrease of H_2O concentration in electrolyte solutions changes the hydrogen-bond equilibrium described in Eq. (1). The reason for this may be a coupling of the anion and cation hydration.

D. SUMMARY

The Hofmeister ion series is directly related to the change in water structure. The influence of ions on the water structure can be described to a first approximation by T_{str}, which has been defined as the temperature at which pure water contains a number of non-hydrogen-bonded OH groups similar to the electrolyte solution under study. Using this definition of T_{str} makes a useful parameter for describing the solubility of electrolyte solutions.

The boundary between structure-making ions [i.e., $T_{str} < T$(solution)] and structure-breaking ions (i.e., $T_{str} > T$) varies somewhat with temperature and ion concentration. The boundary between salt-out or salt-in effects of both types of salts can depend on the hydrogen-bond acceptor or donor strength of the organic solutes.

The position of salts in the Hofmeister series depends mainly on the anions. The structure-maker effect is caused by the orientation of the water dipoles in the Coulombic field around the ions. The structure breaker effect may be induced by the smaller Coulombic field around large anions by comparison with the field strength of the lone-pair electrons.

IV. Water–Organic Solutes–Salt Solutions

A. SOLUBILITY MECHANISMS IN WATER

An idealized two-dimensional model of water is shown in Fig. 14. The hydrogen-bonded aggregates or clusters extend relatively long distances, and orientation defects, O_F, of free OH induce the density maximum at 4°C. Solutes with weak water interactions will easily be dissolved in the defect areas. This has been demonstrated spectroscopically with aqueous solutions of NH_3 (Luck, 1970). Which demonstrate a decrease of free OH by NH_3 addition, but only a small change in the NH frequencies. It is assumed NH_3 is mainly dissolved by hydrogen bonds (OH···N). Schröder (1969) also concluded that gases such as H_2 and N_2 are dissolved in water primarily in the regions between clusters. The anomalous ability of water to form gels using small amounts of solutes is also shown. The defect areas of free OH flicker rapidly, probably with a relaxation time in

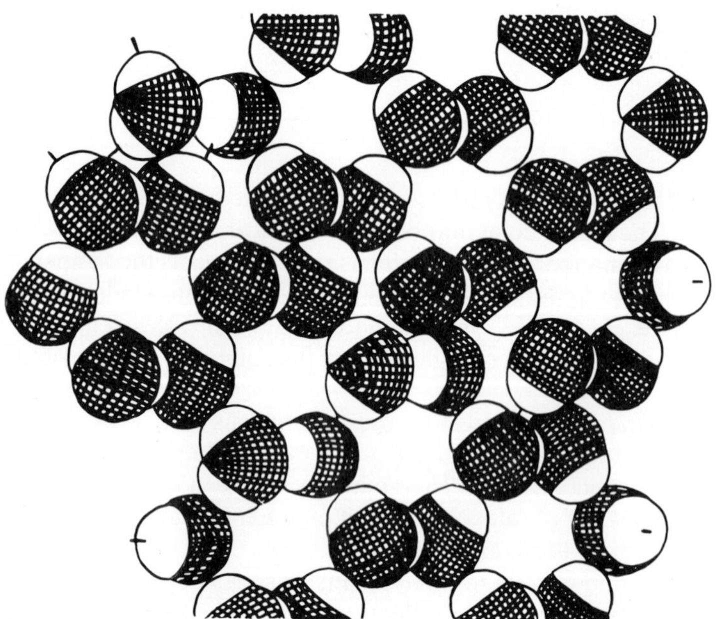

Fig. 14 Idealized model of liquid water consisting on two types of OH groups: hydrogen-bonded and nonbonded. By angle cooperativity, the non-hydrogen-bonded OH are not completely statistically distributed.

water of about 10^{-11} sec (Schröder, 1969). Small amounts of solute may enter the defect areas in water and decrease the flickering rate, thus forming gels. For instance, one part SiO to 500 parts H_2O (Gmelin, 1959), or one part pseudoisocyanine dye to 2000 parts H_2O (Luck, 1976c) can form gels. The properties of solutions will depend on the acceptor and/or donor hydrogen-bond strength of the solute. The acceptor strength can be measured by the frequency shift $\Delta\nu$ of the IR or Raman water bands. For instance, a series of $\Delta\nu$ values is as follows: acetonitrile $<$ dioxane $<$ dimethyl sulfoxide $<$ pyridine $<$ triethylamine (Schiöberg and Luck, 1979). The frequency of the ice I maximum for a linear hydrogen-bonded water in the matrix spectra is between the $\Delta\nu$ of pyridine and triethylamine (Luck, 1978).

The solubility of organic solutes in water is extremely sensitive to ion additions. Measurements of the partition coefficients of solutes in cyclohexane/water allow an estimate of this ionic influence (Luck, 1976d). Assuming the salt-out effect is caused by the insolubility in the ion–hydration shell, an estimate can be made of the size of these hydration shells of ions (i.e., the HN). In partition experiments with *p*-cresol in cyclohexane/water, apparent HNs for Na_2SO_4 at 25°C were calculated to be about 60 at 0.2 M and 20 at 1 M.[8,9] This decrease in concentration may indicate a disturbance of the hydration spheres at higher concentrations. The HN decreases with temperature for 0.2 M Na_2SO_4, from 60 at 25°C to 10 at 90°C.

The spectroscopic T_{str} determinations allow an estimate of the change in the number, N, of hydrogen-bonded water molecules (the cluster size) (Luck, 1976b). Assuming that structure-making ions can, in a simplistic view, cause an increase in N of a hydration sphere around the ions, and that normal water association exists away from the ion-hydration spheres, a spectroscopic estimate of the HNs of ions is possible (Luck, 1974, 1976b, 1976d). Hydration numbers obtained in this way compare favorably with those obtained from partition-coefficient results (Luck, 1976d).

The predictive capability of a model indicates its usefulness and validity. The T_{str} results indicate that the difference N(with electrolyte) $-$ N(pure water) has a small temperature coefficient for NaCl in comparison with other salts (Luck, 1976d). Therefore, salt-out effects from NaCl addition should have a smaller temperature dependence than with other salts. Indeed, HN determinations with *p*-cresol/cyclohexane/water have confirmed this prediction, and are shown in Fig. 15.

Hydrophobic molecules are dissolved in water by dispersion forces. The critical temperature T_c can be viewed as an approximate measure of these forces (Luck, 1979b). Therefore, the solubility of small molecules in water is proportional to T_c, as shown in Fig. 16. Large molecules may

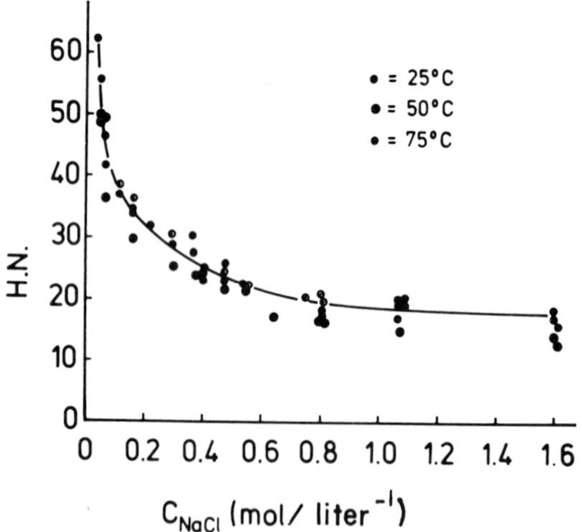

Fig. 15 Apparent hydration numbers of NaCl, estimated by partition coefficients of $CH_3C_6H_4OH$ between cyclohexane–water as function of NaCl in the water phase, establish the spectroscopically predicted temperature insensitivity of the salting-out effect of NaCl.

induce gas hydratelike structures of water depending on their size (Franks and Reid, 1973; Luck, 1976b). Twenty H_2O molecules can form a pentagondodecahedron based on the plain five-membered rings with a cavity in the middle the size of a benzene molecule (Luck, 1976b). This configuration is preferred because of its cavity formation. All the OH and lone-pair electrons face the direction of the pentagondodecahedral surface or point outward. Thus, water turns its hydrophobic sites toward the cavity.

The hydrogen-bond angles of the five-membered rings are about 10°. The hydrogen-bond energy of gas hydratelike structure is therefore slightly smaller than for an icelike six-membered water ring (Luck, 1970). The dispersion energy between water and a hydrophobic guest molecule has to compensate for this energy loss. The high coordination number ($Z = 20$) of these pentagondodecahedrons favors this. Compare this with a Z of about 12 for nonpolar liquids of spherical molecules (Luck, 1979b, 1979c).

The term of "iceberg formation" of water around hydrophobic solutes has been used in the literature. This description is based on small entropy effects. As far as we are concerned, no direct experiments showing a decrease of O_F during the iceberg-formation have confirmed this. The change in entropy during the transfer from icelike nonplanar six-membered rings to the planar five-membered rings in gas hydrates could occur in such a case.

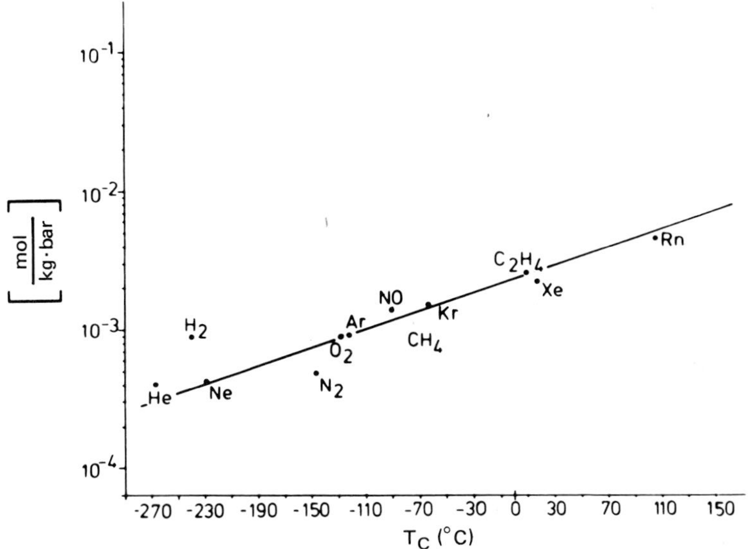

Fig. 16 The solubility of small vapor molecules in water at 60°C depends on their critical temperature, T_c, a measure of the dispersion interaction.

B. Micelle Formation

Molecules with both hydrophilic and hydrophobic parts, like the nonionic PIOP-*n* or detergents, have a different solubility mechanism for each of their parts. There is an energy gain if the hydrophobic parts are associated into a micelle nucleus, because this configuration requires less energy to separate the hydrogen bonds of the solvent water. This is the "hydrophobic bond" and nothing else. The term is misleading because there are no new forces operative. Rather, the formation of hydrophobic bonds by association of hydrophobic groups are only the result of the formation of maximum numbers of hydrogen bonds of water.

Micelle formation is sensitive to the addition of electrolytes. The equilibrium constants K_{mi} of PIOP-9 micelle formation was measured by a UV method (Luck, 1960a, 1960c). The constant K was changed at 20°C by a factor of 1000 by the addition of 0.5 mol/liter of salt in the Hofmeister ion series LiCl < NaCl < BaCl < $NaHCO_3$ < $(NH_4)_2SO_4$ < Na_2SO_4(0.4 M) < Na_2CO_3(0.3 M). The salt effect on micelle formation is used in practice to strengthen detergent efficiency in cleaning processes.

C. ELECTROYTE EFFECTS ON INTERFACES

Interfaces in contact with water are disturbed by ions in a way similar to the micelle formation just described. First, the surface tension of water increases by ion addition in the Hofmeister ion series as a result of a change in the water structure (Luck, 1964):

$$SCN^- < I^- < NO_3^- < Br^- < Cl^- < SO_4^{2-}$$

$$K^+ < Na^+ < Li^+$$

The interaction between water and hydrophobic groups is also affected by ionic addition. This can be demonstrated with force–area (F–A) diagrams of monolayers. Monolayers of water-insoluble $C_{18}H_{37}(OCH_2CH_2)_3OH$ need more force to get the same surface area in the series (Luck and Shah, 1978): $NaClO_4 > NaNO_3 > NaCl > Na_2SO_4$. These ions change F–A diagrams similar to a temperature increase of several degrees. This can be described by the T_{str} (Luck and Shah, 1978).

D. ION SOLUBILITY IN ORGANIC SOLVENTS

The solubility of ions in organic solvents is generally low compared with their solubility in water. Lithium salts perhaps play a special role because of their strong interactions with lone-pair electrons of solutes. Ionic solubility is much smaller in alcohols than in water. With the exception of LiCl, chlorides are less soluble in methanol than in water by a factor of about 0.1, or less soluble in ethanol by a factor of about 0.01 (Kleeberg, 1981). It is probably reasonable to assume that the single OH groups of solvents induces smaller ion solubilities than does the network of hydrogen bonds in water. This view is illustrated in Fig. 17 (Kleeberg, 1981), in which the reciprocal solubility (RS) in mol H_2O/mol salt of saturated solutions is plotted against H_2O/C_2H_5OH or H_2O/acetone solvent mixtures. The RS values vary little for small concentrations of organic solutes until a water–solvent ratio of about 200 : 1 is reached. This is the estimated size of water clusters. At higher concentrations of organic solutes the RS increases or the salt solubility decreases rapidly (note the log scale in Fig. 17). The K_2CrO_4 additions to H_2O/acetone mixtures induce two-phase formation. The organic phase has a lower ion content by a factor of about 10 than does the aqueous phase with lower acetone content. These experiments indicate that ionic solubility needs groups of water molecules (see Section III, D).

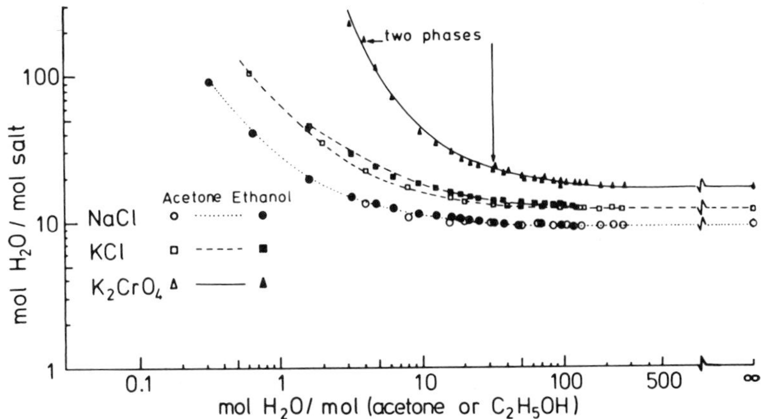

Fig. 17 Reciprocal solubility of NaCl, KCl and K_2CrO_4 in mixtures H_2O/acetone or C_2H_5OH at 23°C (Kleeberg, 1981).

E. ORGANIC HYDRATES

Detailed studies of the upper phase of a two-phase solution containing a high concentration of PIOP-9 above the two-phase forming temperature T_K indicate special effects of the primary hydrates of this type of organic molecules. The organic phase just above T_K has a relatively high water content (about 22 H_2O molecules per ether oxygen atom) that rapidly decreases with temperature to four H_2O molecules per oxygen atom, as shown in Fig. 18. Adding 0.4 *M* Na_2SO_4 to the solution reduces T_K from about 64 to 29°C and the H_2O content to about 12 H_2O molecules per oxygen atom. By increasing the temperature, the water content decreases asymptotically to about two H_2O molecules per oxygen atom. A similar limit value to two H_2O molecules per oxygen atom is observed upon increasing the concentration of structure-making ions at constant temperature (Luck, 1964).

This dihydrate has special properties (Luck, 1964), such as a viscosity maximum, a sound–velocity maximum, and a specific meander or helix X-ray structure. Molecular models have been used to demonstrate that this dihydrate structure may be induced by hydrogen bridges from one oxygen to the next but one with hydrogen-bond angles around $\beta = 0$. A two-dimensional drawing is shown in Fig. 19. The preference of this value of $\beta = 0$ may induce such primary hydrate structures. Also, cooperative effects appear to favor such hydrates. This effect may also play a

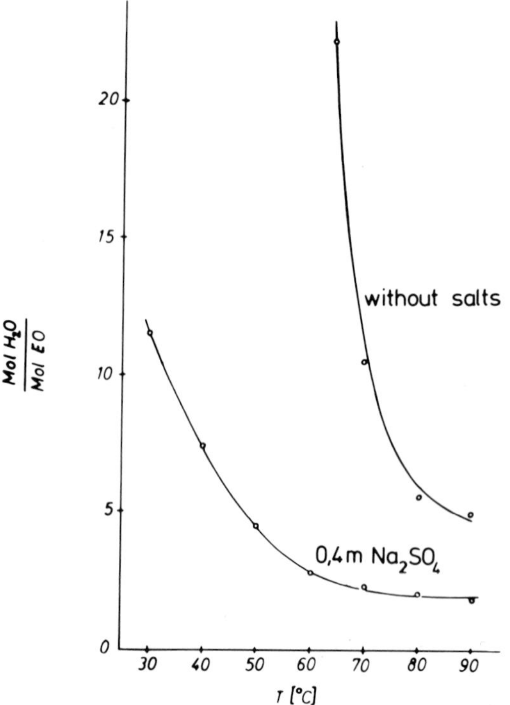

Fig. 18 H_2O content per ether group (EO) of the organic phase above T_K of PIOP-9 with 0.4 M Na_2SO_4 added and without the additive.

role in the action of protective colloids. It seems the preference for $\beta = 0$ and the formation of primary hydrates are important for the secondary and tertiary structures of biopolymers. Additional confirmation of this is still needed.

The higher water content (more than two H_2O molecules per oxygen atom) of the organic phase PIOP-9 may thus be called a secondary hydrate; it dissolves the ions.

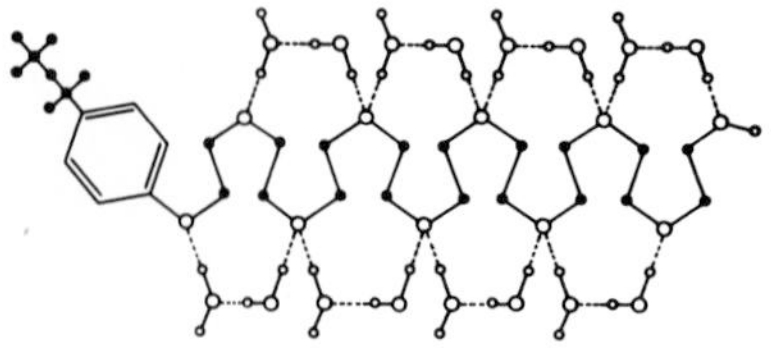

Fig. 19 Model of PIOP-9 dihydrate, O represents oxygen, ○ represents hydrogen and ● represents CH_2; hydrogen bridges with 2 H_2O and $\beta = 0$.

Warner (1962, 1965) has shown many biopolymers have preferred icelike distances for their hydrophilic groups. Some of them are arranged in six-membered rings with icelike distances. This may indicate the importance of primary hydration in biology.

Intramolecular hydrogen bonds as in DNA or cellulose, may influence the structure of primary hydrates. Properties of concentrated solutions of salts and organic solutes suggest a competition exists between the hydration shell of the ions and that of the organic solute. The turbidity point T_K of PIOP-9 has a maximum at 2 *M* KI or at 4 *M* NaI additions, as shown in Fig. 20. Below the maximum, T_K is not strongly influenced by PIOP-9 concentration, whereas above the maximum it does depend on the PIOP-9 concentration (Luck, 1979a). Also, the water content of the organic phase above T_K decreases with increasing ion concentrations, as shown in Fig. 21.

F. COACERVATES

Bungenberg de Jong called the two-phase formation with one organic-rich phase and one a water-rich phase *coacervation* and the organic phase *coacervate* (Bungenberg de Jong and Kruyt, 1930).

The water content of the PIOP-9 coacervate depends strongly on ionic addition to the PIOP-9 solution below its turbidity point T_K, as shown in Fig. 22, in which the Hofmeister ion series can again be recog-

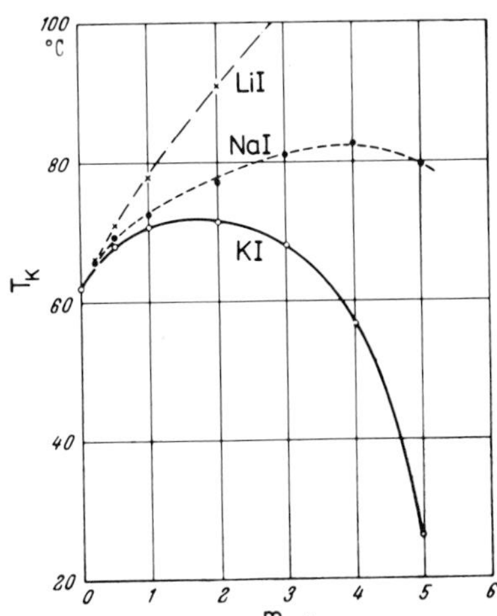

Fig. 20 Turbidity point T_K of PIOP-9 as function of concentration of different salts.

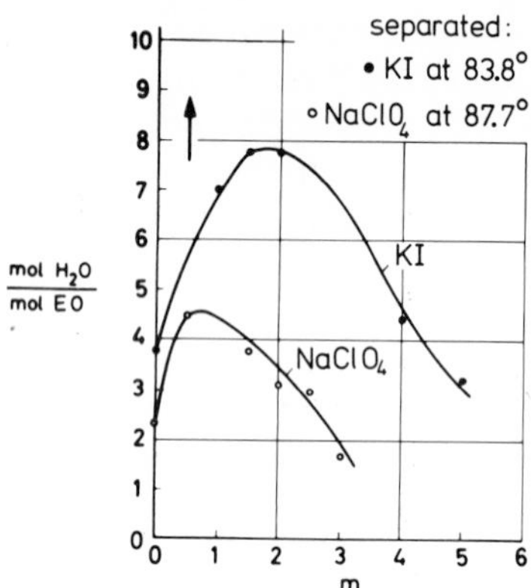

Fig. 21 Water content per ether group (EO) of the organic phase at about T_K of PIOP-9 as function of the salt concentration of the original solution (the KI experiment separated at 83.7°C; the $NaClO_4$ experiment separated at 87.7°C).

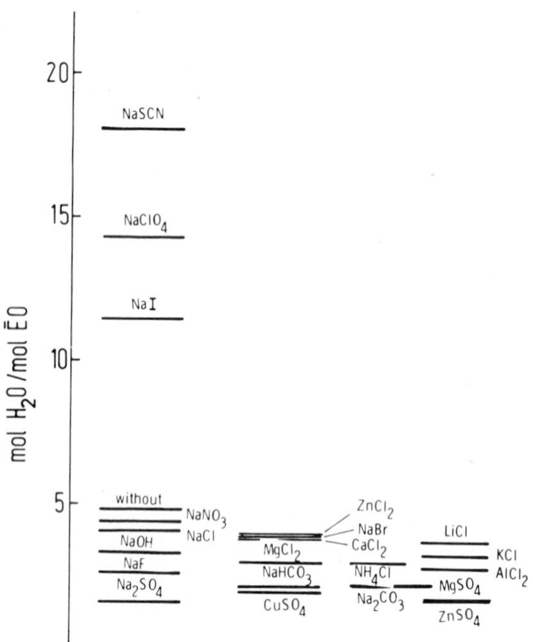

Fig. 22 Water content per ether group of the organic phase at $T = 80°C$ for PIOP-9 for different salt additives at 0.5 M salt content in the homogeneous solution below T_K.

nized. The content of PIOP-9 in the aqueous phase also changes by ionic addition in the same experiment (see the log scale in Fig. 23) (Luck *et al.*, 1980b). Gelatine–chondroitine sulfate also form coacervates in certain concentration regions (Kleeberg, 1981; Luck, 1979a). The ratio of gelatine to chondroitine sulfate in the coacervate can be changed by a factor of three by the presence of NaCl, $NaHPO_4$, or carboxylates (Kleeberg, 1981; Luck, 1979a). These and other experiments indicate cartilage may have properties similar to coacervates (Kleeberg and Luck, 1968).

Coacervates of mixtures of organic solutes have different parameters. Organic additives to aqueous solutions of PIOP-9 affect the formation of a coacervate differently, as shown in Fig. 24. Water-soluble molecules

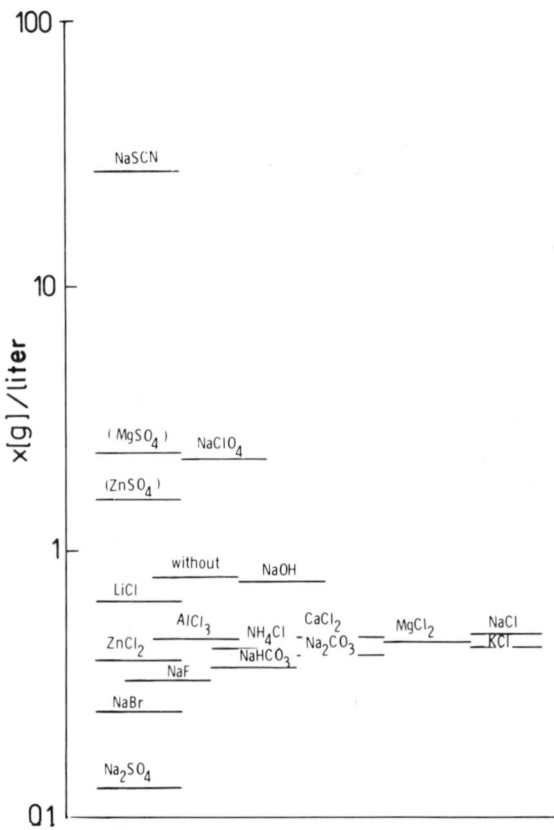

Fig. 23 Content of PIOP-9 in the water phase at 80°C for different salt additives (0.5 M salt content in the homogeneous solution below T_K).

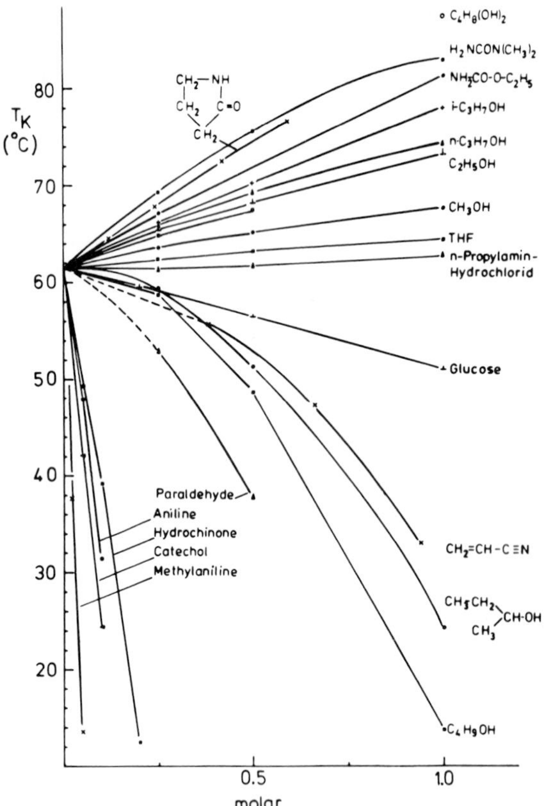

Fig. 24 Influence of different organic additives at the turbidity point of 10 g PIOP-9 per liter H_2O.

may protect hydrophobic PIOP-9 groups, and nonpolar or nonsoluble solvents may protect the hydrophilic groups of PIOP-9.

Bungenberg de Jong's coacervate model proposes a primary hydration shell of the organic solute with a diffuse secondary shell (Bungenberg de Jong and Kruyt, 1930). Jong assumes that at the coacervate formation there exists a connection between different primary hydration shells. The PIOP-9 experiments discussed previously extend this model so the secondary hydration shells form the coacervate. This model resembles the ion pair-formation model (Frank and Wen, 1957; Lilley, 1973).

As with some coacervates, there are some biochemical processes with structured states that favor higher temperature (Lauffer, 1975). Examples of this effect include polymerization of tobacco mosaic protein, gel formation of sickle-cell hemoglobin at $T > 38°C$, formation of insoluble myosin in H_2O at $T > 23°C$ at pH = 7.4, actin association at $T > 25°C$, formation of poly-L-proline flocculates in water at $T > 25°C$, formation of insoluble tropocollagen in water at $T > 37°C$, pseudopods formation of *amoeba dubia* at $T > T_c$, egg cell division at $T > T_c$, favored antibody formation at $T > 25°C$, etc. (Lauffer, 1975). Lauffer (1975) has termed these endothermic processes "entropy-driven processes". He assumed that the entropy of water increased during these processes. These entropy-driven processes are influenced by ions in the Hofmeister ion series. The Hofmeister series also affects the helix coil transformation or denaturation of DNA, ribonuclease, or collagen gelatine (Luck, 1976b).

Different hydration steps for polymers are also observed by nmr methods (Franks, 1975; Luck, 1976b).

G. SUMMARY

The water solubility of molecules with weak hydrogen bonds is primarily determined by interaction with non- or weakly hydrogen-bonded OH groups. Molecules with strong basic groups can disturb the water hydrogen-bond system. Small hydrophobic molecules can be dissolved in water by dispersion interaction and by forming clathratelike structures, which may also occur around hydrophobic moieties of organic molecules. Association of hydrophobic moieties (micelle formation) lead to a minimum of broken hydrogen bonds (hydrophobic bond).

Solubility in water (as with micelle formation or interactions between water at interfaces, or coacervate formation) is sensitive to salt addition. The ionic solubility mechanism in water seems to depend on the formation of ion hydrates; this ion hydrate phase disturbs the solubility of other solutes.

Hydrophilic organic molecules appear to form hydration shells with hydrogen-bond angles (β) near O.

V. Examples of Aqueous Systems

A. THE 6-NYLON (POLYAMIDE) FIBERS

The water content of 6-nylon fibers plays an important role in the infusion of acid dyes (Luck, 1960b, 1965b, 1972). These dyes do not diffuse significantly at room temperature and normal air humidities. Fibers that the dyeing process is conducted in water can be stored at least one year in air without noticeable dye transport from the outer fiber areas. Only by heating the air to 150°C will a diffusion velocity of the acid dyes in 6-nylon be established, comparable to the dyeing process in water at 60°C. The diffusion coefficient D of water in 6-nylon depends on the relative humidity of the air (Luck, 1972), as is shown in Figs. 25 and 26. This may be of biological importance for skins of animal living in arid areas.

The water content in the 6-nylon fiber depends on the water activity, which can be changed by the addition of various ions, as shown in Fig. 27 (Luck, 1965b, 1976b).

Ions have the following effects on dyed textiles:

(1) They change the water content of the fiber,
(2) They show salt-out effects on dyes,
(3) They show salt-out effects of leveling agents, which increase their efficiency, and
(4) The dye salt-out effects during the washing processes increase the wash-fastness effect of the dyes (Luck, 1966).

The properties of fibers usually depend on their pretreatment. For instance, HCl or H_2SO_4 have partition coefficients for 6-nylon–H_2O at pH 4 of about 200 : 1. The result is that an acid-treated 6-nylon fiber (6–6-nylon is similar) has to be exposed 200 times to equilibrium with water for 50% of the absorbed acid to efflux from the fiber (at a 1 : 1 volume ratio).

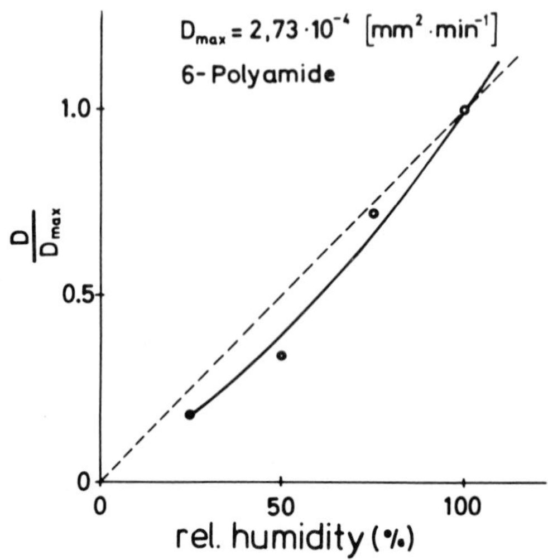

Fig. 25 Apparent diffusion coefficient of water in 6-polyamide as function of relative humidity.

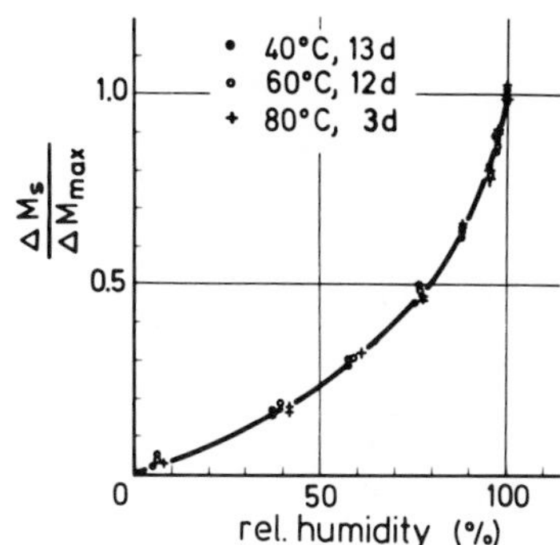

Fig. 26 Equilibrium water uptake of 6-polyamide at different humidities and different temperatures (×d represents time of measurement).

B. COLLAGEN AND CARTILAGE

Samples of collagen, cartilage, or gelatine show a shift of the water band maxima depending on the relative humidity of the air (Kleeberg, 1981; Kleeberg and Luck, 1977), as seen in Fig. 28. Below the primary hydrate completion at low relative humidity (<50%), the frequencies of the water-band maxima are constant, but some polymer IR bands do shift.

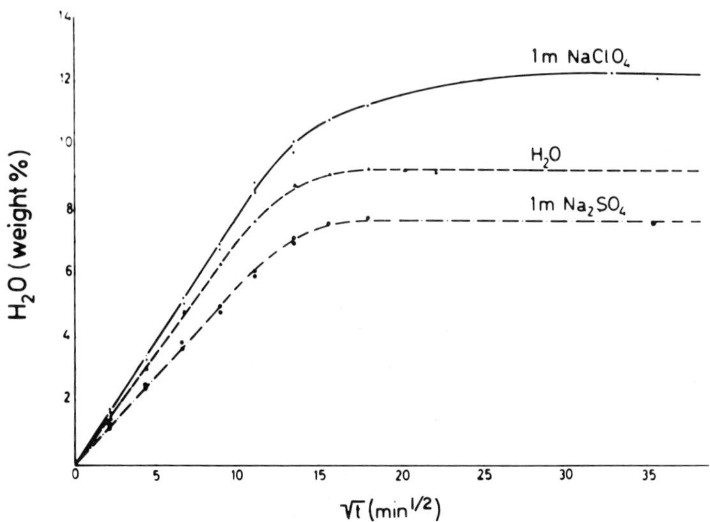

Fig. 27 Water absorption of 6-polyamide as function of time t in presence of liquid water or electrolyte solutions at 70°C (determined by overtone IR spectra).

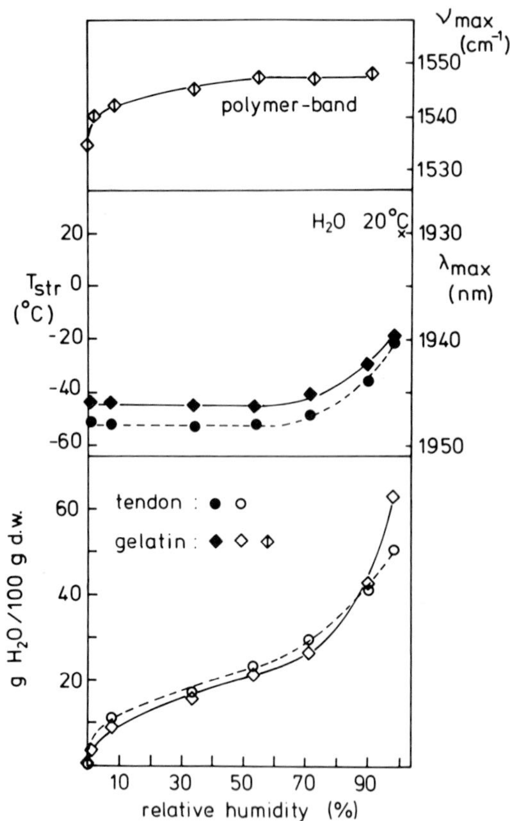

Fig. 28 Influence of relative humidity on the properties of tendon and gelatine. Top: Change of the frequency maximum for the polymer band 1550 cm^{-1}. Middle: Change of the frequency maximum for the water combination band. Bottom: The water desorption isotherm.

Both facts indicate the formation of a primary hydrate that may prefer some polymeric configuration. Above the primary hydrate completion (i.e. rh > 50%), the water IR maxima shift slowly against the bulk water, as seen in Fig. 28. But different spectra taken at different, neighboring relative humidities suggest three types of water: Primary hydrate, bulk water, and a third type with properties between the first two. The third type generally has a greater similarity to bulk water than to the hydrate phase, but as 100% relative humidity is approached, bulk water properties appear. The amount of this intermediate water is influenced by the pres-

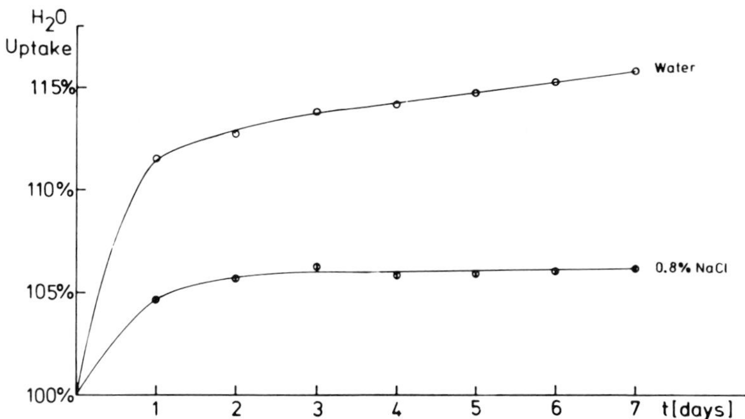

Fig. 29 Water uptake of bovine nasal cartilage in the presence of liquid water and 0.8% NaCl solution. T = 4°C. Fresh weight = 100%.

ence of ions similar to the experiment shown in Fig. 25. This is shown in Fig. 29.

The spectroscopic experiments described here are focused on the combination water band, $(\nu_3 + \nu_2)$, at 1900–2000 nm, because the bending vibration of water (ν_2) is different from alcohol-bending vibrations. Thus, the biopolymer OH groups do not exhibit absorption in this ν region. These experiments tend to support the view of biologists and physicians who differentiate between bound and free water (Hazlewood, 1973, 1979). Only relatively small differences in the properties of the two waters have been found, though. Therefore, the terms "hydrated water" (instead of bound water) and "liquidlike water" (instead of free water) are preferred. Bulk water has a strong hydrogen-bonded or associated system similar to the hydrates of biopolymers. This probably can be explained by the fact the acceptor strengths of the two do not differ much. Many biological processes are extremely sensitive to the water content. For instance, seeds that did not grow in air humidity $< 100\%$ rh for hundreds of years start to grow as soon as the rh reaches 100%. Food stability depends on water activity. Salt or sugar additives reduce the water activity in food and, consequently, increase its stability. A reduction of humidity below some threshold percentage stops the growth of microorganism on food. This discussion suggests that biological processes, such as the dye diffusion process discussed previously, require the presence of liquidlike water.

C. SUMMARY

Fibers or biopolymers exhibit different properties of water at low and high concentrations. Spectroscopic investigations establish small but distinct differences between hydrate water at low humidities, and liquidlike water at high humidities.

VI. Water in Desalination Membranes and the Desalination Mechanism

A. WORKING HYPOTHESIS

The curve at the bottom of Fig. 18 suggest a hypothesis for the mechanism of the desalination process with membranes. If, above the two-phase forming temperature T_K, the organic phase of PIOP-9, is separated from the aqueous phase and reheated, two phases are again formed. Because the newly formed aqueous phase should be free of ions, this procedure provides a method of desalination. The ionic content of the newly formed aqueous phase, however, is similar to the original solution below T_K prior to the phase separation. This means the secondary hydration sphere of PIOP-9 dissolves ions, which are carried with the organic phase during the separation process. The primary hydration water does not dissolve ions, and suggests the following hypothesis (Luck *et al.*, 1979, 1980a): "Desalination membranes should contain mainly primary hydrate water, the special structure of which (due to the polymer structure) weakens the ion solubility. The existence of secondary hydration shells with ion solubility should be prevented by steric effects." In other words, desalination membranes should not contain the water-types needed by the processes mentioned in the last section. The validity of this hypothesis was first suggested in experiments with ethylenoxide–propylene oxide copolymers, which also form two phases above T_K. Further heating of the separated polymer phase separated into an organic aqueous phase with a *reduced* ion content and a low concentration of polymers (Luck, 1973, Luck *et al.*, 1980a). The propylene oxide derivatives have a reduced water affinity that hinders the formation of secondary hydrates.

B. A SPECTROSCOPIC METHOD FOR MEMBRANE STUDYING

A second verification of the hypothesis described previously uses the spectroscopic method. The first question was whether the water structure in efficient desalination membranes is different from liquid water. The temperature dependence of the water infrared combination band, (ν_3 + ν_2), together with ice at −10°C is presented in Fig. 30, the positions of the water band for several organic base–water solutions are also included. An increasing frequency shift, $\Delta\nu$, is noted with decreasing temperature. This temperature dependence of the spectra of liquid H_2O can be described by three subbands with maxima at 5297 cm^{-1} (position of free OH), 5200 cm^{-1} (partition of unfavored hydrogen-bond angles around 110°), and 5038 cm^{-1} (linear hydrogen bonds). In liquid water at room temperature the three subbands have an intensity ratio of about 1 : 3.3 : 3.5, indicating a low content (about 12%) of free OH. (Luck *et al.*, 1980a; Siemann, 1976).

With decreasing temperature, the hydrogen-bond equilibrium is shifted to the hydrogen-bonded OH groups. As a result, a shift to smaller frequencies is observed, indicating a stronger hydrogen-bonded state. In

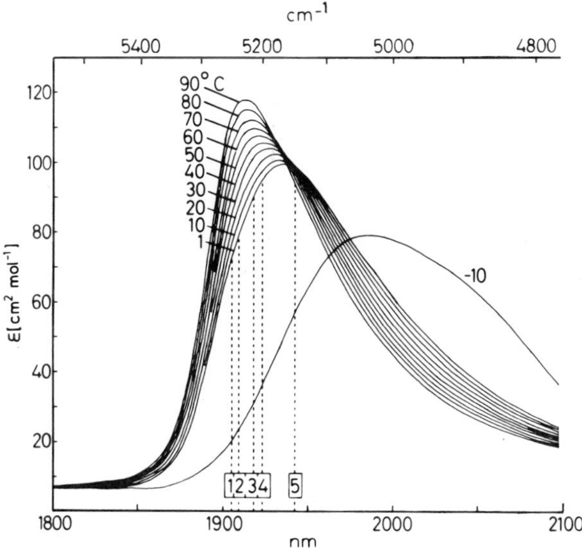

Fig. 30 Temperature dependence of the (ν_3 + ν_2) combination band of liquid water. Dotted lines are the positions of band maxima for water in the presence of different organic bases −10° ice spectrum: (1) acetronitrile, (2) acetone, (3) diethylether, (4) tetrahydrofuran, and (5) dimethylsulfoxide.

Fig. 30, the position of water maxima in the presence of different bases (dotted lines) is shown (Luck *et al.*, 1979). Thus, $\Delta\nu$ of the water band maxima, ν_{max}, is proportional to the average of the hydrogen-bond interaction energy.

Another example is shown in Fig. 31, where ν_{max} of aqueous mixtures of $HO(C_2H_4O)_9H$ is plotted as a function of the water content (Luck *et al.*, 1980a). The unusually strong hydrogen bonding of the dihydrate (2 H_2O per ether oxygen) induces a minimum of ν_{max}.

C. Cellulose Acetate Membranes

Homogenous cellulose acetate membranes (grade of acetylation 39, 1%) cast from acetone solutions of Celite K-700 (Bayer Leverkusen) were used to study the occluded water structures in desalination membranes using IR spectroscopy.* The spectra of water in these cellulose acetate membranes is shown in Fig. 32. Their position is shifted to higher frequencies compared to liquid water. The value of ν_{max} at 11% rh corresponds to ν_{max} of liquid water at 110°C, and ν_{max} (98% rh) corresponds to about 80°C

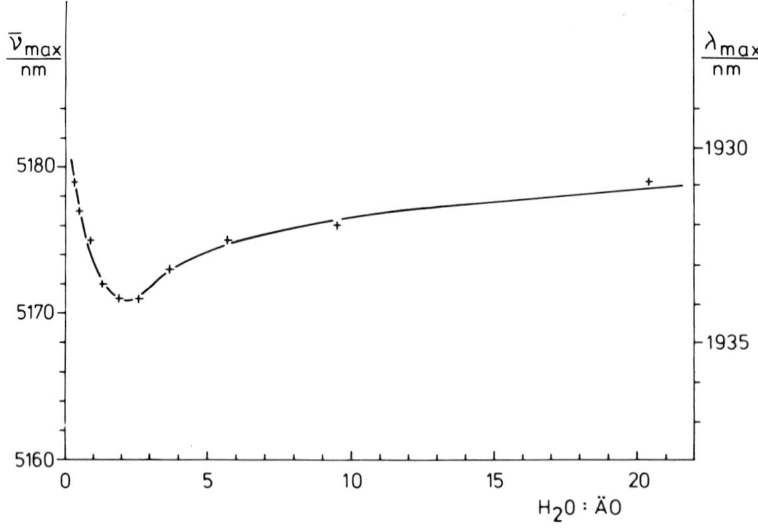

Fig. 31 Band maxima for water in the presence of different amounts of HO—$(C_2H_4O)_9H$. Note the preference for the dihydrate with two H_2O molecules per ether group.

* Dr. Pusch, Max Planck Institut für Biophysik, Frankfurt, Germany, supplied these membranes and is thanked by the author.

for ν_{max} of liquid water. The ratios of the intensities of the three water subbands in cellulose acetate are 1 : 1.5 : 0.5 at 11% rh; and 1 : 2.4 : 1.9 at 98% rh. Comparing this with the pure water ratio of intensities 1 : 3.3 : 3.5, a relatively high content of weak hydrogen-bonded or free OH water is found in the cellulose acetate.

A comparison of the spectra of membrane water with other water bands $2\nu_3;\nu_3;\nu_1;\nu_2$ also indicates a weaker hydrogen-bond system (Luck *et al.*, 1980a; Siemann, 1976). It could be concluded that the hydrogen-bond interaction between water and cellulose acetate is much weaker than that between water molecules themselves, as a result of the weaker hydrogen-bond acceptors of the cellulose acetate. The intensity of the membrane–water interaction in the region of the linear hydrogen bonds is very weak. The content of linear hydrogen bonds in liquid water is about 40% at room temperature, whereas membranes contain a low percentage of liquidlike water. Fundamental IR spectra of water in membranes are complicated by the self-absorption of polymers in this region (Alexander *et al.*, 1971). A compensation technique with dried membranes in the second light beam of a double-beam spectrophotometer induces errors caused by small differences in membrane quality and by energy losses of the instrument (broader slit). In Fig. 33, the IR fundamental spectra is shown for different rh values, with the resultant frequency shifts moving in the direction of

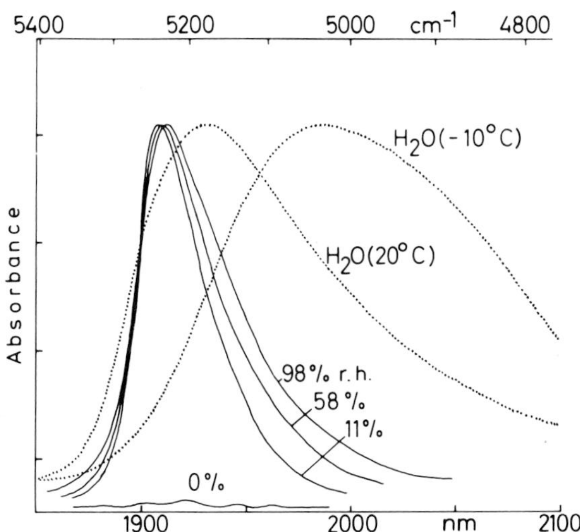

Fig. 32 Full lines: Combination band of water in cellulose acetate membranes of different relative humidities at 20°C. Dotted lines: Liquid water at 20°C and ice at −10°C. All maxima are normalized to equal heights.

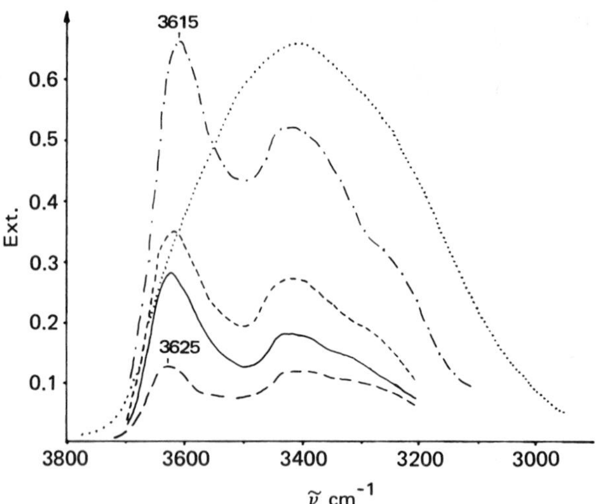

Fig. 33 Fundamental band of water in cellulose acetate membranes at different relative humidities at 20°C. The dotted line is for liquid water at 20°C.

weaker hydrogen bonds than are observed with bulk water. Toprak, Agar, and Falk have recently confirmed these results, observing weaker hydrogen bonds for the fundamental bands of water in cellulose acetate (Toprak *et al.*, 1979).

D. POLYIMIDE MEMBRANES*

Examples of some water spectra in polyimide membranes are depicted in Fig. 34 (Luck *et al.*, 1979). Again, a shift to higher frequencies is observed at various rh values. Because the polyimide polymer has no OH or NH groups, fundamental IR spectra could be taken. Only at 3500 cm^{-1} does a C=O overtone give a small disturbance.

The results for the polyimide membranes also indicate that the occluded water has fewer linear hydrogen bonds of the water–water type. These linear hydrogen bonds would absorb in the region 3300–3200 cm^{-1}, as shown for ice at -10°C in Fig. 34. The desalination characteristics for the polyimide membranes studied, indicated 0.3 liter/m^2 (13.5 liter/m^2) flux for 3.5% NaCl solutions at 100 bar and 100 μm (10 μm) thickness, with an NaCl rejection of $R = 99.0\%$ (99.7%).

* Dr. Walsh, Batelle Institut, Frankfurt, Germany, supplied these membranes and is thanked by the author (Walch *et al.*, 1974).

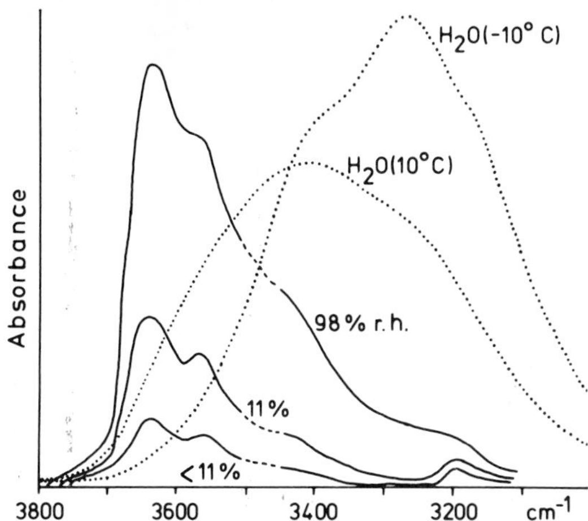

Fig. 34 Full lines: Fundamental IR band for water in polyimide membranes at different relative humidities. Dotted lines: Same band for liquid water at 10°C and ice at −10°C.

E. GLASS MEMBRANES*

Desalination membranes with pore diameters ranging from 26 to 100 Å can be produced as Vycor glass ($B_2O_3/SiO_2/Na_2O$) by dissolving the borate phase from the two-phase glass. The typical desalination performance for such membranes gives a NaCl rejection of about 80% and a flux of about 1.0 $\frac{m^3}{m^2\ day}$. In Fig. 35, the 1900 nm band of water in these glass membranes is shown as a function of rh. At low humidities, the water bands differ strongly from bulk water, indicating that the hydrates are more weakly hydrogen-bonded than is liquid water. At high humidities the water spectra are similar to liquid water, suggesting poor ion rejection. The ν_{max} of membrane water, even at high humidity (98% rh), is still slightly higher (cm^{-1}) than liquid water, indicating on average, a weaker hydrogen-bond network.

The first overtone of water in glass membranes is given in Fig. 36. At low rh values, the SiOH vibration at 1365 cm^{-1} can be observed. The band disappears as the humidity is increased because of increased hydrogen bonding with water. The disappearance of free SiOH by the addition of small amounts of water can also be noticed by the SiOH combination band at 2200 nm (Langer *et al.*, 1979; Luck, 1980a). With the addition of small amounts of water, a peak at 1402–1410 nm (7132–7092 cm^{-1}) ap-

* Dr. Schnabel, Schott & Gen. Mainz, Germany, supplied these membranes and is thanked by the author.

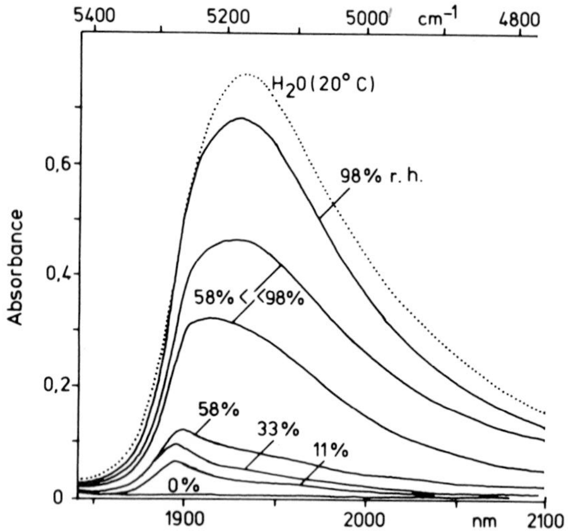

Fig. 35 Combination band of water in glass membranes of different relative humidities. The dotted line is for liquid water at 20°C.

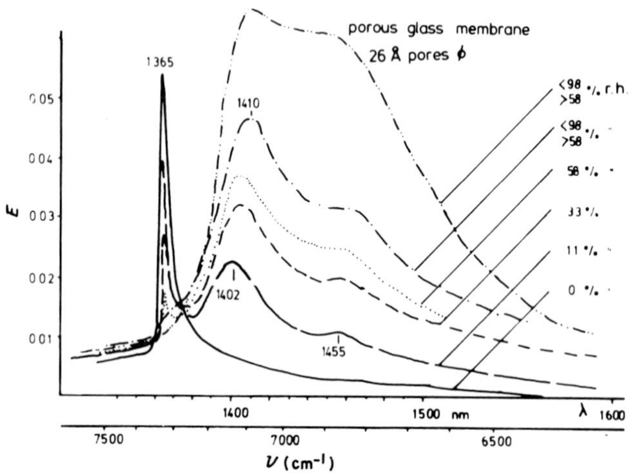

Fig. 36 First overtone band of water in porous glass membranes (26 Å pore diameter) at different relative humidities. The maximum at 1354 nm is assigned to free OH of the membrane SiOH groups.

pears. This peak corresponds to the sharp band at 7125 cm^{-1} for free OH in liquid water at 350°C (Luck, 1963, 1965d, 1976b). In membranes, this peak may be taken as evidence for 1 : 1 complexes of SiOH/H_2O with one free water OH group, and one hydrogen-bonded group. The second maximum at 1455 nm (6872 cm^{-1}) is similar to that of water at 50°C, and can be induced by water hydrogen bonding to SiOH or to H_2O. This second band increases with increasing H_2O content, but even at 98% rh the water bands in glass membranes differ slightly from those of liquidlike water, and the peak at 1410 nm remains.

The NaCl rejection of the glass membranes can be increased to 99% by surface modification of the SiOH groups with $(CH_2)_4SO_3H$. Water spectra of these modified glass membranes, manufactured as hollow fibers (15 μm outer diameter, pore diameter of about 25 Å, and permeation flux of about 60 liter/m^2 day) were measured and are presented in Fig. 37. These spectra differ more from liquidlike water spectra than do the spectra for the membranes presented in Fig. 36. At rh > 70%, the spectra do not differ much from liquidlike water spectra. The water adsorption isotherm flattens to a horizontal line in this rh region, (Belfort and Sinai, 1979), probably by filling the fiber pore. Therefore, the state of water in the membranes cannot be determined spectroscopically at high rh values in hollow fibers. The water spectra for chemically modified and untreated hollow fibers [$(CH_2)_4SO_3H$] at 58% rh are compared in Fig. 38. The water spectra for the treated fibers are shifted further from liquid water spectra than are the unmodified fibers. This suggests that the modified fibers have higher ion rejections. This has been found by Schnabel (R. Schnabel, private communication).

F. MEMBRANE DESALINATION MECHANISM

The working hypothesis that desalination membranes should contain water of a structure different from liquidlike water appears to have been confirmed by the spectroscopic results presented. In contrast to biological processes, the hydrate water in desalination membranes is more weakly bonded than in liquid water. This favors a higher permeation flux. In agreement with this observation, Belfort and Sinai (1979) found, in the same glass membranes, an activation energy ΔH for the microscopic water motion of 1.4 kcal/mol, using pulse nmr. This is much smaller than ΔH for the self-diffusion coefficient of pure water, which is about 5 kcal/mol (Samoilov, 1957, 1961). The formation of extended areas of liquidlike water is disturbed in the membranes discussed in Sections VI,B.–E. The

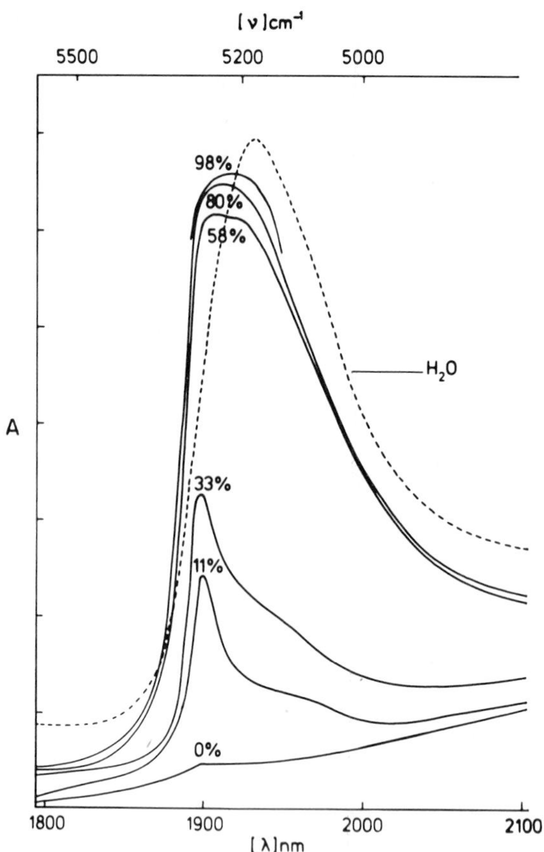

Fig. 37 Combination band of water in porous glass hollow fiber membranes. The dotted line is for liquid water.

water uptake of cellulose acetate at saturation corresponds to about a quadruple H_2O layer. The pores (26 Å) of the used glass membranes seem to have two quadruple H_2O layers, whereas the hollow glass fibers have about a 5 H_2O-per-pore diameter. We concluded in Section IV that NaCl needs a minimum of 9 H_2O per ion pair to be dissolved. These order of magnitude numbers seem to suggest that the water content in the used desalination membranes is too small to dissolve ions. This conclusion is confirmed by the observation that the position of ion in the Hofmeister series determines its rejection by desalination membranes (Hodgson, 1970; Pusch, 1975; Sourirajan, 1971). The order of separation efficiency is $SO_4^- > Cl^- > NO_3^- > I^- > SCN^-$. The reflection coefficients of different ions with cellulose acetate membranes as measured by Pusch (1975), is shown in Fig. 39 as a function of ionic concentration. Thus, SO_4^- would

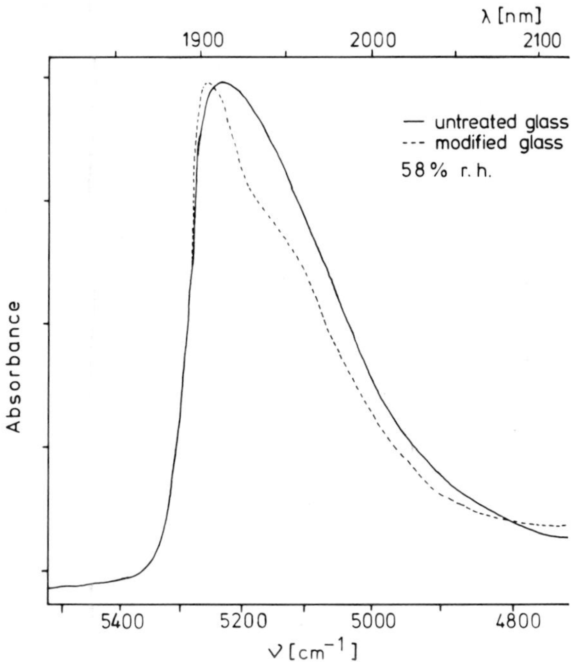

Fig. 38 Combination band of water in hollow fiber porous glass membranes at 58% relative humidity. The full line is for the untreated glass surface, while the dotted line is for the sulfonic acid modified glass surface.

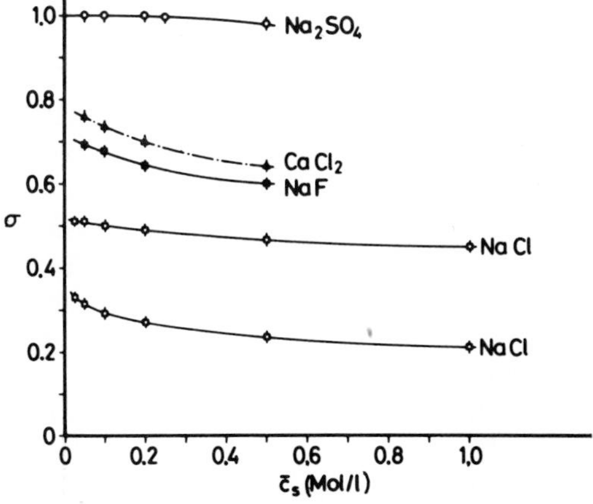

Fig. 39 Reflection coefficients σ of different ions in cellulose acetate membranes as function of salt concentration $\bar{C}_s$ (Pusch, 1975).

need a larger hydration shell than Cl^-; therefore, its rejection by membranes is higher.

The proposed model, in which a differentiation between hydrated and liquidlike water in aqueous systems is made, seems to be a useful concept. Additional experiments are needed to verify this model in more detail. Indication of two types of water in copolyoxamide desalination membranes was also found by Vogl et. al. by Differential Scanning Calorimetry (DSC) technique (Tirell *et al.*, 1979). Hoeve arrived at a similar conclusion, finding that the pores of collagen are too small to form clustered water (Hoeve, 1979). Starkweather (1979) concluded that a monohydrate has about one H_2O molecule per 2 amide groups in 6-6-nylon, and Hauser (1975) using the nmr techniques, suggested the following lipids: (1) the first hydration shell consists of 1 H_2O per lipid group, (2) the "bound shell" consists of 11 H_2O per polymer unit, (3) the "trapped water" consists of about 11 H_2O, and (4) the remainder of the water is bulk water.

G. SUMMARY

The structure of water in efficient desalination membranes differs from that of liquid water. The greater the difference, the better the solute ionic rejection. This confirms the assumed hypothesis that desalination membranes should contain only primary hydration water and not liquidlike water. Membrane mechanisms have been discussed in some earlier papers (Neale and Williamson, 1956; Reid and Breton, 1959; Schultz and Asunmaa, 1970) on the basis of water bound to cellulose or cellulose acetate, but these papers have assumed stronger hydrogen bonding of water to polymers (Neale and Williamson, 1956; Reid and Breton, 1959; Schultz and Asunmaa, 1970). Neal and Williamson (1956) and Schultz and Asunmaa (1970) assumed an icelike water structure inside cellulose. This assumption was suggested by measured entropy effects. But there are other entropy effects possible as icelike structures—for instance, a clathratelike water structure with planar five-membered rings, structures of organic hydrates with favored hydrogen-bond angles or cooperativity. Another factor could be coordination numbers other than 4 in liquid water. Higher dispersion forces, together with coordination numbers higher than 4 may be taken into account too. In our results, we have observed in all efficient desalination membranes a weaker hydrogen-bond system than in liquid water. This would favor an efficient flux of water through the membranes.

Reid and Breton (1959) in their discussion of a difference in ionic rejections, stated that "those ions . . . that combine with the membrane through hydrogen bonding . . . are presumed to be transported across the membrane. . . ." This presumption, however, is in contradiction to the parallelism between the ion rejection and the Hofmeister ions series. Weak hydrogen-bond acceptors like ClO_4^- have a bad rejection coefficient and vice versa.

Acknowledgment

I thank the German Bundesministerium für Forschung und Technologie, Bonn, West Germany and Dechema, Frankfurt, for supporting this work. I thank my coworkers: A. Behrens, C. Buanam-Om, W. Ditter, H. Kleeberg, D. Schiöberg, O. Schrems, and U. Siemann for their activities and discussions, and G. Belfort, R. Schnabel, and W. Pusch for discussions and membranes. I thank Dr. M. Falk for attending our seminar on membrane results at Maria Rain in the spring of 1977, and establishing our results.

References

Alexander, D. M., Hill, D. J. T., and White, L. R. (1971). *Aust. J. Chem.*, **24,** 1143.

Ayers, G. P., and Pullin, A. D. E. (1976). *Spectrochim. Acta* **32A,** 1629, 1641, 1689, 1695.

Behrens, A., and Luck, W. A. P. (1979). "I.R. Matrix Isolation Studies of Self-association of Water and Oxims! Proofs of Cyclic Structures in Matrix and Liquid State?" Presented at Int. Conference Spectroscopy, Frankfurt, *J. Mol. Structure,* **60,** 337.

Belfort, G., and Sinai, N. (1979). "Water in Polymers". Presented at ACS Meeting, Washington, September 1979.

Ben-Naim, A. (1974). "Recent Developments in the Molecular Theory of Liquid Water." In Structure of Water and Aqueous Solutions (W. A. P. Luck, ed.), Chap. 2, Weinheim, Verlag Chemie.

Buanam-Om, C., Luck, W. A. P.,and Schiöberg, D. (1979). *Z. Phys. Chem.* **117,** 19.

Bungenberg de Jong, H. G., and Kruyt, H. R. (1930). *Kolloid Z.* **50,** 39.

Choppin, G. R., and Buijs, K. J. (1963). *Chem. Phys.* **39,** 2042.

Debye, P. (1929). "Polare Molekeln." Dover, Hirzel, Leipzig.

Eucken, A. (1946). Nachr. Akad. Wiss. Goettingen **38, ■■■**.

Eucken, A. (1948). Z. Elektrochem. **52,** 264.

Eucken, A. (1949). Z. Elektrochem. **53,** 102.

Falk, M., and Knop, O. (1973). "Water, A Comprehensive Treatise", (F. Frank, ed.) Vol. 2, p.55. Plenum, New York-London.

Frank, H. S., and Wen, W. V. (1957) *Disc. Faraday Soc.* **24,** 133.

Franks, F. (ed.) (1975). "Water a Comprehensive Treatise," Vol. 4 and 5. Plenum, New York.

Franks, F., and Reid, D. S. (1973). "Water a Comprehensive Treatise", Vol. 2, p. 336. Plenum, New York-London.

Fredin, L., Nelander, B., and Ribbegard, G. (1975a). *Chem. Phys. Lett.* **36,** 375.
Fredin, L., Nelander, B., and Ribbegard, G. (1975). *Chem. Phys. Lett.* **66,** 4065, 4073.
Geiger, A. (1979). Karlsruhe, Film.
Geiger, A., Rahman, A., and Stillinger, F. H. (1979). *J. Chem. Phys.* **70,** 263.
Gmelin, L. (1959). "Handbuch der anorganischen Chemie." Nr. 15, Silicium B, Weinheim, Verlag Chemie, 414, 439, 447, 448, 453, 503.
Halban, H. V., Kortüm, G., and Seiler, M. (1935). *Z. Phys. Chem.,* **173,** 454.
Hallam, H. E. (1973). "Vibrational Spectroscopy of Trapped Species." Wiley, London.
Hasted, J. B. (1973). "Aqueous Dielectrics." Chapman, London.
Hauser, H. (1975). Lipids, in "Water a Comprehensive Treatise" (F. Franks, ed.), Vol. 4. New York, Plenum, p. 209.
Hazlewood, C. F. (ed.) (1973). *Ann. New York Acad. Sci.* **204,** 1.
Hazelwood, C. F. (1979). *In* "Cell-Associated Water" (W. Drost-Hansen and Clegg, J., eds.), p. 165. Academic Press New York.
Hertz, H. G. (1974). "NMR Studies of Aqueous Electrolyte Solutions." *In* Structure of Water and Aqueous Solutions (W. A. P. Luck, ed), Chap. 7. 2, Weinheim, Verlag Chemie.
Hippel, P. H. V., and Wrong, K. Y. (1965). *J. Biol. Chem.* **240,** 3009.
Hodgson, T. D. (1970). *Desalination* **8,** 99.
Hoeve, C. A. S. (1979). "Water in Polymers." Presented at ACS Meeting, Washington, 1979.
Hofmeister, F. (1890). *Arch. Exp. Pathol. Pharmakol.* **25,** 295.
Hofmeister, F. (1891). *Arch. Exp. Pathol. Pharmakol.* **28,** 210.
Hübner, G., Jung, K., and Winkler, E. (1970). "Die Rolle des Wassers in Biologischen Systemen." Akademie Verlag und Braunschweig, Berlin, Vieweg.
Kleeberg, H. (1981). Ph.D. Thesis, Universität Marburg.
Kleeberg, H., and Luck, W. A. P. (1977). *Naturwissenschaften* **64,** 223.
Kleeberg, H., and Luck, W. A. P. (1978). "Is Cartillage a Coacervate?", 6th Colloquium of the Federation of European Connective Tissue Clubs, August, 28–30 (Poster).
Langer, K., Luck, W. A. P., and Schrems, O. (1979). *Appl. Spectrosc.* **33,** 495.
Lauffer, M. A. (1975). Entropy-Driven Processes in Biologie." Springer, Berlin-Heidelberg-New York.
Lilley, T. H. (1973). *In* "Water a Comprehensive Treatise" (F. Franks, ed.), Vol. 3, p. 266. Plenum, New York.
Luck, W. A. P. (1960a). "Spektroskopisch bestimmte Mizellbildung äthoxylierter Octylphenole", Vol. 3. Internat. Kongr. f. Grenzflächenaktive Stoffe, Köln, Bd. I, Sekt. A, S. 264.
Luck, W. A. P. (1960b). Melliand **41,** 315.
Luck, W. A. P. (1960c). *Angew. Chem.* **72,** 57.
Luck, W. A. P. (1963). *Ber. Bunsenges. Phys. Chem.* **67,** 186.
Luck, W. A. P. (1964). *Fortschr. Chem. Forsch.* **4,** 653.
Luck, W. A. P. (1965a). *Naturwissenschaften* **52,** 25, 49.
Luck, W. A. P. (1965b). 100 Jahre BASF, 259.
Luck, W. A. P. (1965c). *Ber. Bunsenges. Phy. Chem.* **69,** 69.
Luck, W. A. P. (1965d). *Ber. Bunsenges. Phys. Chem.* **69,** 626.
Luck, W. A. P. (1966). *Chimia* **20,** 270–271.
Luck, W. A. P. (1967a). *Discuss. Faraday Soc.* **43,** 115.
Luck, W. A. P. (1967b). *Naturwissenchaften* **54,** 601.
Luck, W. A. P. (1970). *J. Chem. Phys.* **74,** 3687.
Luck, W. A. P. (1972). "Handbuch der Mikroskopie in der Technik" (H. Freund, ed.), Umschau-Verlag, Frankfurt, Bd. VI, Teil I, S. 345.

Luck, W. A. P. (1973). DBP Offenlegungsschrift, Nr. 2151/207.
Luck, W. A. P. (1974). "Structure of Water and Aqueous Solutions," p. 222, 248. Verlag Chemie/Physik Verlag, Weinheim.
Luck, W. A. P. (1976a). "The Hydrogen Bond" (Schuster-Zundel-Sandorfy, ed.), Vol. 2, p. 527 Verlag, North Holland.
Luck, W. A. P. (1976b). *Top. Curr. Chem.* **64,** 113.
Luck, W. A. P. (1976c). *Naturwissenschaften* **63,** 39.
Luck, W. A. P. (1976d). "The Hydrogen Bond," (Schuster-Zundel-Sandorfy, ed.), Vol. 3, p. 1369. Verlag, North Holland.
Luck, W. A. P. (1978). *Pro. Colloid Polym. Sci.* **65,** 6,
Luck, W. A. P. (1979a). "Water in Polymers," *In* Am. Chem. Soc. Meeting *Series* 127.
Luck, W. A. P. (1979b). *Angew. Chem.* **91,** 408.
Luck, W. A. P. (1979c). *Angew. Chem. Int. Engl. Ed.* **18,** 350
Luck, W. A. P. (1980a). *Angew. Chem.* **92,** 29.
Luck, W. A. P. (1980b). *Angew. Chem. Int. Engl. Ed.* **19,** 28.
Luck, W. A. P., and Ditter, W. (1967-1968). *J. Mol. Struct.* **1,** 261.
Luck, W. A. P., and Ditter, W. (1968). *Ber. Bunsenges. Phy. Chem.* **72,** 365.
Luck, W. A. P., and Ditter, W. (1969). *Z. Naturforschung* **24b,** 482.
Luck, W. A. P., and Schiöberg, D. (1979). Presented at DFG Seminar Karlsruhe, February.
Luck, W. A. P., and Schrems, O. (1979). Infrared matrix isolation studies of self-association of methanol and ethanol: proof of cyclic dimers, Frankfurt, 1979, Int. Conference Spectroscopy, *J. Mol. Structure,* **60,** 333.
Luck, W. A. P., and Shah, S. S. (1978). *Prog. Colloid Polym. Sci.* **65,** 53.
Luck, W. A. P., and Zukovskij, A. P. (1974). "Molecular Physics and Biophysics of Water Systems" (A. J. Sidorovo, ed.), p. 131. Leningrad University.
Luck, W. A. P., Schiöberg, D., and Siemann, U. (1979). *Ber. Bunsenges. Phys. Chem.* **83,** 1085.
Luck, W. A. P., Schiöberg, D., and Siemann, U. (1980a). *Chem. Soc. Faraday Trans. 2,* **76,** 129.
Luck, W. A. P., Schiöberg, D., and Siemann, U. (1980b). *J. Chem. Soc. Faraday Trans. 2,* **76,** 129.
Némethy, G., and Scheraga, H. (1962). *J. Chem. Phys.* **36,** 3382.
Narten, A. H. (1974). "X-ray and Neutron Diffraction from Water and Aqueous Solutions." *In* Structure of Water and Aqueous Solutions (W. A. P. Luck, ed.), Vol. 1, Verlag Chemie, Weinheim.
Neale, S. M., and Williamson, G. R. (1956). *J. Phys. Chem.* **60,** 741.
Palinkas, G., Riede, W. O., and Heinzinger, K. (1977). *Z. Naturforsch.* **82a,** 1137.
Paquette, J., and Joliceur, C. (1977). *J. Sol. Chem.* **6,** 403.
Philipp, P. R., and Joliceur, C. (1973). *J. Phys. Chem.* **77,** 3076.
Pusch, W. (1975). "Structure of Water and Aqueous Solutions" (W. A. P. Luck, ed.), p. 549, 551. Verlag Chemie/Physik, Weinheim.
Rahman, A., and Stillinger, F. H. (1971). *J. Chem. Phys.* **55,** 3336.
Reid, C. E., and Breton, E. J. (1959). *J. Appl. Polym. Sci.* **1,** 133.
Samoilov, O. Ya. (1957). *Faraday Soc.* **24,** 141.
Samoilov, O. Ya. (1961). "Die Struktur wäßriger Elektrolytlösungen und die Hydration der Ionen." Teubner Verlag, Leipzig.
Schiöberg, D., and Luck, W. A. P. (1979). *J. Chem. Soc. Faraday Trans.* **75,** 762.
Schiöberg, D., Buanam-Om, C., and Luck, W. A. P. (1979). *Spectroscopy Lett.* **12,** 83.
Schnabel, R. (1976). *Int. Symp. Fresh Water Sea 5th* **4,** 409.
Schröder, W. (1969). *Z. Naturforsch.* **24b,** 500.
Schultz, R. D., and Asunmaa (1970). *Rec. Prog. Surface Sci.* **3,** 293.

Siemann, U. (1976). Diplomarbeit, Universität Marburg.
Sourirajan, S. (1971). "Reverse Osmosis," p. 27, 109, 168. Logos, London.
Starkweather, H. W. (1979). "Water in Polymers." Presented at ACS Meeting, Washington.
Stauf, J. (1960). "Kolloidchemie." Springer-Verlag, Heidelberg.
Stillinger, F. H., and Rahman, A. (1972). *J. Chem. Phys.* **67,** 1281.
Stillinger, F. H., and Rahman, A. (1974). *J. Chem. Phys.* **60,** 1545.
Thiel, M. V., Becker, E. D., and Pimentel, G. C. (1957a). *J. Chem. Phys.* **27,** 95.
Thiel, M. V., Becker, E. D., and Pimentel, G. C. (1957b). *J. Chem. Phys.* **27,** 243, 486.
Tirell, D., Grossman, S., and Vogl, O. (1979). "Water in Polymers." Presented at ACS Meeting Washington.
Toprak, C., Argar, J. N., and Falk, M. (1979). *J. Chem. Soc. Faraday Trans. 1* **75,** 803.
Tursi, A. J., and Nixon, E. R. (1970). *J. Chem. Phys.* **52,** 1521.
Warner, D. T. (1962). *Nature* **196,** 1055.
Warner, D. T. (1965). *Ann. New York Acad.* **125,** 605.
Walch, A., Lucas, H., Klimmek, A., and Pusch, W. (1974). *J. Polym. Sci. Polym. Lett.* **12,** 697.
Walch, A., Lucas, H., Klimmek, A., and Pusch, W. (1975). *J. Polym. Sci. Polym. Lett.* **13,** 701.
Zerahn, W. (1976). Diplomarbeit, Universität Marburg.
Zundel, G. (1969). "Hydration and Intermolecular Interaction." Academic Press, New York.

3

Hyperfiltration Membranes, Their Stability and Life

DAVID C. SAMMON
Chemistry Division, AERE Harwell
Didcot, Oxon, United Kingdom

List of Symbols

J	Permeate flux (m day^{-1})	A	Water permeation constant (m day^{-1} bar^{-1})
J_t	Permeate flux at time t (m day^{-1})	P	Applied pressure (bar)
B	Salt permeation constant (m day^{-1})	Π	Osmotic pressure (bar)

ISBN 0-12-085480-5

k	Reaction rate constant (s^{-1})	a, b, τ	Constants in Eq. (4)
m	Slope of log–log plot	x	Membrane thickness (μm)
m_A	Slope for water flux	β	Slope of creep plot—Eq. (5)
m_B	Slope for salt flux	a	Acetyl content—Eqs. (7)–(9)

I. Introduction

In hyperfiltration (HF), the membrane is exposed to a feed liquid at high pressure and the life of the membrane is largely determined by the stability to pressure and to the various constituents of the feed. Temperature exerts a considerable influence on stability in both areas. Membrane life can be defined as the time taken for performance to fall below some chosen value; because membrane replacement contributes significantly to the overall cost of the HF process, there is considerable incentive to be able to predict and increase membrane life.

The factors affecting membrane life can be classified as follows:

(1) feed—chemical constituents (including pH) and temperature;
(2) operating conditions—pressure, water recovery;
(3) membrane type—cellulose acetate, polyamide, and NS and PA series, dynamic, ion-exchange, and glass;
(4) membrane form—asymmetric, composite, and hollow fine fiber;
(5) plant geometry—plate and frame, tubular, spiral wrap, and hollow fine fiber; and
(6) miscellaneous—quality of supervision.

The total picture is complex, but it is convenient to consider the first two categories in detail and include information on the other categories when appropriate and available. The complexity of the overall picture is simplified by the lack of information in many areas. For instance, practically all systematic work has been done on asymmetric membranes made from cellulose acetate. This was the first practical HF membrane and is still used in about half of the world's present-day installed HF capacity. There is much less published information on the DuPont polyamide hollow fine fibers that account for the majority of the remaining installed capacity.

It would be logical to describe the observed effects, possible mechanisms, and how they relate to plant operational experience under various conditions. However, one must accept what is available. Just as the early development of membranes was empirical, so too is much of the work on membrane stability and life. In many instances, several effects were in operation at once, whereas ideally, each effect should be studied in isolation.

II. Mechanical Effects

Before beginning a study of the principal effects controlling membrane life, it is worthwhile to consider the role played by changes elsewhere in the membrane system. Foremost of these are changes in the porous substrate on which asymmetric and composite membranes are normally supported. If this substrate loses its permeability because it is compressed, then the performance of the plant is likely to deteriorate. In multicomponent substrates such as paper tubes inserted into perforated metal tubes, the paper must support the membrane adequately over the hole in the metal tube. In principle, chemical changes in the material of which the substrate is made could also play a role in reducing the effectiveness of the support and result in loss of flux or damage in the membrane. This is unlikely in practice except under unusual operating conditions.

Asymmetric and composite membranes are relatively durable but can be damaged by careless handling during membrane replacement, by poor module design that permits membranes to rub against each other, or by spacers and inadequate feed pretreatment that may result in suspended matter or scale abrading the active surface. In one instance, damage on the membrane next to the inlet on the first module was attributed to ultrasonic effects (Taniguchi, 1977). Pressure reversal may occur on shutdown and cause problems, particularly in tubular systems. Instances have been reported where membranes became detached from the porous substrate.

It is obvious that the effects just described can largely be eliminated by proper plant design and operation.

III. The Effect of Pressure

Although it is often possible to eliminate a deleterious effect by eliminating the cause, pressure-induced effects cannot be eliminated in this manner with pressure-driven processes such as HF. The permeate flux through a membrane is given by

$$J = A(P - \Delta\Pi). \tag{1}$$

Thus, the applied pressure must exceed the difference in osmotic pressure by an amount that is determined by the value of A (see the list of symbols pp. 73–74 for a full definition of the variables) and economic considerations. The capital cost of a plant of given capacity increases as the membrane flux decreases. Because the salt flux is largely independent of pressure, salt rejection increases with pressure, and the required salt rejection (or product water specification) may well define the desired pressure in a given application. Improved membranes with higher A and lower B values will enable operation at lower pressure. For seawater, however, the osmotic pressure is ≃24 bar and with 50% water recovery will reach a final value of 48 bar. Thus, for seawater desalination, the applied pressure must exceed this value. Practices appear to be settling around 27–34 bar for brackish water and 54–68 bar for seawater. Other potential applications, such as the concentration of fruit juices, would require higher pressures to exceed the larger osmotic pressure of these feeds.

A. Membrane Structure

Hyperfiltration membranes comprise a thin desalinating skin, usually termed the "active layer," on a thicker, porous support layer. Except in the case of hollow fine fibers, the membranes are on a pressure-resistant porous substrate. If skin and porous layer are made from one polymer in a single series of operations, the term "asymmetric" membrane is normally used. The term "composite" is applied where active and porous layers are of different polymers and the layers formed in different operations. In the former type, the demarcation between layers is not necessarily sharp, and neither layer can be assumed to be homogeneous throughout its thickness. Porous layers in composite membranes frequently have a gradient of porosity. It is, however, convenient to regard both types as comprising discrete active and porous layers. The former is thin (<1 μm in thickness) and is approximate in properties to a dense, nonporous film, i.e., a film in equilibrium with polymer and water. In water-swollen cellulose acetate, the water content is ~18 wt%. The porous layer has a pore structure that is visible in light and electron microscopes, and in an asymmetric cellulose acetate membrane the pores may occupy ~60% of the volume. Hollow fine fibers also have active and porous layers, but the difference in properties is not as marked. Applied pressure affects both layers, but the magnitude and nature of the effects are different.

B. PRESSURE-INDUCED FLOW THROUGH NONPOROUS WATER-SWOLLEN POLYMER FILMS

The details of this type of flow mechanism are under debate, but various aspects of the problem have been summarized by Meares (1979). One effect is to deswell the polymer and, hence, decrease the permeation rate, but this is probably insignificant at the pressures currently employed in HF. A further effect is the development of a pressure-induced water gradient across the film. Paul and Ebra–Lima (1971) demonstrated the existence of such a concentration gradient across a solvent-swollen layer; the layer was made of several films and, once steady state conditions were reached, the films were separated and the solvent content measured. Rosenbaum and Cotton (1969) demonstrated a similar, though much smaller, gradient in water-swollen cellulose acetate. In both instances the effect of pressure was to reduce the solvent content at the downstream, low-pressure side. For cellulose acetate and an applied pressure of 68 bar the reduction amounted to 6% of the upstream value.

Both effects require the movement of polymer chains and it is possible a slow approach to the equilibrium configuration would result in a permeate flux decreasing with time. However, many of the dense films and active layers of membranes are not fully dense, i.e., not at the equilibrium degree of swell, and a slow approach to equilibrium would also cause flux to change with time. These changes were observed by Baayens and Rosen (1972) and are attributed to viscoelastic creep. The most probable effect is film densification which results in decreased water and salt fluxes and increased salt rejection.

C. EFFECT OF PRESSURE ON THE POROUS LAYER

The porous layer may be compared to a sponge or open-cell foam with the polymer network accounting for about one-third of the volume. The effect of pressure on such a structure is to compress it, particularly when one recalls that membranes are made from nonrigid polymers such as cellulose acetate. As the structure is compressed, the pore volume is reduced and resistance to water flow is increased. The effect this has on the flux through an asymmetric or composite membrane depends on the degree of compression and the contribution the porous-layer resistance makes to the overall resistance of the membrane to water flow. Depending upon the nature of the polymer and extent that it is plasticized by water, elastic and inelastic effects may be observed. The former are likely to occur relatively quickly, but creep may occur over a longer time scale and result in a marked dependence of membrane properties on the duration of exposure to applied pressure. The principal effect is a decrease in permeate flux and this, in association with a constant salt flux, results in decreased salt rejection.

D. EXPERIMENTAL RESULTS

The principal problem encountered in laboratory studies on pressure-induced effects in HF membranes is the elimination of other effects. Hydrolysis can be minimized by careful control of pH, but fouling is more difficult to completely remove. Apart from any species that are present in the feedwater, the two most likely sources of fouling are microbiological contaminants and corrosion products. It is ideal to use a feed made from water free of suspended material, e.g., the permeate from HF, and add salts that inhibit microbiological growth and cause minimal corrosion of the constructional materials chosen. In practice, stainless steel is used for some or all of the high-pressure components. Because most resistant grades are somewhat sensitive to chlorides, it is wise to avoid the use of sodium chloride. Nitrates may be used, provided appropriate action is taken to avoid microbial slimes, although other effects of any additives used for this purpose must be considered. Sulfates may be used, but rejection is high and sensitive to membrane flaws. Other salts such as chlorates and perchlorates may be better choices. In much of the reported work, only water flux is studied, and with the use of suitably purified water it should be possible to eliminate fouling in these studies. However, measurements of salt flux are useful in shedding light on the mechanisms involved.

When membranes are operated under constant conditions, the usual observation is that permeate flux decreases, salt rejection increases or stays constant, and membrane thickness decreases. There is much more information on the first of these than on the other two. The effects of pressure cycling have been studied much less extensively, and are far less understood than is flux decline. Relaxation and hysteresis effects are commonly observed (Henkens and Smit, 1979; Martinez Guerrero and Filipe del Castillo, 1978).

1. *Water Flux Decline*

The observed flux loss slows with increased time; it is usually greater for membranes with higher initial flux, and is strongly dependent on the applied pressure. It is customary to plot log J against log t (time) when a straight line is obtained with slope m that is an empirical measure of membrane stability under the chosen conditions. This relationship can be expressed as

$$J_t = J_1 t^m \tag{2}$$

or

$$\log(J_t/J_1) = m \log t. \tag{3}$$

It is common to use J_1, the flux at one hour, as the reference value; t as the time in hours; and m as the (negative) slope of the log–log plot. If one distinguishes between changes in water flux and salt flux, then m_A may be used for the former and m_B the latter. The slope of the log–log plots for A and J are virtually identical. The implication of the log–log relationship is that if the water flux decreases by 1% in the time interval 1–100 hr, then it will decrease further by 1% in the period 100–10,000 hr, thus, a quasi-steady state is reached. The relationship between the observed change in flux and the calculated value of m is illustrated in Table I and Fig. 1. In both, permeate fluxes are normalized to a value of one at one hour. An example of m varying with applied pressure and the initial flux of the membrane is given in Fig. 2. Many values of m are quoted in the literature, and a representative selection is given in Table II in addition to conditions under which the measurements were made. Different membrane types are also included, and laboratory studies and field operations covered. It is likely in several instances that some of the flux loss is

TABLE I

Relevance of $-m_A$

	F_w at time t relative to value at 1 hr			
$-m_A$	168 hr 1 week	730 hr 1 month	2200 hr 3 months	8800 hr 1 year
0.01	0.95	0.94	0.93	0.91
0.03	0.86	0.82	0.79	0.76
0.07	0.70	0.63	0.58	0.53
0.10	0.60	0.52	0.46	0.40

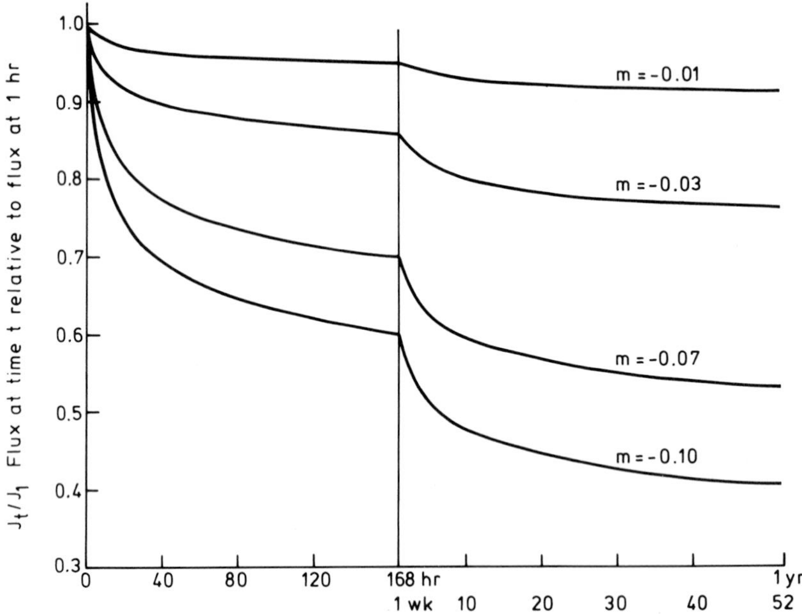

Fig. 1 Flux versus time for different values of *m*.

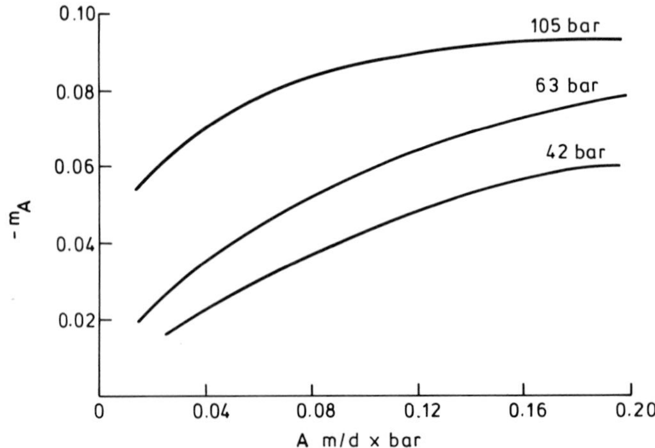

Fig. 2 Flux decline slope, m_A, versus initial value of *A* for homopolymer membranes of cellulose acetate; *A* is measured at 14 bar prior to exposure to higher pressure.

attributable to fouling, and thus, it is best to regard the values for $-m_A$ as upper limits. Improvements are being made to the pressure stability of membranes, but it is clear that $-m_A$ values of slightly greater than 0.01 can be attained for seawater and brackish water membranes. This indicates that about 10% loss of flux occurs between 1 hr and 1 year.

The slope of the log–log plot remains the most widely used measure of flux stability. The plot was shown to be linear from a few minutes to hundreds of hours, for pressures up to 100 bar and flux decreases of >30% of the 1 hr value (Sammon, 1975). Various other equations have been suggested, but it is often difficult to fit the results conclusively to any one equation because there is considerable scatter in the results, and the test may be of relatively short duration. Ohya (1978) suggested the equation

$$1/A = 1/A_0 + at + be^{-t/\tau}. \tag{4}$$

Using published results, it was found that this equation can be used to represent experimental information. Plots of reciprocal A versus time are normally curved for a short period after exposure to pressure. This is represented by the exponential term and attributed to rapid compaction of the porous layer. At longer times, a straight line is obtained, and the slope of this line may be a measure of the compaction of the active layer.

The equations used to represent flux decline are empirical because a good model of the flux decline process does not exist. This is a major drawback in understanding pressure-induced changes. Although much experimental work has been reported, it is preferable not to draw detailed conclusions because it is likely that the important factors will differ from one membrane system to another.

2. *Thickness Changes*

One problem in interpreting water-flux decline measurements is that effects from the active and porous layers may be present. However, changes in overall thickness reflect what is occurring in the porous layer. For both asymmetric and composite membranes, the active layer contributes less than 1% to the overall thickness, whereas in hollow fine fibers the active layer may occupy a larger fraction of the total thickness. Results for asymmetric membranes where exposure to pressure may reduce the thickness from 75–50% of the starting value have been reported.

It is difficult to measure thickness as a function of time under normal operating conditions. Hoernschemeyer *et al.* (1970) measured compressive creep in wet cellulose acetate methacrylate asymmetric membranes and found

TABLE II

Values of Flux Decline Slope ($-m_A$) for Various Membranes

Membrane	Type	Module	Feed	Pressure (bar)	Temperature (°C)	Flux (m/d)	Rejection (%)	$-m_A$	Duration (hr)	Ref.[m]
CA[a]	D[i]	flat cell	3.5% NaCl	100	27	—	—	0.012	>100	1
CA	A[j]	flat cell	1.0% NaCl	140	23–30	0.54	97.7	0.078	900	2
CA	A	flat cell	1.0% NaCl	105	23–30	0.81–0.94	98.2–98.7	0.059	48	2
CA	A	flat cell	1.0% NaCl	70	23–30	0.71–0.76	96.2–98.3	0.040	48	2
CA	A	flat cell	1.0% NaCl	42	23–30	0.48	98.0	0.026	310	2
CA (2.45 ds[b])	A	flat cell	3.5% NaCl	105	—	0.40	98.0	0.08	–	3
CA (blend)	A	flat cell	3.5% NaCl	105	—	0.46	99.5	0.012	–	3
CA (2.63 ds)	A	flat cell	3.5% NaCl	105	—	0.44	99.5	0.03	–	3
CAM[c]	A	flat cell	3.5% NaCl	105	—	0.36	99.7	0.011	–	3
CA/CN[d], CA	C[k]	flat cell	3.5% NaCl	105	—	0.40	99.5	0.03	–	3
CA (2.45 ds)	A	flat cell	1.0% NaCl	56	—	1.04	94.0	0.035	–	3
CA (blend)	A	flat cell	1.0% NaCl	56	—	1.68	95.0	0.039	–	3
CAM	A	flat cell	1.0% NaCl	56	—	1.64	90.0	0.013	–	3
PA-300[e]	C	spiral	seawater	68	25	0.82–1.02	>99.4	0.01	500	4
PA-300	C	spiral	water	27	55	—	—	0.012	200	4
CA	A	spiral	seawater	56	25	—	95	0.029	6000	5
CA	A	spiral	0.2% NaCl	30	25	—	98	0.00	6000	5
CA	A	tubular	seawater	50	25	—	89	0.008	6300	5

CA	F[l]	hollow fiber	seawater	55	25	—	94	0.027	1780	5
CA	F[l]	hollow fiber	0.2% NaCl	22	25	—	97	0.00	1780	5
B-10[f]	F[l]	hollow fiber	seawater	56	20–30	—	—	0.012	2300	6
PBIL[g]	A	flat cell	3.5% NaCl	80	25	—	—	0.07	720	7
PAH[h]	A	flat cell	3.5% NaCl	80	25	—	—	0.002	720	7

[a] CA, cellulose acetate.
[b] ds, degree of acetylation.
[c] CAM, cellulose acetate methacrylate.
[d] CA/CN, porous cellulose acetate–cellulose nitrate support.
[e] PA-300, a polyether–amide membrane.
[f] B-10, aromatic polyamide.
[g] PBIL, polybenzimidazole.
[h] PAH, polyamide hydrazide.
[i] D, dense film.
[j] A, asymmetric membrane.
[k] C, composite membrane.
[l] F, hollow fine fiber.
[m] References, 1 Baayens and Rosen (1972)
2 Merten *et al.* 1966
3 Podall, 1971
4 Riley *et al.* 1976
5 Kunisada and Murayama, 1978
6 Murayama *et al.* 1976
7 Hara *et al.* 1977.

$$\log x = \text{constant} + \beta \log t, \tag{5}$$

where x is membrane thickness and β the slope of the log–log plot analogous to m_A for water flux decline. Values of $-\beta$ from 0.028 to 0.048 were found where the values of $-m_A$ would have been lower. Another significant observation was that much of the compression is elastic. For one specimen of cross-linked cellulose acetate methacrylate compressed at 56 bar for 170 hours, the following results were obtained: initial thickness (zero pressure) 70 μm; initial thickness (56 bar) 30 μm; final thickness (56 bar) 24 μm; and final thickness (zero pressure) 44 μm. Deanin *et al.* (1970) reported similar studies on a stack of 30 dense films (each 25 μm thick) and observed $<$10% compression after 100 hours under a compressive load of 100 bar. As was expected, the degree of compression was much less than for asymmetric membranes and could be further decreased by adding fillers to the dense films.

These experiments, however, do not accurately reproduce the situation under operating conditions when water is flowing through the membrane. The flow-induced pressure gradients modify the compressive-stress profile across the thickness of the membrane, the result being that the membrane is compressed less than in a static experiment under the same applied pressure. This effect is marked for those membranes where the majority of the resistance to flow resides in the porous layer. The pressure and compressive-stress gradients are shown schematically in Fig. 3. These gradients have been discussed by Baayens and Rosen (1972) and by Hoernschemeyer *et al.* (1970). The latter covered the membrane with an impermeable film to increase compression of the active layer. Walmsley *et al.* (1976) have also shown that thickness changes are greater after exposure to pressure in the "no-flow" state than under normal operation. This is indicative of higher compressive stress on the porous layer in the no-flow experiment.

Asymmetric and composite membranes are mounted on a substrate that has little resistance to flow as depicted in Fig. 3. This optimal design, however, is not always realized and will further complicate pressure and compression gradients. In instances where compression of the porous layer contributes to flux decline, the use of a poor substrate (appreciable flow resistance) can lead to lower observed flux decline and somewhat lower starting flux.

It is understood from what has been stated that the effect of compression on membrane performance is dependent on the contribution made by the porous layer to the overall resistance of the membrane. It is also dependent on how flow resistance is increased as thickness is decreased and this depends on the detailed structure of the porous layer.

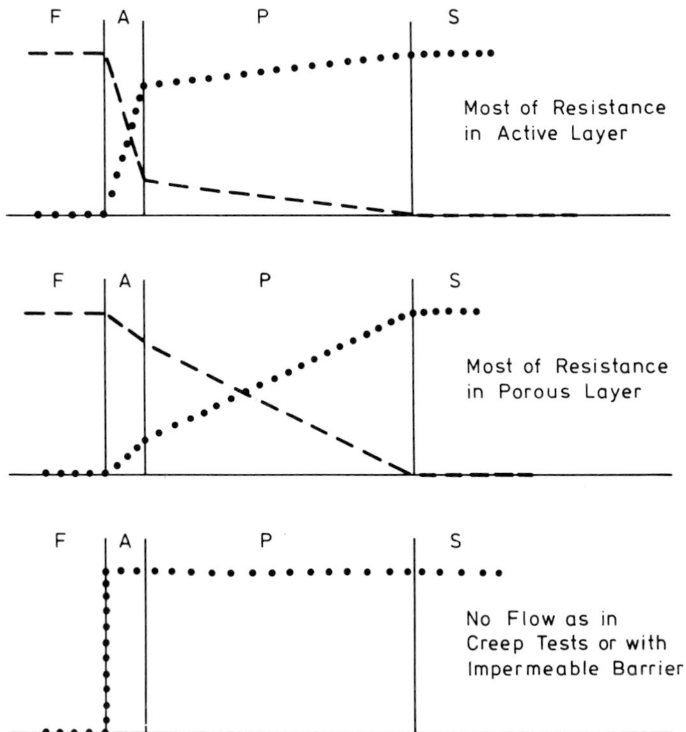

Fig. 3 Gradients of pressure (– – –) and compressive stress (· · ·) in an asymmetric membrane. Symbol F denotes the feed solution, A the active layer, P the porous layer, and S the support.

3. *Changes in Salt Flux*

Although there is little published data on this topic, the conclusion usually drawn is that salt rejection changes little or improves with time. Laboratory tests on flat-sheet homopolymer membranes (Sammon, 1975) are summarized in Fig. 4, and similar results were obtained for blend membranes. In contrast, Kimura and Nakao (1975) obtained values of m_B/m_A that range from 0.62 to 1.35 for tubular membranes with water at 40 bar and intermittent use of 1000 ppm NaCl to measure salt rejection. The measured values of m_B for the two data sets at the same pressure lie in the same range, and the difference in the ratio m_B/m_A arises from the lower m_A values obtained with flat sheets. There may be some contribution because of fouling with the tubular membranes; however, it is not appropriate to compare results from different membrane systems.

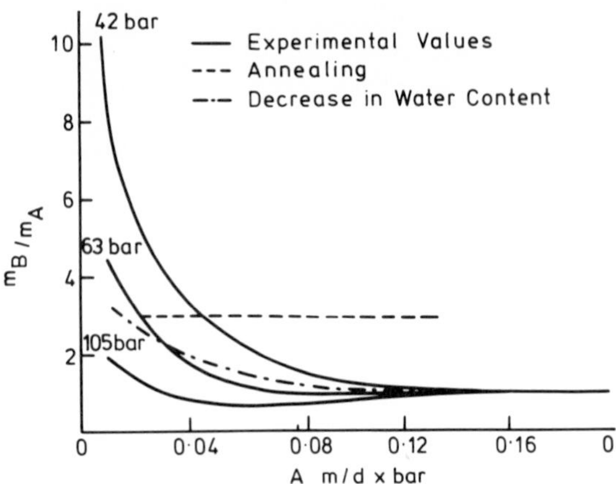

Fig. 4 Ratio m_B/m_A versus A for the homopolymer membranes as in Fig. 2.

E. MECHANISMS

The conclusion of the previous section is also true when trying to understand the processes occurring in the membrane. Some scientists have argued that increased flow resistance of the porous layer is the only phenomenon of importance, whereas others have presented evidence to suggest all changes occur in the active layer. Each hypothesis may be true for the particular system studied. Some mechanisms proposed are listed in Table III; all result in the permeate flux decreasing with time. If only one process occurs, then the value m_B/m_A would give an indication of which one is operating. This assumes the absence of chemical effects and fouling, and in the first example of Table III, that salt transport is independent of water flux. One can understand the various methods in which the active layer may be densified. Annealing causes water and salt flux to decrease, and is considered a relaxation process to a more stable configuration; it is possible pressure may induce a similar change. For the membranes in Fig. 4, this results in the value m_B/m_A becoming approximately 3 at 40 bar. Alternatively, applied pressure can lead to a progressive reduction in the water content of the active layer. If this has an effect similar to increasing the acetyl content (which also decreases the water content), then the ratio would be 2.8. The effect of pore plugging is dependent on pore structure of the active layer and the nature of the plugging species. If a few pores contribute significantly to salt flux, but much less so to water flux, then high values of m_B/m_A can occur.

TABLE III

Mechanisms of Flux Decline

Mechanism	m_B/m_A	Change in salt rejection
Compressing porous layer	0	decrease
Densifying active layer	>1	increase
Thickening active layer	1	none
Pore plugging (active layer)	>1	decrease

F. METHODS OF REDUCING FLUX DECLINE

Fillers are used to strengthen dense films and membranes. Deanin *et al.* (1970) incorporated powders such as aluminum silicate into dense cellulose acetate films and achieved a considerable reduction in thickness changes. Harriott *et al.* (1973) used silica gel as a filler in dense films; water flux increased with little decrease in salt rejection and flux decline was reduced. Merten *et al.* (1966) precipitated inorganic salts in the porous layer of asymmetric membranes, but *m* values were affected little. Baum *et al.* (1972) used an aluminum silicate filler in similar membranes. They obtained a reduction in water-flux change and thickness, but salt rejection was impaired. Salt rejection was restored by coating with a dilute solution of cellulose acetate in acetone to give 96–98% rejection of 3.5% NaCl at 100 bar and a flux of ≤0.5 m/d. During a 300 hr period, $-m_A$ values of 0.017–0.030 were obtained. Aluminum oxide was also shown to reduce flux decline (Goossens and van Haute, 1976); water flux was increased, but salt rejection was slightly reduced. It is significant, however, that the lowest value of $-m_A$ reported is >0.04 at 80 bar.

Another approach is that of strengthening the polymer film with cross linking. Some success was achieved with cross-linked cellulose acetate methacrylate (see Table I). Many blend membranes have better compaction resistance than homopolymer, but this might be due to differences in structure rather than differences in creep resistance. One objective of the search for noncellulosic membranes is the possibility of identifying a polymer that has high creep resistance and good salt rejection.

In composite membranes, the active and porous layers are usually made from different polymers; one chosen for its desalinating properties and another for formation of a stable porous layer. This appears to offer the best chances of long-term low flux decline. Because the layers are separately optimized, it is possible to reduce thickness changes and increase the active-layer stability.

All membranes considered so far are made from organic polymers. Porous glass membranes also exhibit salt rejection (Kraus *et al.*, 1966), and are extremely stable against compression. Some indication of the stability is given in tests at pressures up to 300 bar (Ermakova *et al.*, 1977). Water flux is found to be directly proportional to applied pressure in marked contrast to polymeric membranes where compression results in considerable curvature of the flux–pressure graph at pressures <100 bar (Merten *et al.*, 1966).

Although the present polymeric membranes are satisfactory for many purposes, the useful life is considerably shorter at temperatures above 30–35°C. At higher temperatures, e.g., above 60°C, one must use dynamic membranes or porous glass if the chemical stability is adequate. The former have been tested to 190°C and show good salt rejection to 60°C (Elmer, 1978), whereas the latter can be operated near 100°C (El Nasher, 1976). Flux decline parameters for both of these systems are not available, but it is likely that $-m_A$ values are far less than those of current polymeric membranes at 60–100°C.

IV. Chemical Effects

The chemical effects governing membrane life are interactions between the membrane material and species to which it is exposed. In true chemical effects, a chemical reaction occurs. In this section it is also convenient to include other interactions between membrane and species in the feed. The sorption of organic solutes is one example.

The species involved may be feed components, and it may be possible to remove the harmful species except where these are important constituents of a process stream. Adjustment of pH is one of the most common changes in feed composition. Other harmful species may be added in pretreatment, as when chlorine is used to control biological growth, but they may reduce membrane life. It is frequently necessary to clean membranes and it is clear that the chemicals used for this must be chosen carefully.

All commercially important HF membranes are polymeric, and one can view two chemical reaction types. The first, severing the polymer chain, results in a lower mean MW, leading to a loss of strength. For flat sheet membranes this effect is minimal over the range studied, but for membranes inside tubes, it is advantageous to use a polymer of higher

MW e.g., 398–10 instead of the 398–3 cellulose acetate. Tubular membranes undergo stretching forces with applied pressure, and the role of longer polymer chains is to impart increased resistance to damage because of stretching. Chain scission is not an important factor with cellulose acetate membranes and feeds, although it may become relevant in other applications such as the treatment of radioactive wastes. Ali and Clay (1979) showed chain scission occurs on cellulose acetate membranes exposed to ionizing radiation. These membranes lost strength and eventually fractured with a marked increase in flux. It is likely that chain scission may be occurring in membranes exposed to chlorine.

The second chemical change of considerable importance involves chemical alteration of units on the polymer chain. Because salt rejection is dependent on the chemical nature of the polymer, it is understood that chemical changes in the polymer can have either a beneficial or detrimental effect on performance. The best known example is the hydrolysis of cellulose acetate resulting in the replacement of acetate with hydroxyl. In the early development of HF, it was demonstrated that salt rejection increased with the acetate-to-hydroxyl ratio (Lonsdale *et al.,* 1965), and thus, the deleterious effect of hydrolysis is readily explained. The loss of salt rejection because of hydrolysis is one of the most important limitations of cellulose acetate as a membrane material. It has been studied systematically and is the principal theme of this section, not only because of its practical importance, but because it has features in common with changes involving other chemical reactions.

Both chain scission and chemical nature change are permanent effects, in that the effect remains when the agent causing it is removed. Temporary effects have also been reported (Duvel *et al.,* 1972) and are the result of chemical affinity of certain feed constituents for the membrane material. Various organic solutes are found to reduce the water flux reversibly.

Porous glass membranes represent a significantly different system. Although the use of glass in contact with water is widespread, attack on the glass can be measured, but this occurs at such a slow rate it may be neglected. An enormous difference exists between glassware, where the surface area exposed to water is the geometrical area in contact with the liquid, and porous membranes, where the walls of the pores have a large surface area in contact with the liquid. Dynamic membranes share some advantages with porous glass, but have one considerable advantage: easy replacement. This means a fair amount of chemical degradation can be tolerated because a short life is not disadvantageous with these membranes. The chemical nature of dynamic membranes is likely to give advantages of chemical stability in some applications. This is also true of polymeric ion-exchanging membranes such as sulfonated polysulfone. Chemical degradation may be expected, as with other polymers, and loss of ion-exchange capacity may occur as a result of the formation of relatively permanent bonds between the functional groups and feed constituents.

A. HYDROLYSIS OF CELLULOSE ACETATE MEMBRANES

Vos *et al.* first reported systematic studies on the hydrolysis of asymmetric HF membranes made from cellulose acetate. In the first of two papers (Vos *et al.*, 1966a), samples of asymmetric membrane were immersed in buffer solutions with different pH values and analyzed for acetyl content. The experiments covered a pH range of 2.2–10 and a temperature range of 23–95°C.

Hydrolysis rate studies on esters such as cellulose acetate have been made for many systems; the overall reaction rate constant k_1 may be treated as the sum of three terms:

$$k_1 = k_{H^+}[H^+] + k_{OH^-}[OH^-] + k_{H_2O}. \tag{6}$$

The rate constants of acid-catalyzed, base-catalyzed and spontaneous reactions are k_{H^+}, k_{OH^-}, and k_{H_2O}, respectively. The rate of change of acetyl content with time yields the results summarized in Fig. 5.

In the second paper (Vos *et al.*, 1966b), membranes were exposed to alkaline brine under HF conditions, and the changes in membrane constants for water (A) and salt permeation (B) were studied. These changes were correlated with previous measurements of hydrolysis rate constants and known effects of acetyl content on salt and water permeation (Lonsdale *et al.*, 1965). Over the range of acetyl content (a) from 33.6 to 39.8%, A and B can be expressed as follows:

$$\ln A = -10.7 \ln a + \text{constant}, \tag{7}$$

$$\ln B = -29.6 \ln a + \text{constant}. \tag{8}$$

The first-order rate constant, k_1, for hydrolysis is

$$\ln a/a_0 = -k_1 t, \tag{9}$$

where a_0 and a are the acetyl contents at time zero and time t, respectively. Combining Eqs. (7), (8), and (9) yields

$$\ln A = 10.7\, k_1 t + \text{constant}, \tag{10}$$

$$\ln B = 29.6\, k_1 t + \text{constant}. \tag{11}$$

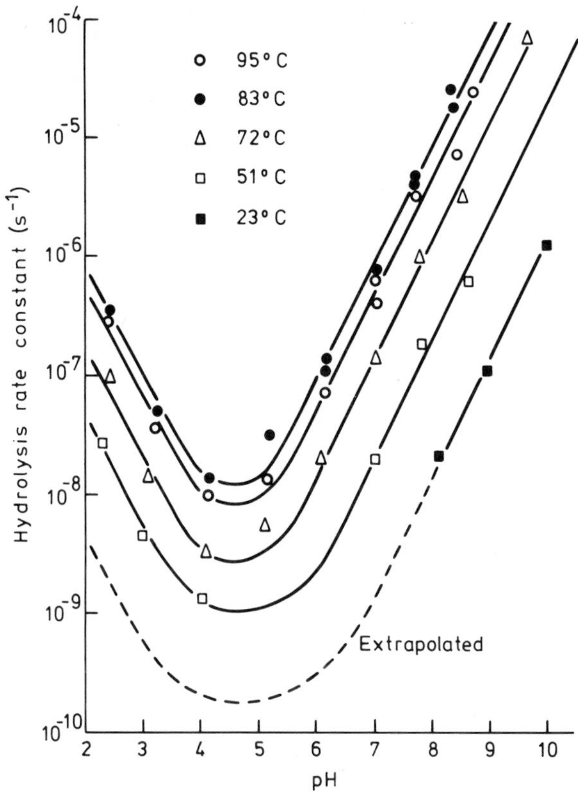

Fig. 5 Hydrolysis rate constant for cellulose acetate versus pH and temperature.

These equations indicate salt rejection will decrease with time, and because k_1 is dependent on pH and temperature, as shown in Fig. 5, the rate of salt-rejection loss will also depend on these parameters.

The linear dependence of ln A and ln B predicted in these equations is observed in HF experiments, but an induction period occurs before the linear relationships are established. These induction periods are attributed, in part, to the existence of pH gradients across the membrane.

Apparent rate constants calculated from the HF data are a factor of 3–10 lower than those predicted from the data involving loss of acetyl content. This measure of agreement is probably as good as can be expected from the simple treatment of data employed.

Further work (Sammon *et al.*, 1976) on similar cellulose acetate membranes describes the mechanism of hydrolysis and its effect on HF properties. Hydrolysis experiments similar to those previously discussed were performed using alkaline buffer solutions on homopolymer and blend membranes with a range of values of salt rejection. Pseudo–first-order kinetics are observed for the change in acetyl content with immersion time in the hydrolyzing medium (carbonate buffer, pH 10), as shown in Fig. 6. Considerable differences are observed for membranes with different properties. In HF experiments, an induction period was observed (Phase I), followed by a period during which log A/A_0 and log B/B_0 increased linearly with time (Phase II). In some cases, A and B decreased (Phase III) and was attributed to collapse of the weakened porous layer. At the beginning of Phase II, the pH gradient across the membrane is less

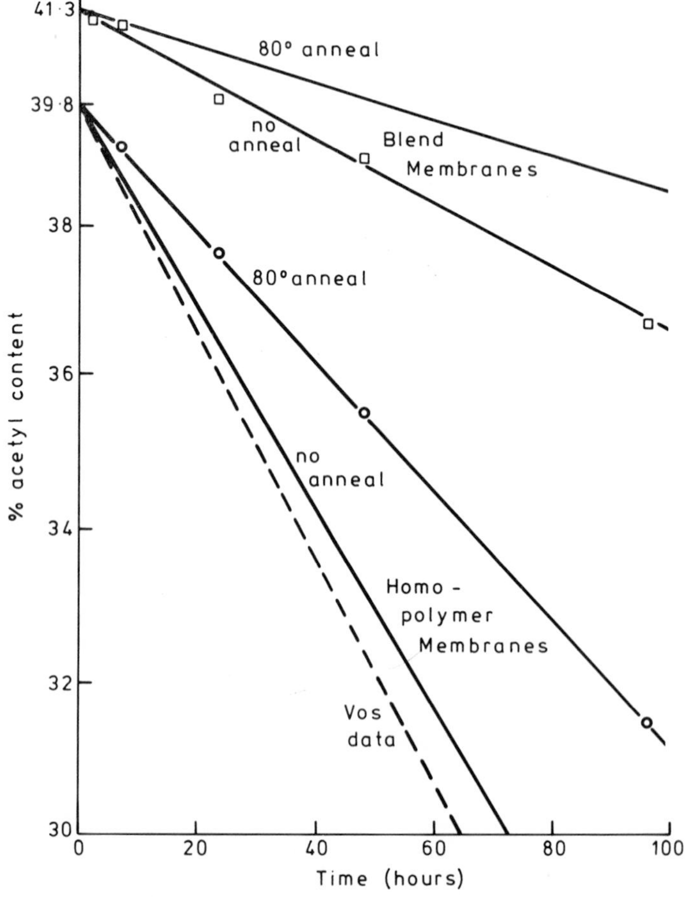

Fig. 6 Acetyl content versus time for cellulose acetate membranes immersed in carbonate buffer at pH 10 and 25°C.

than 0.1 pH and salt rejection is less than 90%. This is postulated to represent homogeneous hydrolysis of the active layer and the remainder of the membrane. In Phase I, however, the existence of pH gradients across the membranes suggests attack is not homogeneous. At any point within the active layer the rate of attack is determined by the local pH, which in turn is controlled by the salt rejecting properties of the active layer and the concentration gradients across the active layer. The known distribution of ionic species between feed solution and membrane results in the hydroxyl ion concentration in the feed side of the active layer being lower for membranes of higher salt rejection. Thus, the initial rate of hydrolysis is lower. This is an over-simplified model, but it underlines the role exercized by the salt-rejecting properties of the active layer. At the end of Phase I, membranes have too low a salt rejection for most purposes, and the change in properties during Phase I defines the useful lifetime of the membrane. Unfortunately, changes in properties during this phase cannot be predicted and the operating mechanisms are poorly understood. A further complication is that the rate of hydrolysis increases with increasing ionic strength of the feed.

B. EMPIRICAL TESTS ON CHEMICAL STABILITY

Because the systematic studies outlined above cannot be used to estimate membrane life, one must use empirical measurements. One study of pH stability is given in Fig. 7 (Spatz and Friedlander, 1978). In order to accelerate the tests, a temperature of 55–65°C was used and was assumed to increase the reaction rates by a factor of 8–16 over those at

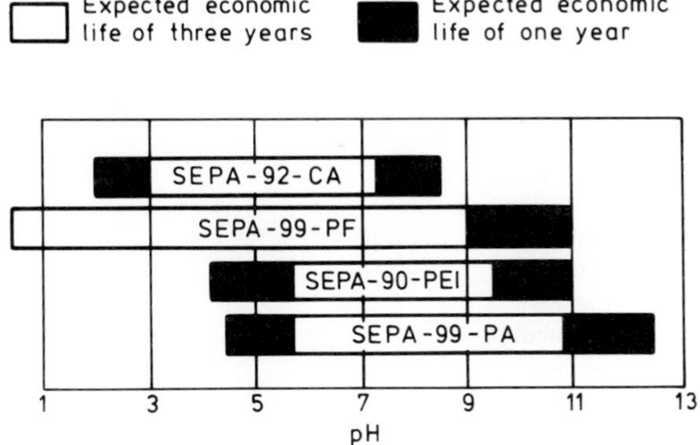

Fig. 7 Hydrolytic stability of SEPA membranes. Membrane materials are cellulose acetate (CA), polyfurane (PF), polyethylene-imine (PEI), and polyamide (PA).

25°C. The economic life of the membrane was defined as the point when solute passage reaches twice the initial value. Similar tests were conducted with 1.0–1.5 ppm chlorine in the feed. After three weeks at 60°C (equivalent to 8–10 months at 25°C), solute passage increased by less than a factor of 2 for cellulose acetate membranes and by 15–30 for the others.

C. EFFECT OF ORGANIC SOLUTES

Duvel *et al.* (1972) felt the development of HF for waste-water treatment was hampered by the loss in flux when water containing soluble or colloidal organics was treated. In a series of short-term tests at 40 bar and 40°C using tubular membranes, they measured flux loss resulting from the presence of individual solutes, mostly at 10^{-2} M. Solutes containing fewer than four carbon atoms did not produce significant flux change. Appreciable decreases occurred with solutes containing five or more carbon atoms, but the decrease was dependent on the solutes' chemical natures. The results obtained for straight chain primary alcohols and phenol are summarized in Table IV. The effect of increasing chain length and concentration is shown. Flux loss was found to be unrelated to solute removal, and the losses were completely restored on removing the solute from the feed. A larger loss in flux was observed with a wastewater from azo-dye manufacture. The flux loss is considered a result of the organic solute blocking the water transport by being adsorbed in the membrane or at the membrane–water interface.

The process is not chemical degradation, but the presence of such

TABLE IV

Effect of Alcohols on Permeate Flux

Solute	Concentration M	% Decrease in flux
1-pentanol	10^{-2}	8.6
1-hexanol	10^{-2}	18.9
1-heptanol	10^{-2}	39.4
1-heptanol	10^{-3}	7.3
1-octanol	10^{-3}	15.1
phenol	10^{-2}	22.9

Note: Only flux decreases >6.4% were considered significant. The values reported are averaged for two membranes.

organics in waste water is likely to affect the stability of membrane performance.

D. POROUS GLASS MEMBRANES

The excellent pressure stability of these membranes has been mentioned (Section III. F.) and is in marked contrast to their chemical stability. Ballou *et al.* (1971) observed considerable losses in salt rejection during a period of 100 hr, but noted that a decrease in pH and salt content and an addition of aluminum chloride to the feed improved the stability of the salt rejection. In a later paper, Ballou *et al.* (1973) showed that intermittent treatment with 30 g/l $AlCl_3 \cdot 6H_2O$ every 100 hr was sufficient to maintain the initial rejection over a period of 450 hr with a high-rejection membrane. Salts other than $AlCl_3$ were found to be less effective. The loss in rejection is attributed to a slow dissolution of silica from the glass; the rate of dissolution can be measured and is reduced by the $AlCl_3$ treatment. It is clear that commercial exploitation of this membrane type will require more detailed life studies, and the development of stable or more easily stabilized membranes.

E. STABILITY AT ELEVATED TEMPERATURES

The upper temperature limit for currently available membrane systems is quoted at 30–40°C, and because the usual practice is to operate within prescribed pH limits, the temperature ranges are largely controlled by the need to minimize flux decline. The effect of elevated temperature in accelerating hydrolysis of cellulose acetate can be seen in Fig. 5 where pH 8 and 23°C is equivalent to pH 6.5 and 51°C. Tests at 60°C were used by Spatz and Friedlander (1978) to estimate membrane life at 25°C as a function of pH (see Fig. 7). Riley *et al.* (1976) reported that the performance of PA–300 was maintained over a period of 1000 hr at 55°C. Membranes have also been evaluated at 75°C for use in washwater recycling in a space environment (Goldsmith *et al.*, 1973). For polybenzimidazole hollow fibers and cellulose acetate blend membranes in spiral-wrap modules, rejection was maintained over a period of 900 hr at pH 6.5–7.5. Flux was, however, reduced to about 25% of the starting value. At higher temperatures, porous glass and dynamic membranes have been evaluated, but not over extended periods. Brandon *et al.* (1975) state that dynamic membranes can be used at pH 4–11 and at >85°C. This degree of stability combined with relatively easy replacement makes dynamic membranes the most logical prospect at elevated temperatures.

F. STABILITY AT ZERO PRESSURE

During shutdown, the membranes may be more vulnerable to chemical attack than during normal operation. In the former state, the concentration gradient across the membrane disappears and the entire thickness of the active layer experiences the same conditions as the feed side during operation. The same considerations apply during cleaning, flushing, storage, and transit. Although most HF membranes suffer damage when allowed to dry, some wet–dry stable asymmetric membranes have been developed (Kesting, 1973); the PA–300 composite membrane can also be wet–dry cycled with no effect on performance (Riley *et al.*, 1976).

V. Microbiological Attack

Cantor and Mechalas (1969) exposed asymmetric cellulose acetate membranes to microbial species from three sources: biologically degraded membranes, surface soil, and lake-bottom mud. The membranes were tested under HF conditions; salt rejection was constant at first but then decreased rapidly to zero at times between 200 and 300 hr. This failure is associated with the appearance of etched areas on the active surface, and ultimately, holes may form. Cellulose triacetate was found to be unaffected under conditions when cellulose acetate membranes failed.

VI. Membrane Regeneration

Most studies of membrane regeneration are associated with cleaning to remove fouling. The purpose has been to reverse the deleterious effects of pressure, although in principle, more complex techniques may be developed to counter chemical attack. Hamer and Kalish (1969) showed that heat treatment at temperatures close to those used in preparation restored 60% of the flux loss and all of the salt rejection loss. Treatment by swelling agents or solvents, followed by heat treatment, was also effective. Cantor *et al.* (1970) studied the effect of dilute acetic acid solutions (2–4% at ≃50°C) and produced similar recoveries in performance. Complete flux recovery was reported by Higley and Saltonstall (1971), using 30% acetic acid at ambient temperature with little or no loss of salt rejection. Periodic regenerations of this type were used to maintain high fluxes over ≃1700 hr in membranes operated at 56 bar. The process was successfully used on homopolymer and blend membranes in flat sheet and tubular form. The use of guar gum to repair damaged membranes was also demonstrated.

The ultimate in regeneration is complete replacement of the membrane *in situ*. This is readily achieved for dynamic membranes that are always formed *in situ;* the replacement of cellulose acetate membranes has been described (Belfort, *et al.* 1973), and cost calculations given.

VII. Conclusion

Most of the work discussed previously is related to the use of HF for the production of drinking water. Satisfactory membrane life (⩾3 years) has been achieved by control of pH to limit hydrolysis (in cellulose acetate), by removal of added chlorine (particularly for polyamides), and by the development of more stable membranes that give adequate salt rejection at lower pressures than those employed in the early days of HF. As with the development of the membranes, these improvements were brought about largely by empirical development. Extension of the use of the process to a wide range of effluents and process streams will make even greater demands on membrane stability and stimulate more work on the mechanisms involved in chemical degradation and flux decline.

References

Ali, S. M., and Clay, P. G. (1979). *J. Appl. Polym. Sci.* **23,** 2893–2897.

Baayens, L., and Rosen, S. L. (1972). *J. Appl. Polym. Sci.* **16,** 663–670.

Ballou, E. V., Wydeven, T. and Leban, M. I. (1971). *Env. Sci. Tech.* **5**(10), 1032–1038.

Ballou, E. V., Leban, M. I., and Wydeven, T. (1973). *J. Appl. Chem. Biotechnol.* **23,** 119–130.

Baum, B., Margosiak, S. A., and Holley, W. H., Jr. (1972). *Ind. Eng. Chem. Prod. Res. Dev.* **11**(2), 195–199.

Belfort, G., Littman, F. E., and Bishop, H. K. (1973). *Water Res.* **7,** 1547–1559.

Brandon, C. A., El Nasher, A., and Porter, J. J. (1975). *Am. Dyest. Rep.* **64**(10), 20–41.

Cantor, P. A., and Mechalas, B. J., (1969), *J. Polym. Sci. Part C.* **28,** 225–241.

Cantor, P. A., Higley, W. S., and Saltonstall, C. W., Jr. (1970). Office of Saline Water Research and Development Progress Report, No. 601. Washington, D.C.

Deanin, R. D., Baum, B., Margosiak, S. A., and Holley, W. H., Jr. (1970). *Ind. Eng. Chem. Prod. Res. Dev.* **9**(2), 172–175.

Duvel, W. A. Jr., Helfgott, T., and Genetelli, E. J. (1972). *AIChE Symp. Ser.* **68**(124), 250–261.

Elmer, T. H. (1978). *Am. Ceram. Soc. Bull.* **57**(11), 1051–1053.

El Nasher, A. M. (1976). *NWSIA J.* **3**(2), 23–27.

Ermakova, T. P., Ananich, N. I., and Polyakov, G. V. (1977). *J. Appl. Chem. USSR* **50**(5), 975–979.

Goldsmith, R. L., Hossain, S., and Tan, M. (1973). Office of Saline Water Research and Development Progress Report, No. 877. Washington, D.C.

Goossens, I., and van Haute, A. (1976). *Desalination* **18,** 203–214.

Hamer, E. A. G., and Kalish, R. L. (1969). Office of Saline Water, Research and Development Progress Report, No. 471.

Hara, S., Mori, K., Taketani, Y., Noma, T., and Seno, M. (1977). *Desalination* **21,** 183–194.

Harriott, P., Wu, J., and Klunker, F. (1973). Office of Saline Water Research and Development Progress Report No. 846. Washington, D.C.

Henkens, W. C. M., and Smit, J. A. M. (1979). *Desalination* **28,** 65–85.

Higley, W. S., and Saltonstall, C. W., Jr., (1971). Office of Saline Water Research and Development Progress Report, No. 694. Washington, D.C.

Hoernschemeyer, D. L., Saltonstall, C. W. Jr., Schaeffler, O. S., Schoellenbach, L. W., Secchi, A. J., and Vincent, A. L. (1970). Office of Saline Water, Research and Development Progress Report No. 556. Washington, D.C.

Kesting, R. E. (1973). *J. Appl. Polym. Sci.* **17,** 1771–1784.

Kimura, S., and Nakao, S. (1975). *Desalination* **17,** 267–288.

Kraus, K. A., Marcinkowsky, A. E., Johnson, J. S., and Shor, A. J. (1966). *Science* **151,** 194–195.

Kunisada, Y., and Murayama, Y. (1978). *Desalination* **27,** 333–344.

Lonsdale, H. K., Merten, U., and Riley, R. L. (1965). *J. Appl. Polym. Sci.* **9,** 1341–1362.

Martinez Guerrero, J., and Filipe del Castillo, L. (1978). *Desalination* **26,** 141–151.

Meares, P. (1979). *Ber. Bunsenges. Phys. Chem.* **83,** 342–351.

Merten, U., Lonsdale, H. K., Riley, R. L., and Vos, K. D. (1966). Office of Saline Water Research and Development Progress Report No. 208. Washington, D.C.

Murayama, Y., Kasamatsu, T., and Gaydos, J. G. (1976). *Desalination* **19,** 439–446.

Ohya, H. (1978). *Desalination* **26,** 163–174.

Paul, D. R., and Ebra-Lima, O. M. (1971). *J. Appl. Polym. Sci.* **15,** 2199–2210.

Podall, H. E. (1971). *AIChE Symp. Ser.* **67**(107), 260–266.

Riley, R. L., Fox, R. L., Lyons, C. R., Milstead, C. E., Seroy, M. W., and Tagami, M. (1976). *Desalination* **19,** 113–126.

Rosenbaum, S., and Cotton, O. (1969). *J. Polym. Sci.* **Al**(7), 101–109.

Sammon, D. C. (1975). *NATO Adv. Study Inst. Ser.* **E13,** 63–90.

Sammon, D.C., Stringer, B., and Stephen, I. G. (1976). *Int. Symp. Fresh Water Sea 5th* **4,** 179–188.

Spatz, D. D., and Friedlander, R. H. (1978). *Water Sewage Works* **125**(2), 36–40.

Taniguchi, Y. (1977). *Desalination* **20,** 353–364.
Vos, K. D., Burris, F. O., Jr., and Riley, R. L. (1966a). *J. Appl. Polym. Sci.* **10,** 825–832.
Vos, K. D., Hatcher, A. P., and Merten, U. (1966b). *Ind. Eng. Chem. Prod. Res. Dev.* **5**(3), 211–218.
Walmsley, D., Stringer, B., and Russell, P. J. (1976). *Int. Symp. Fresh Water Sea 5th* **4,** 209–218.

4

Polarization Phenomena in Membrane Processes

G. JONSSON

C. E. BOESEN

Instituttet for Kemiindustri, The Technical University of Denmark
Lyngby, Denmark

List of Symbols

- C solute concentration (mol/cm^3)
- C_b solute concentration in bulk solution (mol/cm^3)
- C_g solute concentration in gel layer (mol/cm^3)
- C_p solute concentration in permeate (mol/cm^3)
- C_w solute concentration at membrane-solution interface (mol cm^{-3})
- D diffusion coefficient (cm^2/sec)
- d_h hydraulic diameter of flow channel (cm)
- d particle diameter (cm)
- f flow friction factor

SYNTHETIC MEMBRANE PROCESSES

ISBN 0-12-085480-5

F Faraday constant (96,487 coul/eq)
h half-channel height (cm)
i current density (amp/cm^2)
$i_{\lim}$ limiting current density (A/cm^2)
J_D Chilton–Colburn factor
J_{dif} solute diffusion flux (eq/cm^2 sec)
J_{el} solute transference flux (eq/cm^2 sec)
$J_{m(el)}$ solute transference flux in the membrane (eq/cm^2 sec)
J_m solute flux in the membrane (eq/cm^2 sec)
J_s solute flux (mol/cm^2 sec)
J_v volume flux (cm/sec)
$J_{v_{\lim}}$ limiting flux in UF (cm/sec)
k mass-transfer coefficient (cm/sec)
L length of the membrane (cm)
l_p membrane permeability (cm/sec atm)
M concentration polarization modulus, C_m/C_b
P permeability (cm^3 sec/g)
ΔP pressure difference (atm)
P_g permeability of the gel layer (cm^2/sec atm)
R true retention, $1 - (C_p/C_m)$
R_m hydraulic resistance of the membrane (atm sec/cm)
Re Reynold number, $ud_h\rho/\mu$
S observed retention, $1 - (C_p/C_b)$
Sc Schmidt number, $\mu/\rho D$
Sh Sherwood number, kd_h/D
$\overline{\text{Sh}}$ average Sherwood number
t_g thickness of the gel layer (cm)
t_m transference number in the membrane
t_s transference number in solution
u channel velocity parallel to membrane (cm)
v channel velocity perpendicular to membrane (cm)
x coordinate along membrane in flow channel
y coordinate perpendicular to membrane
$\dot{\gamma}_w$ fluid shear rate at membrane (sec^{-1})
δ concentration boundary layer thickness (cm)
ε eddy diffusivity coefficient (cm^2/sec)
ε porosity of the gel layer
ξ dimensionless coordinate, Eq. (15) and Eq. (17)
μ viscosity (g/cm sec)
σ density (g/cm)

I. Introduction

In any membrane process an accumulation or depletion of solutes takes place near the membrane surface because of the permselectivity of the membrane. Thus a concentration gradient is formed at which steady state is established by the diffusion of solute back through the stagnant film layer. This phenomenon is called *polarization*.

In general, the polarization phenomenon reduces the separation efficiency. With increasing polarization, the flux of the more permeable solute decreases and the flux of the less permeable solute increases.

In forced convection, the velocity boundary layer gradually grows in thickness and usually approaches a constant value toward the far downstream end of the channel. In turbulent flow, the thickness assumes the constant value much faster than in laminar flow. The concentration profile develops in a manner similar to the longitudinal velocity, but the concentration boundary layer is much thinner than the corresponding velocity boundary layer. Therefore, the concentration gradient across the concentration boundary layer is steeper than the velocity gradient. In the bulk solution the concentration profile is almost uniform in the direction normal to the membrane surface, so transport of any species in the bulk solution occurs by convective motion in the longitudinal direction parallel to the membrane. Within the boundary layer, the transport is in the direction normal to the membrane primarily because of diffusion, convection, and transference. Thus the longitudinal motion of the bulk stream influences the mass transport in the normal direction only by maintaining a thin stagnant film layer.

When the polarization is sufficiently high that significant density distribution occurs near the membrane surface, free convection induced by buoyancy effects also influences the mass transfer.

II. Reverse Osmosis and Ultrafiltration

When water permeates selectively through the membrane, the retained solute accumulates at the solution–membrane interface. The solute is then transported back from the membrane by diffusion and consequently a concentration gradient is formed in the boundary layer. This concentration build up at the membrane surface is termed concentration polarization.

Concentration polarization results in decreasing product rate and solute retention and eventually to precipitation or gelation of certain components on the membrane surface. One of the main requirements in process design is to obtain a fluid flow pattern on the membrane feed side, which minimizes concentration polarization.

A. THE FILM-THEORY MODEL

Figure 1 shows the concentration profile in the stagnant boundary layer. Longitudinal mass transport within the boundary layer is assumed negligible, so mass transport within the film is one-dimensional. In the steady state the solute flux is constant throughout the film and equal to the solute flux through the membrane J_s.

A material balance for the solute in a differential element gives the equation

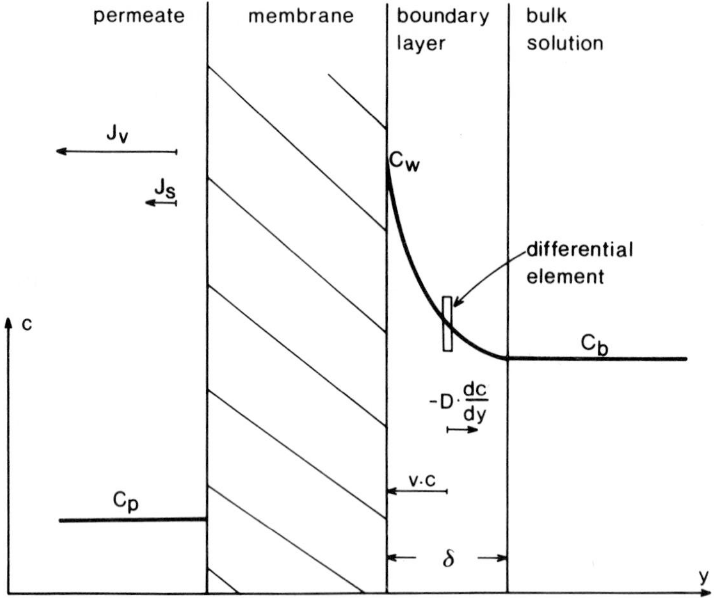

Fig. 1 Schematic of the film-theory model.

$$J_s = C_p J_v = CJ_v - D(dC/dy). \tag{1}$$

The first equality refers to the product condition with the boundary conditions

$$C = \begin{cases} C_b & \text{at} \quad y = 0, \\ C_w & \text{at} \quad y = \delta. \end{cases}$$

Integration of Eq. (1) gives

$$(C_w - C_p)/(C_b - C_p) = \exp(J_v/k), \tag{2}$$

where the mass-transfer coefficient k by definition is equal to D/δ.

Equation (2) can be rearranged to give a relation between the observed retention $S \equiv 1 - (C_p/C_b)$ and the concentration polarization M:

$$M \equiv C_w/C_b = 1 - S + S \exp(J_v/k). \tag{3}$$

Thus the concentration polarization can be calculated from the measurement of the retention and the permeate flux, when the mass-transfer coefficient for the given reverse osmosis (RO) module is known. The film-theory model contains many simplifying assumptions known to be incorrect, but the effect of these assumptions on the film-theory predictions are often found to be small. The film model is applicable in turbulent flow beyond the entrance region and in other flow situations resulting in a constant concentration along the membrane surface. It has, however, been used in laminar flow to give average M values over a membrane area by using the appropriate value for k. It is common practice to introduce the dimensionless Sherwood Number Sh defined as $\mathrm{Sh} \equiv kd_h/D$.

1. *Turbulent Flow*

The Sh number can be related to the Re and Sc numbers by means of the Chilton–Colburn analogy (Brian, 1966)

$$J_D \equiv (k/u) \cdot \mathrm{Sc}^{2/3} = f/2, \tag{4}$$

and using the Blasius formula (Bird *et al.*, 1960),

$$f = 0.0791 \cdot \mathrm{Re}^{-1/4} \tag{5}$$

which gives the relation

$$\mathrm{Sh} = 0.04 \cdot \mathrm{Re}^{0.75} \cdot \mathrm{Sc}^{0.33}. \tag{6}$$

This expression links the concentration polarization to flow friction. More directly, Sh was determined empirically from heat- and mass-transfer data in channel flow (e.g., Dittus and Boelter, 1930):

$$\mathrm{Sh} = 0.023 \cdot \mathrm{Re}^{0.80} \cdot \mathrm{Sc}^{0.33}. \tag{7}$$

Sherwood *et al.* (1965) expanded the film model to include eddy diffusivity by changing the last term in Eq. (1) from DdC/dx to $(D + \varepsilon)\, dC/dx$, where ε is the eddy diffusion coefficient. This leads to a slightly different film model equation

$$(C_w - C_p)/(C_b - C_p) = \exp[f(\varepsilon)J_v/k], \tag{8}$$

where $f(\varepsilon)$ indicates a factor that is a function of the eddy diffusivity. Gill *et al.* (1969) compared M calculated by the film model, the eddy diffusivity model, and the equations of change. For the Sc numbers and permeate fluxes normally found in RO and ultrafiltration (UF), the agreement was very good.

2. *Laminar Flow*

a. In fully developed laminar flow, under conditions where the concentration boundary layer is developing, the Sh number is given by (Sieder and Tate, 1936),

$$\overline{\mathrm{Sh}} = 1.86(\mathrm{Re} \cdot \mathrm{Sc} \cdot d_h/L)^{0.33}. \tag{9}$$

Here $\overline{\text{Sh}}$ is the average value of Sh from the front edge of the membrane up to L, where L is the length of the membrane.

According to Blatt *et al.* (1970), the length of the "concentration entrance region" where Eq. (9) can be used is approximately

$$L = \dot{\gamma}_w h^3/10D. \tag{10}$$

For normal conditions this gives L values that are longer than any channel of interest.

b. If both the velocity and the concentration profiles are developing, then the following relationship is applicable (Gröber *et al.*, 1961):

$$\overline{\text{Sh}} = 0.664(\text{Re} \cdot d_h/L)^{0.50}\text{Sc}^{0.33}. \tag{11}$$

According to Gröber *et al.* (1961), the velocity profile is completely developed at a distance from the channel inlet given by the equation

$$L = 0.029 \cdot \text{Re} \cdot d_h. \tag{12}$$

B. Laminar Flow Conditions

In the steady-state system with feed solution in laminar flow between two parallel, flat plate membranes as shown in Fig. 2, a solute material balance on a differential volume element yields the following partial differential equation:

$$\frac{\partial}{\partial x}(uC) + \frac{\partial}{\partial y}\left(vC - D\frac{\partial C}{\partial y}\right) = 0, \tag{13}$$

where u and v are fluid-velocity components in the x and u directions, respectively. Both diffusion and convection of solute in the transverse

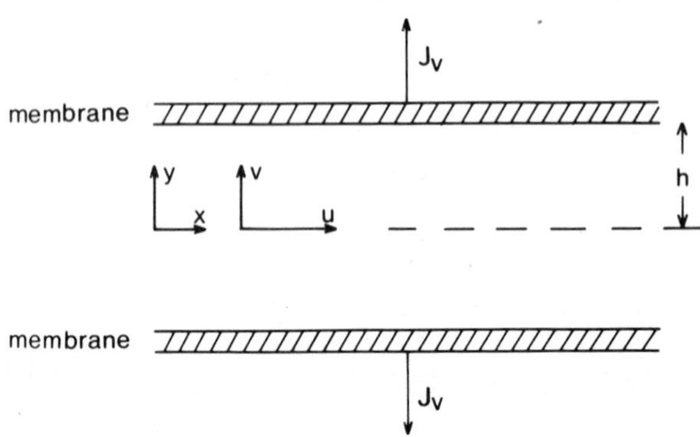

Fig. 2 The two-dimensional system for laminar flow between flat parallel membranes.

direction are included, but diffusion in the longitudinal direction is assumed negligible. The boundary conditions are

$$C(0,y) = C_b,$$

$$\frac{\partial C}{\partial y}(x,0) = 0, \text{ and}$$

$$J_v C_p = J_v C(x,h) - D\left[\frac{\partial C}{\partial y}(x,h)\right]. \tag{14}$$

To solve the differential equation, one needs an equation for the velocity-field profile, but it is not possible to establish this equation for the case when the permeate velocity, J_v, varies with the x position.

However, assuming that J_v is constant, $R = 1$, the velocity profile can be expressed by a set of equations developed by Berman (1953), describing the velocity field in laminar channel flow with liquid removal through the channel walls. Using this expression in a truncated form, Dresner (1964) obtained an approximative solution that is valid near the channel entrance:

$$M(R = 1) = \begin{cases} \xi + 6 - 5\exp[-(\xi/3)^{1/2}] & \text{for} \quad \xi \geq 0.02, \\ 1 + 1.536(\xi)^{1/3} & \text{for} \quad \xi < 0.02, \end{cases} \tag{15}$$

where

$$\xi = J_v^3 xh/3uD^2.$$

This solution has been extended to the interval $0.8 \leq R < 1$ by Johnson *et al.* (1966):

$$M = M(R = 1) - (\xi + 5)\frac{R}{1 - R}$$
$$\times (1 - \exp[-(\xi)(1 - R)]) + 5\exp[-\xi(1 - R)]. \tag{16}$$

Fisher *et al.* (1964) modified Dresners solution to apply for tubular membranes and found that Eq. (15) could be used when the factor ξ was calculated from the equation:

$$\xi = J_v^3 xr/4uD^2. \tag{17}$$

C. Measurements of Concentration Polarization

Direct measurement of the concentration profile can, in principle, be made by microelectrode or optical measurements. Both have been applied successfully to RO studies. Goldsmith and Lolachi (1969) developed an Ag–AgCl microelectroprobe and used this to measure the build up of concentration profile in batch cells and concentration polarization in laminar flow. Hendricks and Williams (1971) developed a conductivity microprobe and used this to measure the concentration profile in laminar flow. Johnson (1970) used a Mach–Zendner interferometer to measure the concentration profile in free convection on a vertical membrane. Additionally, Lim *et al.* (1971) sampled and analyzed the gel layer formed during UF of whey and Dejmek *et al.* (1973) followed the cumulative build up of a ^{131}I–labeled casein. All direct measurements of the concentration polarization are more or less restricted to batch cells and special channel flow experiments under laminar flow conditions.

Indirect measurements of the concentration polarization in both laminar and turbulent flow can be obtained from the film-theory model. Defining the true retention, $R \equiv 1 - C_p/C_w$, Eq. (2) may be rearranged to give

$$\ln\left(\frac{1 - S}{S}\right) = \ln\left(\frac{1 - R}{R}\right) + \frac{J_v}{k}. \tag{18}$$

Because the mass-transfer coefficient is unknown, an appropriate value of k must be used. In turbulent flow the use of Eq. (7) leads to

$$\ln\left(\frac{1 - S}{S}\right) = \ln\left(\frac{1 - R}{R}\right) + \text{constant}\,\frac{J_v}{u^{0.80}}. \tag{19}$$

This means that a plot of $\ln[(1 - S)/S]$ vs. $J_v/u^{0.80}$ should give a straight line, intersecting the ordinate axis at $\ln[(1 - R)/R]$. When R is determined, the concentration polarization can be calculated from

$$M = (1 - S)/(1 - R). \tag{20}$$

Jonsson and Boesen (1977) analyzed how well the method works: The main problem is to do the measurements in such a way that the true retention is truly constant. Normally, the extrapolation is done by measuring the permeate flux and concentration with varying flow rates, keeping the pressure and bulk concentration constant. This means C_w decreases with increasing u, so the method normally gives true retentions that are too high because of the decreasing retention with increasing concentration.

It was found that the turbulent data could be represented by the equation

$$\text{Sh} = A \cdot \text{Re}^{0.80} \cdot \text{Sc}^{0.33}, \tag{21}$$

where the constant A was independent of the solute mobility and membrane retention but varied with the channel height and distance from the inlet zone. Generally, the experimental Sh values were 25% higher than predicted from Eq.(7).

The laminar data could be represented by Eq. (9), except that the constant was found to be about 1.6. Equation (16) was in reasonable agreement with the experimental laminar data and gave a good description of the increase in concentration polarization with increasing channel length.

D. GEL FORMATION

In UF, the retention of low molecular weight (MW) compounds is generally low and concentration polarization behavior is likely to be determined by the properties of the macromolecules in solution.

As the diffusivity of the macromolecules is low and the permeate flux is high in UF, concentration polarization can be extremely high, especially as the retention is high. From the film model, Eq. (2), one has

$$M = \frac{C_w}{C_b} \approx \frac{C_w - C_p}{C_b - C_p} = \exp\left(\frac{J_v}{k}\right) \gg 1. \quad (22)$$

As macromolecules normally have a certain concentration at which they behave as gels, the concentration polarization will have a maximum value dependent on the bulk concentration. For a given value of k, determined by the flow conditions, the permeate flux will be independent of the pressure when the concentration at the membrane surface equals the gel concentration C_g as shown in Fig. 3. By increasing the pressure above this point a temporary flux increase results in a gel formation at the membrane surface. Steady state is attained when the hydraulic resistance of the gel layer has decreased the permeate flux to the original value:

$$J_{v\lim} = \frac{\Delta P}{R_m + (t_g/P_g)} = k \ln(C_g/C_b). \quad (23)$$

Thus the limiting flux increases with decreasing bulk concentration and increasing flow rate, as shown in Fig. 4.

The gel concentration C_g at which a particular macromolecule displays anomalous rheological properties depends on the size, shape, and degree of solvation of the solute. For highly structured spheroidal macromolecules, such as proteins, gellike properties may begin to develop at concentrations much higher than for rigid chain, solvated macromolecules such as polysaccharides.

From Eq. (23) it is seen that under gel-polarized conditions, the per-

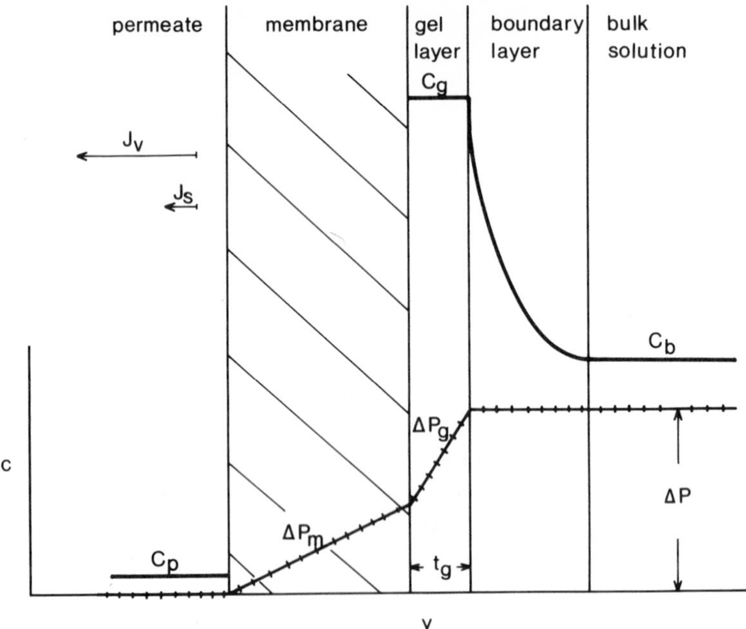

Fig. 3 Schematic of the gel formation model.

meate flux decreases linearily with the logarithm of the bulk concentration and equals zero when $C_b = C_g$. This has been experimentally verified as shown in Fig. 5, and the extrapolated C_g-values appear reasonable (Blatt *et al.*, 1970).

Porter (1972) compared experimental flux data for UF of human albumin solutions with Eq. (23), determining the gel concentration from a plot as in Fig. 5 and calculating the mass transfer coefficient from Eqs. (7) and (9). The agreement between theoretical and experimental UF fluxes was within 15 to 30%.

Because the concentration at the membrane surface can be more than 10 times the bulk concentration in a highly polarized system, the diffusivity and viscosity may vary appreciably across the boundary layer, giving an uncertainty in the calculation of the mass transfer coefficient.

Probstein *et al.* (1978) recently analyzed this problem in UF of bovine serum albumin solutions at high polarization in laminar channel flow. They found that the appropriate diffusivity defining the limiting flux was that of the gelling concentration, rather than that of the bulk concentration.

In an earlier paper (Shen and Probstein, 1977) they found that the

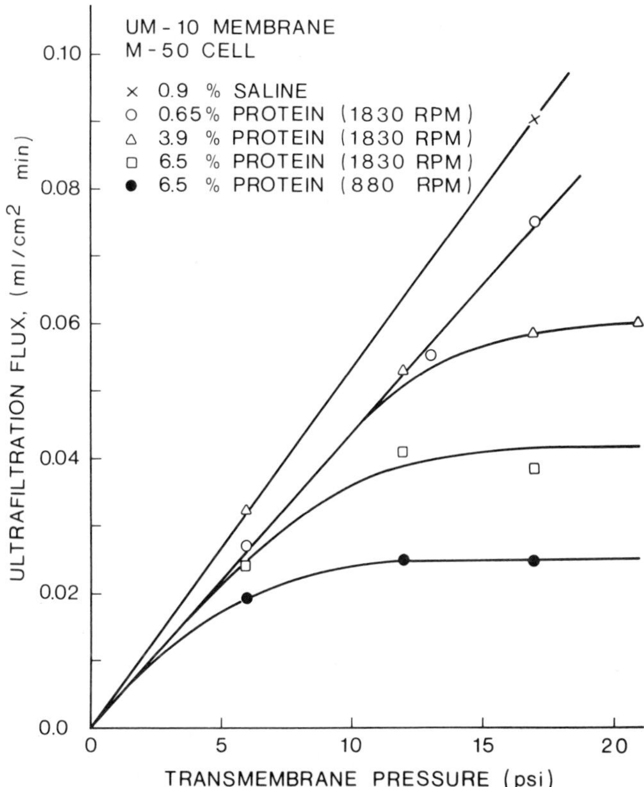

Fig. 4 Flux–pressure relationships for bovine-serum albumin solutions in a stirred batch cell. (Reprinted from Blatt *et al.*, 1970, courtesy of Plenum Press.)

concentration dependence of the viscosity had only minor effect on the limiting flux.

Although the gel model has shown great utility, the assumption that a gel layer with well-defined gel concentration C_g and variable thickness t_g determined by the pressure difference across the gel layer is not generally acceptable. Nakao *et al.* (1979) measured directly the concentration of the gel layer for polyvinylalcohol and ovalbumin by scratching out the gel layer from the membrane after measuring steady state fluxes. It was found that C_g was not constant but that it increased with increasing bulk concentration and decreasing feed velocity.

Vilker *et al.* (1981) measured the osmotic pressure of bovine serum albumin at concentrations up to 475 g/liter. They found that these highly concentrated solutions had osmotic pressures comparable to the normally

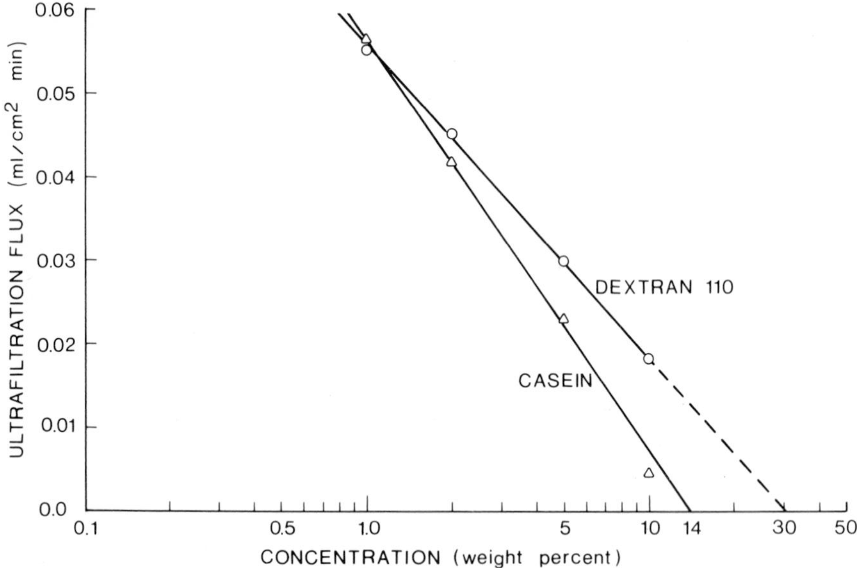

Fig. 5 Flux–concentration relationships for casein and dextran 110 solutions in thin-channel flow cells. (Reprinted from Blatt *et al.*, 1970, courtesy of Plenum Press.)

applied pressures in UF. The osmotic pressure increased parabolically with concentration because of increased importance of the second and third virial coefficients. It is likely that UF flux may be limited by the osmotic pressure in addition to the formation of a gel layer, depending on the nature of the solute and the operating conditions.

Wales (1981) argues that for most lyophilic systems it is not likely there exists a pressure gradient across the polarization layer. For such materials as agar, pectin, gelatin, and some proteins, which might be denatured at the membrane to give true gels, it would be expected to find gel-controlled polarization layers.

Trettin and Doshi (1981) showed that the best method to distinguish between gel-limited and osmotic-pressure-limited UF is from unstirred batch cell experiments using different pressure levels. For bovine serum albumin they found increasing wall concentrations to 6.9 atm, indicating osmotic-pressure-limited UF below this pressure.

For colloidal suspensions, experimental flux values are often one–two orders of magnitude higher than predicted from the theory.

One explanation could be that the steady-state situation shown in Fig. 3 is unattainable. If the thickness of the gel layer is sufficiently high, it migrates with the flowing bulk solution because of the high shear rate and Eq. (23) will give far too low flux values. The permeability of the gel layer can be approximated by the Kozeny–Carman relation for porous solids (e.g., Blatt *et al.*, 1970):

$$P \approx (d^2/180\mu)\varepsilon^3/(1 - \varepsilon)^2, \tag{24}$$

where d is the diameter of the particles and ε the porosity of the gel layer.

Porter (1972) presents flux data for UF of solutions of styrene–butadiene polymer latex with an average particle size of 0.19 μm and a gel concentration determined to be 75%. If the gel layer is the limiting resistance to flow, one can calculate the thickness from the Poiseuille equation, using the permeability from Eq. (24) and appropriate values of flux and pressure drop. This gives a value of about 100 μm.

Porter argues, however, if the gel layer is not the limiting resistance to flow, the flux should be invariant with concentration and proportional to the transmembrane pressure drop. But because this is not the case, he concludes that, for colloidal suspensions, mass transfer from the membrane into the bulk stream is driven by some force other than the concentration gradient.

The "tubular-pinch" effect has been observed for many colloidal suspensions giving an enhanced mass transfer of particles moving away from the wall. Brandt and Bugliarello (1966) made a direct photographic observation of the phenomenon and found the process was accelerated by increasing the flow rate and decreasing the average concentration. The physical cause of the tubular-pinch effect has been postulated to occur because of a transverse force arising from either a slip–spin force (Rubinow and Keller, 1961) or a slip–shear lift force (Saffman, 1956). The different analyses of the phenomenon indicate that the migration velocity is proportional to the flow rate raised to the second power and the ratio of the particle size to the channel dimension raised to the fourth power. Therefore, high flow rates in narrow channels are advantageous in depolarizing the membrane surface via the tubular-pinch effect.

In a dissertation by Madsen (1977), other effects that may partially explain the high fluxes obtained experimentally are further discussed.

Because a gel layer acts as a membrane in series with the primary membrane, the influence on the retention of a permeable solute is expected, depending on the mechanical properties of the gel layer. If the mobility of the low MW solute in the gel layer is relatively high, the concentration polarization of the solute will increase with accompanying decrease in retention. However, if the solute mobility in the gel layer is low, it becomes virtually impossible for solute molecules to pass through the layer and reach the primary membrane, at which point the solute retention increases. Nakao and Kimura (1981) determined the transport coefficients for vitamin B_{12} in an ovalbumin gel layer by comparing the retention flux data for the UF membrane alone with those where the gel layer was in series with the membrane. They found the reflection coefficient was much higher in the gel layer than in the membrane, giving a considerable increase in the retention at a given flux level. If the gel layer has polyelectrolyte character, a high retention of electrolytes is usually seen, because of the Donnan exclusion. Thus, gel layer formation normally reduces the separative capability of a membrane, so if the purpose of the UF is to effect a separation between larger and smaller molecules, the efficiency of the process is seriously compromised.

As many wastewaters contain high MW polyelectrolytes, dynamically formed membranes having high salt retention can be prepared on UF membranes, giving high flux desalination membranes (e.g., Perona *et al.*, 1967).

E. Fouling

In the literature discussing water and wastewater treatment by membrane methods, one often encounters the term "membrane fouling." The definition is not precise but normally is concerned with a long-term flux decline and, eventually, retention decrease as a result of accumulation of some fouling material. Because gel formation, membrane compaction, and membrane hydrolysis result in similar phenomena, it is often impossible to distinguish between them. The main difference between gel formation and fouling is that the gel layer is formed on the membrane surface because the gel concentration is reached, whereas the fouling layer is formed by another mechanism and is more closely bound to the membrane surface.

Fouling may be caused by a variety of compounds. These foulants may be classified as (Belfort *et al.*, 1976)

(1) dissolved organics, including humic substances, biological slimes and macromolecules;

(2) dissolved inorganics, including inorganic precipitates such as $CaSO_4$, $CaCO_3$, $Mg(OH)_2$, $Fe(OH)_3$, and other metal hydroxides; and

(3) particulate matter.

Matthiason and Sivik (1980) wrote an extensive review discussing fouling of RO and UF membranes.

1. *The Fouling Mechanism*

Jackson and Landolt (1973) studied the rate of fouling by deposition of iron hydroxide on tubular reverse osmosis membranes. They postulated that the fouling occurred from a two-step nucleation–growth mechanism. This viewpoint is probably useful for other foulants, too. In the nucleation phase, foulants are deposited in pores and surface cavities of the membrane. The attachment is caused by the mechanical force acting from the convection of foulants to the membrane surface and van der Waals' forces of attraction. The number of nucleation sites is dependent on the relative size of the foulants and pores in the membrane as well as surface charges. When sufficient amounts of foulant are trapped on the membrane surface, they act as nuclei from which growth proceeds by a polymerization reaction analogous to those in flocculation. Large flocculated particles build up on the membrane surface forming a thin porous layer. The rate of growth depends on the number of nuclei, the rate of polymerization reactions, and the transport of foulants to the membrane surface.

In a review article, Gregor and Gregor (1978) explained that fouling is normally caused by materials that have large surface areas and are hydrophobic, therefore repelling water. When a hydrophobic substance is in an aqueous environment, it can reduce its total energy by reducing the area exposed to water. Therefore, it will be held to the surface of the membrane by the elimination of repulsive interactions with the surrounding water. As fouling materials normally bear a negative charge, hydrogen bonding involving these charges will further contribute to the fouling of the membrane surface.

Jonsson and Kristensen (1980) distinguish between two types of fouling: Irreversible adsorption of hydrophobic substances to the membrane surface and reversible sorption of different compounds within the membrane phase. The irreversible adsorption mechanism depends mainly on the transport of foulants to the membrane surface, which is a function of foulant concentration in the bulk solution and of the total volume of permeate. Therefore, it depends strongly on the elapsed time and the hydrodynamics of the membrane model. Further, membranes require a mechanical or chemical cleaning to restore the pure water permeability. The reversible sorption mechanism results in smaller l_p values compared with water, because of increased frictional resistance and a decrease in the water content of the membrane. Because of the existance of a dynamic equilibrium between the bulk solution and the membrane phase, the change in l_p is established quickly and is almost independent of time.

2. *How to Prevent Fouling?*

Because of the different nature of foulants, a general answer to the question of how to prevent fouling is not possible, but the main approaches to the prevention of fouling are concerned with (a) the hydrodynamics of the membrane module, (b) the pretreatment of the feed solution, and (c) the properties of the membrane.

a. *Hydrodynamics of the Membrane Module*

Fouling generally decreases with decreasing concentration polarization, so that a high flow velocity and a high Re number is advantageous. Sheppard *et al.* (1972) found that with river water feed, the rate of flux decline varied with the feed velocity raised to the $(-1/2)$, when it was less than some threshold velocity. Above this value, the accumulation of particulates on the membrane was prevented and the rate of flux decline was significantly less than predicted from the half power relationship.

Boen and Johannsen (1974) tested five RO units including tubular, hollow fiber, and spirally wound concepts for treatment of secondary sewage effluents. All unit membranes were subject to fouling, showing that periodic cleaning procedures are necessary.

b. *Pretreatment of the Feed Solution*

Depending on the concentration and foulant type in raw water, combinations of the following pretreatments can be used:

(1) filtration,
(2) chemical clarification,
(3) pH adjustment,
(4) chlorination, and
(5) adsorption on active carbon.

Kuiper *et al.* (1974) studied how the pretreatment of river Rhine water influences the membrane fouling. The results indicated that a chemical clarification was necessary and that the floc removal should be sufficiently effective to reduce the iron content to normal drinking water levels (<0.05 ppm). Without chemical clarification the flux decline was high and all membranes were completely covered with a dark fouling layer containing 70% organic matter, 10–15% silica, and 5–10% iron.

Jackson and Landolt (1973) found that traces of oil aggravated the fouling considerably, so oil contamination from high pressure pumps should be carefully avoided.

Belfort and Marx (1978) presented another method—the fixed protective cover method—to reduce submicrometer colloidal fouling of RO membranes. They used irradiated polycarbonate membranes with a mean pore diameter in the range of 0.2–0.9 μm as protective cover above the RO membrane. This resulted in 78% increased fluxes at 43% decreased salt rejections over the unprotected membrane. The mechanism of protection was explained by the presence of submicrometer pinholes in the surface skin, which were prevented from being plugged by the colloids in the feed. It seems the microfilter acts as a depth filter, giving a greater surface area for the accumulation of colloids. Simultaneously the microfilter increased the polarization of dissolved molecules, so the method is not applicable to feeds containing gel-forming macromolecules.

c. *Properties of the Membrane*

As previously mentioned, the pore size distribution of the membrane influences the fouling tendency. Dense membranes are normally less exposed to fouling than open membranes. Nonfouling membranes are based on the assumption that negatively charged colloids in the feed solution are the major fouling factor. A negatively charged membrane should repel the colloids and ensure fouling-free service. Fisher and Lowell (1970) have tested a cellulose acetate–hydrogen succinate membrane on secondary effluent and found no flux decline under the first six days. Madsen (1977) reported that a hydrophilic cellulose acetate membrane needs much less flow over the membrane than a hydrophobic membrane with the same performance to give the same UF results in the same system.

This agrees with the fouling mechanism described by Gregor and Gregor (1978). Thus an obvious technique to avoid hydrophobic interactions is to create an extremely hydrophilic membrane, which has a strong affinity for water. Such a membrane remains wetted even in the presence of hydrophobic particles, therefore these particles cannot adhere to the surface by excluding water.

3. *Cleaning Methods*

When the product flux has decreased to unacceptable values, the membrane must be cleaned. The cleaning method and frequency depend on the type of foulant and the chemical resistance of the membrane. Generally, it is easier to clean a membrane that is slightly contaminated.

The cleaning method may be divided into (a) hydraulic cleaning, (b) chemical cleaning, and (c) mechanical cleaning.

a. Hydraulic Cleaning

Sometimes a depressurizing followed by flushing with water at high linear velocity is sufficient to remove a fouling layer (Kuiper *et al.*, 1974). Then chemical cleaning is necessary only at rare intervals.

b. Chemical Cleaning

The following types of chemicals have been used in cleaning:

(1) acids (HNO_3, H_3PO_4, citric acid),
(2) bases (NaOH),
(3) complexing agents (EDTA),
(4) enzymes,
(5) detergents, and
(6) disinfectants (NaOCl, H_2O_2).

c. Mechanical Cleaning

In the tubular module concept, *in situ* mechanical cleaning can be performed by the passage of oversized sponge balls through membrane tubes (Sachs and Zisner, 1977).

III. Electrodialysis

One of the major factors limiting the practical and economical application of electrodialysis (ED) is polarization. In the polarized boundary layers, conditions of concentration and composition may occur that differ significantly from those in bulk solutions. Because these conditions are generally deleterious, it is necessary to inquire into methods to limit polarization.

A. General Description

To illustrate the polarization phenomenon in ED, consider the simple model shown in Fig. 6. A cation-selective membrane is placed between two electrodes and the system immersed in a NaCl solution. When an EMF is applied across the membrane Na^+ ions move from right to left.

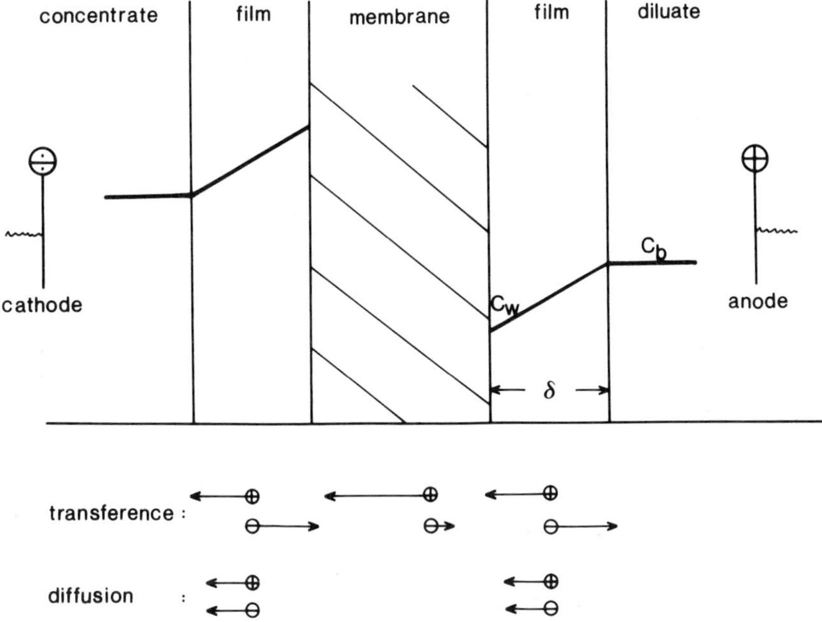

Fig. 6 Schematic of the steady-state situation at a cation selective membrane during ED.

The flux of cations through the membrane and the boundary layers because of transference are, respectively,

$$J_{\mathrm{m(el)}} = t_{\mathrm{m}} i/F \tag{25}$$

and

$$J_{\mathrm{(el)}} = t_{\mathrm{s}} i/F. \tag{26}$$

Because the transport in the membrane is faster than in the boundary layers ($t_{\mathrm{m}} > t_{\mathrm{s}}$), a region will be formed next to the membrane on the anode side, in which the ions are removed faster than they are supplied. Thus, a concentration gradient in the boundary layer is established, so that the ions needed for steady-state transport through the membrane are supplied by diffusion. On the cathode side, a similar but opposite concentration gradient is established.

The flux of ions resulting from diffusion can be expressed in terms of Fick's first law

$$J_{(\mathrm{dif})} = -D\ dC/dy. \tag{27}$$

At steady state the combined electrical and diffusive flux through the boundary layers equals the flux through the membrane

$$J_{\mathrm{m}} = t_{\mathrm{m}} i/F = -D\ (dC/dy) + t_{\mathrm{s}} i/F. \tag{28}$$

Since the concentration gradient is linear (Spiegler, 1971), the current density i can be isolated from Eq. (28) to give

$$i = DF(C_{\mathrm{b}} - C_{\mathrm{w}})/(t_{\mathrm{m}} - t_{\mathrm{s}})\delta. \tag{29}$$

The slope of the concentration profile in Fig. 6 characterizes the rate at which Na^+ ions cross the stagnant film by diffusion. If the EMF is increased, the rate of ion transport must also increase, and the concentration at the membrane surface must fall to provide the necessary increase in driving force.

If the EMF is raised still further, a point is reached when the concentration at the membrane surface falls to zero. At this point, a limiting condition has been reached because the concentration gradient cannot become any steeper.

Thus, there is a limiting current density i_{lim} described by setting $C_{\mathrm{w}} = 0$ in Eq. (29), in which case

$$i_{\mathrm{lim}} = DFC_{\mathrm{b}}/(t_{\mathrm{m}} - t_{\mathrm{s}})\delta. \tag{30}$$

From Eq. (30) it is seen that the limiting current density divided by the product normality $i_{\mathrm{lim}}/C_{\mathrm{b}}$, normally called the polarization parameter, is primarily a function of the boundary layer thickness δ. So the study of classical polarization is primarily a study of hydrodynamics.

Spiegler (1971) analyzed the total potential drop across the system shown in Fig. 6. This is composed of

(1) ohmic drops in the bulk solutions, membrane and boundary layers and

(2) membrane potential and junction potentials.

The ohmic potential drops (1) should vanish rapidly when interrupting the current, whereas the other potential drops relax more slowly because the relaxation depends on ionic diffusion. Spiegler showed that the total potential drop is the sum of two terms. One is linear in i and the other a logarithmic function of i. It is of interest that ohmic drops in the boundary layers, which are of dissipative nature, are contained in the second term together with the membrane and junction potentials.

When the voltage across an ED stack is raised, the current initially increases roughly in proportion to the voltage. The apparent resistance of the stack increases however, and a point is reached when large voltage increments cause only small current increases, as shown in Fig. 7. True plateaus, similar to those observed at metal–solution interfaces under hydrodynamically similar conditions, are rarely seen, as should be expected from Eq. (30).

In the absence of these plateaus, an alternative parameter, the limiting current density determined from a "Cowan plot" (Cowan and Brown, 1959) is used in ED technology to designate the upper limit at which the unit will demineralize satisfactorily. This type of limiting current is defined as the current corresponding to the minimum of a plot of the apparent resistance, defined as cell voltage divided by cell current, of the stack versus reciprocal current, as shown in Fig. 8.

Forgacs *et al.* (1972) showed that true plateaus in current–voltage curves can be obtained if the current is corrected for efficiency. This indicates that part of the electric current being used for desalination

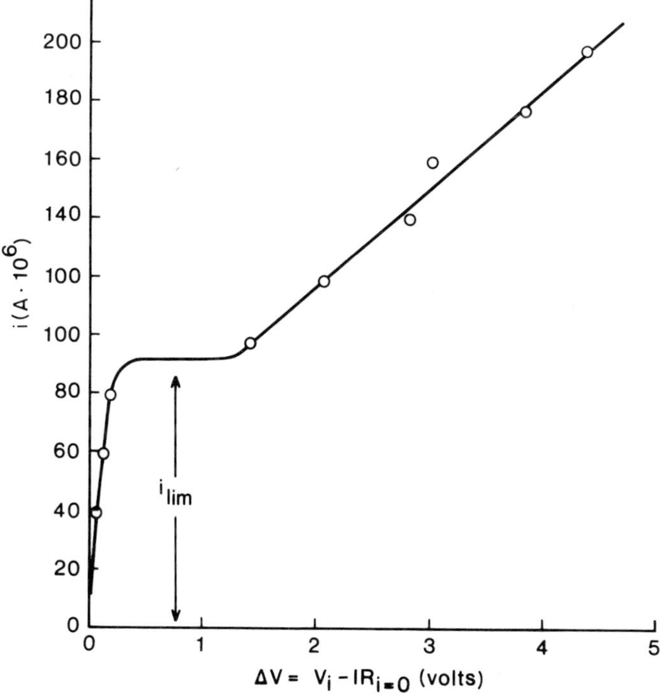

Fig. 7 The current–voltage curve for a cation-selective membrane between 0.01 *N* NaCl solutions. (Reprinted from Peers, 1956, courtesy of The Chemical Society.)

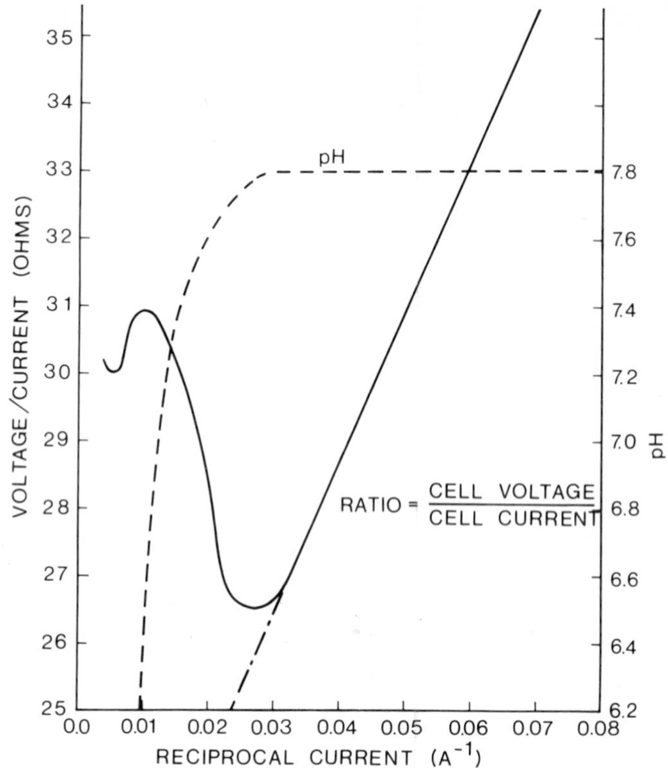

Fig. 8 Apparent resistance versus reciprocal current for an experimental ED cell. (Reprinted from Cowan and Brown, 1959, courtesy of the American Chemical Society.)

reaches a plateau, whereas the remaining portion of the current does not, as shown in Fig. 9.

B. THE EFFECTS OF POLARIZATION

The immediate effect of polarization is to increase the electrical resistance of the apparatus, which leads to an increase in power consumption. If this were the only problem, it would be of little significance.

1. *The "Water-Splitting" Phenomenon*

As seen from Fig. 9, an increase in current beyond the limiting current density results from a further increase in EMF. This is because of

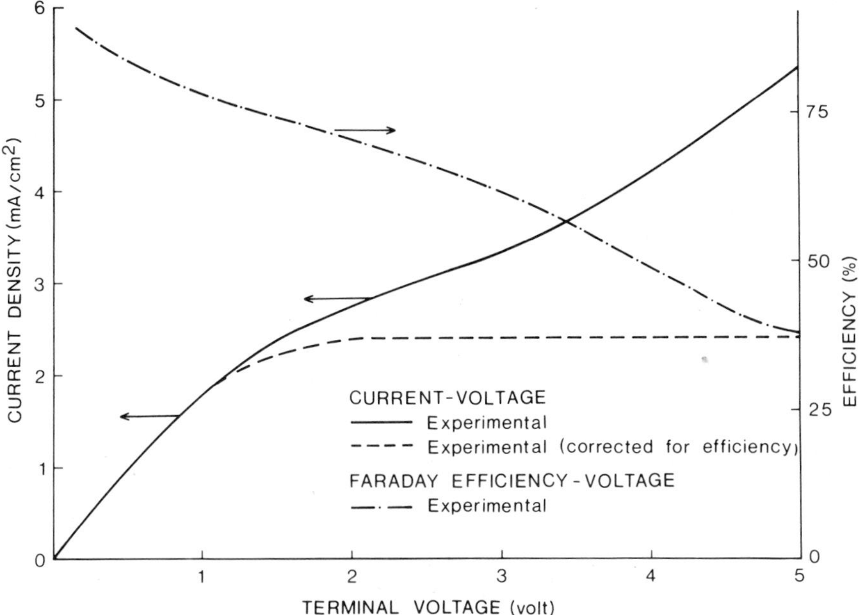

Fig. 9 Current density and Faraday efficiency versus terminal voltage for an anion selective membrane between dilute KCl solutions. (Reprinted from Forgacs *et al.*, 1972, courtesy of Elsevier Scientific Publishing Company.)

another mechanism of ion transfer. One explanation might be an increase in coion transport, which would decrease the transport number in the membrane and thereby increase the limiting current density from Eq. (30).

Another explanation could be that the contribution of H^+ to ion transport becomes significant when the concentration of Na^+ ions in the depleted boundary layer falls to zero. Unlike the Na^+ concentration, the H^+ concentration at the membrane surface can be supplemented by dissociation of water, so a supply of H^+ is always available. Because of the high mobility of H^+ in cation-selective membranes, and OH^- in anion-selective membranes, the "water-splitting" phenomenon is expected to start before the concentration in the depleted boundary layer reaches zero.

Normally, the mobility of anions is somewhat higher than that of cations in aqueous solutions. As the transport number in both anion- and cation-selective membranes approaches one, the term $t_m - t_s$ is greatest for the cation-selective membrane. Assuming similar film thickness for both types of membranes, the limiting current density is reached first at the cation-selective membrane, so one would expect this to be the critical membrane in ED. Figure 10 shows the pH change of a NaCl solution in

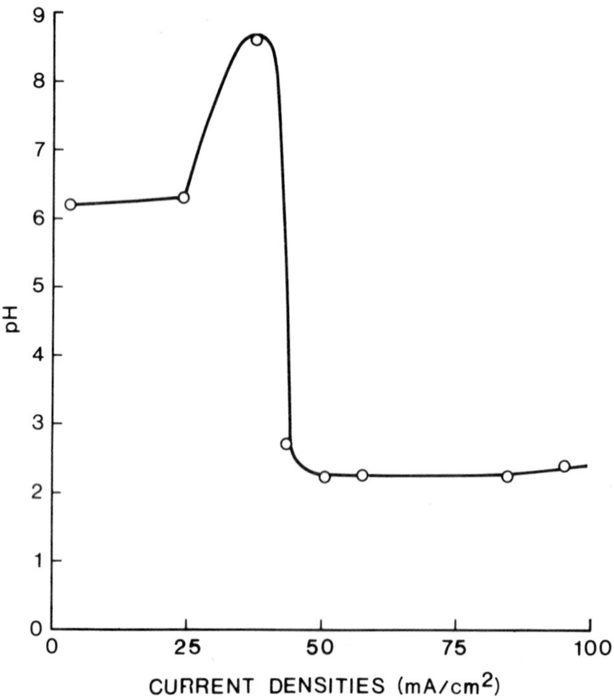

Fig. 10 pH of the diluate with increasing current densities in a multicompartment cell operated with 0.3 *N* NaCl solutions. (Reprinted from Yamabe and Seno, 1967, courtesy of Elsevier Scientific Publishing Company.)

the desalting compartment, separated by a cation- and anion-selective membrane, with increasing current density. When i_{lim} for the cation-selective membrane is reached, transport of H^+ ions begins to leave OH^- ions behind in the desalting compartment, so the solution becomes slightly alkaline. Above i_{lim} for the anion-selective membrane, OH^- transport through this membrane starts and the solution suddenly becomes intensively acidic. This shows the relation between the water-splitting phenomenon and the limiting current density and shows that the phenomenon is much more pronounced at the anion-selective membrane. Oda and Yawataya (1968) analyzed the phenomenon of neutrality disturbance on several types of membranes and solutions. Normally, concentration polarization will only result in a minor transport of H^+ ions through cation-selective membranes. If the solution contains ions that have a tendency to hydrolyze (Mg^{2+}, Ca^{2+}), these ions are capable of becoming an acceptor of the OH^- ion generated from the dissociation of water, thereby increasing the transport of H^+ through the membrane. Another factor appears to be the structure of the membrane. With increasing resistance of a series of cation-selective membranes, the pH change occurred at a lower current density for simulated seawater, indicating that the reduced mobility of the counter ions favors the transport of the high-mobility H^+ ions.

Forgacs *et al.* (1975) and Boari *et al.* (1973) determined the concentration at the membrane–solution interface in natural convection systems by two different methods. Both found that side effects had already set in at a moderate degree of electrolyte depletion at the membrane–solution interface.

The pH changes are, in general, believed to occur more readily with anion-selective membranes having strongly basic ion exchange groups than with ordinary cation-selective membranes. This is probably because the fouling and poisoning phenomenon overshadow the normal behavior of an anion-selective membrane (Kressman and Tye, 1969).

2. *The Fouling Phenomenon*

Fouling of ion-selective membranes is a major problem in ED, caused by the precipitation of colloids on the membrane surfaces. Because most of the colloids present in natural waters are negatively charged, it is normally the anion-selective membranes that are affected.

Korngold *et al.* (1970) examined the process of anionic fouling on different anion-selective membranes using sodium humate as the fouling agent. Fouling is caused by H^+ ions generated, by even minimal polarization, at the surface of the anion-selective membrane in the dialysate compartment. It is an autocatalytic process because the precipitate forms a composite sandwich membrane, which works in its closed direction, generating more H^+. Thus fouling will increase more rapidly after an incubation period. Figure 11 is a typical fouling curve showing an increase in voltage drop across the anion-selective membrane with time. By reversing the current, the voltage immediately decreases to its original value and remains low as long as fouling does not begin at the other side of the membrane. Reversing the current to its original direction gives an almost instantaneous voltage rise back to the value before the first reversal, and the fouling-time curve continues unchanged.

This shows clearly that the fouled membrane is effectively bipolar and so rectifies the electric current as a type of ionic transistor.

3. *The Poisoning Phenomenon*

Poisoning is defined as the adsorption of strongly held ions. Kobus and Heertjes (1972) studied the poisoning of commercial anion-selective

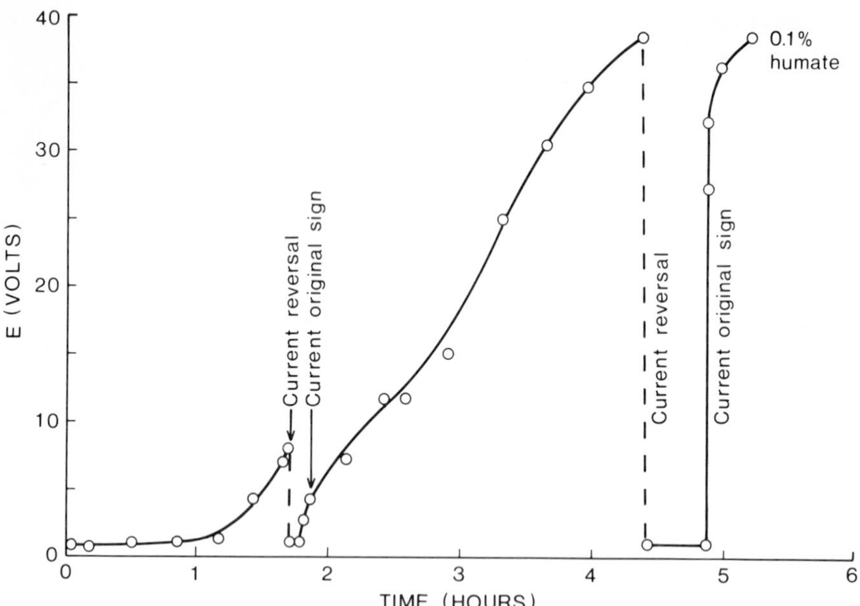

Fig. 11 Fouling of anion selective membranes by Na-humate containing 0.1 *N* KCl solutions, represented by the voltage increase with time at constant current density. (Reprinted from Korngold *et al.*, 1970, courtesy of Elsevier Scientific Publishing Company.)

membranes by sodium dodecyl sulfate (SDS). The membranes had a high affinity for the DS^- ions, which were exchanged stoichiometrically with the original counter ions in the membranes. During this poisoning, the water content decreased and the membranes showed a strong increase in resistance and a decrease in the permselectivity, tending toward zero.

During ED of a NaCl solution containing small amounts of SDS, the DS^- ions were transferred into the anion-selective membrane. Because of the low mobility of these ions, they formed a "soap layer" with low permselectivity and high resistance in the membrane on the diluate side. Although the mechanism is different, this layer acts like a fouling film layer, only it seems the water-splitting phenomenon is not associated with the poisoning phenomenon. This can be explained because the soap layer acts like a neutral film, while the fouling layer acts like a cation-selective membrane, as shown in Fig. 12. The current density at which the concentration at the membrane–film interface equals zero and water splitting starts is therefore much smaller for fouling than for poisoning phenomena (Grossman and Sonin, 1973).

Another phenomenon, scaling, is the precipitation of crystalline inorganic compounds such as $CaCO_3$, $MgCO_3$, $Mg(OH)_2$, and $CaSO_4$, nor-

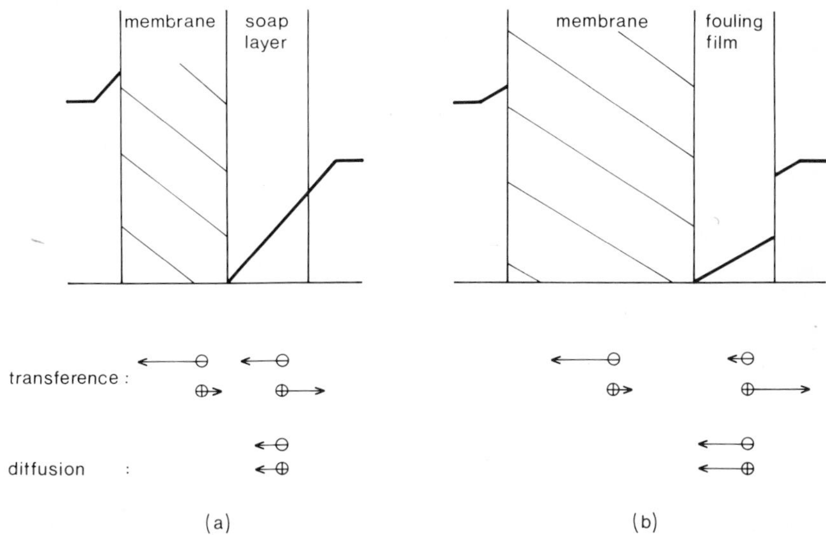

Fig. 12 Schematic of concentration profiles and ionic fluxes for an anion-selective membrane that has been poisoned (a) or fouled (b).

mally on the brine side of the anion-selective membranes. Scaling alone is relatively harmless, but there seems to be a mutual interaction of scaling and fouling that increases the water splitting and resistance of the anion-selective membranes (Korngold *et al.*, 1970).

C. HYDRODYNAMICS

The design of ED plants is to a great extent a matter of experience and rule of thumb because hydrodynamics in an ED apparatus has to date, eluded formal description. Care must be taken in the design to improve mixing and turbulence in the liquid streams. The problem is that the cell thickness must be kept small, because it is the main cause of ohmic resistance in the cell pair. As a result, the liquid velocity is limited by the high pressure loss. Thus most commercial apparatus operate at Reynold's numbers of approximately 100.

Belfort and Guter (1972) made an extensive hydrodynamic study on various commercial types of turbulence promotors. It was found that the limiting current density showed a definite plateau in the region of Reynold's numbers of 90 to 200. This is believed to be because a friction-dependent laminar flow is converted to an inertia-dependent laminar flow and eventually to nonsteadiness and vortex shedding. Turbulence does not yet occur, as is commonly believed and misunderstood. Because the turbulence promoters always increase the hydraulic pressure drop and the electrical resistance of the stack, these effects must adequately be balanced by an increase in limiting current density and coulomb efficiency.

Kedem (1975) showed that suitably shaped ion-conducting spacers can simultaneously increase the limiting current density and decrease the resistance of the dialysate compartments. The spacers consist of interdigiting anion and cation fibers, which are in direct contact with membranes of the same ion selectivity. The mode of action of such spacers is that both ions enter the resin phase at a distance from the membranes and can migrate in this continuous phase to the proper membrane. Near the interface, salt is removed by the current and must be supplied by diffusion through the unstirred layers. The surface for diffusion is, however, larger than the membrane area, and salt depletion takes place at every distance from the membranes in the whole cross section of the dialysate compartment. Simultaneously, the presence of the fibers decreases the total ohmic resistance, and protons and hydroxyls formed at the junction may re-exchange for salt cations and anions during migration toward the membrane surfaces.

Another method of decreasing the resistance of the dialysate compartments is to employ ion exchange resins between the membranes. Korngold (1975) analyzed the different features of the types of ion exchange resins that can be introduced in the diluate and brine compartments. By introducing anion exchange resin next to the anion-selective membrane and cation exchange resin next to the cation-selective membrane in the diluate compartments, polarization near both membranes decreased markedly and a higher electrical efficiency was obtained. By introducing anion exchange resins in the brine compartments, the concentrated layer was near the cation-selective membrane, which decreased the danger of scaling on the surface of the anion-selective membrane.

D. PROCESS VARIATIONS

Several process variations have been investigated to minimize the polarization phenomena.

A high-temperature operation has several beneficial effects on the ED process (Leitz *et al.*, 1974). With increasing temperature, resistances and fluid viscosity are lowered and the rate of diffusion is increased. Thus the limiting current density increases with temperature [Eq. (30)].

Whey and similar solutions are usually treated at higher temperatures. Such solutions are viscous when cold and cannot be handled without excessive pressure loss, while their conductivities in the cold are low in relation to their salt content.

With strong fouling solutions like whey, where the removal of foulants is impossible, the transport depletion process has proved to be an alternative to conventional ED (Lang and Huffman, 1969). In this process, the anion-selective membrane is replaced by a nonselective membrane like cellophane. This reduces the maximum current efficiency to about 60%, but because no polarization takes place at this membrane, the fouling problems concerned with the anion-selective membrane disappear.

References

Belfort, G., and Guter, G. A. (1972). *Desalination* **10,** 221–262.

Belfort, G., and Marx, B. (1978). *Int. Symp. Fresh Water Sea 6th* **4,** 183–192.

Belfort, G., Alexandrowicz, G., and Marx, B. (1976). *Desalination* **19,** 127–138.

Berman, A. S. (1953). *J. Appl. Phys.* **24,** 1232–1235.

Bird, R. B., Stewart, W. E., and Lightfoot, E. N. (1960). "Transport Phenomena". Wiley, New York.

Blatt, W. F., Dravid, A., Michaels, A. S., and Nelsen, L. (1970). *In* "Membrane Science and Technology" (J. E. Flinn, ed.), pp. 47–97. Plenum, New York.

Boari, G., Lacava, G., Merli, C., Passino, R., and Tiravanti, G. (1973). *Int. Symp. Fresh Water Sea 4th* **3,** 169–180.

Boen, D. F., and Johannsen, G. L. (1974). "Reverse Osmosis of Treated and Untreated Secondary Sewage Effluent" (Environmental Protection Technology Series EPA-670/2-74-007), U.S. E.P.A., Cincinnati, Ohio.

Brandt, A., and Bugliarello, G. (1966). *Trans. Soc. Rheol.* **10,** 229–251.

Brian, P. L. T. (1966). *In* "Desalination by Reverse Osmosis" (U. Merten, ed.), pp. 161–202. M.I.T. Press, Cambridge, Massachusetts.

Cowan, D. A., and Brown, J. H. (1959). *Ind. Ing. Chem.* **51,** 1445–1448.

Dejmek, P., Hallström, B., Klima, A., and Winge, L. (1973). *Lebensm. Wiss. Technol.* **6,** 26–29.

Dittus, F. W., and Boelter, L. M. K. (1930). *Univ. Calif. Berkeley Publ. Eng.* **2,** 443.

Dresner, L. (1964). *Oak Ridge Nat. Lab. Rep.* **3631**.

Fisher, B. S., and Lowell, J. R., Jr. (1970). "New Technology for Treatment of Wastewater by Reverse Osmosis" (Water Pollution Control Research Series 17020 DUD 09/70).

Fisher, R. E. Sherwood, T. K., and Brian, P. L. T. (1964). *MIT Desalination Res. Lab. Rep.* **295-5**.

Forgacs, C., Ishibashi, N., Leibovitz, J., Sinkovic, J., and Spiegler, K. S. (1972). *Desalination* **10,** 181–214.

Forgacs, C., Leibovitz, J., O'Brien, R. N., and Spiegler, K. S. (1975). *Electrochimica Acta.* **20,** 555–563.

Gill, W. N., Derzansky, L. J., and Doshi, M. R. (1969). *OSW Res. Dev. Prog. Rep.* **403**.

Goldsmith, H., and Lolachi, L. (1969). *OSW Res. Dev. Prog. Rep.* **527**.

Gregor, H. P., and Gregor, C. D. (1978). *Sci Am.* **239**(1), 88–101.

Gröber, H., Erk, S., and Grigull, U. (1961). "Fundamentals of Heat Transfer." McGraw-Hill, New York.
Grossman, G., and Sonin, A. A. (1973). *Desalination* **12,** 107–125.
Hendricks, T. J., and Williams, F. A. (1971). *Desalination* **9,** 155–180.
Jackson, J. M., and Landolt, D. (1973). *Desalination* **12,** 361–378.
Johnson, A. R. (1970). "Concentration Polarization in Reverse Osmosis under Natural Convection." Ph.D. Thesis, Stanford University.
Johnson, J. S., Dresner, L., and Kraus, K. A. (1966). *In* "Principles of Desalination" (K. S. Spiegler, ed.), pp. 346–439. Academic Press, New York.
Jonsson, G., and Boesen, C. E. (1977). *Desalination* **21,** 1–10.
Jonsson, G., and Kristensen, S. (1980). *Desalination* **32,** 327–339.
Kedem, O. (1975). *Desalination* **16,** 105–118.
Kobus, E. J. M., and Heertjes, P. M. (1972). *Desalination* **10,** 383–401.
Korngold, E., De Körösy, F., Rahav, R., and Taboch, M. F. (1970). *Desalination* **8,** 195–220.
Korngold, E. (1975). *Desalination* **16,** 225–233.
Kressman, T. R. E., and Tye, F. L. (1969). *J. Electrochem. Soc.* **116,** 25–29.
Kuiper, D., Van Hezel, J. L., and Bom, C. A. (1974). *Desalination* **15,** 193–212.
Lang, E. W., and Huffman, E. L. (1969). *OSW Res. Dev. Prog. Rep.* **439**.
Leitz, F. B., Accomazzo, M. A., and McRae, W. A. (1974). *Desalination* **14,** 33–41.
Lim, T. H., Dunkley, W. L., and Merson, R. L. (1971). *J. Dairy Sci.* **54,** 306–311.
Madsen, R. F. (1977). "Hyperfiltration and Ultrafiltration in Plate-and-Frame Systems." Elsevier, Amsterdam.
Matthiason, E., and Sivik, B. (1980). *Desalination* **35,** 59–103.
Nakao, S., and Kimura, S. (1981). *In* "Synthetic Membranes: Hyper- and Ultrafiltration Uses" (A. F. Turbak, ed.), pp. 119–132. American Chemical Society, Washington, D.C.
Nakao, S., Nomura, T., and Kimura, S. (1979). *AIChE J.* **25,** 615–622.
Oda, Y., and Yawataya, T. (1968). *Desalination* **5,** 129–138.
Peers, A. M. (1956). *Discuss. Faraday Soc.* **21,** 124–125.
Perona, J. J. *et al.,* (1967). *Environ. Sci. Technol.* **1,** 991–996.
Porter, M. C. (1972). *Ind. Eng. Chem. Prod. Res. Dev.* **11,** 234–248.
Probstein, R. F., Shen, J. S., and Leung, W. F. (1978). *Desalination* **24,** 1–16.
Rubinow, S. I., and Keller, J. B. (1961). *J. Fluid Mech.* **11,** 447–459.
Sachs, S. B., and Zisner, E. (1977). *Desalination* **20,** 203–215.
Saffman, P. G. (1956). *J. Fluid. Mech.* **1,** 540–553.
Shen, J. S., and Probstein, R. F. (1977). *Ind. Eng. Chem. Fundam.* **16,** 459–465.
Sheppard, J. D., Thomas, D. G., and Channabasappa, K. C. (1972). *Desalination* **11,** 385–398.
Sherwood, T. K., Brian, P. L. T., Fisher, R. E., and Dresner, L. (1965). *Ind. Eng. Chem. Fundam.* **4,** 113–118.
Sieder, E. N., and Tate, G. E. (1936). *Ind. Eng. Chem.* **28,** 1429–1435.
Spiegler, K. S. (1971). *Desalination* **9,** 367–385.
Trettin, D. R., and Doshi, M. R. (1981). *In* "Synthetic Membranes: Hyper- and Ultrafiltration Uses" (A. F. Turbak, ed.), pp. 373–409. American Chemical Society, Washington, D.C.
Vilker, V. L., Colton, C. K., and Smith, K. A. (1981). *J. Colloid Interface Sci.* **79,** 548–566.
Wales, M. (1981). *In* "Synthetic Membranes: Desalination" (A. F. Turbak, ed.), pp. 159–170. American Chemical Society, Washington, D.C.
Yamabe, T., and Seno, M. (1967). *Desalination* **2,** 148–153.

5

Mathematical Modeling of Fluid Flow and Solute Distribution in Pressure-Driven Membrane Modules

CLEMENT KLEINSTREUER

GEORGES BELFORT

Rensselar Polytechnic Institute
Troy, New York

List of Symbols

The symbols defined are commonly used in membrane modeling. To simplify comparison in certain case studies, we followed the author's notation and gave the definition in the text.

ISBN 0-12-085480-5

A	membrane permeability, cross-sectional area
c	solute concentration
D/Dt	Stokes' derivative
D	effective diffusion coefficient
D_{AB}	binary diffusion coefficient
$\vec{f}$	body force per unit mass
F	external force
h	channel half-height, characteristic height
j, J, v_w	volumetric flow rates, ultrafiltration rate, volume flux
$\vec{J}_c$	solute mass flux
J_w	solvent flux
J_v	product flux
k	mass-transfer coefficient, distribution coefficient
K	hydraulic permeability
l	length scale parameter
L	channel or tube length
L_p	hydraulic membrane conductivity
$\dot{m}$	mass flow rate
n	exponent, particle density function, porosity
N_A, N_B	molar flux of constituent A or B
p	thermodynamic pressure
Δp	hydrostatic pressure difference
$\vec{q}_H$	heat flux
r	membrane rejection coefficient, radius
r, R	resistance layers
Re	Reynolds number
Sc	Schmidt number
Sh	Sherwood number
S_c	net source of solute c
t	time
T	temperature, characteristic time
$\vec{T}_{ext}$	external torque vector
u, v	velocity components in axial and normal direction
$\bar{u}$	average axial velocity
u_e	velocity at boundary layer edge, outer flow
U	internal energy
$\vec{v}$	velocity vector
v_w, j, J	wall velocity, volumetric flux, permeate flux, suction velocity
V	volume
x, y, z	cartesian coordinates
x_A	mole fraction of constituent A
x_0, x_2	mole fraction at location 0 or 2

Greek Symbols

α	inverse Péclét number
Δ	difference
δ	boundary layer or film thickness
$\overleftrightarrow{\delta}$	unit tensor
∇	del operator
η	viscosity (non-Newtonian)
$\overleftrightarrow{\dot{\gamma}}$	shear rate tensor
ρ	fluid density
ψ	stream function
$\overleftrightarrow{\pi}$	total stress tensor
ν	kinematic viscosity
σ	reflection coefficient
$\overleftrightarrow{\tau}$	stress tensor
$\vec{\Omega}$	angular-velocity vector

Subscripts

α	indicator of component considered
c	cake layer
CL	centerline
f	fluid
m	membrane
o	condition at location o (bulk or entrance)
p	permeate, particle
s	solute
t	turbulent
w	condition at (membrane) wall, water
y	y direction

I. Introduction and Overview

A. Background

The important and critical role that fluid mechanical phenomena play in the optimal operation of mass-transfer systems is well known (Bird *et al.*, 1960). More specifically, it has been shown that for each membrane separation process "one or more of the critical technical problems which lead to increased fresh water costs is associated with a fluid mechanical phenomenon" (Probstein, 1972). Consequently, it is evident that a self-consistent hydrodynamic theory for the performance of each membrane process considered here as a function of flow conditions, module geometry, and membrane characteristics is needed. This would provide a rational basis for optimizing the design and operation of the electrodialysis (ED), hyperfiltration (HP), and ultrafiltration (UF) processes.

In this chapter we present an updated review of the critical fluid engineering problems associated with each of these mass-transfer membrane processes and where possible the solutions and their consequences with respect to maximizing the performance/cost ratio. As with Probstein's earlier review, future research directions with respect to hydrodynamic analysis will also be recommended here.

B. APPROACH

As of 1982, the general approach in describing the dynamics of mass-transfer membrane processes has been to develop a mathematical description of the fluid (momentum) and the solute (mass) transport phenomena, together with reasonable boundary conditions and assumptions. Analytical and numerical techniques are used to obtain a solution to the problem statement. Comparison of the solution with experimental observations establishes the validity of this simulation procedure and allows for a parametric sensitivity analysis. Often, simplifications and assumptions are made to make the problem mathematically tractable. Unfortunately this may result in solutions that have limited practical value. They do, however, provide a basis for further incorporation of more complicated and realistic phenomena.

An important analogy between heat-transfer and isothermal mass-transfer processes, such as the membrane processes considered here, is that capital costs decrease with reduced transfer area. Therefore, it is usually desirable to operate at the highest transfer rates per unit transfer area, i.e., at the highest transfer fluxes. Because the rate of salt removal or ion transfer from the feed is proportional to the current density in ED, it is normal to attempt operation at the highest current density possible. For the pressure-driven membrane processes it is also normal to operate at the highest permeation flux possible per unit driving force for the same solute retention. Increasing the transfer fluxes requires increased power consumption for higher current densities in ED and higher applied pressures in UF and HF. Clearly, some optimum balance between increased performance and increased power consumption results in an optimum transfer flux for a given process. Generally, this economically optimal transfer flux is not attainable because of limitations such as the inherent properties of the membrane (hydrolysis, compaction, creep) or polarization phenomena resulting from different solute-transfer rates.

C. POLARIZATION

Polarization phenomena are characterized by an increase or decrease near the membrane–solution interface in solute concentration above or below that of the bulk solution concentration, respectively. This change in concentration in the laminar boundary layer is because of differential transfer rates in the membrane, boundary layer, and bulk solution. In ED as the current density is increased, the anion concentration in the dialysate laminar boundary layer adjacent to the anion-exchange membrane decreases until it reaches zero. With a further increase in the current density, water splitting occurs and the current is carried by hydronium and hydroxide ions resulting in extreme pH changes near the membrane surface. Unwanted scale formation and deposition on the brine side of the anion-exchange membrane may then result, causing a decrease in process removal efficiency. In pressure-driven membrane processes, an analogous but different type of concentration polarization occurs. For this case, solute molecules are carried toward the membrane by convective drag. Because they are partially or completely retained by the membrane they begin to concentrate in the laminar sublayer adjacent to the membrane–solution interface and diffuse back into the bulk solution. Increased concentrations of solute near the membrane surface can result in high osmotic pressures and, hence, reduced effective driving force (pressure) for HF and sometimes UF (Smith, 1981, private communication). When the solution contains foulants including dissolved macromolecules, colloidal species, and inorganic precipitates, deposition onto the membrane surface may result. This introduces an additional resistance to permeation and may increase concentration polarization, thereby reducing the efficiency of the process.

Because the unwanted concentration polarization effects occur within the mass boundary layer near the membrane surface, they can be directly reduced by manipulation of the cell geometry (design) and/or the hydrodynamics of the system and by a propitious mix of pretreatment and periodic membrane cleaning. Much of the early work was conducted with plane unobstructed cells in fully developed laminar and turbulent flow.

Mixing promoters or eddy inducers are often used in commercial systems to disturb the solute concentration profile in the boundary layers, resulting in better performance at lower relative Reynold's numbers. Empirical studies determining the effects of these promoters in the ED process, for example, have been conducted (Belfort and Guter, 1972). Detailed analytical studies, however, have yet to appear in the literature.

To fully understand the critical hydrodynamic problems associated with the cost of producing water using membrane processes and to provide rational solutions it is imperative to quantitatively describe fluid and solute behavior within the membrane modules. In Section II, various submodels and their applications are reviewed. Steady incompressible laminar–turbulent flow and solute transfer models are discussed in Section III. In Section IV, mathematical modeling needs are outlined.

II. Field Equations and Common Submodels

In the review of mathematical models we concentrate on the fluid flow and solute distribution within a given membrane module rather than discussing the entire system (Fig. 1). The hydrodynamic subsystems of feed and permeate flow for common membranes are depicted in Fig. 2. Most of the simulation models examined steady-state tubular or channel-type membrane flows; a few modeling efforts focus on transient transmembrane flows occurring in unstirred batch cell systems (Fig. 3). Hence, the dynamic regimes for momentum and mass transfer can be grouped into nonflow batch cell in addition to laminar or turbulent flow in conduits with "porous walls." These porous walls are semipermeable membranes that are the microporous convective sieve type or the diffusive type. Their utilization primarily depends on solute characteristics and the type of separation desired. In UF, suspended colloidal matter and macromole-

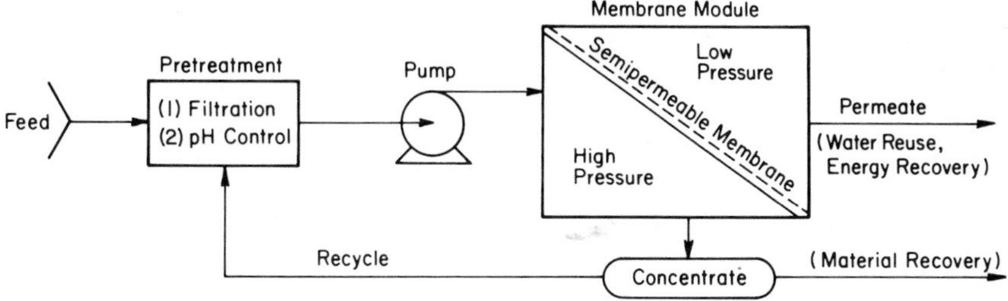

Fig. 1 Schematics of membrane separation system.

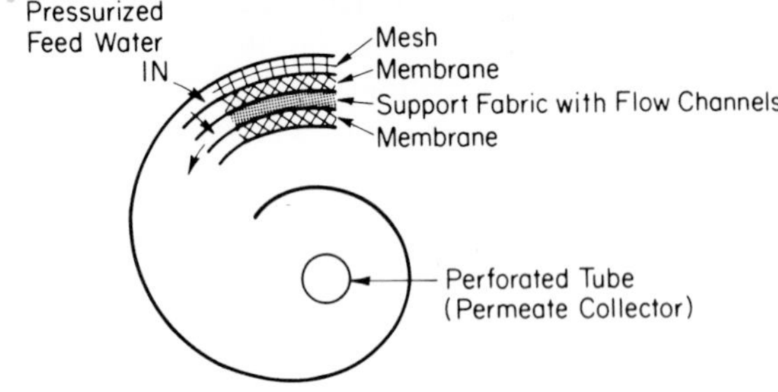

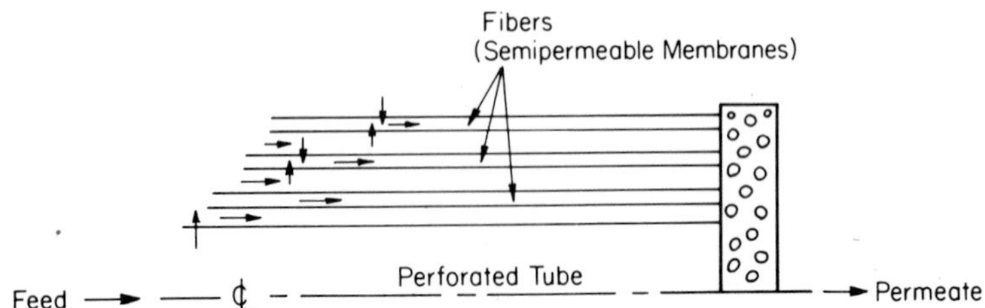

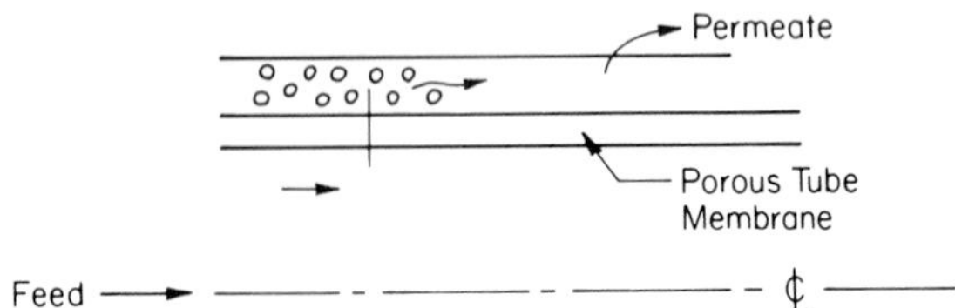

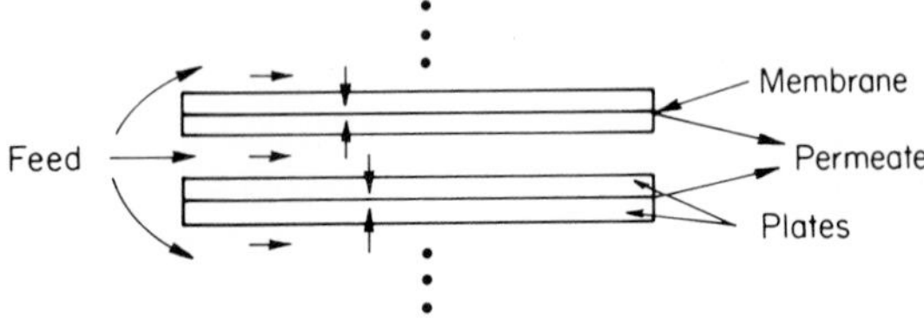

Fig. 2 Flow system schematics for common membrane elements.

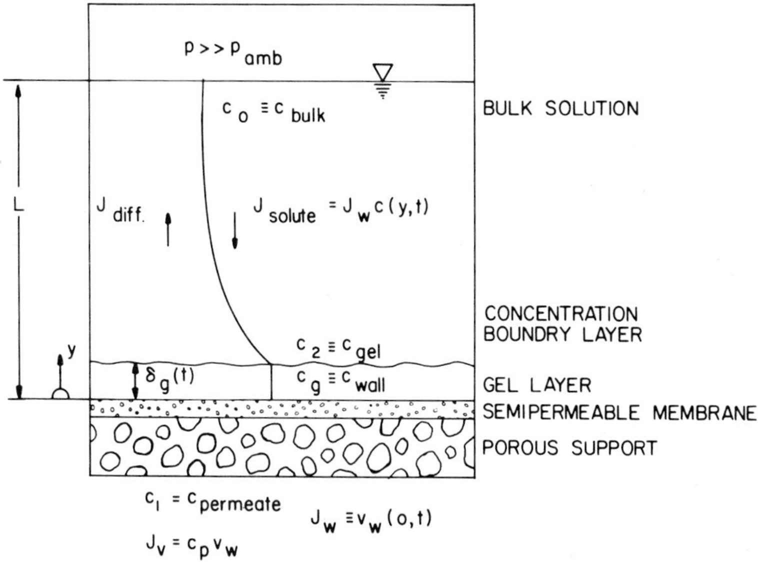

Fig. 3 Unstirred batch cell system.

cules in solution are mechanically retained by membranes with pores only large enough for the carrier fluid and nonretained solutes. In HF, membranes probably do not have definable pores, only spaces between polymer fibers. Most of the solutes (in particular ions) are rejected when the carrier fluid passes through the barrier.

In this section we outline relevant modeling approaches and introduce the field equations for a comprehensive description of momentum, heat and mass transfer in pressure-driven membrane modules. Simplified versions of these equations and appropriate solution techniques most frequently employed are also given. Associated with the governing equations are mathematical submodels that are used to gain closure and express the necessary boundary conditions. From a physical point of view, auxiliary models simulate phenomena such as concentration polarization or membrane fouling that prevent steady, optimal operating conditions. The submodels discussed represent the following phenomena: Turbulence (Boussinesq's hypothesis and Prandtl-Von Karman's theory), membrane surface roughness and finite slip velocity based on pipe flow analyses (Darcy–Weisbach equation), flux resistance because of deposition layers (Darcy's law and the Carman-Kozeny equation), solute flux through the concentration boundary layer (Nernst's stagnant film concept), and transfer across particular membranes (irreversible thermodynamics).

A. GENERAL MODELING APPROACH AND FIELD EQUATIONS

The main objectives of process simulation are to have a cost-efficient tool for experimental guidance, a means of testing hypotheses, predicting design parameters, and optimizing and controlling the process. The specific modeling approach to be chosen depends largely on the objectives, type of membrane system, availability of data sets, and mathematical complexity of the governing equations. Three basic modeling methodologies can be distinguished: Empirical, stochastic, and deterministic (Kleinstreuer, 1983). The macroscopic description of membrane filtration can be provided on an empirical basis through statistical analysis of measured data. This method, a type of lumped parameter or black box approach, has not been successful and obscures the physical significance of the design parameters. Stochastic models, usually random terms added to deterministic equations, should closely represent membrane separation processes because the nature of most transport phenomena (e.g., particle deposition and reentrainment) is random. Reliable data sets (i.e., time series) are required, however, to find or test the probability density function for the random variables. Furthermore, once a random variable is introduced, the results using this model will be expressed in terms of probabilities rather than fixed numbers. The deterministic approach is exclusively employed in modeling membrane transport and is presented in this chapter. A generalized system conceptualization and mathematical modeling framework are given in Fig. 4.

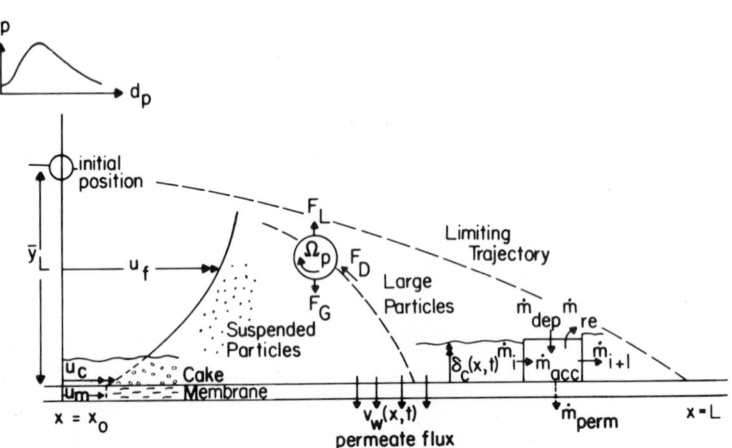

Fig. 4a Problem conceptualization for dilute suspension flow and fluid–particle systems.

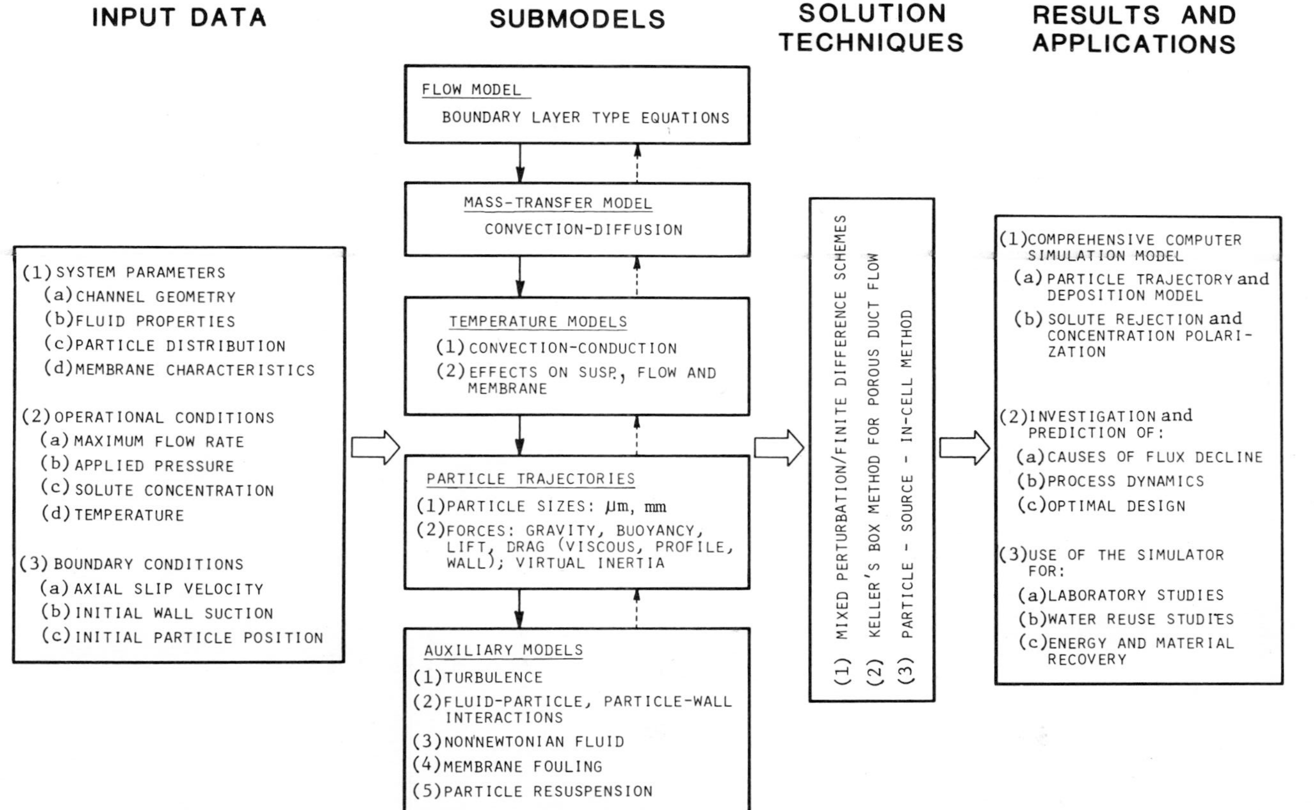

Fig. 4b Mathematical modeling framework.

1. *Governing Equations for Momentum, Heat, and Mass Transfer*

Of interest here are the equations of motion and continuity, species (or solute) mass-transport equation, energy equation and equation of state and species flux expressions. Larger particles that do not follow the fluid motion are usually represented by single or monodispersed spheres. Their trajectories are computed based on force and torque balances (see Section II.C.) replacing the mass-transport equation. Hence the field equations in terms of fluxes for a generalized membrane system are given by (Kleinstreuer, 1983)

$$\text{continuity} \quad \frac{\partial \rho}{\delta t} + \nabla \cdot \rho \vec{v} = 0, \tag{1}$$

$$\text{motion} \quad \frac{\partial}{\partial t} \rho \vec{v} + \nabla \cdot \rho \vec{v}\vec{v} = -\nabla p - \nabla \cdot \vec{\vec{\tau}} + \sum \vec{f}, \tag{2}$$

$$\text{mass transport} \quad \frac{\partial c}{\partial t} + \nabla \cdot \vec{v} c = -\nabla \cdot \vec{\mathrm{j}}_{\mathrm{c}} + S_{\mathrm{c}}, \text{ and} \tag{3}$$

$$\text{internal energy} \quad \frac{\partial}{\partial t} \rho U + \nabla \cdot \rho U \vec{v} = -\nabla \cdot \vec{q}_{\mathrm{H}} - \vec{\vec{\pi}} : \nabla \vec{v}. \tag{4}$$

The stress tensor $\vec{\vec{\tau}}$ in addition to the mass and heat flux ($\vec{j}_{\mathrm{c}}$ and $\vec{q}_{\mathrm{H}}$) must be specified for the solute–fluid system in terms of the principal variables $\vec{v}$, c and $U = c_{\mathrm{p}}T$ to gain closure (see the subsequent discussion). An equation of state $\rho = \rho[p, c, U(T)]$ is necessary when the assumption of incompressible flow does not hold. Suitable submodels for $\vec{\vec{\tau}} = \eta \cdot \cdot \vec{\vec{\gamma}}$ have to be found when the suspension flow exhibits rheological properties. The energy dissipation term, $\vec{\vec{\pi}} : \nabla \vec{v}$ with the total stress tensor $\vec{\vec{\pi}} = p\delta + \vec{\vec{\tau}}$, couples the energy equation with the momentum equation explicitly. For isothermal, viscous flow with constant fluid and solute parameters, Eqs. (1)–(3) reduce to

$$\text{continuity} \quad \frac{\partial u}{\partial x} + \frac{\partial v}{\partial y} = 0, \tag{5a}$$

$$x \text{ momentum} \quad \frac{\partial u}{\partial t} + u \frac{\partial u}{\partial x} + v \frac{\partial u}{\partial y} = -\frac{1}{\rho} \frac{\partial p}{\partial x} + \nu \left(\frac{\partial^2 u}{\partial x^2} + \frac{\partial^2 u}{\partial y^2} \right), \tag{5b}$$

$$y \text{ momentum} \quad \frac{\partial v}{\partial t} + u \frac{\partial v}{\partial x} + v \frac{\partial v}{\partial y} = -\frac{1}{\rho} \frac{\partial p}{\partial y} + n \left(\frac{\partial^2 v}{\partial x^2} + \frac{\partial^2 v}{\partial y^2} \right), \text{ and} \tag{5c}$$

$$\text{solute transport} \quad \frac{\partial c}{\partial t} + u \frac{\partial c}{\partial x} + v \frac{\partial c}{\partial y} = D_{\mathrm{y}} \frac{\partial^2 c}{\partial y^2} + S_{\mathrm{c}} \tag{6}$$

subject to the essential boundary conditions

$$\text{at solid wall} \quad u = v = 0, \qquad \frac{\partial c}{\partial y} = 0,$$

$$\text{at membrane} \quad u = u_{\text{slip}}, \qquad v = v_{\text{w}},$$

$$\alpha \frac{\partial c}{\partial y} = rc(x;y_{\text{wall}}), \tag{7}$$

and the nonessential boundary condition

$$\text{at boundary-layer edge} \quad u = u_{\text{e}}, \qquad \frac{\partial c}{\partial y} = 0.$$

The initial conditions require that the flow field and the solute distributions are known throughout the domain at time $t = 0$. The solution of this equation system constitutes formidable numerical problems. In the case $v \ll u$, the computational problem can be reduced to solving a set of boundary layer-type equations

$$\frac{\partial u}{\partial x} + \frac{\partial v}{\partial y} = 0, \tag{8a}$$

$$\frac{\partial u}{\partial t} + u \frac{\partial u}{\partial x} + v \frac{\partial u}{\partial y} = -\frac{1}{\rho}\frac{\partial p}{\partial x} + \nu \frac{\partial^2 u}{\partial y^2}, \text{ and} \tag{8b}$$

$$\frac{\partial c}{\partial t} + u \frac{\partial c}{\partial x} + v \frac{\partial c}{\partial y} = D_{\text{y}} \frac{\partial^2 c}{\partial y^2} + S_{\text{c}}, \tag{9}$$

subject to appropriate initial and boundary conditions. If measured data for the pressure gradient are available, the system can be solved numerically with available routines (Cebeci and Bradshaw, 1977). It has to be noted, however, that these equations are valid only for dilute laminar suspension flow in conduits with semipermeable walls where suction (i.e., the permeation flux) is small compared with the axial bulk velocity. Additional terms for turbulent flow should be included in Eqs. (8b) and (9). If non-Newtonian behavior in the gel and fouling layer becomes significant, or if the flow is turbulent, Eqs. (2) and (3) will remain the governing equations with specific submodels for $\vec{\vec{\tau}}$, $\vec{j}_{\text{c}}$, $\vec{q}_{\text{H}}$, and $\Sigma\,\vec{f}$.

Fluid flow and solute transport models published in the open literature are generally steady state. This poses a severe shortcoming for (their) practical applications (Sections III and IV). Assuming steady, axisymmetric laminar flow with constant fluid and material properties, Eqs. (5a)–(5c) and (6) can be simplified to

$$\frac{\partial u}{\partial x} + \frac{\partial v}{\partial y} = 0, \tag{10a}$$

$$u\frac{\partial u}{\partial x} + v\frac{\partial v}{\partial y} = -\frac{1}{\rho}\frac{\partial p}{\partial x} + \nu\left(\frac{\partial^2 u}{\partial x^2} + \frac{\partial^2 u}{\partial y^2}\right), \tag{10b}$$

$$v\frac{\partial v}{\partial y} = -\frac{1}{\rho}\frac{\partial p}{\partial y} + \nu\frac{\partial^2 v}{\partial y^2}, \text{ and} \tag{10c}$$

$$u\frac{\partial c}{\partial x} + \frac{\partial c}{\partial y} = D\frac{\partial^2 c}{\partial y^2} + S_c. \tag{11}$$

The associated boundary conditions for system (10) are: A prescribed velocity profile at the tube or channel entrance, the "no-slip" condition at the "walls" ($u|_{wall} = 0$), no solvent flux across solid walls ($v|_{wall} = 0$), and suction or permeate flux v_w at the membrane surface ($v|_{wall} = v_w$). For Eq. (11) the following conditions must be satisfied: (1) "Initial" condition $c(x = 0,y) = c_o$, the solute concentration in the feed stream; (2) at the membrane wall $D\,\partial c/\partial y|_w = rv_w c_w(x)$, where r is the solute rejection efficiency and $c_w(x)$ is the solute concentration along the membrane surface; and (3) at the impermeable wall $\partial c/\partial y|_w = 0$. If appropriate, a symmetry condition at the conduit centerline ($\partial c/\partial y|_{CL} = 0$) could also be imposed on Eq. (11). System (10) can be transformed to one equation in using the stream function approach where $u = \partial\psi/\partial y$ and $v = -\partial\psi/\partial x$ so that continuity is automatically satisfied and Eqs. (10b) and (10c) collapse into

$$\frac{\partial\psi}{\partial y}\frac{\partial}{\partial x}\nabla^2\psi - \frac{\partial\psi}{\partial x}\frac{\partial}{\partial y}\nabla^2\psi = \nu\,\nabla^4\,\nu. \tag{12}$$

The gel polarization model, discussed in Section II.B., takes the mathematical form (Merten, 1963)

$$v_w\,\Delta c = A(\Delta p - \Delta\pi)\,\Delta c = -D\,\partial c/\partial y|_{\text{membrane}}. \tag{13}$$

Hence, system (10) or Eq. (12) and Eq. (11) are coupled via the velocity field as well as through the concentration polarization, Eq. (13).

Methods of solution for these equations are summarized subsequently. Details and merits of frequently used solution procedures are also discussed.

2. *Solution Techniques*

Because analytical solutions to the modeling equations [Eqs. (10a)–(10c) and (11)] subject to appropriate boundary conditions are not available, asymptotic and approximate solution methods are usually used. The representation of physical systems is sometimes simplified to obtain a more tractable mathematical problem.

A perturbation solution of a simplified equation of motion describing laminar flow between two porous plates (or in a porous tube) and constant wall velocity (permeate flux) was given by Berman (1953). Approximate solutions of problem-specific equations were reported by Gill *et al.* (1965) employing a series expansion, Kozinski *et al.* (1970) using Bessel functions, and Leung and Probstein (1979) resorting to the Von Kármán integral method. Traditionally, the assumption of constant permeate flux v_w is suspended *after* the momentum equation is solved and replaced by a phenomenological relationship [i.e., Eq. (13)] as discussed by Merten (1963), Dandavati *et al.* (1975), and Spiegler and Laird (1980). This simplification is only justified if wall suction does not disturb the bulk flow. The no slip condition is usually invoked for the longitudinal velocity at the walls. Beavers and Joseph (1967) and Sparrow *et al.* (1972), however, investigated the effect of a thin moving layer in the porous walls. The computed velocity field is then inserted into the convection–diffusion equation to obtain the dissolved and/or suspended species distribution using again approximation techniques (e.g., Gill *et al.*, 1965; Hung and Tien, 1976; Johnson and McCutcham, 1972; Leung and Probstein, 1979; Sherwood *et al.*, 1965) or a finite difference method (e.g., Brian, 1965; Kleinstreuer and Paller, 1983; Singh and Laurence, 1979). This procedure allows a separate treatment of momentum and mass transfer. Kleinstreuer *et al.* (1983a) employed the integrated compartment or finite volume method (Rich, 1974) to solve for solute and solvent transfer across red blood cell membranes as well as membranes.

The convenient reduction of the motion equations to one ordinary differential equation (ODE) is based on the conditions of similarity. This implies that Eqs. (11b) and (11c) are parabolic and the (axial) velocity profiles differ only by a scaling factor. The magnitude and variability of the wall flux (v_w) for example, can destroy similarity. Furthermore, in the entrance region of laminar internal flows, such similar velocity distributions do not exist. Weissberg (1959) found a nonsimilar solution for laminar flow in the entrance region of a porous pipe.

An approximate solution method more powerful than the similarity theory is the method of integral relationships (MIR). The integral method is often used for calculating laminar or turbulent boundary layer flow parameters (thin shear layers) and boundary layer-type flows (jets and wakes). The unique feature of the method is that a suitable velocity and/or concentration profile has to be postulated and the governing partial differential equation (PDE) is then converted into an ODE for a system parameter by integrating the governing PDE across the boundary layer subject to associated boundary conditions. In Section III., the MIR is discussed for the solution of the convection–diffusion equation describing polarization in laminar UF (Leung and Probstein, 1979).

When a mathematical problem statement, i.e., the modeling equations, cannot be solved by analytical, asymptotic or approximate solution techniques, then numerical schemes, notably finite difference methods, must be employed (e.g., Ames, 1978; Cebeci and Bradshaw, 1977; Roache, 1976). For example, Brian (1965, 1966) solved the mass-transport equation [Eq. (12)] for an RO system using a finite difference method. His results can be compared with the work of Sherwood *et al.* (1965) and Leung and Probstein (1979) for the limiting case of constant solution diffusivity and permeate flux as well as a linear osmotic pressure concentration variation. Brian noted that the generation of the finite difference results required considerably less computer time than finding the infinite series solution. Singh and Laurence (1979) used a backward finite differencing scheme to solve Eq. (12). They imposed symmetry conditions to obtain a tridiagonal matrix problem that was solved with standard routines. Kleinstreuer and Paller (1983) advanced their work for the asymmetric case of a plate-and-frame membrane module. The flexibility of finite difference or finite element schemes make numerical modeling a powerful tool in the simulation of practical membrane processes.

In the remainder of this section we discuss submodels for dissolved and suspended solutes, trajectories of large particles such as spherical colloids and impurities, development of colloidal or gel layers (membrane fouling), turbulence, and finite slip velocities at membrane surface.

B. MASS-TRANSFER SUBMODELS

In the previous section we reviewed the generalized field equations, associated boundary conditions, reduced sets of modeling equations, and appropriate solution methods for dilute suspension flows in pressure-driven membrane units. In order to achieve completeness, problem-specific submodels have to be provided. These auxiliary relationships should reflect salient transport phenomena such as mass transfer across deposition layers and membranes, heat transfer across system boundaries, and momentum transfer in bulk and boundary-layer flow.

1. *Film-Theory Model*

Because it is difficult to observe conditions in the immediate region of an interface between phases, submodels for mass transfer were developed (Danckwerts, 1951; Higbie, 1935; Nernst, 1904; Whitman, 1923). The three simplest models are the stagnant film, penetration–surface renewal, and turbulent boundary-layer model (Sherwood *et al.*, 1975). It should be noted that the film concept was originally developed for turbulent flow fields where the region outside the film or interface was assumed to be at a constant uniform value of temperature or concentration. Hence, empirical fitting of these models to different systems, such as the unstirred batch cell, should be regarded with care when case studies are investigated.

The film theory model employs a lumped parameter approach to estimate the limiting permeate velocity for membrane flow systems (Belluci *et al.*, 1979; Brian, 1965; Trettin and Doshi, 1980). In general, the concentration boundary layer is idealized as a thin liquid film in which possible eddy motion (from the turbulent bulk flow) is assumed to be negligible. The mass transport within this film takes place under steady laminar conditions with Fickian molecular diffusion occurring only normal to the membrane surface. Figures 5a and 5b illustrate the build-up of

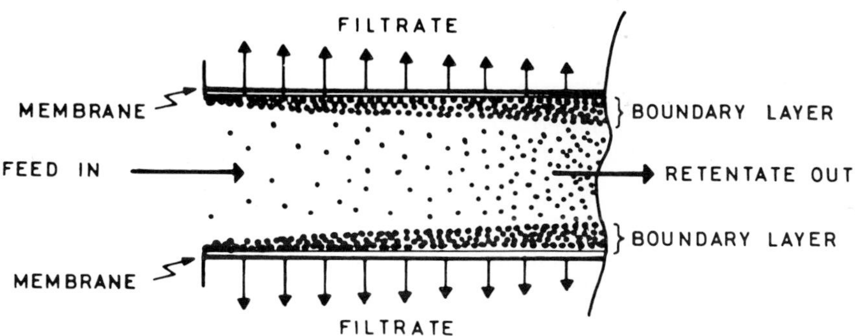

Fig. 5a Schematics of concentration boundary-layer development.

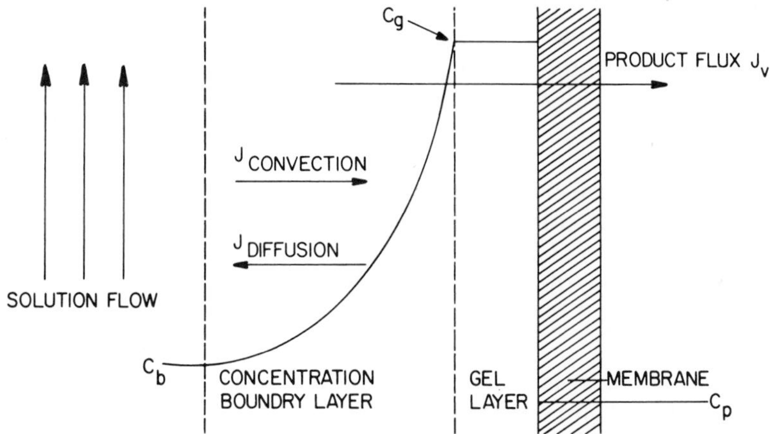

Fig. 5b Concentration profile for a gel-polarized UF membrane.

solute at the membrane surface. Of particular interest is the solute mass transfer written in terms of the molar flux N_A and the mole fraction x_A relative to stationary coordinates (Bird *et al.*, 1960):

$$\vec{N}_A = -\rho D_{AB} \nabla x_A + x_A(\vec{N}_A + \vec{N}_B). \tag{14a}$$

The second term in Eq. (14a) is the molar flux of species A resulting from the bulk fluid motion. For one dimensional, steady, isothermal mass transfer without chemical reactions, Eq. (14a) yields with $N_B = 0$ and N_A being constant:

$$N_A \int_0^\delta dy = \rho D_{AB} \int_{x_0}^{x_2} \frac{dx_A}{1 - x_A},$$

or (14b)

$$N_A = \frac{\rho D_{AB}}{\delta} \ln \frac{1 - x_2}{1 - x_0}.$$

Equation (14b) derived from the film model suggests two debatable dependencies: (1) $N_A \propto D^{1.0}$, although experiments show that $N_A \propto D^n$ where $0.5 \leq n \leq 1.0$, and (2) $N_A \propto$ ln(concentration ratio), which was questioned by Trettin and Doshi (1980). Another fundamental relationship based on the film theory has to be derived before the "film concept" can be discussed in connection with existing membrane filtration models. In equimolar counter diffusion (as in distillation) it is assumed that $N_B = -N_A$, such that integration of the Eq. (14a) for a one-dimensional case gives

$$N_A = -(D_{AB}/\delta\rho)(x_2 - x_0). \tag{14c}$$

Equation (14c) is based on the stagnant film theory and holds for binary molecular diffusion with high mass-transfer rates at constant temperature and pressure. For Eq. (14c) the same assumptions are made except that the two-phase film is slowly moving. For low mass-transfer rates a binary mass-transfer coefficient can be defined (Bird *et al.*, 1960)

$$N_A = -(k_c/\rho)(x_2 - x_0), \tag{14d}$$

where the difference in molar concentrations $\Delta x \propto c_{\text{wall}} - c_{\text{bulk}}$ constitutes the "driving force" for the (molar) solute flux N_A. Comparison of Eq. (14d) with Eq. (14c) yields an expression for the mass-transfer coefficient as a function of the (unknown) film thickness

$$k_c^o = D_{AB}/\delta. \tag{14e}$$

If component A is transported through a nondiffusing fluid, i.e., in comparing Eqs. (14b) and (14d), we obtain

$$k_c = (D_{AB}/\delta) f(x_0, x_2). \tag{14f}$$

When the concentration of A is small, the function $f(x_0, x_2)$ is approximately equal to one, as found in dilute liquid suspension flows. In practice, the film thickness δ is replaced by the mass-transfer coefficient k_c to correlate experimental data suggested by dimensional analysis: $k_c l/D_{AB} = f(\text{Re}, \text{Sc})$, where Re and Sc are the Reynolds and Schmidt numbers, respectively.

In summary, the film theory assumes that resistance to mass transfer exists in a fictitious film in which molecular diffusion normal to the interface takes place. For stagnant-layer diffusion, as well as for equimolar counter diffusion, the film theory model predicts that the convective mass-transfer coefficient is directly related to the molecular mass diffusivity and inversely proportional to the imaginary film thickness.

A more realistic dependence of $k \propto D^n$ is achieved with the penetration model developed by Higbie (1935) and Danckwerts (1951). The simplest version of the penetration theory pictures a small fluid element of uniform solute concentration c_o being brought into contact with a phase boundary of concentration c_A for a fixed time t (Sherwood *et al.*, 1975):

$$N_A = 2\,(c_A - c_o)(D/\pi t)^{1/2}. \tag{15}$$

Equating (14d) and (15) yields the time-smoothed mass-transfer coefficient, neglecting convective flux, as

$$k_c = 2D^{1/2}(\pi t)^{-1/2}. \tag{16}$$

As it is difficult to obtain a value for the film thickness δ for Eq. (14e), the exposure time t as required in Eq. (16) is also seldom known.

The application of these submodels as expressed through Eqs. (14)–(16) are found in the literature, e.g., Brian (1965), Kimura and Sourirajan (1967, 1968), Shephard and Thomas (1971), Belluci *et al.* (1979), Leung and Probstein (1979), and Trettin and Doshi (1980). The film theory model is widely used in describing the volumetric permeate flux (permeate or wall velocity) v_w for all membrane modules and fluid flow fields. The model is also indirectly used as a Neumann-type boundary condition for the mass-transport equation representing dilute suspension flows in porous ducts.

Kimura and Sourirajan (1967) developed a correlation between the Sherwood number ($\text{Sh} = 4hk/D$) and concentration polarization (c_w/c_o) in solving Eq. (11), and incorporated the thin film concept for laminar flow in RO units. Their equation reads for ideal membranes and insignificant osmotic pressure:

$$\ln(c_w/c_{CL}) = 1.538(v_w^3 hx/3D^2\bar{u}_o)^{1/3}. \tag{17a}$$

TABLE I

Channel half-height b (cm)	Average, U_{ao} (cm/sec)	Flux, w (cm/sec)	Distance $X = 1$ cm Diffusion coefficient for ln (c_w/c_o)		
			1.1	2	10
0.02	50	10^{-2}	1.9×10^{-5}	7.8×10^{-6}	6.9×10^{-7}
0.02	50	10^{-3}	6.1×10^{-7}	2.5×10^{-7}	2.2×10^{-8}
0.02	50	10^{-4}	1.9×10^{-8}	7.8×10^{-9}	7.0×10^{-10}
0.02	100	10^{-2}	1.3×10^{-5}	5.5×10^{-6}	4.9×10^{-7}
0.02	100	10^{-3}	4.4×10^{-7}	1.8×10^{-7}	2.6×10^{-8}
0.02	100	10^{-4}	1.3×10^{-8}	5.5×10^{-9}	4.9×10^{-10}
0.02	200	10^{-2}	1.0×10^{-5}	3.9×10^{-6}	3.5×10^{-7}
0.02	200	10^{-3}	3.2×10^{-7}	1.3×10^{-7}	1.1×10^{-8}
0.02	200	10^{-4}	1.0×10^{-8}	3.9×10^{-9}	3.5×10^{-10}
0.05	50	10^{-3}	9.6×10^{-7}	3.9×10^{-7}	3.5×10^{-8}
0.05	100	10^{-3}	6.7×10^{-7}	2.8×10^{-7}	2.5×10^{-8}
0.05	200	10^{-3}	4.9×10^{-7}	2.0×10^{-7}	1.7×10^{-8}

[a] From Madsen (1976).

Typical operating parameters for this relationship based on a theoretical sensitivity analysis are given in Table I. Brian (1966) approximated Eq. (17a) for cases of low permeate flux (i.e., $\mathrm{Re}_w \ll 1.0$) to

$$c_w/c_o = 1 + 1.538(v_w^3 hx/3D^2\bar{u}_o)^{1/3}. \tag{17b}$$

Additional simplifications yielded a practical expression for estimating concentration polarization:

$$c_w/c_o = 1 + (v_w^2 h^2/3D^2). \tag{17c}$$

Sherwood *et al.* (1965) generated a composite graph showing the theoretical development of concentration polarization for various Péclét numbers in the entrance and main region of an RO module (Fig. 6). The entrance-region solution is based on Dresner's work (Dresner, 1964). Fisher *et al.* (1964) and Brian (1965) computed a family of asymptotic curves for laminar RO flow in tubular and channel-type configurations.

Trettin and Doshi (1980) compared the results of the film theory to the direct integration of the solute transport equation for UF in an unstirred batch cell (Fig. 3). For the one-dimensional system, Trettin and Doshi (1980) considered the solute mass balance equation without reac-

Theoretical Diffusion Coefficients According to Eq.(17a)[a]

Distance					
$X = 10$ cm Diffusion coefficient for ln (c_w/c_o)			$X = 100$ cm Diffusion coefficient for ln (c_w/c_o)		
1.1	2	10	1.1	2	10
6.1×10^{-5}	2.5×10^{-5}	2.2×10^{-6}	1.9×10^{-4}	7.8×10^{-5}	7.0×10^{-6}
1.9×10^{-7}	7.8×10^{-7}	7.0×10^{-8}	6.1×10^{-6}	2.5×10^{-6}	2.2×10^{-7}
6.1×10^{-7}	2.5×10^{-8}	2.2×10^{-9}	1.9×10^{-7}	7.8×10^{-8}	7.0×10^{-9}
4.4×10^{-5}	1.8×10^{-5}	1.6×10^{-6}	1.3×10^{-4}	5.5×10^{-5}	4.9×10^{-6}
1.3×10^{-6}	5.5×10^{-7}	4.9×10^{-8}	4.4×10^{-6}	1.8×10^{-6}	1.6×10^{-7}
4.4×10^{-8}	1.8×10^{-8}	1.6×10^{-9}	1.3×10^{-7}	5.5×10^{-8}	4.9×10^{-9}
3.2×10^{-5}	1.3×10^{-6}	1.1×10^{-7}	1.0×10^{-4}	3.9×10^{-5}	3.5×10^{-6}
9.6×10^{-7}	3.9×10^{-7}	3.5×10^{-8}	3.2×10^{-6}	1.3×10^{-6}	1.1×10^{-7}
3.2×10^{-8}	1.3×10^{-8}	1.1×10^{-9}	1.0×10^{-7}	3.9×10^{-8}	3.5×10^{-9}
2.9×10^{-6}	1.2×10^{-6}	1.1×10^{-7}	9.6×10^{-6}	3.9×10^{-6}	3.5×10^{-7}
2.1×10^{-6}	8.7×10^{-7}	7.8×10^{-8}	6.7×10^{-6}	2.8×10^{-6}	2.5×10^{-7}
1.5×10^{-6}	6.2×10^{-7}	5.5×10^{-8}	4.9×10^{-6}	2.0×10^{-6}	1.7×10^{-7}

tion as

$$\frac{\partial c}{\partial t} + v\,\frac{\partial c}{\partial y} = D\,\frac{\partial^2 c}{\partial y^2}, \tag{18a}$$

subject to the initial and boundary conditions

$$c(o,y) = c_o \quad \text{for} \quad y > 0, \qquad c(t,o) = c_2, \quad \text{and} \quad c(t,\infty) = c_o \quad \text{for all} \quad t;$$

$$v(c_2 - c_p) = -|v_w|(c_2 - c_p) = D\;\partial c/\partial y|_{y=0}.$$

In applying the integral method, a trial solution for $c(y,t)$, satisfying all boundary conditions, is postulated as

$$c = c_o + (c_2 - c_o)(1 - y/\delta)^n, \qquad n > 0. \tag{18c}$$

Insertion of this solution into Eq. (18a) and integration across the concentration layer thickness δ yields an ordinary differential equation for $\delta(t)$ that can be solved with the initial condition $\delta(t = 0) = 0$ to

$$\delta = \left[2n(n + 1)D\left(\frac{c_o - c_p}{c_2 - c_p}\right)\right]^{1/2} t^{1/2}. \tag{18d}$$

From Eq. (18d) an expression for the "wall" velocity v_w can be obtained

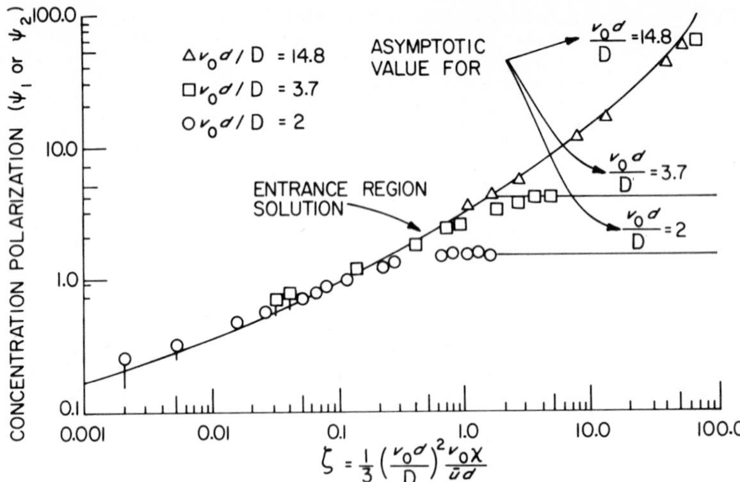

Fig. 6a Concentration polarization between plane-parallel membranes for laminar solution flow $R = 1$.

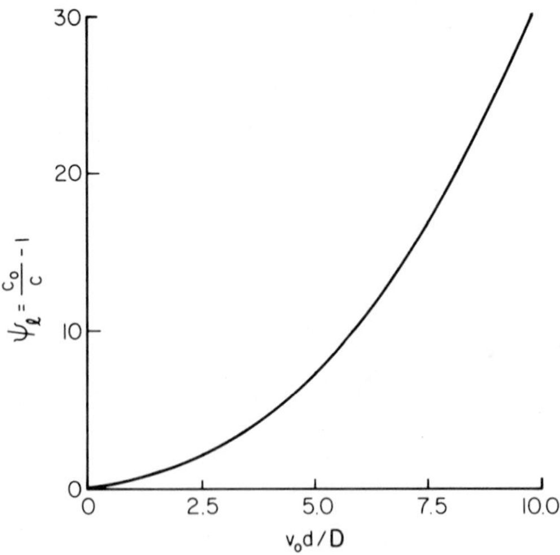

Fig. 6b Asymptotic concentration polarization between plane parallel membranes in the laminar case as a function of Péclét number $v_0 d/D$.

by using the boundary condition of incomplete solute rejection:

$$-|v_w|(c_2 - c_p) = -Dn(c_2 - c_o)/\delta, \tag{19a}$$

so that

$$|v_w| = \frac{c_2 - c_o}{c_2 - c_p}\left(\frac{c_2 - c_p}{c_o - c_p}\right)^{1/2}\left(\frac{nD}{2(n+1)t}\right)^{1/2} \tag{19b}$$

The value of the coefficient n is determined using a moment technique where Eq. (18a) is multiplied by y, and the trial solution for $c(y,t)$ Eq. (18c), is inserted. The resulting equation is integrated over the concentration layer thickness δ, which yields a formula for $n = n(c_2/c_o, c_p/c_o)$. Equation (19) is now compared with the unstirred batch cell analogy of the film theory. The steady-state version of Eq. (18a) is integrated to obtain

$$1 - \frac{c_2 - c_o}{c_2 - c_p} = \frac{c_o - c_p}{c_2 - c_p} = \exp\left[-\frac{|v_w|\delta}{D}\right]. \tag{20}$$

The quantity D/δ is expressed as the unsteady-state mass-transfer coefficient [see Eq. (14e) and (16)]

$$k = D/\delta = (D/\pi t)^{1/2}. \tag{21}$$

Combining Eq. (21) with Eq. (20) yields an expression for the permeate velocity

$$|v_w| = (D/\pi t)^{1/2} \ln(c_2 - c_p)/(c_o - c_p). \tag{22}$$

Equation (19), which was obtained with the integral method, and Eq. (22), based on both the diffusive film and the penetration model, are key equations, and are compared in terms of the constant flux parameter

$$v_w = |v_w|(4Dt)^{1/2}/D = \text{constant}. \tag{23}$$

According to Trettin and Doshi (1980), the film-theory model consistently underpredicts the integral method solution with progressively better agreement as the value of c_2/c_o approaches unity. The film theory deviates from the more exact integral method by more than 25% for all values of c_2/c_o greater than 4.0. For the experimental procedure the authors used bovine serum albumin (BSA) as a solute in a saline buffer solution (pH = 7.4; BSA gelling concentration $c_2 \approx 0.585$ g/cm^3), and in an acetate buffer solution (pH = 4.7; BSA gelling concentration $c_2 \approx 0.340$ g/cm^3). A comparison of the film-theory model, the integrated method solution, and the experimental data are shown in Figs. 7a and 7b. Here, $\lim_{T\to\infty} \Delta V/\sqrt{T} = 2AV_w\sqrt{D/4}$ is plotted versus c_o, the bulk concentration, for various gelling concentrations and diffusivities. The total perme-

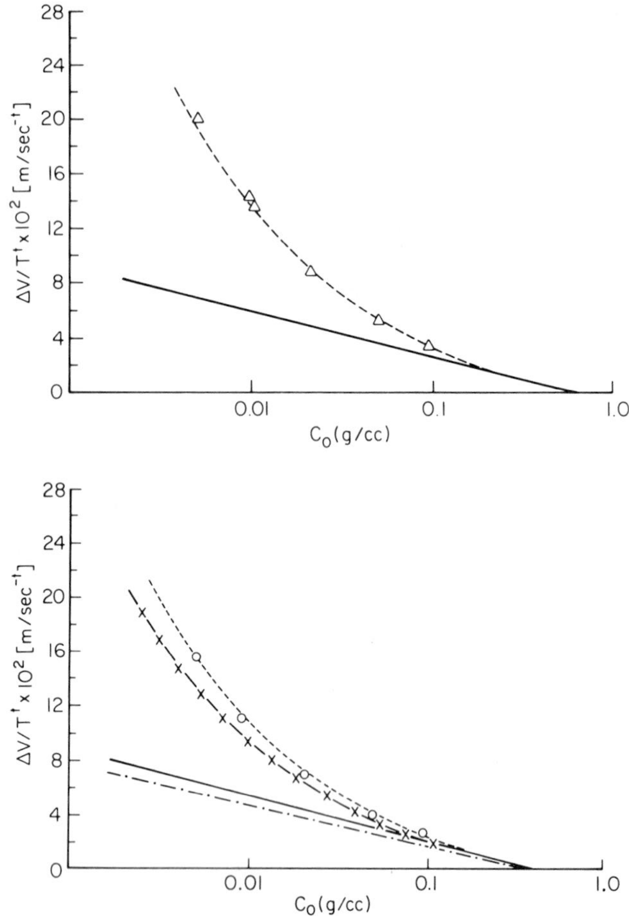

Fig. 7 Comparison of theoretical and experimental results.

ate is defined as $\Delta V = \int_0^T A v_w \, dt$. It is not surprising that the film-theory model deviates significantly from the integral method solution, because the unstirred batch system resembles more an Arnold diffusion cell than the film-concept model. It is unfortunate that a number of researchers mix results from equimolecular counter diffusion (film theory) with parameters of the penetration model (Danckwert's surface renewal concept) and then fit the resulting expression to a particular system that has little in common with the applied theory.

2. *Concentration Polarization and Transmembrane Fluxes*

In the previous section, concepts, applications, and limitations of the film theory and the surface renewal models were discussed. With these submodels, important system parameters such as the concentration boundary layer thickness δ, overall mass-transfer coefficient k, and solute concentration $c_2 \equiv c_{\text{wall}}$ can be evaluated. The convective flow of solute components to the membrane surface is greater than that resulting from diffusion backflow to the bulk solution until steady state is reached (Fig. 5). This well known phenomenon is called concentration polarization. It is also observed in RO modules with sufficiently prefiltered feed streams. It often causes a serious problem because of its negative influence on the transmembrane flux. Detrimental aspects associated with concentration polarization are (Matthiasson and Sivik, 1980):

(1) An increase in chemical potential at the surface, which reduces the driving force for filtration.

(2) Precipitation or formation of a gel on the membrane surface if the wall concentration of solute reaches the saturation concentration. This increases the hydrostatic resistance.

(3) High concentration of solute at the membrane interface, which increases the risks for changes in membrane material composition because of chemical attack.

(4) The deposition of solute on the surface, which can change the separation characteristic of the membrane.

As a consequence of these factors, the transmembrane fluxes in commercial plants may be as low as 10% of the transmembrane fluxes for pure water (Matthiasson and Sivik, 1980). Therefore, it is extremely important to make some efforts to reduce the concentration polarization. It is not always possible, however, to explain the flux behavior solely as a consequence of concentration polarization. In addition, fouling also occurs. In UF processes the occurrence of concentration polarization is for the same reason as in RO. The mechanism of mass transfer is governed by the same principles. Another important factor must also be taken into account, i.e., the characteristic of concentrated macrosolutes. The properties of the most important macromolecular solutions in this case are

(1) high concentration-dependent viscosity,
(2) possible non-Newtonian fluid behavior,
(3) low and concentration-dependent self-diffusivity,

(4) low osmotic pressure, and
(5) gel can be formed at high concentrations.

In summary, when polarization occurs, the solute concentration at the membrane–solution interface rapidly reaches a constant value (or gel point, i.e., 20–60% solute by volume), which is virtually independent of bulk solution concentration, operating pressure, fluid flow conditions or membrane characteristics. This gel layer imposes a second resistance to permeation. The cake formation stops when the convective flux of solute toward the membrane ($J_w c_o$) is exactly equal to the back transfer rate of solute ($D\, dc/dy$). Hence, the UF rate is virtually independent of the applied pressure and actual membrane permeability (except at low pressures). The UF rate is directly proportional to the back transfer rate of solute in the polarization boundary layer (not the gel layer). It increases with increasing fluid velocity, decreasing channel dimension normal to the membrane, and decreasing bulk solution solute concentration. Because of the extremely low diffusion coefficients of macromolecules and colloids in solution, the minimization of polarization c_2/c_o (i.e., maximum UF rates J_w) is far more critical for UF than for RO where ionic species are separated.

In many experiments the influence of concentration polarization has been observed, mostly in the form of a decrease in transmembrane flux and changes in rejection characteristics. Many theoretical analyses have been conducted to obtain a model that describes these phenomena and predicts the transmembrane flux (Blatt *et al.*, 1970; Merten, 1966, Michaels, 1968; Porter, 1972; Shen and Probstein, 1977). The number of direct or indirect studies to test the validity of these models are limited. The most widely used approach to model transfer of solute and solvent across membranes is of the lumped parameter-type. The mass flux expression in terms of customary membrane parameters such as reflection coefficient, permeability coefficient, and hydraulic conductivity, is based on the linear transport equations of irreversible thermodynamics (Dresner and Johnson, 1981; Kedem and Katchalsky, 1958). Assuming the membrane is homogeneous and mechanical equilibrium prevails, the volume flux of solvent for a two-component system is given by

$$J_w = -L_p(\Delta p - \sigma\, \Delta\pi). \tag{24}$$

The hydraulic conductivity of the membrane L_p is directly proportional to the membrane permeability constant and inversely proportional to solvent viscosity and membrane thickness. The driving potential for J_w is the applied pressure difference Δp that might be reduced by the reflection coefficient $\sigma = (\Delta p/\Delta\pi)|J_v = 0$ but primarily by the osmotic pressure

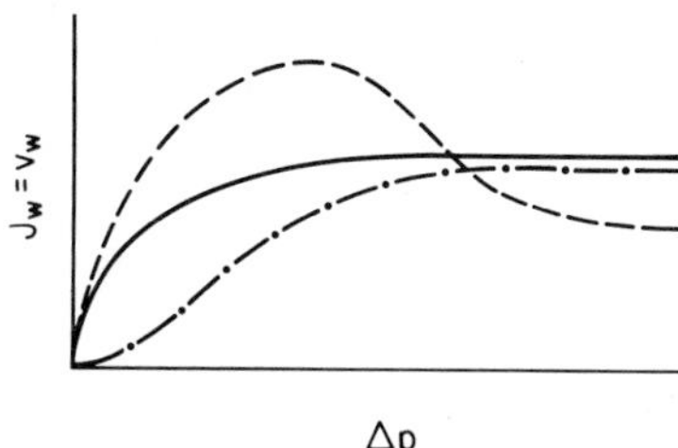

Fig. 8 Variation of transmembrane flux with applied pressure.

$\Delta\pi = \Delta\pi(c_2)$ where $c_2 = c_w$ is the solute concentration at the membrane surface (Fig. 8). The corresponding expression for solute mass flux is

$$J_s = J_w(1 - \sigma)\bar{c}_s - P_s\,\Delta c_s. \tag{25a}$$

Equation (25) is often simplified and written in the form

$$J_s = \sigma_s J_v c_o, \tag{25b}$$

where $\sigma_s = (c_p/c_o)(1 - r)$ can be interpreted as the fraction of the total liquid flowing through membrane pores large enough to pass solute molecules. The rejection coefficient r is defined as $r = 1 - (c_p/c_o)$. For diffusive-type (RO) membranes, the solute mass flux is

$$J_s = (k_s D_s/h)(c_o - c_p), \tag{25c}$$

where k_s is the distribution coefficient between membrane and solution and D_s is the solute diffusivity in the membrane. Of fundamental interest is to find c_p to evaluate the rejection coefficient of a particular membrane. Clearly, for sieve-type membranes $r \neq r(\Delta p)$ as for diffusive-type membranes. From mass conservation,

$$J_s = c_p J_v. \tag{25d}$$

Within the steady-state polarized boundary layer the transport equation (Eq. 11) can be written as

$$J_w\, c - D\, dc/dy = 0; \qquad J_w \equiv v_w. \tag{26a}$$

For the case of a stagnant film of thickness δ (film theory) the polarization modulus is then (Fig. 5)

$$c_2/c_o = \exp[\,j_w\,\delta/D] \qquad \text{or} \qquad v_w \equiv J_w = K\,\ln(c_2/c_o), \tag{26b}$$

which can be directly compared with Eq. (14b) of Section II.B.1.

It is obvious, an accurate representation of transmembrane fluxes is important in connection with membrane design, precoating and cleaning, membrane fouling, and optimal process operation. The paper by Derjaguin *et al.* (1980) is a convenient carrier to exemplify for RO the mechanisms of transmembrane fluxes and dependencies of the rejection coefficient r. More detailed investigations and more complex submodels can be found in Kedem and Katchalsky (1958), Sourirajan (1970), and Spiegler and Laird (1980).

The membrane selected is of the diffusive type. Hence, it is considered to be a quasi-isotropic medium pierced by randomly distributed "fine pores" whose effective radius is much smaller than the actual range of the surface forces. The solute distribution in fine pores is determined by different components of surface forces that are expressed as gradients of scalar potentials. The mass flux of component α, J_α, is driven through the membrane because of the following mechanisms

$$\vec{J}_\alpha = c_\alpha \vec{v}_\alpha - D(\Delta c_\alpha + c_\alpha \cdot \Delta\phi_x + Z_\alpha c_\alpha \Delta\psi) \cdot \tag{27}$$

$$\begin{Bmatrix}\text{Mass}\\\text{Flux}\end{Bmatrix} = \begin{Bmatrix}\text{Solute}\\\text{Convection}\end{Bmatrix} - \begin{Bmatrix}\text{Molecular}\\\text{Diffusion}\end{Bmatrix} + \begin{Bmatrix}\text{Diffuse}\\\text{Adsorption}\end{Bmatrix} + \begin{Bmatrix}\text{Electrostatic}\\\text{Interactions}\end{Bmatrix}$$

For a one-dimensional case assuming electrical neutraliy, the flow region is subdivided into six zones, as shown in Fig. 9. The governing equation for solute mass flux reads

$$J_s = v \cdot c - D\left(\frac{dc}{dy} + c\,\frac{d\phi}{dy}\right) = \text{constant.} \tag{28}$$

The main goal is to find c_p to determine the membrane selectivity (or rejection) coefficient $r = 1 - (c_p/c_o)$. Because $\phi(y)$ and $D(y)$ are usually known, Eq. (28) is integrated piecewise with the appropriate boundary conditions along the y-axis, $y_o \le y \le y_\psi$, so that only the extreme (constant) values, i.e., ϕ_m, D_o, and D_m appear in the final equation. This restriction could be lifted by extending the finite difference grid into the membrane. The total integral of Eq. (28) becomes

$$c(y) = J\exp\left[v\int_{y_o}^{y}\frac{dy}{D(Y)} - \phi(y)\right] \times \left\{B - \int_{y_o}^{y}\exp\left[-v\int_{y_o}^{y}\frac{dy}{D(y)} + \phi(y)\right]\frac{dy}{D(y)}\right\}, \tag{29}$$

where B is an integration constant. From continuity it follows that

$$c_p = J/v = \text{constant} \qquad \text{and} \qquad c_o = JB$$

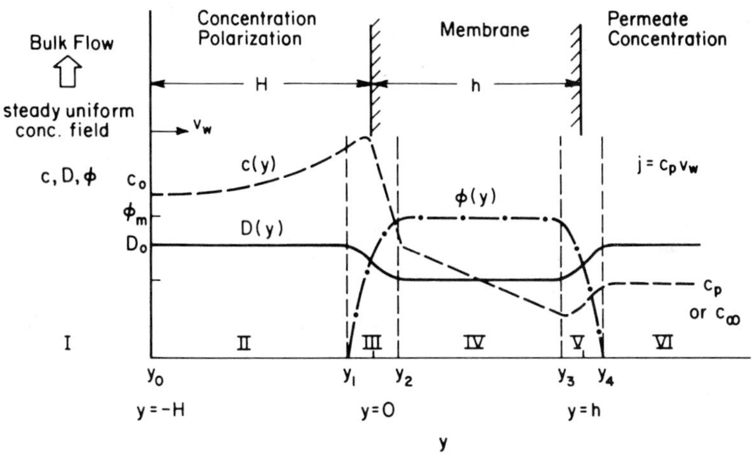

Fig. 9 Schematics for solution flow through RO membrane.

or

$$c_o = \frac{J}{v} + \frac{J}{V}[\exp(\phi_m) - 1] \exp\left(\frac{-vH}{D_o}\right)\left[1 - \exp\left(\frac{-vh}{D_m}\right)\right]. \tag{30}$$

Hence,

$$r = 1 - \frac{1}{1 + [\exp(\phi_m) - 1][1 - \exp(-vh/D_m)] \exp(-vH/D_o)}. \tag{31}$$

With regard to Eq. (31) several conclusions can be drawn. The value of r does not depend on the initial concentration of solution c_o, which is because of the assumption $\phi = \phi(c)$. When $\phi \to \infty$, $r \to 1$ and $c_p \to 0$; conversely, when $\phi \to 0$, $r \to 0$, i.e., no separation of the solution takes place. It is interesting that r decreases when H (concentration polarization layer thickness) increases because of a lack of bulk flow agitation. High values of D_o have a positive effect on r, whereas the selectivity decreases when D_m increases and the permeate velocity v decreases. This is explained by the diffusion D_o on the feed side of the membrane that is directed opposite to the suspension flow, whereas the diffusion inside the membrane D_m enhances the solute transfer. Equation (31) also indicates $r(v)$ possesses a maximum. As the solution velocity v decreases, the selectivity drops because of the predominant effect of the diffusion mechanism through the membrane (i.e. the internal Péclét number $Pe_m = (vh/D_m) \to 0$). An increase of v would improve r if the influence of concentration polarization characterized by the external Péclét number ($Pe_H = (vH/D_o)$)

could be neglected. Differentiating Eq. (31) with respect to v and equating the derivative to zero yields an expression for the optimal permeate flux v_o and a formula for the maximum selectivity r_o:

$$v_o = (D_m/h)\ \ln(1 + \omega), \tag{32a}$$

$$r_o = [(\gamma - 1)\omega/\omega(\gamma - 1) + (1 + \omega)^{(1+\omega)/\omega}], \tag{32b}$$

where $\omega = \mathrm{Pe_m}/\mathrm{Pe_H}$ and $\gamma = e^{\phi_m}$. A plot of $r(v)$ for various Péclét number ratios from Eq. (31) is given in Fig. 10. Equation (31) reduces to $r_o = 1 - (1/\gamma)$ when $\omega \to \infty$ or $H \to 0$, i.e., when concentration polarization is suppressed. In Fig. 10 it can be seen that the selectivity approaches $r_o = 0.96$ for a Péclét ratio of $\omega = 8$, which corresponds to $H/h = 0.25$. For lower values of the immiscible film thickness H, the membrane selectivity is practically unaffected, as indicated by the locus of the maximum r_o (dotted line) in Fig. 10. An analysis of the function $\partial\psi/\partial\gamma = f(v)$ from Eq. (31) shows the greatest differences in r values occur as $v \to 0$ and as $v \to \infty$ when the absolute values of ψ decrease. Hence, for the separation of multicomponents in r_o, a suitable membrane material has to be chosen so the values of ϕ_α differ from one another as much as possible.

3. *Membrane Fouling*

Membrane fouling was mentioned in the previous section in connection with enhanced concentration polarization and gel layer formation. In general, fouling is an accumulation of material on the membrane surface

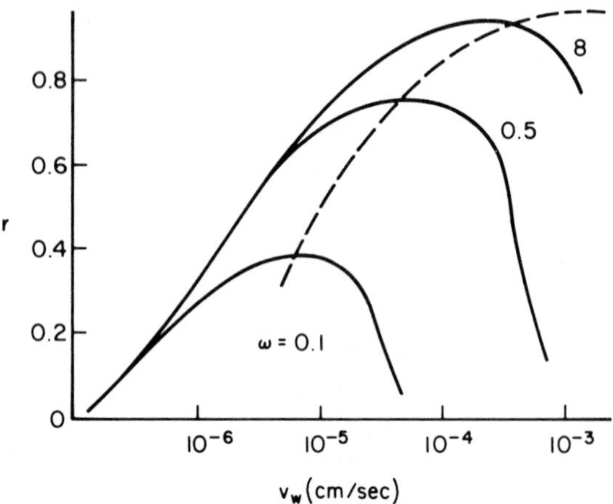

Fig. 10 Dependences of the selectivity coefficient r on the solution flow rate v_w through the membrane.

that causes a decrease in permeate flux. The phenomenon could be classified as particulate, corrosion, chemical reaction, and biological fouling. In some cases, fouling is an irreversible adsorption of macromolecules whereas gel formation (caused by concentration polarization) is a more reversible adsorption without the strong intramolecular forces. The gel layer may be still strongly attached to the membrane surface, however, which makes membrane cleaning difficult (Section III.C).

The interactions responsible for fouling could be viewed as purely chemical in nature. Then hydrolysis of the membrane material (Kesting, 1971), which is a well-known problem with CA membranes, should be considered. It can also involve a chemical reaction between a solute and the membrane as described by Jonsson and Kristensen (1980) for a case of sulfite liquors showing a severe flux decline. Substituted phenols have been identified as a possible cause (Matthiasson and Sivik, 1980).

Purely physical interactions occur by way of compaction—a mechanical effect. Most of the problems encountered in membrane filtration, however, have physico–chemical reasons (Kesting, 1971). They stem from distribution differences of electrons in the atoms or molecules in the solute–membrane system.

Although membrane fouling is associated with UF of colloidal and/or macromolecular solutions, a similar phenomenon might occur in RO units because of the deposition of corrosion particles and other feed constituents refractory to pretreatment (Table II). It is tempting to use the thin film (or gel polarization) model discussed earlier to describe the main effect of membrane fouling, which is permeation flux-decline. Experiments indicate, however, that these submodels predict flux values for the UF of colloidal suspensions one or two orders of magnitude too low. Green and Belfort (1980) termed this phenomenon "flux-paradox for colloidal suspensions." Blatt *et al.* (1970) explained that colloidal species form a much less resistive cake than the considerably smaller macromolecules. This implies the flux is not limited by the hydraulic resistance of the fouling layer. An alternate hypothesis for the observed higher UF rates with colloidal solutions suggests that particle-lift forces augment the diffusive transport of colloids away from the membrane (Green and Belfort, 1980; Madsen, 1976; Porter, 1972). Hence it is relevant to review some modeling aspects of large particle behavior in viscous internal flow fields before discussing existing submodels simulating membrane fouling.

a. Simulation of Particle Trajectories

Particles that do not follow the fluid motion because of their own force fields are not modeled by the mass-transport equations reviewed in

TABLE II

Descriptions of Fouling Phenomena and Foulants[a]

Foulant	Source of foulant	Author
Heavy metal oxides bacterial slimes $CaSO_4$ $CaCO_3$		Leiserson (1973)
Organic and inorganic colloids		
Iron	Water, sewage treatment including iron coagulation	Kuiper *et al.* (1973)
		Cruver and Nusbaum (1974)
		McCutchan and Johnson (1970)
		Grover and Delve (1972)
Corrosion products	Stainless steel test loop	Carter *et al.* (1974)
		Agrawal *et al.* (1972)
Microbial slime	Waste water from sulfite pulping of wood	Wiley *et al.* (1972)
	Alum-treated sand-filtered primary effluent	Feuerstein *et al.* (1971)
$CaSO_4$	Sulfuric acid, pH-adjusted primary sewage	Feuerstein (1971)
	Polluted surface waters	Beckman *et al.* (1973)
Casein	Whey	Lim *et al.* (1971)
Polyhydroxy aromatics	Waste water	Cruver and Nusbaum (1974)
Organic acids and polysaccharides	Polluted surface water	Beckman *et al.* (1973)
Protein	Milk	Glover and Brooke (1974)
Organic material	Simulated brackish water	Minturn (1973)
Organics	Plating wastes	Bevege *et al.* (1973)
Calcium phosphate complex	Whey	Hayes *et al.* (1974)
Ca, P, organic material	Trickling filter effluent	Bashow *et al.* (1972)
Pectin and insoluble celluloselike material	Mandarin juice	Watanabe *et al.* (1979)
Dissolved organic material	Secondary sewage effluent	Winfield (1979)
Oil	Oily bilge water	Jackson *et al.* (1973)
		Bhattacharyya *et al.* (1979)
Calcium salt and humic acid	Surface water and sewage	Sammon and Stringer (1975)

[a] From Matthiasson and Sivik (1980).

Sections II.A and II.B. The pathlines or trajectories of particles in viscous fluid can be described by Newton's Second Law of Motion (Fig. 4a). This law is important in many areas of fluid–particle interactions such as UF, flow field fractionation, deep bed filtration, sedimentation and fluidized bed phenomena. A comprehensive review of small particle motion and drops and bubbles in a viscous fluid at low Reynolds numbers (creeping flow approximation) is in Happel and Brenner's (1965) text. Extensive reviews can also be found in the literature of Goldsmith and Mason (1967), Brenner (1970), Caswell (1977), and Leal (1980). The text of Clift *et al.* (1978) is a concise overview of practical formulas for evaluating integral properties of viscous flow at all Reynolds numbers past spherical and nonspherical particles. In notes by Pruppacher (1980) and Esmail (1980), a brief discussion and an updated bibliography on numerical and empirical results of flow fields around spheres at intermediate Reynolds numbers are presented. Expressions for integral properties, such as the total drag, could be extracted from these papers for the force balance from which the particle trajectory is determined. Typical forces and torques acting on a representative particle are summarized in Table III as compiled by Rajagopalan and Tien (1979). The accurate prediction of particles in space and time depends mainly on: (1) The degree of inertia in a Newtonian fluid (Reynolds number), (2) the rheological fluid properties, and (3) the changing shape of the particle. Experimental observations showed that even small departures from the assumptions of "zero" Reynolds number, Newtonian rheology, and fixed shape lead to "preferred" particle orientations and/or positions in the flow field. A well-known observation is the tubular pinch effect measured by Segre and Silberberg (1962) but noticed much earlier by Poiseuille, where particles migrate laterally toward the solid boundaries for viscous fluid duct flow and to the centerline in viscoelastic fluids. A deformable drop, however, migrates laterally toward the centerline in a unidirectional Newtonian shear flow at low Reynolds numbers. Lateral migration phenomena are a special class of weak inertia effects on (single) particle dynamics. Segre and Silberberg (1962) showed that a neutrally buoyant sphere would translate across streamlines of a Poiseuille flow to an equilibrium position about 60% of the distance from the centerline to the tube walls. A detailed theoretical analysis of lateral migration in creeping shear flows was reported by Ho and Leal (1974) and later by Vasseur and Cox (1976). Their predicted migration velocity, equilibrium positions and trajectories are in excellent quantitative agreement with observations of Halow and Wills (1970). A practical application of the Segre–Silberberg effect to membrane fouling in UF will be discussed with other existing math models.

In summary, for a comprehensive simulation of fouling layer forma-

TABLE III

Forces and Torques Acting on the Suspended Particles in Spherical Coordinates[a]

Inertial force and torque

$\underline{f}^I = m(D\underline{u}/Dt)$

$\underline{t}^I = \underline{O}$

Gravitational force and torque

$\underline{f}^G = \frac{4}{3}\pi a_p^3 g(\rho_p - \rho_f) = \frac{4}{3}\pi a_p^3(\rho_p - \rho_f)g(-\cos\theta\underline{e}_r + \sin\theta\underline{e}_\theta)$

$\underline{t}^G = \underline{O}$

Surface force and torque

Molecular dispersion force and torque (London force and torque)

$\underline{f}^{Lo} = [-2H\alpha(\delta;\ a_p,\lambda_e)a_p^3/3\delta^2(2a_p + \delta)^2]\underline{e}_r$, where $\alpha(\delta;a_p,\lambda_e)$ is the retardation correction factor

$\underline{t}^{Lo} = \underline{O}$

Double layer interaction force and torque

$\underline{f}^{DL} = \{[\nu a_p\kappa(\zeta_c^2 + \zeta_p^2)/2][[2\zeta_c\zeta_p/(\zeta_c^2 + \zeta_p^2)] - e^{-\kappa\delta}][e^{-\kappa\delta}/(1 - e^{-2\kappa\delta})]\}\underline{e}_r$

$\underline{t}_{DL} = \underline{O}$

Drag forces and torques

Resulting from the translation of the particles

$(\underline{f}^D)^t = -6\pi\mu a_p[\underline{u}_r f_r^t(\delta^+)\underline{e}_r + \underline{u}_\theta f_\theta^t(\delta^+)\underline{e}_\theta]$

$(\underline{t}^D)^t = 8\pi\mu a_p^2 u_\theta g_\theta^t(\delta^+)\underline{e}_\phi$

Resulting from the rotation of the particle

$(\underline{f}^D)^r = 6\pi\mu a_p^2\omega f_\theta^r(\delta^+)\underline{e}_\theta$

$(\underline{t}^D)^r = -8\pi\mu a_p^3\omega g^r(\delta^+)\underline{e}_\phi$

Resulting from the fluid velocity in the presence of the stationary particle

$(\underline{f}^D)^m = 6\pi\mu a_p\{-Ay^2 f_r^m(\delta^+)\underline{e}_r + [Byf_{1\theta}^m(\delta^+) + Dy^2 f_{2\theta}^m(\delta^+)]\underline{e}_\theta\}$

$(\underline{t}^D)^m = 8\pi\mu a_p^3[Bg_{1\phi}^m(\delta^+) + Dyg_{2\phi}^m(\delta^+)]\underline{e}_\phi$

[a] From Rajagopalan and Tien (1979).

tion resulting from large (colloidal) particles, it is necessary to simultaneously model the forces acting on the particles, fluid flow patterns, particle interactions, membrane–solute interactions, and transmembrane fluxes. The type of modeling equations involved are best illustrated with a problem conceptualization. Newton's Second Law of Motion and the principle of mass conservation are used for simulating the transport of colloidal matter as developed by Kleinstreuer *et al.* (1978) based on Saffman's "dirty gas" equation. It basically takes fluid dynamics, lift, buoyancy, and drag forces into account:

Particle transport $$N^b\left[\frac{\partial \vec{v}_p}{\partial t} + (\vec{v}_f \cdot \nabla)\vec{v}_p\right] = N^b\vec{g} + kN^b(\vec{v}_p - \vec{v}_f) \tag{33a}$$

Particle continuity $$\frac{\partial N^b}{\partial t} + \nabla \cdot (N^b\,\vec{v}_p) = N_s + D\,\nabla^2 N^b, \text{ where} \tag{33b}$$

$N^b \equiv [N^b(\vec{x},t)_{i,j}]$ is the number density of particles i and characteristics j. The velocity vector $\vec{v}_f$ could be obtained from Eqs. (8a and b) provided the particles do not disturb the fluid flow field. If statistical data sets are available, the particle velocity vector could be decomposed into a deterministic and random part. Once the velocity distributions are known, a realistic submodel for fouling layer growth could be constructed.

Such a project is presently underway using a simplified form of Eqs. (33a and b) and Newton's Second Law of Motion for the particle trajectories (Kleinstreuer *et al.* 1983b). The change in particle size distribution on the angström to micrometer scale because of internal and external processes is described by

$$(\partial n_p/\partial t) + (\vec{u}_f \cdot \nabla)n_p = -\nabla \cdot \vec{j} + \sum S_p, \tag{33c}$$

where $n_p = n_p(V_p,\vec{r},t)$ is the particle size distribution function and the number concentration is given by $N = \int_{\Delta V} n \, dV_p$, i.e., the number of particles per unit fluid volume in the particle size volume range ΔV, where $\nabla \cdot \vec{j} = \nabla \cdot D \nabla n_p - (\nabla \cdot \vec{v}_p)n_p$ is the change in particle flux because of diffusion and convection, $\vec{v}_p = (\vec{F}_{ext}/f)$ the particle migration velocity (settling, phoresis, lift, etc.), and $\sum S_p = S_{conv} + S_{coag} - S_{sink}$ represent the growth rates. A typical boundary condition for ideal membranes is

$$n_p(x, y = y_w) = \alpha_o \left.\frac{\partial n_p}{\partial y}\right|_{wall}.$$

The equation of motion for particles on the micrometer to millimeter scale can be written as

$$\frac{d(m\vec{v})}{dt} = \sum \vec{F}_{ext} = \vec{F}_{grav} - \vec{F}_{buoy} - \vec{F}_{D\ prof} - \vec{F}_{D\ visc} \pm \vec{F}_{VI\ acc} \pm \vec{F}_{VI\ visc} \pm \vec{F}_{lift} \pm \vec{F}_{elect}. \tag{33d}$$

The force field may consist of gravity, buoyancy, profile (or pressure), viscous drag, virtual inertia because of particle acceleration in irrotational fluid, virtual inertia because of vorticity diffusion (Basset, 1961), lift because of particle rotation (fluid inertia), and far field as well as near field electric forces. In addition, a torque balance is necessary to compute the particle rotation that is required for calculating the lift force among others

$$\frac{d(I\vec{\Omega}_p)}{dt} = \sum \vec{T}_{ext} \tag{33e}$$

The equation for two-dimensional particle trajectories can be easily deduced as

$$dy/dx = v/u, \qquad y = \int_{x_0}^{x} v/u \, dx. \tag{33f}$$

The modeling equations for particle deposition are (Fig. 4a)

$$Dm_{\text{cake}}/Dt = \sum S_c = \dot{m}_{\text{depos}} - \dot{m}_{\text{re}} - \dot{m}_{\text{perm}}, \tag{33g}$$

where $m_{\text{cake}} \propto \delta_c(x,t)$ is the thickness of deposition layer (cake); $\dot{m}_{\text{depos}} \propto \int_{t_0}^{t} \int_{p}^{y_{\min}} n_p(d, y;t)(h - y)\, dy\, dt$ is the deposition of particle that is dependent on particle size distribution, limiting trajectory, time, and channel geometry; $\dot{m}_{\text{re}}$ is proportional to London–Van der Waals forces, rheological properties of the cake, and shear stresses; and $\dot{m}_{\text{perm}}$ is proportional to particle and membrane characteristics.

This fluid–particle system is described by a set of coupled transient nonlinear PDEs. The (strong) coupling between fluid flow and particle system is established through the rheological fluid properties, boundary conditions at the wall (permeation flux across cake and membrane plus axial slip velocity), fluid dynamic forces exerted on the particle(s), and the associated limiting particle trajectories.

Three solution methods should be considered: (1) A comprehensive numerical solution technique using, for example, Keller's box method, (2) a mixed approximation scheme using a perturbation method for the fluid flow problem and an iterative integration method for particle trajectory and deposition, and (3) the "particle source in cell" (PSIC) model as proposed by Migdal and Agosta (1967) and applied to gas-droplet flows by Crowe *et al.* (1977).

Some obvious problem areas include: (1) The availability of data sets for model input, calibration, and verification, (2) the determination of the significance of internal and external transport mechanisms, forces, etc., and (3) the selection of a computationally efficient (numerical) solution technique.

b. Submodels Simulating Membrane Fouling

Even when the fluid flow and particle velocity fields are known, for example, by solving Eqs. (8) and (33), suitable equations for fouling and fluxes across the solute layer and membrane have to be solved. The deposition layer and membrane can be regarded as multiresistance layers in series. The energy loss because of the resistance, $h_f = (\Delta p/\delta)$, is computed with the Carman–Kozeny equation based on Darcy's Law and the Darcy–Weissbach equation for pipe flow losses. Thus (Carman, 1937; Kozeny, 1927)

$$h_f = f'(L/\phi_s d_p)[(1 - n)/n^3](\bar{v}_s^2/g), \tag{34}$$

where the friction factor f' has been correlated with the Reynolds number $\mathrm{Re}_{dp} = \phi_s d_p \bar{v}_s/\nu$ to

$$f' = 150[(1 - n)/\mathrm{Re}_{dp}] + 1.75, \tag{35}$$

the Ergun equation. The Carman–Kozeny equation is often rewritten in terms of the hydraulic resistance R_c of the cake (or fouling layer) as

$$R_c = 180(1 - n)^2 h_c/n^3 d_p^2, \tag{36}$$

where h_c is the cake thickness. The Carman–Kozeny equation predicts the head loss across stationary beds, in which voids are unobstructed. Before solidification, gel layers act more like fluidized beds. Hence, the standard filtration equation must be augmented where appropriate.

Another effect, the lateral migration phenomenon (Segre and Silberberg, 1962), was integrated into the standard filtration theory by Green and Belfort (1980). They proposed an expression for the dynamic growth of the cake thickness based on Eq. (34) as

$$\delta_c = \frac{c_s}{\rho_p(1 - n)} \int_o^t (j - v_L)\, dt, \tag{37}$$

where c_s, ρ_p and n are properties of the cake, j the permeation flux discussed as follows, and v_L the lateral drift velocity proportional to the square of the average axial velocity, as discussed by Brenner (1966).

For resistance layers in series

$$j = \frac{\Delta p}{\mu(R_m + R_c)} \quad \text{or} \quad j = \frac{\Delta p}{(\delta_m/K_m) + (\delta_c/K_c)}, \tag{38}$$

where $K \propto d^2 n^3/180(1 - n)^2$ (Carman–Kozeny) is the hydraulic permeability of the membrane (subscript m) or cake layer (subscript c).

Blatt *et al.* (1970) derived an alternative expression for the permeation flux j that follows directly from the film theory model (Section II.B.1) and is repeated for comparison:

$$j = k_s \ln(c_{depos}/c_{bulk}), \tag{39}$$

where $k_s = B(\dot{\gamma}_{wall} D_s^2/L)^{1/3}$ is a mass-transfer coefficient based on Leveque's work, $\dot{\gamma}_w$ the fluid shear rate at the membrane surface, B a constant, L the flow channel length, D_s the solute diffusivity, and c_{depos} the (constant) gel layer concentration. Without the deposition layer, $j_s = (D_s/\delta)\ln(c_{wall}/c_{bulk})$, where (c_w/c_b) adjusts itself to any imposed j_s.

Based on Eqs. (37) and (38), the permeation flux $j(t)$ can be calculated with an iterative procedure as outlined by Green and Belfort (Fig. 11). This approach considers *transient* changes (e.g., deposition layer

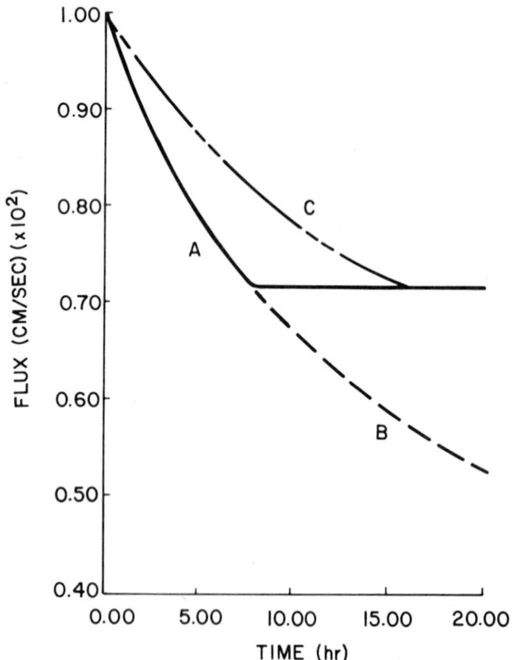

Fig. 11 Flux-decline plots using iterative procedure described in Green and Belfort (1980).

growth or decline or permeation fluxes) as observed in *laminar* boundary-layer flow. This modeling concept is the first step in the simulation of membrane fouling in laminar UF of colloidal solutions. It can be expected that the concentrated macrosolute and colloidal particle solution at the membrane surface might exhibit rheological properties as discussed in Section II.C.

A mathematical model for the rate of fouling-layer formation under *turbulent flow* conditions was proposed by Gutman (1977) based on the work done by Carter and Hoyland (1976) and Kimura and Nakao (1975). The basic equation for the fouling-layer formation is a first-order kinetic model

$$dm/dt = r_d - r_e, \tag{40}$$

where m is the mass of foulant, directly proportional to δ, the fouling-layer thickness. Empirical relationships are proposed for r_d (deposition rate) and r_e (reentrainment rate)

$$r_d = \begin{cases} (k + \frac{1}{2}j)c_o & \text{for } j < 2k, \\ jc_o & \text{for } j > 2k, \end{cases} \tag{41a}$$

$$r_e = m\beta/\theta = m(a/\sigma)\tau_w; \qquad \tau_{wall} = \tau_w(\bar{u}, j). \tag{41b}$$

A second basic expression relates the flux through the coated membrane j_c to the hydraulic resistance of the unfouled membrane R_m, and the resistance of the fouling layer R_f as

$$j_c/j = R_m/(r_m + R_f). \tag{42}$$

Again, the flux of water through the membrane j is a function of applied pressure drop Δp, osmotic pressure $\Delta\pi$, and membrane resistance R_m. Hence,

$$j = (\Delta p - \Delta\pi)/R_m. \tag{43}$$

Equations (41) and (42) inserted into Eq. (40) and integrated, form with Eq. (43), the tailored equation for the rate of fouling layer formation.

C. FLUID FLOW SUBMODELS

We pointed out in Section II.A that a family of submodels (i.e., auxiliary expressions and conditions) in addition to the governing equations of Section II.A.1 are necessary for accurate and comprehensive modeling of fluid flow and solute distribution in membrane units. The fluid flow submodels to be discussed include (1) non-Newtonian models, because concentrated macrosolutes and colloidal particle solutions at the membrane surface might exhibit rheological properties, (2) turbulence models primarily for momentum transfer, and (3) models for incorporating finite slip velocities at and inside the membrane that might affect the overall system performance (Fig. 4a).

1. *Submodels for Non-Newtonian Behavior of Fouling Layers*

Concentrates near the membrane might behave as "true gels," i.e., they distort elastically under shear and rupture at a characteristic shear stress (Blatt *et al.*, 1970). In other cases, they behave as "Bingham–

pseudoplastics," i.e., a threshold yield stress is required before they begin to flow as highly viscous fluids. A suitable model in the latter case can be constructed by writing the momentum flux (or shear stress) tensor, $\vec{\vec{\tau}} = \eta \cdot \dot{\vec{\vec{\gamma}}}$ in Eq. (2) in the form (Bird *et al.*, 1977)

$$\begin{aligned} \eta &= \infty && \text{for} \quad \tau \leq \tau_o, \\ \eta &= \mu_o + (\tau_o/\dot{\gamma}) && \text{for} \quad \tau \geq \tau_o, \end{aligned} \tag{44a}$$

where τ_o is the yield stress that might contribute to the explanation of turbulent bursts observed by Thomas *et al.* (1973) and $\dot{\vec{\vec{\gamma}}} = \nabla\vec{v} + (\nabla\vec{v})^T$ is the rate of deformation tensor.

Depending on the solute characteristics, a power law might be sufficient, i.e.,

$$\eta = m\dot{\gamma}^{n-1}, \tag{44b}$$

where m and n are constants that must be determined experimentally. Note that the zero shear rate region, $\eta\ (\dot{\gamma} \approx 0) \approx \eta_o$, is not described by the power law.

2. *Turbulence Submodels*

The various models for turbulent momentum and mass transfer may be divided into three general classes: (1) Model based on the film theory, (2) models based on the surface renewal concept, and (3) zero- or multiequation models in boundary-layer theory. The first two approaches are most frequently employed in membrane process simulation, but they have distinct drawbacks (Section II.B). For a mathematically rigorous approach, time-averaged transport equations and suitable turbulence models must be used. Semiempirical approaches using prescribed (turbulent) velocity profiles and mass-transfer coefficients are also discussed.

The principal idea in turbulence description is to decompose the instantaneous variable f into a time-smoothed and a randomly fluctuating part:

$$f = \bar{f} + f', \qquad \text{where} \quad \bar{f} = (1/\Delta T) \int_{t_0}^{t_0+\Delta T} f\, dt. \tag{45}$$

If Eq. (45) is inserted into the governing transport equations (Section II.A) and the time-averaging procedure is carried out, additional terms, "apparent stresses" or Reynolds transport terms, are generated, e.g., $\tau_t = -\rho\overline{u'v'}$ and $N_t = \overline{-c'v'}$. The conservation equations do not provide information regarding the magnitudes of these statistically correlated terms. It is, therefore, necessary to simulate their effects by introducing turbulence models based on either a statistical (Frost and Moulden, 1977; Monin and Yaglom, 1971; Seinfeld, 1975) or deterministic description, the latter favored by many researchers (Bradshaw, 1976; Launder and Spalding, 1972). The submodels for the turbulent stresses are expressed in terms of transport parameters such as the kinematic eddy viscosity, the mixing length, and kinetic turbulence energy and/or frequency of the kinetic energy. Boussinesq's hypothesis is most frequently employed:

$$\frac{\tau}{\rho} = (\nu_{\text{mol}} + \nu_{\text{turb}}) \frac{d\bar{u}}{dy} = \nu_{\text{e}} \frac{d\bar{u}}{dy}, \tag{46a}$$

$$N = (D_{\text{mol}} + D_{\text{turb}}) \frac{d\bar{c}}{dy} = D_{\text{e}} \frac{d\bar{c}}{dy}, \tag{46b}$$

where ν_e and D_e are the effective transport coefficients. The hypothesis is turned into a turbulence model when the coefficients ν_t and D_t are expressed in terms of one or more transport parameters. For example, Prandtl's mixing length formula

$$-\overline{u'v'} = l_{\text{m}}^2 \left|\frac{\partial u}{\partial y}\right| \frac{\partial u}{\partial y} \tag{47}$$

includes an (alternative) expression for ν_t, that is $\nu_t = l_m^2 |\partial u/\partial y|$.

An alternative equation proposed by Braun (1977) is even simpler and may cover the entire flow domain:

$$\tau_{\text{t}} = \alpha\rho(u_{\text{max}} - u)(u - u_{\text{c}}), \tag{48}$$

where α is a parameter reflecting the scale of turbulence (like l_m), ρ is the density, and u_c is a cutoff velocity much like u^+ for the laminar viscous sublayer to be discussed later. The unique features of Eq. (48) are (1) it reflects the *nonlocal* characteristics of turbulence, (2) it represents fluid as well as turbulence scale (or system) parameters, and (3) it is a polynomial of order two because the shear stress has two maximal roots. Equations (46) and (47) or (48) together with mass-transfer analogies (Sherwood *et al.*, 1975; Spalding, 1977) can be combined to relatively simple but powerful turbulent flow representations.

A contrasting approach to solving the time-averaged transport equations directly after suitable turbulence submodels are established is to postulate a turbulent velocity profile for the boundary layer flow: Either the $\frac{1}{7}$-law (Schlichting, 1979) or the law of the wall (White, 1974). For the latter approach, the boundary layer is decomposed into several regions: The linear sublayer $0 \leq y^+ \leq 5$, the buffer zone $5 \leq y^+ \leq 50$, the log–law region $50 \leq y^+ \leq 500$, and the outer layer $y^+ > 500$, which is influenced by the turbulent bulk flow and hence the pressure gradient. Here, the inner variables are defined as $u^+ = \bar{u}/u^*$, where $u^* = \sqrt{\tau_w/\rho}$ is the friction velocity, $y^+ = u^* y/\nu$ a Reynolds number, and $\varepsilon^+ = 1 + \nu^+/\nu$ the dimensionless effective viscosity for the inner region when $\tau \approx \tau_w$. Cebeci and Bradshaw (1977) discuss two formulas for the mean velocity distributions along porous surfaces with mass transfer v_w:

$$2/v_w^+[(1 - v_w^+ u^+)^{1/2} - 1] = (1/\kappa) \ln y^+ + c \tag{49a}$$

$$u^+/\tfrac{1}{2}[(1 + v_w^+ u^+)^{1/2} + 1] = (1/\kappa) \ln y^+ + c. \tag{49b}$$

The latter equation, proposed by B. E. Launder, collapses to the logarithm law for $v_w^+ \equiv v_w/u^* = 0$.

Turbulence concepts in experimental studies on membrane separation processes were applied by a number of researchers (e.g. Brosh and Winograd, 1974; Goldsmith, 1971; Johnson and McCutchan, 1972; Kinney and Sparrow, 1970; Michaels, 1968; Sheppard *et al.*, 1972; Thomas *et al.*, 1973). Representative for most investigations is a semiempirical expression for turbulent flow (based on Froessling's equation and the film theory model) confirmed by Goldsmith (1971):

$$kD_H/D = 0.0096 \cdot \mathrm{Re}^{0.913} \cdot \mathrm{Sc}^{0.346}. \tag{50a}$$

The maximum flux obtained in the turbulent system was around 3.6 $\times 10^{-4}$ cm/sec for an approximately 1% carbowax 20,000 MW solution. An alternative to the approach given with Eq. (50a) is the derivation of a semiempirical expression for the solute transport parameter $D_{AM}/K\delta$. The solvent water transport through an RO membrane could be written as (Kimura and Nakao, 1975; Sourirajan, 1978)

$$N_b = (D_{AM}/K\delta)[(1 - X_{A3})/X_{A3}](c_2 X_{A2} - c_3 X_{A3}), \tag{50b}$$

where D_{AM} is the (turbulent) solute diffusivity, δ the effective film thickness, K the equilibrium constant relating solute concentration in the membrane phase and in the solution phase in equilibrium with the membrane phase, and X_A and c are the mole fraction of solute and the molar density of solution, respectively. Laboratory case studies can be found in Kimura and Nakao (1975), Sourirajan (1978), and Tweddle *et al.* (1980).

If the membrane surface or impermeable channel walls are rough, the turbulent velocity profiles have to be extended to accommodate roughness effects that play an important role in boundary-layer stability and hence, membrane unit operation. In general, an extra term that correlates to the roughness height k is added to the velocity profile for the inner region, i.e., $u^+ = u^+(y^+) + f(k^+)$, where the dimensionless roughness height $k^+ = u^+ k/\nu$ is another Reynolds number. The outer region is unaffected from the surface roughness.

Depending on the value and distribution of k, it is possible the viscous (laminar) sublayer of the turbulent boundary layer might "drown" the roughness effect so that hydrodynamically smooth flow occurs. Meyer (1980) and Schlichting (1979) give an updated outline of internal turbulent flows over rough surface. Rekin (1976) determined the turbulent viscosity above a permeable plate by using the magnitude of the maximum turbulent friction in the boundary layer. Simple dependences are obtained for the velocity and friction stress distribution. Epifanov and Gus'kov (1979) correlated roughness heights with boundary-layer parameters and energy losses for turbulent flow on porous walls with blowing.

3. *Finite Slip Velocity at Membrane Surfaces*

The conventional boundary condition for Newtonian fluid flow over a membrane surface is that $u = 0$ at the wall, the "no-slip" condition. Beavers and Joseph (1967) pointed out there is a migration of fluid tangent to the boundary within the porous matrix. They relate the slip velocity to the bulk flow by the ad hoc boundary condition

$$du/dy|_{y=0} = \beta(u_B - Q). \tag{51a}$$

Singh and Laurence (1979) studied the influence of slip velocity at a membrane surface on UF performance for tubular and channel flow systems. They write the slip–flow boundary condition of Beavers and Joseph (1967) for tubular membranes as

$$u(x, r_w) = -\frac{\sqrt{k}}{\alpha}\frac{\partial u}{\partial r}. \tag{51b}$$

When the membrane permeability $k = 0$, Eq. (51b) reduces to the no-slip condition appropriate to a solid wall. A slip coefficient θ equal to $\sqrt{k}/\alpha r_w$ is introduced and the dramatic effect on the laminar velocity profile for various wall Reynolds numbers ($Re_w = 2v_w r_w/\nu$) is shown.

III. Comprehensive Case Studies

Field applications of membrane systems to reclamation of municipal–industrial wastewater and to water purification (desalination) were reported for RO units only (e.g. Argo and Montes, 1979; Gaddis *et al.*, 1979; Schippers *et al.*, 1978; Wojcik, 1980). Realistic laboratory experiments with HF modules using municipal or industrial wastewater were performed to investigate causes of membrane fouling (Kimura and Nakao, 1975), potential removal mechanisms for fouling layers (Winfield, 1979), and a representative fouling index (Reed, 1979; Schippers and Verdoŭw, 1980). Existing mathematical models fall short in the attempt of a comprehensive and accurate simulation of real world systems, mainly because of the complexity of the governing equations discussed in Sections (II.A and II.B). Hence, some models are based on highly empirical formulas using a lumped parameter approach. In particular, models are presently employed to "summarize," or interpret, specific laboratory case studies and to "predict" trends for the individual system or problem area. Nevertheless, it should be evident that many mathematical models are helpful in the understanding membrane separation processes and in the representation of hypotheses toward a generalized theory. One of the more comprehensive modeling studies is concentrating on hollow-fiber RO systems because such units are used worldwide for desalination of brackish water. Laminar, dilute suspension flows in tubular or slit-type UF modules is another area for which comprehensive modeling work exists. We shall discuss the system conceptualization and the results for these two case studies. Section III closes with a review of physical and mathematical modeling results on the reduction of polarization layers that plague RO and UF systems.

A. HOLLOW-FIBER REVERSE OSMOSIS SYSTEM

The modeling study for radial flow and solute transfer in hollow-fiber RO units is based on investigations by Kabadi *et al.* (1979) and earlier works by Dandavati *et al.* (1975), Bansal and Gill (1973), Srinivasan *et al.* (1967), and Gill *et al.* (1965). Hollow-fiber RO membranes are attractive because of high productivity per unit module volume resulting from a large surface-to-volume ratio. In addition, concentration polarization is apparently unimportant in radial flow hollow-fiber systems as stated by Kabadi *et al.* (1979), based on their observations and the work by Orofino (1977). In hollow-fiber RO systems, high-pressure solutions exit from a central feeder (porous tube) and flow around tiny hollow fibers that are densely packed in a shell. The solution, minus the rejected solutes, permeates the fiber walls radially. Inside the fibers the permeate flows axially toward the atmospheric pressure-end of the module. The concentrate is collected from the outside shell wall (Fig. 2). The flow fields in the shell-side fiber bundle and inside a typical fiber are three-dimensional. This complex flow field is conceptualized as (1) radial flow through a shell packed with quasi-parallel fiber tubes and (2) Poiseuille-type flow inside the fibers (i.e., laminar porous tube flow with mild injection). Experiments were carried out by Kabadi *et al.* (1979) for applied pressures p_f ranging from 200 to 400 psia, feed rates F varying from 75 to 380 cm/sec, and feed concentrations, c_o up to 0.034 gm NaCl/gm H_2O. The objective was to simulate the performance of a hollow-fiber unit as measured by the productivity ϕ

$$\phi = 1 - y_T \int_0^1 V_1(z, y_T)\, dz, \tag{52}$$

and the dimensionless overall product concentration $\theta_p = c_p/c_o$

$$\theta_p = \frac{\int_0^1 \int_1^{y_T} y v_w \theta_s\, dy\, dz}{\int_0^1 \int_1^{y_T} y v_w\, dy\, dz} \approx \frac{1}{\phi}\left[1 - y_T \int_0^1 V_1(y_T)\theta_1(y_T)\, dz\right], \tag{53}$$

where $V_1 = v_1/v_o$ is the dimensionless radial velocity in the shell, $y_r = R_1/R_o$ the radii ratio of the fiber bundle to the central feeder, v_w the fiber wall permeation velocity, $y = r/R_o$ the dimensionless radial coordinate, z the axial coordinate, $\theta_1 = c_1/c_0$ the concentration on the shell side, and $\theta_s = c_3/c_o$ the concentration in the fiber.

Thus, the principal unknowns are the radial velocity in the fiber-rod packed shell, the associated solute concentration, and the solute concentration inside the fibers. The modeling equations are basically steady-state expressions for radial pressure drop, fluid continuity and solute transfer in the fiber packed shell, and axial pressure drop and fluid velocity inside a representative fiber. The coupling between shell and fiber is achieved via balances for the permeate flux and the solute mass flux at the (fiber) membrane wall.

The last two relationships can be written for direct comparison with results of Section II.B as

$$j_w = K_1(\Delta p - \Delta\pi) + K_3\, \Delta p \tag{54a}$$

$$j_s = K(c_1 - c_3) + K_3\, \Delta p c_1. \tag{54b}$$

The pressure dependence of K_1 (solvent permeability coefficient) and K_3 (membrane pore coefficient) resulting from membrane compaction was found experimentally. They both decrease as the feed pressure increases. The membrane diffusion coefficient K also decreases with increasing pressure. All K values are, in addition, dependent on the solute concentration as mentioned in Section II.B. As the feed concentration increases, K_1 and K are found to decrease whereas K_3 increases (Applegate and Antonson, 1971). The problem oriented equations for pressure gradients, radial velocity profile, and solute concentration distributions in shell and

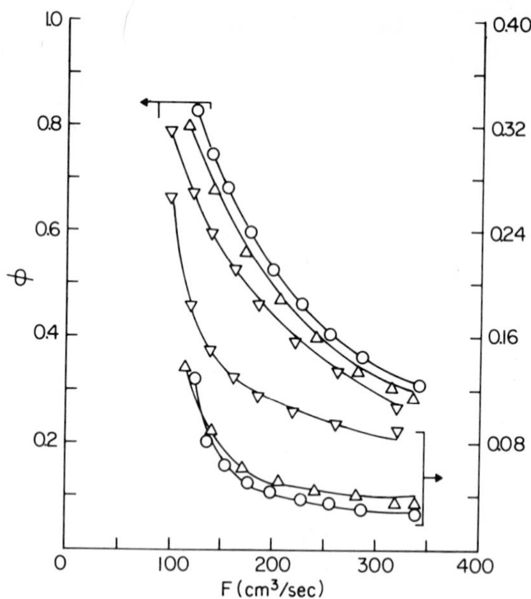

Fig. 12a Variation of dimensionless system productivity with feed rate.

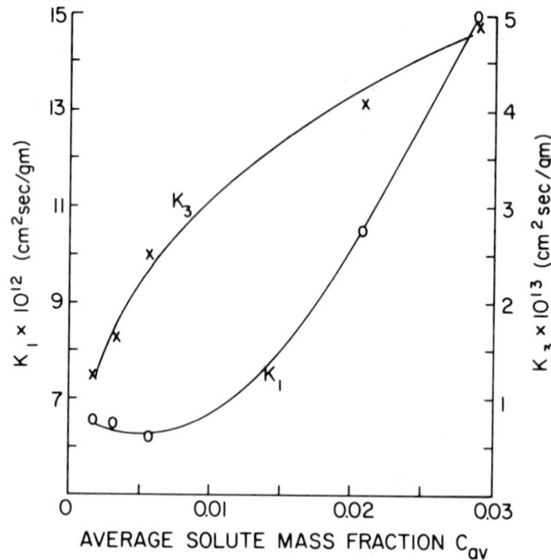

Fig. 12b Variation of K_1 and K_3 with average solute mass fraction.

fiber were solved with a Runge–Kutta routine keeping K_1–K_3 as adjustable parameters to match experimental results (Fig. 12a). Such excellent agreement between theoretical and experimental dependence of system productivity on feed rate and initial solute concentration was not achieved for higher feed concentrations and applied pressures. Still, the accuracy for the parameters ϕ and θ_p using a well-calibrated simulator was reported to be within 8% for ϕ and 17% for θ_p over the entire range of operating conditions (i.e., 100 cm^3/sec $\leq F \leq$ 400 cm^3/sec, 200 psig $\leq \Delta p \leq$ 400 psig and 900 ppm $\leq c_o \leq$ 34,000 ppm).

Provided that the underlying assumptions for momentum and mass transfer are reasonable, this approach provides insight to the pressure and (average) concentration dependence, especially of K_1 and K_3. It was found that K was insignificant compared with K_3. It appears that the salt transport through the membrane was primarily by pore flow. Kabadi *et al.* (1979), in contrast to Applegate and Antonson (1971), observed that the solvent permeability coefficient K_1 first decreases with increasing $c_{av} = (c_F + c_R)/2$ as expected, but increases with further increase in c_{av}, where c_F is the feed and c_R the reject concentration (Fig. 12b). The authors point out that the problem could be solved with an improved solubility-diffusion imperfection model (e.g., Kedem and Katchalsky, 1958; Sherwood *et al.*, 1967), a more accurate simulation of the transport mechanisms, and the implementation of K_1 and K_3 functions that vary *locally* with shell-side concentration c_1. It was demonstrated, however, that the assumptions $K/v_w \ll 1.0$ and $c_3 \ll c_1$ for high-feed concentrations made earlier by Dandavati *et al.* (1975) are invalid.

B. STRAIGHT CONDUIT ULTRAFILTRATION SYSTEMS

Separation of macromolecular solutions by UF was discussed in Section II.B. Tubular membranes, two-sided channel membranes, or plate-and-frame modules are conceptualized as straight conduit with porous walls (Fig. 2). The differences between the systems can be accommodated with simple coordination transformations and modifications of the boundary conditions. The basic assumptions include steady, two-dimensional, laminar, or turbulent flow without external forces and entrance or end effects. The fluid properties are constant and the constituents (dissolved or suspended matter) follow the motion of the carrier fluid. Again, modeling efforts concentrate on solutions for the momentum and mass-transport equation subject to appropriate boundary conditions.

Berman (1953) found the basic solution for Eqs. (10a)–(10c) in terms of a power series in $\lambda = y/h$ (transverse coordinate) by using a perturbation technique:

$$u(x,\lambda) = [\bar{u}(0) - v_w x/h][\tfrac{3}{2}(1 - \lambda^2)][1 - (\mathrm{Re}_w/420)(2 - 7\lambda^2 - 7\lambda^4], \quad (55a)$$

$$v(\lambda) = v_w[(\lambda/2)(3 - \lambda^2) - (\mathrm{Re}_w\lambda/280)(2 - 3\lambda^2 + \lambda^6)], \text{ and} \quad (55b)$$

$$p(0,\lambda) - p(x,\lambda) = -(\mu k/h^2)[\bar{u}(0)x - (v_w x^2/2h). \quad (55c)$$

The perturbation parameter is the wall Reynolds number $\mathrm{Re}_w = v_w h/\nu$. It is necessarily constant and less than one in Berman's derivation.

In modeling applications, terms in Eqs. (55a) and (55b) following the factor Re_w are neglected so that familiar parabolic velocity profiles emerge. Otherwise, the axial velocity profile differs from the Poiseuille parabola by flattening out at the center of the channel and steepening in the region close to the walls, depending upon the degree of uniform suction. Singh and Laurence (1979) employed the truncated Berman solution but varying the wall Reynolds number up to $\mathrm{Re}_w = 2$. Variations of Berman's solution were developed by other authors. Beavers and Joseph (1967) and later Singh and Laurence (1979) demonstrated the pronounced effect of a finite slip-velocity on axial velocity distributions in symmetric channels and tubes. The velocity profiles approach plug flow with increasing slip coefficient θ. Meyer (1980) illustrated the deformation of the laminar velocity profile in a channel because of wall roughness. For a channel with one rough wall the velocity distribution is drastically skewed toward the rough surface. This, together with superimposed suction, might have a measured effect on productivity in plate-and-frame UF modules. Kleinstreuer and Paller (1983) followed Berman's approach to derive the velocity and pressure field for one-sided porous channels (Fig. 13) as found in plate-and-frame units (Madsen, 1977).

For the case of turbulent flow passed membranes, Merkine *et al.* (1971) developed an expression for the velocity fields in terms of inner variables (Section II.C.2) that match data sets collected by Weissberg and Berman (1955) well. Their Reynolds numbers ($\mathrm{Re} = \bar{u}r/\nu$) range from 27,500 to 50,000 with suction ratios $v_w/\bar{u}$ varying from 0 to 0.01. Most modelers employ the momentum equation or postulate directly a Poiseuille solution as an auxiliary equation to the solute transport equation. The paper by Leung and Probstein (1979), based on similar investigations by Sherwood *et al.* (1965), Brian (1966), and Probstein *et al.* (1978) might serve as a representative case study. Leung and Probstein (1979) simulated UF of dilute suspensions in a flow system consisting of two parallel porous plates a distance of $2h$ apart. They selected the Von Kármán–Pohlhausen integral method to solve the solute transport equation, and prepared the modeling equations and boundary conditions

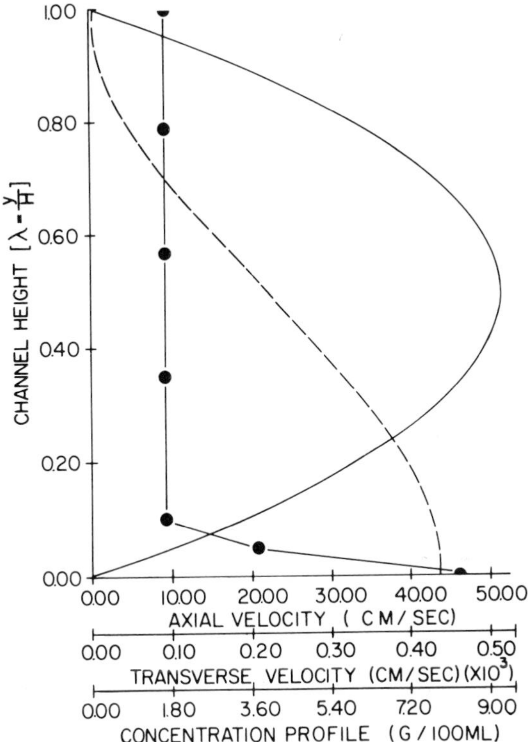

Fig. 13 Velocity distributions and concentration profile versus channel height at mid-channel for plate-and-frame membrane.

accordingly. They used the conversation form of Eq. (11) and the condition of zero mass flux along the concentration boundary-layer edge. The features of the method of integral relationships (MIR) are outlined in Section II.A.2. Hence, the well-posed problem

Convection–diffusion equation $$\frac{\partial(uc)}{\partial x} \div \frac{\partial(vc)}{\partial y} = \frac{\partial}{\partial y} D(c) \frac{\partial c}{\partial y} \tag{56a}$$

bulk feed concentration at channel inlet $c(x = 0) = c_o$ (56b)

complete rejection of macromolecules at membrane surface $D\, \partial c/\partial y|_{y=0} = v_w c(y = 0)$ (56c)

no mass flux at Boundary-layer edge $y = \delta$ $D\, \partial c/\partial y|_{y=\delta} = 0$ (56d)

has the solution with the postulated quadratic profiles for u, v, and c:

$$\gamma\bar{\delta}^2[1 - (\bar{\delta}/5)] = 4S/(1 - S), \tag{57}$$

where $S(x) = -\int_0^x v_w(x)\,dx$ is the fraction of permeate that has been ultrafiltrated from the channel inlet to the point x. Equation (57) contains the additional unknowns $\gamma = (c_w/c_o) - 1$, the degree of polarization; and $\bar{\delta} = \delta(x)/h$, the nondimensional concentration boundary-layer thickness. Because $\bar{\delta} \ll 1$ and $S \ll 1$ the resulting equations for the three unknowns γ, v_w, and $\bar{\delta}$ read

$$\gamma\bar{\delta}^2 \approx 4S, \tag{58a}$$

$$v_w = A[\Delta p - \pi(c_w)], \text{ and} \tag{58b}$$

$$D\,\partial c/\partial y|_{y=0} = v_w c(y = 0) \qquad \text{viz.} \qquad \bar{\delta} = -(2D_w/hv_w)(\gamma/(1 + \gamma). \tag{58c}$$

Equation (58a) is an approximation of Eq. (57), Eq. (58b) is discussed as Eq. (13) in Section II.A.1, and Eq. (58c) deduced from Eq. (56c) is the boundary condition for complete rejection of solute at the membrane surface. After empirical relationships for $\pi(c_w)$ and $D_w(c)$ were specified, the system of equations was solved simultaneously using a Runge–Kutta routine.

The results compare favorably with flux data collected by Huffman (1970) and in-house measurements (Figs. 14a and b). Figure 14a shows the expected dependence of the volumetric permeate flux v_w with the applied pressure Δp. This graph should be viewed with Figs. 15a and b, which are based on pilot plant measurements that are self-explanatory. Figure 14b indicates the expected model prediction of flux decline along the channel because of solute build-up at the membrane surface. The type of postulated velocity profile used to satisfy Eq. (55a) has a noticeable influence on the dependence $v_w(x)$. The uniform velocity profile is somewhat unrealistic because it masks important transport phenomena occurring in the boundary layer at the membrane walls. In imposing the nonessential boundary condition Eq. (55c), a certain mathematical simplicity is gained, but model flexibility is reduced. The series solution of Eq. (55a) by Sherwood *et al.* (1965) and the numerical solution by Brian (1966) do not have this drawback. They considered an RO system as discussed by Dresner and Johnson (1981) rather than a UF system, however.

A comparison between pilot plant studies and model simulations for plate-and-frame-type membranes (Fig. 16) was undertaken by Kleinstreuer and Paller (1983). Figure 17 depicts model predictions for whey separation based on input data given in Table IV and measurements published by Madsen (1976). It should be noted that information on the test conditions and functional forms and values of key parameters were either scarce or not reported. For example, no "initial," or entrance, conditions were described and expressions for the diffusivity and osmotic pressure of whey had to be obtained elsewhere (Sourirajan, 1978). The effective

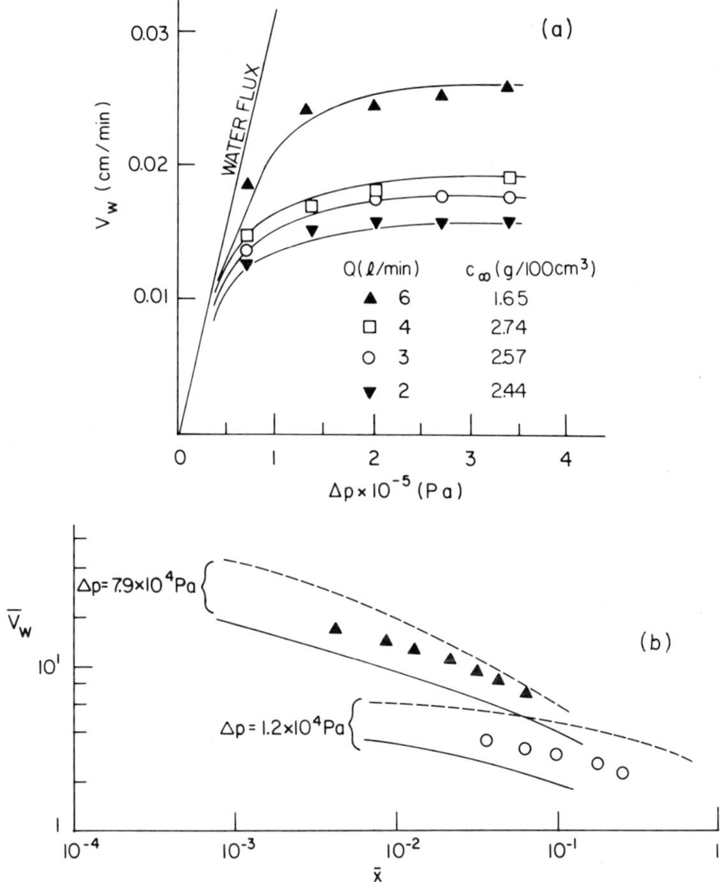

Fig. 14 Comparison of integral solution with flux data for laminar ultrafiltration of pH 5 bovine serum albumin. (See text for discussion.)

local membrane permeability denoted by k^i_{eff} was deduced from the measured pure water flux, which explains the similar trend between pure water flux and computer prediction. The simulation predicts a sharp flux-decline at the beginning of the membrane channel as expected and documented elsewhere. Because of the enhanced membrane permeability, the permeation flux increases, and predicted and measured permeate flux values agree favorably. Based on the sparse data given, we feel the predicted trend in the initial phase is more realistic than the first two data points given by Madsen (1976). For example, according to the experiment

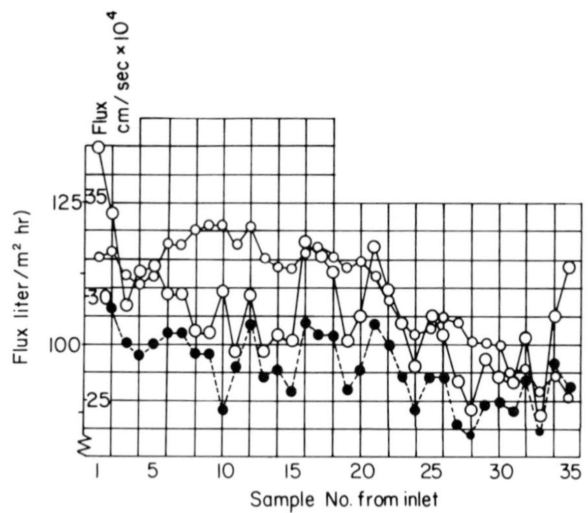

Fig. 15a Distilled water fluxes on 70-cm thin-channel test cell.

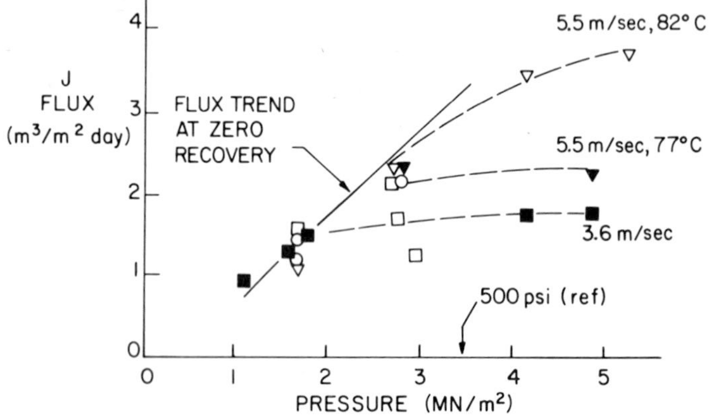

Fig. 15b Field results—hyperfiltration of preparation range water washer—moderate concentration.

the initial membrane flux for the given applied pressure difference and volumetric flow rate is maintained in the first 10 cm of the module before a moderate flux decline occurs. This is rather unrealistic unless abnormal behavior of $k^i_{\mathrm{eff}}(x_i)$ and $\pi^i(c_w)$ account for these (experimental) results.

The contrasting real world system and smooth modeling results given so far set the stage for a discussion of research needs in mathematical modeling of pressure-driven membrane transport processes (Section IV).

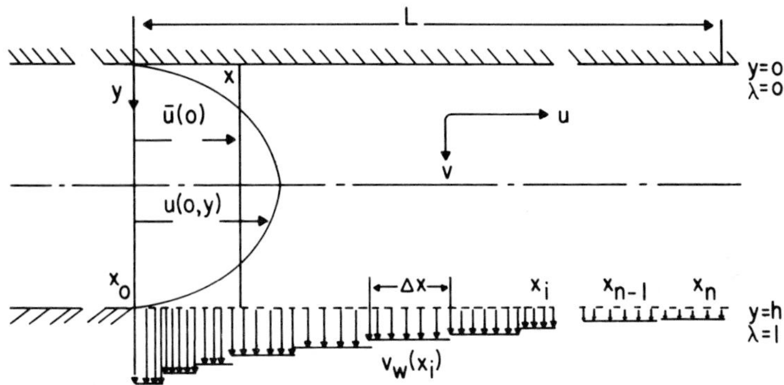

Fig. 16 Coordinate system and segmentation of plate-and-frame membrane unit into *n*-permeable-walled subchannels.

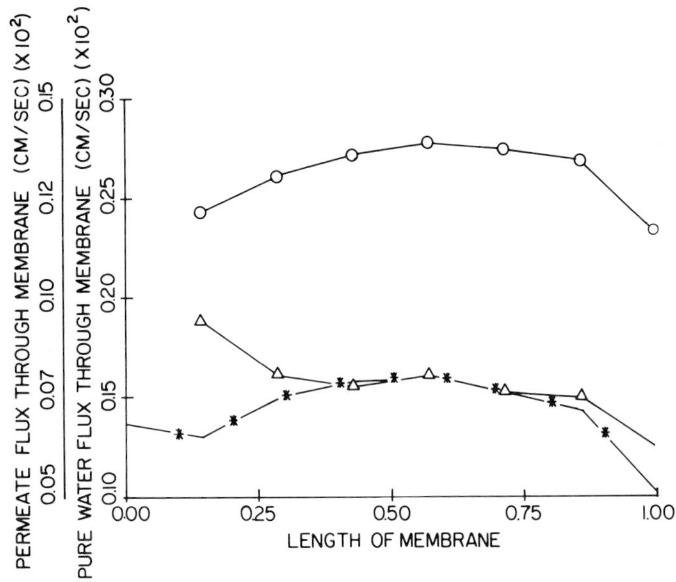

Fig. 17 Comparison of observed and predicted data for flux decline.

C. Reduction of Polarization Layers

From the theoretical analysis presented previously, it is clear that to reduce the degradative effect of polarized layers adjacent to the membrane surface either the convection causing the solute build-up should be reduced or the back-transport of the solute to the bulk solution should be enhanced.

TABLE IV

Input Data for Fig. 17 (Solution: Whey with Total Solid Concentration of 6.2% at pH = 5.0)[a]

Channel length	L	$= 70.0$ cm
Channel height	h	$= 0.10$ cm
Initial axial mean velocity	$\bar{u}_o$	$= 100.0$ cm/sec
Initial transverse velocity	v_{wo}	$= 9.44 \times 10^{-4}$ cm/sec
Applied pressure difference	Δp	$= 4.0 \times 10^5$ Pa
Initial diffusivity	D_o	$= 2.628 \times 10^{-6}$ cm^2/sec
Initial concentration	C_o	$= 2.2$ g/100 cm^3
Initial membrane permeability	k_o	$= 2.36 \times 10^{-9}$ cm/Pa sec
Kinematic viscosity	ν	$= 0.0198$ stokes

[a] From Kleinstreuer and Paller (1983).

One method to reduce convection is to choose a relatively low-flux membrane with a high packing density to maintain a high product volume rate. Hollow-fine fibers are an example of this type of membrane system.

One of the most common methods used to reduce the mass-boundary layer thickness and enhance the back diffusion of solute is to use tangential feed flow across the membrane surface. In most cases the turbulent flow regime is used to accomplish this while in some cases the high shear stress laminar flow regime is used. Theoretical and experimental analyses indicate that inertial effects in low Reynolds number viscous flow can induce lateral migration of solid particles away from the membrane surface, resulting in enhanced back transport (Green and Belfort, 1980).

Different types of mixing promoters have also been inserted in the feed flow channel to enhance back transport. These include static mixers, spiral wires, and fluidized spheres (Blatt *et al.* 1970; Csurny *et al.*, 1973; Hamer, 1969; Hamer and Kalish, 1969; Hiddink and de Boer, 1979; Hiddink *et al.*, 1980; Lai, 1971; Lolachi, 1973; Pitera and Middleman, 1973; Thomas *et al.*, 1971; Van der Waal *et al.*, 1977). The objective in using these mixing promoters is to obtain higher membrane fluxes without the promoters at the same or lower pumping energy rate.

Other methods reported to be successful in increasing the mass-transfer coefficient include feed-flow pulsing (Kennedy *et al.*, 1973, 1974), and periodic reversal of feed-flow direction (Goel and McCutchan, 1976; Thayer *et al.*, 1975). Experimental results with blood flow down a porous regenerated cellulose tube has shown that periodic feed pulsing keeps the red blood cells away from the membrane wall as with the tubular pinch effect (Bauser *et al.*, 1981). Using a trajectory analysis on a graphics computer, Belfort and co-workers have shown that under certain circumstances particle build-up occurs at the feed entrance of the membrane channel and moves along the flow path. Clearly, periodic reversal of the feed-flow direction would help reduce this degradative build-up.

Rather than move the fluid extensively, several workers designed modules to move the membrane system by rotation. Flat rotating disclike membrane modules have been developed (Croopnick, 1975, private communication), and tubular rotating membranes with an outside annulus (Taylor vorticies) flow have also been studied (Lopez-Leiva, 1979). None of these rotating modules have been commercialized. One problem is their high operating costs.

To take advantage of the entrance region in which the mass and viscous boundary layers are both growing, some workers have suggested that a module with very short membrane path lengths be used. One way to construct such a system is to alternate membrane sections with nonporous sections along the feed-flow path (Shaw *et al.*, 1972). Clearly, capital costs in constructing such a module could be a limiting factor.

Radovich showed that by electrically depolarizing a solute build up, improved fluxes can be obtained (Radovich and Sparks, 1980). Both operating and capital costs to incorporate this method into a viable module may pose problems for its commercialization.

Besides the studies by Cleaver and Yates (1973, 1975, 1976) in which reentrainment of deposited particles from nonporous ducts is modeled, we are unaware of any attempts to model particle reentrainment from porous ducts.

IV. Conclusions and Future Work

Reverse osmosis membrane units can be economically applied for desalination of water and the removal of microbial and dissolved impurities. Ultrafiltration modules are successfully used in the separation of macromolecules and colloidal particles from wastewater streams. Three major critical points must be addressed, however, before using membrane technology rather than conventional separation techniques: (1) Capital investment and energy requirements, (2) expenditures for maintenance and optimal process control, and (3) the lifetime of a particular membrane under mechanical, chemical, and thermal stress. Models reviewed in this chapter address some of the technical aspects of these problem areas. A topic that deserves additional attention is membrane fouling, i.e., causes of deposition layers and means of reduction or removal. Sections II and III indicate that for present mathematical models the nonlinear momentum-transfer equations are still solved in their reduced form with approximate solution techniques, whereas finite difference techniques are already employed for solving the complete solute diffusion–convection equation. This mixed approach involves mathematical assumptions (e.g., constant transport parameters, very low wall Reynolds numbers, decoupling of momentum from mass-transfer equations, etc.) that often misrepresent the physical phenomena. Kleinstreuer *et al.* (1983c) are presently modifying a finite difference scheme for momentum, mass and heat transfer in porous wall conduits that was applied to external boundary-layer flow problems.

One of the more successful models is the theory for thin channel design as proposed by Brian (1965) and Kimura and Sourirajan (1968). Strathmann and Keilin (1969) confirmed Brian's result [Eq. (17b)] experimentally at low flow rates. Although several predictive computer simulation models for (laminar) boundary-layer type suspension flows in plate or tubular membrane units exist (e.g., Gill *et al.,* 1965; Green and Belfort, 1980; Gutman, 1977; Kleinstreuer and Paller, 1983; Leung and Probstein, 1979; Singh and Laurence, 1979), their application is restricted to certain solutes, flow patterns, and membrane characteristics. Important membrane coefficients and process parameters need to be kept constant for the sake of mathematical simplicity. Hence, one basic research task is to develop a flexible, comprehensive computer model that is capable of simulating the dynamic interactions of various solutes (dissolved or colloidal matter) with the carrier fluids (water) and specific membranes (sieve or diffusive type). As a first step the following shortcomings of existing (sub)models should be eliminated:

(1) Some models are based on highly empirical formulas with constant coefficients and parameters (lumped parameter approach).

(2) Most models are constructed only for one particular membrane type and problem area.

(3) Of all the well-known factors (forces and fluxes) important for fouling layer formation only one or two are explicitly considered in existing models.

(4) For all practical purposes RO and UF are time-dependent processes that are usually only considered for models of unstirred batch-cell membrane units. Another term that is often neglected is the pressure gradient, $p = p(x,y;t)$, and cannot be replaced by a measured (constant) pressure drop.

Other fundamental research areas that should be investigated more thoroughly are the various submodels for turbulent suspension flow, concentration polarization, fouling layer formation, resuspension, and transmembrane transport (see additional discussions by Dresner and Johnson, 1981; Eriksson, 1980; Madsen, 1976; Matthiasson and Sivik, 1980). It is again the (extended) flexibility and generality of the proposed mathematical model and computer code that will enable researchers to test and improve existing submodels and to postulate new hypotheses for unresolved discrepancies between theory and observation. For example, even though there is a good qualitative agreement of the gel-polarization model with experimental data, there are some unexplained experimental observations that are not predicted by the theory. Most common observations are (Matthiasson and Sivik, 1980)

(1) Slow decline of permeate flux with time,

(2) reduction in feed solution-concentration is not followed by increase in permeate flux,

(3) permeability loss resulting from macrosolute polarization or fouling is not restored in spite of chemical cleaning,

(4) changes in solute-rejection behavior of UF membranes exposed to macrosolute solutions.

Of course, a middle ground must be found between the development of an accurate, comprehensive simulation and its usefulness in terms of easy handling and low operating costs.

Acknowledgment

The authors would like to thank Ms. Lorraine McGrath for her excellent job in word processing, typing the equations, and finalizing the manuscript.

References

Agrawal, J. P., Antonson, C. R., and Rosenblatt, N. W. (1972). *Desalination* **11,** 71–90.

Ames, W. F. (1978). "Numerical Methods for Partial Differential Equations." Academic Press, New York.

Applegate, L. E., and Antonson, C. R. (1971). *Polym. Prepr.* **385.**

Argo, D. G., and Montes, J. G. (1979). *J. Water Pollut. Control Fed.* **51**(3), 590–600.

Bansal, B., and Gill, W. N. (1973). *AIChE J.* **19**(4).

Bashow, J. D., Lawson, J. K., and Orofino, T. A. EPA Report No. EPA-R2-72-103, Dec., 1972.

Basset, A. B. (1961). "A Treatise on Hydrodynamics." Dover, New York. (Reprint)

Bauser, H., Chmiel, H., and Waliztsa (1981). *Proc. 3rd Symp. Synth. Membranes Sci. Ind., Tubingen,* September 7–9.

Beavers, G. S., and Joseph, D. D. (1967) *J. Fluid Mech.* **30,** 197–207.

Beckman, J. E., Bevege, E. E., Cruver, J. E., Kremen, S. S., and Nusbaum, I. OSW Res. Develop. Progress Rept 882, Sept. 1973.

Belfort, G. (1977). *Desalination* **21,** 285–300.

Belfort, G., and Guter, G. A. (1972). *Desalination* **10,** 221–262.

Belluci, F., Carrieri, C., and Drioli, E. (1979). *Gazz. Chim. Ital.* **109,** 499.

Berman, A. S. (1953). *J. Appl. Phys.* **24,** 1232–1235.

Bevege, E. E., Cruver, J. E., Kilbridge, J. G., Kremen, S. S., and Riedinger, A. B. OSW Res. Develop. Progress Rept 883, July, 1973.

Bhattacharyya, D., and Grieves, R. B. NTIS Publ. PB 297 209, May, 1979.

Bird, R. B., Stewart, W. E., and Lightfoot, E. N. (1960). "Transport Phenomena." Wiley, New York.

Bird, R. B., Hassager, O., Armstrong, R., and Curtiss, C. (1977). "Dynamics of Polymeric Liquids," Vol. 2. Wiley, New York.

Blatt, W. F., Dravid, A., Michaels, A. S., and Nelson, L. (1970). "Membrane Science and Technology," pp. 47–97. Plenum Press, New York.

Bradshaw, P. (1977). *Aeronaut. J.* **76,** 403.

Braun, W. H. (1977). "Representation of Turbulent Shear Stress by a Product of Mean Velocity Differences" (NASA-TP-1029).

Brenner, H. (1966). "Advanced Chemical Engineering Series," Vol. 4, p. 377. Academic Press, New York.

Brenner, H., (1970). *Prog. Heat Mass Transfer* **6,** 509.

Bresler, E. H. *et al.* (1976). *Biophys. Chem.* **4,** 229.

Brian, P. L. T. (1965). *Ind. Eng. Chem. Fundam.* **4,** 439–445.

Brian, P. L. T. (1966). "Desalination by Reverse Osmosis," (U. Merten, ed.). M.I.T. Press, Massachusetts.

Brosh, A., and Winograd, Y. (1974). *Trans. ASME J. Heat Transfer* **74,** 338–342.

Carman, P. C. (1937). *Trans. Inst. Chem. Eng. London* **15,** 150–166.

Carter, J. W., and Hoyland, G. (1976). *5th Int. Symp. Fresh Water Sea* **4,** 21–29.

Carter, J. W., Hoyland, G., and Hasting, A. P. M. (1974). *Chem. Eng. Sci.* **29,** 1651–1658.

Caswell, B. (1977). *ASME AMD* **22,** 19.

Cebeci, T., and Bradshaw, P. (1977). "Momentum Transfer in Boundary Layers." McGraw-Hill, New York.

Cleaver, J. W., and Yates, J. (1973). *Coll. Interface Sci.* **44**(3), 464–474.

Cleaver, J. W., and Yates, J. (1975). *Chem. Eng. Sci.* **30,** 983–992.

Cleaver, J. W., and Yates, J. (1976). *Chem. Eng. Sci.* **31,** 147–151.

Clift, R. *et al.* (1978). "Bubbles, Drops and Particles." Academic Press, New York.

Crowe, C. T., Sharma, M. P., and Stock, D. E. (1977). *J. Fluids Eng.*

Cruver, J. E., and Nusbaum, I. (1974). *J. WPCF* **46,** 301.

Csurny, J., Johnson, J. S., Jr., Kraus, K. A., Philips, H. O., Sisson, W. G., and Westmoreland, C. G. (1973). Biennial Progress Report for the period March, 1968–March, 1970 (G. E. Moore and J. S. Johnson, Jr., eds.), p. 270. Oak Ridge Nat'l. Lab., Oak Ridge, Tennessee.

Danckwerts, P. V. (1951). *Ind. Eng. Chem. Fundam.* **43,** 1460.
Dandavati, M. S., Doshi, M. R., and Gill, W. N. (1975). *Chem. Eng. Sci.* **30,** 877–886.
Derjaguin, B. V., Churaev, N. V., and Martynov, G. A. (1980). *J. Colloid Interface Sci.* **75**(2), 419–433.
Doshi, M. R., Dewan, A. K., and Gill, W. N. (1971). *AIChE Symp. Ser.* **68,** 323–339.
Dresner, L. (1964). Rept. No. 3621. Oak Ridge, Tennessee, Oak Ridge Nat'l. Lab.
Dresner, L., and Johnson, J. S. (1981). "Principles of Desalination" (K. S. Spiegler and A. D. K. Laird, eds.), 2nd edition. Academic Press, New York.
Epifanov, V. M., and Gus'kov, V. I. (1979). Translated by Plenum Publ. Co. from Inzhenero-Fizicheskii Zhurnal, Vol. 37, 412–418.
Eriksson, P. (1980). Reverse osmosis, ultrafiltration and mass transfer in turbulent duct flow. Ph.D. thesis, Lund University, Sweden.
Esmail, R. H. (1980). *Am. Meteorol. Sc.* **4,** 905.
Feuerstein, D. L. Report to EPA on Project 17040 EFQ, Contract 14-12-885, Feb. 1971.
Feuerstein, D. L., and Burzstynski, ■■, (1971). Amer. Inst. Chem. Eng. Sympos. Series **67, 107,** 568.
Fisher, R. E., Sherwood, T. K., and Brian, P. L. T. (1964). M.I.T. Desalination Res. Lab. Rept. No. 295-5.
Frost, W., and Moulden, T. H. (1977). "Hand Book of Turbulence," Vol. 1. Plenum Press, New York.
Gaddis, J. L., Brandon, C. A., and Porter, J. J. (1979). EPA Report No. EPA-600/7-79-131.
Gill, W. N., Tien, C., and Zeh, D. W. (1965). *Ind. Eng. Chem. Fundam.* **4,** 433–439.
Gill, W. N., Derzansky, L. J., and Doshi, M. R. (1971). "Surface and Colloid Science," Vol. 4, p. 261. Wiley, New York.
Glover, F. A., and Brooker, B. E. (1974). *J. Dairy Res.* **41,** 80–93.
Goel, V., and McCutchan, J. W. (1976). *5th Int'l. Symp. Fresh Water Sea* **4,** 315–324.
Goldsmith, H., and Mason, S. (1967). *Rheology* **49,** 257.
Goldsmith, R. L. (1971). *Ind. Eng. Chem. Fundam.* **10,** 113–120.
Green, G., and Belfort, G. (1980). *Desalination* **35,** 129–147.
Grover, J. R., and Delve, M. H. (1972). *Chem. Eng.* **257,** 24–29.
Gutman, R. G. (1977). *Chem. Eng. U.K.,* **322,** 510–513, 521–523.
Halow, J. S., and Wills, G. B. (1970). *AIChE J.* **16,** 281.
Hamer, E. A. G. (1969). U.S. Patent No. 3,425,562-4.
Hamer, E. A. G., and Kalish, R. L. (1969). Paper presented 2nd O.S.W. Symp. on Reverse Osmosis, Miami, Florida.
Happel, J., and Brenner, H. (1965). "Low Reynolds Number Hydrodynamics." Prentice-Hall, New Jersey.
Hayes, J. F., Dunkerley, J. A., Muller, L. L., and Griffin, A. T. (1974). *Aust. J. Dairy Technol.* **29,** 132–140.
Hiddink, J., and de Boer, R. (1979). Paper presented 2nd Int'l. Congress on Engineering and Food, Helsinki, Aug. 27–31.
Hiddink, J., de Boer, R., and Nooy, P. F. C. (1980). *J. Dairy Sci.* **63,** 204–214.
Higbie, R. (1935). *Trans. AIChE* **31,** 365.
Ho, B. P., and Leal, L. G. (1974). *J. Fluid Mech.* **63,** 365.
Hung, C. C., and Tien, C. (1976). *Desalination* **18,** 173.
Jackson, J. M., and Landolt, D. (1973). *Desalination* **12,** 361–378.
Johnson, R. A. (1974). *AIChE J.* **20,** 966.
Johnson, J. S., and McCutchan, J. W. (1972). *Desalination* **10,** 147–156.
Jonsson, G., and Kristensen, S. (1980). *Desalination* **32,** 327–339.
Kabadi, V. N. *et al.* (1979). *Chem. Eng. Commun.* **3,** 339–365.

Kedem, O., and Katchalsky, A. (1958). *Biochim. Biophys. Acta* **27,** 229.
Kennedy, T. J., Monge, L. E., McCoy, B. J., and Merson, R. L. (1973). *Chem. Eng. Prog. Symp. Ser.* **69,** 81.
Kennedy, T. J., Merson, R. L., and McCoy, B. J. (1974). *Chem. Eng. Sci.* **29,** 1927.
Kesting, R. E. (1971). "Synthetic Polymeric Membranes," pp. 12–52, 227–270. McGraw-Hill, New York.
Kimura, S., and Nakao, S. I. (1975). *Desalination* **17,** 267.
Kimura, S., and Sourirajan, S. (1967). *AIChE J.* **13,** 497.
Kimura, S., and Sourirajan, S. (1968). *Ind. Eng. Chem. Process Des. Dev.* **7,** 197.
Kinney, R. B., and Sparrow, E. M. (1970). *J. Heat Transfer* **92,** 117–125.
Kleinstreuer, D. (1983). "Mathematical Modeling of Engineering Systems." Wiley (Interscience), New York. (In preparation.)
Kleinstreuer, C., and Paller, M. (1983). *AIChE J.* Vol. 29, No. 4, 529–533.
Kleinstreuer, C. *et al.* (1978). "Math Modeling and Computer Simulation of Biota Entrainment near Power Plant Intake Structures (ORNL-TM). Oak Ridge Natl. Lab., Oak Ridge, Tennessee.
Kleinstreuer, C. *et al.* (1983a). *Comp. Biomed. Res.* **16,** 29–39.
Kleinstreuer, C. *et al.* (1983b). *Chem. Eng. Comm.* (submitted).
Kleinstreuer, C. *et al.* (1983c). "Numerical Solution of Turbulent Flow FIelds with Mass Transfer in Porous Ducts (RPI Rept.). Department Chemical and Environment Engineering, RPI, Troy, New York. (In preparation.)
Kozeny, G. (1927). *Sitzber. Akad. Wiss. Wien Math. Naturw. Kl. Abt IIa,* p. 136.
Kozinski, A. A., Schmidt, F. P., and Lightfoot, E. N. (1970). *Ind. Eng. Chem. Fundam.* **9**(3), 502–505.
Kuiper, D., Bom, C. A., van Hezel, J. L., and Verdouw, J. Proceedings 4th Int. Symposium Fresh water from the sea, Heidelberg, Aug. 1973, **4,** 207–215.
Lai, J. (1971). Ph.D. Thesis, Montana State University, Bozeman, Montana.
Launder, B. E., and Spalding, D. B. (1972). "Lecture in Mathematical Model in Turbulence." Academic Press, New York.
Leal, L. G. (1980). *Ann. Rev. Fluid Mech.* **12,** 435–476.
Leiserson, L. Proceedings of workshop symposium Membranes in Separation Process, Case Western Reserve University, May, 1973, 143–146.
Leung, W. F., and Probstein, R. F. (1979). *Ind. Eng. Chem. Fundam.* **18,** 274–278.
Lim, T. H., Dunkley, W. L., and Merson, R. L. (1971). *J. Dairy Sci.* **54,** 306–311.
Lolachi, H. (1973). O.S.W. Res. Dev. Rept. 843. U.S. Dept. of the Interior.
Lopez-Leiva, M. (1979). "Ultrafiltration in Rotary Annular Flow." Ph.D. Thesis, Lund University, Ystad, Sweden.
Madsen, R. F. (1976). Ph.D. Thesis, Technical Univ. of Denmark, Copenhagen.
Matthiasson, E., and Sivik, B. (1980). *Desalination,* **35,** 59–103.
McCutchan, J. W., and Johnson, J. S. (1970). *J.A.W.W.A.* **62,** 346.
Merkine, L., Solan, A., and Winograd, Y. (1971). *Trans. ASME J. Heat Transfer* **12,** pp. 242–244.
Merten, U. (1963). *Ind. Eng. Chem. Fundam.* **2,** 229–232.
Merten, U. (ed.) (1966). "Desalination by Reverse Osmosis." M.I.T. Press, Massachusetts.
Meyer, L. (1980). *Int. J. Heat Mass Transfer* **23,** 591.
Michaels, A. S. (1968). *Chem. Eng. Prog.* **64,** 31–43.
Migdal, D., and Agosta, V. D. (1967). *Trans. ASME J. Appl. Mech.* **36,** 4.
Minturn, R. E. OSW Res. Develop. Progress Rept 897, Oct. 1973.
Monin, A. S., and Yaglom, A. M. (1971). "Statistical Fluid Dynamics." MIT Press, Massachusetts.

Nernst, W. (1904). *Z. Phys. Chem.* **47,** 52.

Orofino, T. A. (1977). "Reverse Osmosis and Synthetic Membranes (S. Sourirajan, ed.), pp. 313–341. NRC, Ottawa, Canada.

Peri, C., and Dunkley, W. L. (1971). Part 1 and Part 2, *J. Food Sci.* **36,** 25–30, and *J. Food Sci.* **36,** 395–396.

Peri, C., and Pompei, C., Int. Symposium on heat and mass transfer problems in food engineering, Wageningen, Oct., 1972.

Pitera, E. W., and Middleman, S. (1973). *Ind. Eng. Chem. Proc. Des. Dev.* **12,** 52.

Porter, M. C. (1972). *Ind. Eng. Chem. Prod. Res. Dev.* **11,** 234–248.

Probstein, R. F. (1972). *Trans. ASME J. Basic Eng.* June, 286–313.

Probstein, R. F., Shen, J. S., and Leung, W. F. (1978). *Desalination* **24,** 1–16.

Pruppacher, H. R. (1980). *Am. Meteorol. Sci.,* April, 903.

Radovich, J. M., and Sparks, R. E. (1980). "Ultrafiltration Membranes and Applications" (A. R. Cooper, ed.), pp. 249–268. Plenum Press, New York.

Rajagopalan, R., and Tien, C. (1979). "Progress in Filtration and Separation" (J. Wakeman, ed.). Elsevier, New York.

Reed, R. H. (1979). M.S. Thesis. Department of Chemical and Environmental Engineering, RPI, Troy, New York.

Rekin, A. D. (1976). Translated by Plenum Publ. Co. from *Inzhenerno-Fizicheskii Zhurnal,* Vol. 30, No. 6, 1009–1016.

Rich, L. G. (1974). "Environmental Systems Engineering." McGraw-Hill, New York.

Roache, P. J. (1976). "Computational Fluid Dynamics." Hermosa, New York.

Sammon, D. C., and Stringer, B. (1975). *Process Biochem.* March 4–12.

Schippers, J. C., and Verdouw, J. (1980). *Desalination* **32,** 137–148.

Schippers, J. C., Bonn, C. A., and Verdouw, J. (1978). *Int. Symp. Fresh Water Sea,* **3,** 363.

Schlichting, H. (1979). "Boundary Layer Theory," 7th edition. McGraw-Hill, New York.

Segre, G., and Silberberg, A. (1962). *J. Fluid Mech.* **14,** 115.

Seinfeld, J. H. (1975). "Air Pollution." McGraw-Hill, New York.

Shaw, R. A., Delucia, R., and Gill, W. N. (1972). *Desalination* **11,** 189–205.

Shen, J. S., and Probstein, R. F. (1977). *Ind. Eng. Chem. Fundam.* **16,** 459–465.

Shephard, J. D., and Thomas, D. G. (1971). *AIChE J.* **17,** 910.

Sheppard, J. D. *et al.* (1972). *Desalination* **11,** 385.

Sherwood, T. K., Brian, P. L. T., and Fisher, R. E. (1963). M.I.T. Desalination Res. Lab. Rept. 295–291.

Sherwood, T. K., Brian, P. L. T., Fisher, R. E., and Dresner, L. (1965). *Ind. Eng. Chem. Fundam.* **4,** 113–118.

Sherwood, T. K., Brian, P. L. T., and Fisher, R. E. (1967). *Ind. Eng. Chem. Fundam.* **1,** 2.

Sherwood, T. K. *et al.* (1975). "Mass Transfer." McGraw-Hill, New York.

Singh, R., and Laurence, R. L. (1979). *Int. J. Heat Mass Transfer* **22,** 721–729.

Sourirajan, S. (1970). "Reverse Osmosis." Logos, London.

Sourirajan, S. (1978). *Pure Appl. Chem.* **50,** 593–615.

Spalding, D. B. (1977). "GENMIX—A General Computer Program for 2-D Parabolic Phenomena." *In* Heat and Mass Transfer Series. Pergamon, Oxford.

Sparrow, E. M. *et al.* (1972). *Trans. ASME J. Basic Eng.*, June, 314–320.

Spiegler, K. S., and Laird, A. D. K. (eds.) (1980). "Principle of Desalination." Academic Press, New York.

Srinivasan, S., Tien, C., and Gill, W. N. (1967). *Chem. Eng. Sci.* **22,** 417.

Strathmann, H., and Keilin, B. (1969). *Desalination* **9,** 179.

Thayer, W. L., Pageau, L., and Sourirajan, S. (1975). *Can. J. Chem. Eng.* **53,** 422–426.

Thomas, D. G., Griffith, W. L., and Keller, R. M. (1971). *Desalination* **9,** 23.

Thomas, D. G., Gallagher, R. B., and Johnson, J. S., Jr. (1973). EPA Report EPA-R2-73-228. Office of Research and Monitoring, U.S. EPA, Washington, D.C.

Trettin, D. R., and Doshi, M. R. (1980). *Ind. Eng. Chem. Fundam.* **19**(2), 189.

Tweddle, T. A. *et al.* (1980). *Desalination* **32,** 181–198.

Van der Waal, M. J., Van der Velden, P. M., Koning, J., Smolders, C. A., and van Swaay, W. P. M. (1977). *Desalination* **22,** 465–483.

Vasseur, P., and Cox, R. G. (1976). *J. Fluid Mech.* **78,** 385.

Vilker, L. V., Colten, C. K., and Smith, K. A. (1981). *AIChE J.*

Watanabe, A., Yoshio, O., Kimura, S., Keiji, U., and Kimura, S. (1979). *Nippon Shokuhin Kogyo Gakkaishi* **26,** 260–265.

Weissberg, H. L. (1959). *Phys. Fluids* **2,** 510.

Weissberg, H. L., and Berman, A. S. (1955). *Proc. Heat Transfer Fluid Mech. Inst.* **14,** 1–30.

White, F. M. (1974). "Viscous Fluid Flow." McGraw-Hill, New York.

Whitman, W. G. (1923). *Chem. Metall. Eng.* **29,** 146–148.

Williams, F. A. (1969). *SIAM J. Appl. Math.* **17,** 59–73.

Winfield, B. A. (1979). *Water Res.* **13,** 561–564.

Wojcik, C. K., Lopez, J. G., and McCutchan, J. W. (1980). *Desalination* **32,** 353–364.

6

Electrodialysis—Membranes and Mass Transport

E. KORNGOLD

Division of Membranes and Ion Exchanges, Applied Research Institute, Research and Development Authority, Ben-Gurion University of the Negev, Beer-Sheva, Israel

ISBN 0-12-085480-5

List of Symbols

A Surface area of membrane (cm^2)
a Activity in solution
b Empirical number
C_s Solution concentration
$\bar{C}_r$ Concentration of the dissociated polymer
D Diffusion coefficient (cm^2/sec)
e Membrane thickness (cm)
E Energy (W hr/m^3 or kW hr/m^3)
F Faraday constant, 96,500 (C/g eq)
i Electrical current density (mA/cm^2)
I Current intensity (A)
j Activity coefficient
J_e Ion flux by electrotransport (g eq/sec cm^2)
J_D Ion flux by diffusion (g eq/sec cm^2)
K Empirical number
$\bar{M}_r$ Capacity of ion-exchange resin
ΔN Number of gram moles
P Pressure drop meter head of water
Q Flow rate (m^3/hr)
r Specific resistance (Ω cm^2)
R_{cp} Electrical resistance of a cell pair (Ω)
R Resistance (Ω)
R' Gas constant 8.3147 joules K mole
T Absolute temperature (K)
t_s Transport number of ion in solution
$\bar{t}_m$ Transport number of ion in membrane
U Linear velocity (cm/sec)
u_c Electrochemical mobility cm^2/volt sec
V Voltage (volt)
v Cell-pair voltage (volt)
Z Valence
δ Thickness of boundary layer (cm)
ρ_r Specific conductivity of resin (mho/cm)
ρ_s Specific conductivity of solution (mho/cm)
μ Chemical potential
η Efficiency

I. Principle of Electrodialysis

Electrodialysis (ED) is based on the electromigration of ions through cation- or anion-exchange permselective membranes that permit the passage of positive or negative ions, respectively.

Ion selectivity is the result of the high electrical mobility of the counterions (ions attached to the ion-exchange polymer) in the membrane. The counterions are easily replaced by other ions of the same charge that migrate to their respective electrode when an electrical potential is applied.

In an ED stack, cation- and anion-exchange membranes are alternated between two electrodes, thus forming a repeating cell-pair pattern (positive and negative). One cell in each pair will contain a concentrated solution (brine, *B*), and the other will contain a dilute solution (diluate, *D*), as shown in Fig. 1.

In industrial units, several hundred cell pairs can be assembled between two electrodes. The diluate and the brine streams are removed separately from the apparatus after ED. This system can be used for desalination, electrolyte concentration, separation of nonelectrolytes from electrolytes, and separation of electrolytes exhibiting different electromigration velocities in solution or in a membrane.

II. Mass Transfer through Permselective Membrane

Mass transfer through permselective membranes consists of two steps (as shown in Fig. 2):

(1) the reduction of salt concentration in the solution by electrotransport of ions from the boundary layer near the membrane, and
(2) the diffusion of ions to the partially desalinated boundary layer.

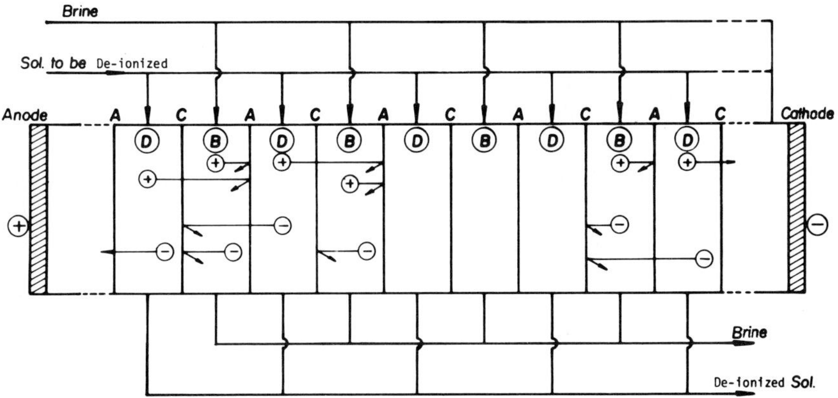

Fig. 1 Electrodialysis Unit

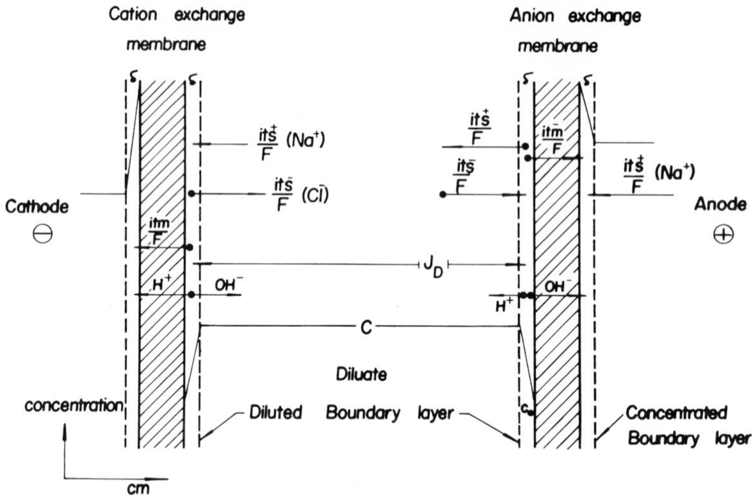

Fig. 2 Mass Transfer in Electrodialysis

The kinetics of the first step is given by the Nernst equation

$$J_e = (\bar{t}_m - t_s)i/F, \tag{1}$$

where (see the list of symbols p. 192 for a full definition of the variables) J_e is the flux of ions by electrotransport, i the current density, F the Faraday number, t_s the transport number in solution, and $\bar{t}_m$ the transport number in membrane.

The second step is given by Fick's First law:

$$J_D = D(C - C_0)/\delta, \tag{2}$$

where J_D is the flux of ions by diffusion, D the diffusion coefficient, C the concentration of the solution, C_0 the concentration of the solution at the boundary layer, and δ the thickness of the boundary layer.

The thickness of the boundary layer δ is a function of the linear velocity of the solution in the cell and the geometry of the spacer (Belfort and Gutter, 1972; Solan *et al.*, 1971; Winograde *et al.*, 1972).

Under steady-state conditions,

$$J_e = J_D. \tag{3}$$

From Eqs. (1)–(3), the following can be derived:

$$i = DF(C - C_0)/\delta(\bar{t}_m - t_s). \tag{4}$$

Increasing the voltage of the stack raises the current density. The flux of ions by electrotransport is also increased until the concentration of the solution in the boundary layer approaches zero ($C_0 \simeq 0$). Under these conditions the flux of ions by diffusion is maximal:

$$J_{D(max)} = DC/\delta, \tag{5}$$

and

$$i_{(max)} = DFC/\delta(\bar{t}_m - t_s). \tag{6}$$

A further increase in J_D can be achieved only by decreasing δ. This can be achieved by raising the linear velocity of the solution in the cell to a level at which the pressure drop across the cells will not cause internal leakage.

When

$$J_e = J_{D(max)}, \tag{7}$$

the ED unit is operating at the highest value of mass transfer. A further increase in the stack voltage will raise the current density. Most of this additional current, however, will cause dissociation of water rather than mass transfer from the diluate to the brine cells.

When the concentration of the solution in the boundary layers decreases, the electrical resistance of the cell pair increases. When the concentration in the boundary layers is low, the water dissociates, causing scaling and fouling on the anion-exchange membranes (Grossman and Sonin, 1972; Korngold *et al.*, 1970). Therefore, it is important that the current density be prevented from approaching the limiting current-density value. This value can be obtained by plotting cell–pair resistance versus current density. The polarization concentration phenomenon has been investigated by Cowan and Brown (1959), Cooke (1965), Spiegler (1971), Belfort and Gutter (1968), and Forgacs *et al.* (1972), among others.

The introduction of screens and mixing promoters in the diluate cell improves turbulence and mass transfer to the boundary layers and, consequently, ED performance (Belfort and Gutter, 1972; Sonin, 1968; Sonin *et al.*, 1971; Winograde *et al.*, 1972). Pressure drop of the mixing promoters must be kept to a minimum with the objective of avoiding mechanical stress on the membrane and keeping the pumping energy requirement as low as possible. It thus becomes evident that the role of hydrodynamics in ED mass transport is associated with problems of water dissociation, electrical resistance, scaling, fouling, mechanical stress on the membranes, and the pumping energy requirement.

A practical equation for limiting current density, derived from empirical results and suggested by several authors (Davis and Lacey, 1970; Mason-Rust, 1970; Maurel, 1972; Winograde, 1972) is

$$i/c_{av} = KU^b, \tag{8}$$

where i is the current density, U the linear solution velocity in stack, K an empirical number, generally between 50 and 200, b an empirical number, generally between 0.5 and 0.9, C_{av} the average concentration as expressed, in the following equation:

$$C_{av} = (C_1 - C_2)/[2.3 \log(C_1/C_2)]. \tag{9}$$

Equation (8) can be approximately (Maurel, 1972)

$$i/c_{av} = 145^{0.6}. \tag{10}$$

A. LIMITING CURRENT DENSITIES FOR WATER DESALINATION

Equations (6), (8), and (10) give the maximal current density permissible in an ED unit. Cost analysis for water desalination, however, results in other values for optimal current densities. According to economic calculations based on several factors such as membrane investment and replacement, electrical energy requirements, and investment in and depreciation of equipment, the optimal current density is a function of the water concentration. In the range of 15–40 meq/liter, the most economical current density is twice as high as the maximum current density for operation without water polarization. For more concentrated solutions, of 100–150 meq/liter, the maximum permissible current density and the most economical current density are identical, and at high concentrations, 400–600 meq/liter, the latter is one-half to one-fifth the former.

III. Manufacture of Permselective Membranes

A. INTRODUCTION

Permselective membranes for ED contain either groups of positive ions (anion-exchange membranes) or negative ions (cation-exchange membranes). In an applied electric field and in aqueous solution, an anion-exchange membrane permits the passage of anions only; a cation-exchange membrane permits the passage of cations only. The most important characteristics of permselective membranes used for ED are

(1) low electrical resistance,
(2) good permselective qualities for cations or anions,
(3) good mechanical properties,
(4) good form stability (contraction or expansion of the membrane must be minimal in transition from one ionic form to the other or from concentrate to diluate), and
(5) high chemical stability.

It is difficult to optimize these properties, and only a small number of companies produce permselective membranes commercially.

B. PRODUCTION OF PERMSELECTIVE MEMBRANES

Two types of membranes are usually variable—heterogeneous and homogeneous.

1. *Heterogeneous Membranes*

These membranes are manufactured by mixing a commercial ion exchanger with a solution of binder polymer, such as polyvinyl chloride (PVC), rubber, or polyvinylidene fluoride. The mixture is heated and poured under pressure onto a plastic mesh or cloth (such as polypropylene or PVC). To ensure continuous contact between the ion-exchange grains, the concentration in the polymer must be at least 50–70%. Cation-exchange membranes are obtained if the ion exchanger is cationic, and anion-exchange membranes if the ion exchanger is anionic. In general, the ion exchanger is ground to a powder before introduction to the production process. The ratio of ion exchanger to binder polymer determines the electrical and mechanical properties of the membrane. The higher the ratio between the ion exchanger and the binder polymer, the lower the electrical resistance, but worse mechanical properties result.

The ion exchangers used in ED membranes are usually made of a copolymer of styrene and divinylbenzene (DVB). The cation-exchange group is introduced into the copolymer by sulfonation with concentrated sulfuric acid at 60–90°C. The chemical reactions involved in the process are given as follows.

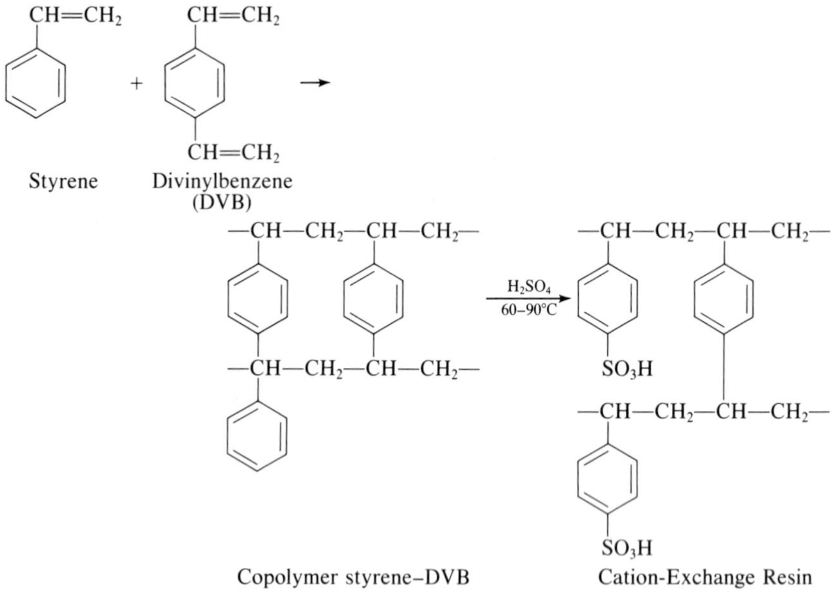

$$\underset{\text{Styrene}}{C_6H_5CH{=}CH_2} + \underset{\text{Divinylbenzene (DVB)}}{CH_2{=}CH{-}C_6H_4{-}CH{=}CH_2} \rightarrow$$

$$\underset{\text{Copolymer styrene–DVB}}{\text{—CH—CH}_2\text{—CH—CH}_2\text{—}} \xrightarrow[60\text{–}90°C]{H_2SO_4} \underset{\text{Cation-Exchange Resin}}{\text{—CH(C}_6\text{H}_4\text{SO}_3\text{H)—CH}_2\text{—}}$$

Copolymer styrene–DVB

Cation-Exchange Resin

The anion-exchange group is introduced into the polymer by chloromethylation and amination with a triamine [such as $(CH_3)_3N$].

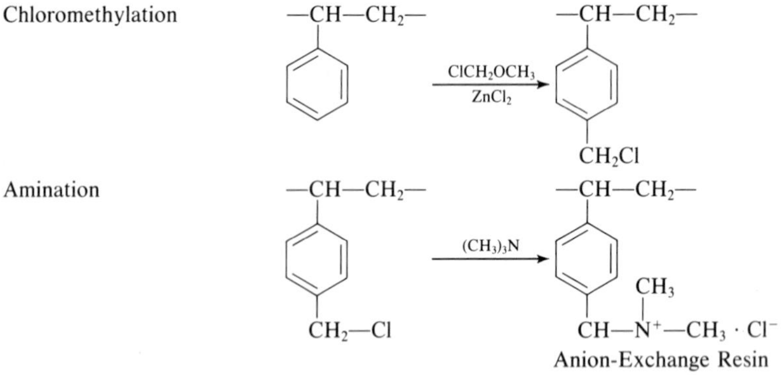

Chloromethylation

$$\text{—CH(C}_6\text{H}_5\text{)—CH}_2\text{—} \xrightarrow[ZnCl_2]{ClCH_2OCH_3} \text{—CH(C}_6\text{H}_4\text{CH}_2\text{Cl)—CH}_2\text{—}$$

Amination

$$\text{—CH(C}_6\text{H}_4\text{CH}_2\text{—Cl)—CH}_2\text{—} \xrightarrow{(CH_3)_3N} \text{—CH(C}_6\text{H}_4\text{CH—N}^+(\text{CH}_3)_2\text{—CH}_3 \cdot \text{Cl}^-\text{)—CH}_2\text{—}$$

Anion-Exchange Resin

The membranes manufactured by Ionac Co. (N.J.) appear to be produced by this method (Sybron Corporation).

2. *Homogeneous Membranes*

These membranes consist of a continuous homogeneous film onto which an active group (cationic or anionic) is introduced. These membranes can be reinforced or nonreinforced. An example of a nonreinforced homogeneous membrane is produced by the American Machine and Foundry Co. (Connecticut), which is based on graft copolymerization of styrene in a polyethylene film. Another method of producing homogeneous membranes is by the sulfochlorination of a polyethylene film (Korngold, 1970; Körösy and Shorr, 1963), an active group SO_2Cl is bound to the polyethylene film. A cation-exchange membrane is then obtained by hydrolysis and the anion-exchange membrane by amination and quaternization.

The chemical reactions involved in the production of polyethylene membranes are given as follows.

Sulfochlorination

$$\underset{\text{Polyethylene}}{—CH_2—(CH_2)_n—CH_2—} + SO_2 + Cl_2 \rightarrow —\underset{\substack{|\\ SO_2Cl}}{CH}—CH_2— + HCl$$

Basic hydrolysis

$$—\underset{\substack{|\\ SO_2Cl}}{CH}—CH_2—CH_2— + 2NaOH \rightarrow \underset{\text{Cation-exchange membrane}}{—\underset{\substack{|\\ SO_3Na}}{CH}—CH_2—CH_2—} + NaCl + H_2O$$

Amination

$$—\underset{\substack{|\\ SO_2Cl}}{CH}—CH_2—CH_2— + \underset{\substack{|\\ NH_2}}{C}—R_1—\underset{\substack{|\\ R_2}}{N}—CH_3 \rightarrow —\underset{\substack{|\\ SO_2—NH—R_1—\underset{\substack{|\\ R_2}}{N}—CH_3}}{CH}—CH_2—CH_2— + HCl$$

Quaternization

$$—\underset{\substack{|\\ SO_2—NH—R_2—\underset{\substack{|\\ R_2}}{N}—CH_3}}{CH}—CH_2—CH_2— + CH_3Br \rightarrow \underset{\text{Anion-exchange membrane}}{—\underset{\substack{|\\ SO_2—NH—R_1—\overset{\substack{CH_3\\ |}}{\underset{\substack{|\\ R_2}}{N^{+}}}—CH_3\ Br^-}}{CH}—CH_2—CH_2—}$$

A homogeneous reinforced membrane may be produced by pouring styrene and DVB onto a matrix or cloth, then carrying out the copolymerization. The remainder of the chemical reactions are the same as those for ion exchangers. A plasticizer must be mixed with the polymers to improve the mechanical properties. The production of membranes by Ionics Inc. (Massachusetts) is apparently based on this method (Hodgen *et al.*, 1973).

It is also possible to pour a solution of sulfochlorinated polyethylene onto a screen cloth. The reinforced Neginst membranes are produced in this way (Korngold and Körösy, 1973).

C. PROPERTIES OF COMMERCIAL MEMBRANES

The properties of some commercially manufactured membranes are given in Table I. A brief summary of the companies referred to is as follows:

1. *Ionac Membranes: Ionac Chemical Co., Birmingham, New Jersey*

The Ionac Company produces five types of heterogeneous reinforced membrane, two cation-exchange membranes, and three anion-exchange membranes. The MC-3470 cation-exchange membrane and the MA-3475 anion-exchange membrane have excellent mechanical properties and display high stability in various chemicals, including chlorine. Although their electrical resistance is relatively high (especially in dilute solutions), these membranes may still be classified as among the best on the market. The IM-12 anion-exchange membrane is produced in such a way that it is less sensitive to organic materials. In addition, it can be used, according to the manufacturer's specifications, at higher current densities than other membranes.

2. *A.M.F. Membranes: American Machine and Foundry Co., Stamford, Connecticut*

These membranes have good chemical, electrical, and mechanical properties, but because they are not reinforced, they undergo shape changes while in use, causing many technical problems. This is especially true of the cation-exchange membrane.

3. *Ionics Membranes: Ionics Inc., Watertown, Massachusetts*

Ionics membranes are homogeneous, reinforced, and have very good mechanical, chemical, and physical properties. They are not sold separately, but as part of a complete ED installation. Ionics membranes have

TABLE I

Properties of Commercially Produced Membranes

Manufacturer	Name of membranes	Membrane	Thickness (mm)	Capacity (meq/gm)	Electrical resistance (Ω cm^2 in 0.1 N NaCl)	Reinforcement
Ionac Chemical Co. New Jersey	Ionac	MC-3142	0.15	1.06	9.1	Yes
		MC-3470	0.35	1.05	10.5	Yes
		MA-3148	0.17	0.93	10.1	Yes
		MA-3475	0.40	1.13	23	Yes
		IM-12	0.13	—	4	Yes
American Machine and Foundry Connecticut	A.M.F.	C-60	0.30	1.5	6	No
		A-60	0.30	1.6	5	No
Ionics Inc. Massachusetts	Nepton	CR61 AZL 183	0.60	2.7	9	Yes
		AR 111 BZL 183	0.60	1.8	14	Yes
Asahi Glass Co. Ltd. Tokyo, Japan	Selemion	CMV	0.15	1.4	6.1	Yes
		AMV	0.14		4.0	Yes
Tokuyama Soda Ltd. Tokyo, Japan	Neosepta	CL 25 T	0.16	1.8–2.0	3.5	Yes
		AV 4 T	0.15	1.5–2.0	4.0	Yes
Asahi Chemical Industry Co. Ltd. Tokyo, Japan	A.C.I. or Acipex	DK 1	0.23	2.6	6.5	Yes
		DA 1	0.21	1.5	4.5	Yes
Ben-Gurion University of the Negev, Research & Development Authority Beersheva, Israel	Neginst	NEGINST-HD	0.35	0.8	12	Yes
		NEGINST-HD	0.35	0.8	10	Yes
		NEGINST-HC	0.2	1.6	6	No
		NEGINST-HC	0.2	1.7	8	No

been supplied for about 25 years to some 500 installations the world over, establishing Ionics as a leader in this field.

4. *Japanese Membranes: Asahi Glass Co., Ltd., Tokyo; Asahi Chemical Industry Co., Ltd., Tokyo; Tokuyama Soda Co., Ltd., Tokyo, Japan*

The permselective membranes produced and marketed by these three independent Japanese companies are fairly similar. All are thinly reinforced and relatively low priced when sold in large quantities. These membranes have lower electrical resistance than the American counterparts but are mechanically weaker and vulnerable to damage under dry conditions. They may also be used for purposes such as separation by diffusion (especially between acid and salt) (Nishiwaki and Itoi, 1969) or ion exchange.

5. *Neginst Polyethylene Membranes: Research and Development Authority, Ben-Gurion University of the Negev, Beer-Sheva, Israel*

These membranes are aliphatic and are, therefore, less sensitive to fouling (Korngold *et al.*, 1970). The form, stability and mechanical properties of the reinforced membranes are very good. The electrical resistance is higher than that of the Japanese membranes and similar to that of the American membranes.

D. PRODUCTION OF SPECIAL ANION-EXCHANGE MEMBRANES

In the conventional ED plant, the permissible current density at the anion-exchange membrane is smaller than that at the cation-exchange membrane (Korngold, 1973, 1974; Korngold *et al.*, 1970), largely because of the risk of precipitation. The anion-exchange membrane is more sensitive than the cation-exchange membrane to organic materials and, as a consequence, its electrical resistance may rise during operation. To overcome this problem, a number of companies produce special anionic membranes. These membranes are characterized by the fact that they can be used at a higher current density without water polarization, or splitting, and large organic anions do not drastically increase electrical resistance of the membranes. In general, their permselectivity is lower than the regular anionic membranes.

The Ionics Company (Hodgen *et al.,* 1973) produces macroreticular membranes that are less sensitive than regular membranes to traces of detergents. They are produced by mixing into the solution of copolymers, styrene and DVB, an organic solvent that dissolves in the copolymer mixture but not in the material obtained after copolymerization. When this material diffuses out of the membrane after the reaction, it leaves large pores through which large anionic molecules can penetrate, thus preventing a steep increase in the membrane's electrical resistance. The disadvantage of this type of membrane is that its electrical resistance is higher than that of a regularly produced membrane. Ionac Co. has developed a technique in which passive salts (potassium iodide or sodium iodide) are added to the solvent (dimethyl formamide) of the binder polymer. On completion of the membrane formation, the salts diffuse out of the membrane, leaving large pores through which organic anions can penetrate (Sybron Corporation, 1972). Partial penetration of large organic anions into the membrane prevents the formation of a thin boundary layer with a high electrical resistance on the membrane's surface and thus prevents a sharp increase in the membrane's electrical resistance.

Another method (Kusomoto *et al.,* 1973) of producing an anion-exchange membrane that is insensitive to traces of organic anions is used by the Tokuyama Soda Co., Japan. The anion-exchange membrane is coated with a thin layer of cation-exchange groups causing electrostatic repulsion on organic molecules. The coating is done by weak sulfonation of the membrane surface followed by chloromethylation and amination with trimethylamine. Another type of special anionic membrane is made from a polyethylene matrix (Korngold, 1972). Ion exchanger powder with a high percentage of swelling is introduced into the sulfochlorinated polyethylene solution, the screen is coated, and amination and quaternization are carried out as described above. Because the ion exchangers are highly swollen, their pores are large and allow the passage of large organic anions through the membrane. If the ion exchanger is cationic, a sharp drop in the permselectivity of the membrane occurs, thus preventing both water dissociation and the complete removal of ions from the unstirred boundary layer near the membrane (Korngold, 1973, 1974). An anion exchanger with a high swelling rate also causes a drop in permselectivity, but not to the same degree as in the case for a highly swollen cation exchanger. The greater the quantity of large molecules in the water, the more important it is to reduce the permselectivity of the membrane and prevent polarization and the resulting side effects (scaling).

Further efforts have been made to produce an aliphatic anionic membrane, in which the sensitivity to large molecules is low. Comparison of different membranes to this polyethylene membrane with respect to sensitivity to different organic materials (Korngold *et al.*, 1970) has shown that the latter membrane is more effective than other commercial membranes.

IV. Membrane Permselectivity

The concentration of mobile electrolyte (coion) in the pores of an ion exchanger that is in equilibrium with an external solution is lower than the electrolyte concentration in the solution. This phenomenon was discovered by Donnan and is called Donnan exclusion (Donnan, 1934). The relationship between the electrolyte concentration in the ion-exchange polymer and that in the solution can be obtained by assuming the chemical potential of the ion-exchange polymer μ_r is equal to the chemical potential of the solution μ_s with which it is in equilibrium. The chemical potentials of a cation-exchange resin of capacity M_R and of an NaCl solution of known concentration can be expressed by the following equations

$$\mu_r = \mu^\circ + R'T \ln(\bar{a}_R + \bar{a}_{Cl^-})\bar{a}_{Na^+}, \tag{11}$$

$$\mu_s = \mu^\circ + R'T \ln a_{Cl^-} a_{Na^+}, \tag{12}$$

where μ° is the standard chemical potential, R' the gas constant, T the absolute temperature, a_{Cl^-} the activity of Cl^- in solution, a_{Na^+} the activity of Na^+ in solution, $\bar{a}_{Na^+}$ the activity of Na^+ in the polymer, $\bar{a}_{Cl^-}$ the activity of Cl^- in the polymer, and $\bar{a}_R$ the activity of the polymer.

Activity can be expressed as a function of the concentration C and the activity coefficient j,

$$a = Cj, \tag{13}$$

where C_{Na^+}, C_{Cl^-}, and j_s indicate the ion concentration and activity coefficient in solution, and $\bar{C}_{Na^+}$, $\bar{C}_{Cl^-}$, $\bar{j}_{Cl^-}$, and $\bar{j}_{Na^+}$ the ion concentration and activity coefficients in the polymer. If we assume that the chemical potential of the resin is equal to the chemical potential of the solution, then Eq. (14) can be obtained from Eqs. (11)–(13):

$$(\bar{C}_r\bar{j}_r + \bar{C}_{Cl^-}\bar{j}_{Cl^-})\bar{C}_{Na^+}\bar{j}_{Na^+} = C_{Na^+}C_{Cl^-}j_s^2, \tag{14}$$

where $\bar{C}_r$ is the concentration of the dissociated polymer, and $\bar{j}_r$ the activity coefficient of the polymer.

We can substitute into Eq. (14) as follows: (1) If there is no association with the counterion, $\bar{M}_r = \bar{C}_r$; (2) for reasons of electroneutrality, $C_{Cl^-} = C_{Na^+} = C_s$, and $\bar{C}_{Cl^-} = \bar{C}_{Na^+} = \bar{C}_{NaCl}$; and (3) if the resin has high capacity and the concentration of NaCl in the polymer is very low, $\bar{C}_r \gg \bar{C}_{Cl}$ and $\bar{C}_{Cl}$ can therefore be disregarded. Eq. (14) can then be expressed as

$$\bar{M}_r \bar{j}_r \bar{C}_{NaCl} \bar{j}_{NaCl} = C_s^2 j_s^2 \quad (15)$$

$$\bar{C}_{NaCl} = C_s^2 j_s^2 / \bar{M}_r \, \bar{j}_r \bar{j}_{NaCl}. \quad (16)$$

If we assume, as an approximation, that in the diluate solution $j_{NaCl} = 1$ and the activity coefficient in the resin is approximately equal to unity, a modified equation is obtained:

$$\bar{C}_{NaCl} = C_s^2 j_s^2 / \bar{M}_r. \quad (17)$$

The electrical conductivity of ion-exchange polymers consists of the counterion conductivity of the polymer and the conductivity of electrolyte diffused into the pores of the ion-exchange resin. In the case of cation-exchange polymers, the conductivity of cations in the polymer will be the result of mobility of the polymer cations and the electrolyte cations diffused into the cationic exchanger; the conductivity of the anions will only be because of the electrolyte diffused into the polymer. The ratio of the two conductivities will determine the ratio between cation and anion electrotransport. As electrolyte diffusion into the pores of the resin is low in the diluate solution, cations can be transported through the resin by electrical force. The same is true for anion-exchange polymers. If we assume that the approximation made by Eq. (17) is valid, the degree of permselectivity of a membrane can be estimated by calculating the electrolyte concentration in the ion-exchange polymer.

When a membrane separates a diluate from a concentrate, there will be a concentration gradient of diffused electrolytes across the membrane. High permselectivity can therefore be obtained even with high electrolyte concentration on one side of the membrane, as long as a low concentration is present on the other side. It has been shown that the ED process can be used with brine solutions of up to 3–4 *N* (Asahi Glass Co. Ltd., 1970).

A. Measurement

Permselectivity of a membrane can be measured in one of two ways: (1) by measuring the transport number or (2) by measuring the potential.

1. *Transport Number*

This method measures the increase in the concentration of certain ions as a result of ED and the amount of current passing through the unit. The ratio of the two values (transformed to the same units) will give the transport number of a membrane.

2. *Potential Measurement*

The potential between two solutions of different concentrations separated by a permselective membrane is measured.

For monovalent ions, the potential V is given by the equation

$$V = (2\bar{t}_m - 1)(2.3R'T/F)\log(a_1/a_2), \tag{18}$$

where $\bar{t}_m$ is the transport number of the counterion, R' the gas constant, T the absolute temperature, F the Faraday constant, and a_1a_2 are the activities of the two solutions separated by the membrane.

If the membrane is completely selective, $\bar{t}_m = 1$, the membrane potential V_0 will be

$$V_0 = (2.3R'T/F)\log(a_1/a_2). \tag{19}$$

Equation (20) is obtained from Eqs. (18) and (19),

$$V/V_0 = 2\bar{t}_m - 1, \tag{20}$$

$$\bar{t}_m = (V + V_0)/2V_0. \tag{21}$$

Therefore, V_0 can be calculated from Eq. (19). The potential V can be measured and the transport number of a membrane can be obtained by direct potential measurement.

V. Diffusion

Diffusion of electrolytes from the concentrate to the diluate cells is carried out in three steps (Fig. 2):

(1) diffusion from the bulk stream through the boundary layer to the membrane–solution interface,

(2) diffusion inside the membrane, and

(3) diffusion from the membrane–solution interface through the boundary layer on the other side of the membrane to the bulk stream.

If the diffusion inside the membrane is much slower than that in the boundary layers, the total rate of diffusion is controlled by the diffusion in the membrane. If the opposite is true, then the diffusion in the boundary layers is rate determining (film-controlled diffusion).

The diffusion from the brine to the diluate increases with current density because the latter increases the concentration in the boundary layer of the brine and decreases the concentration in the boundary layer of the diluate. The net result is greater difference between the concentrations on the two sides of the membrane.

A. Determination of Diffusion Flux

Diffusion flux can be determined by placing a membrane in a bath between two electrolyte solutions of different concentrations, continuously stirring both liquids vigorously (to reduce boundary or film resistances) until equilibrium is reached (about 1 hr), and then titrating the liquids. The diffusion constant can be determined according to Fick's First Law:

$$D = eJ_D/(C_1 - C_2), \tag{22}$$

where J_D is the diffusion flux, e the membrane thickness, D the diffusion coefficient, and C_1, C_2 the concentrations of the two cells.

The specific diffusion flux for a particular membrane J_{DS} can be defined as the flux through the membrane where the difference in concentration is 1 mol. Equation (22) will then be

$$J_{DS} = D/e. \tag{23}$$

The gross average diffusion flux can be determined according to the equation

$$J_D = \Delta N/At, \tag{24}$$

where ΔN is the number of moles that diffuse from one cell to the other, A the surface area of the membrane, and t the time of diffusion.

The specific diffusion flux for commercial membranes is given in Table II.

Diffusion inside the membrane is influenced by the following factors:

(1) conductivity of the membrane,

(2) type and concentration of the electrolyte absorbed by the membrane, and

(3) temperature.

TABLE II

Specific Ionic Diffusion Flux

Membrane	Type	Specific diffusion flux (meq/hr cm^2) NaCl	HCl
Ionac MA-3475	Anionic[a]	0.0155	0.143
AMF A-63	Anionic[a]	0.0105	0.134
Cellophane (0.22 mm)	Neutral[a]	0.15	0.42
Selemion CMV-10	Cationic[b]	0.027	—

[a] Korngold (1973).
[b] Prigent (1968).

B. INFLUENCE OF ELECTRICAL CONDUCTIVITY ON THE DIFFUSION CONSTANT

According to the Nernst–Einstein equation for an ideal solution, the diffusion constant is related to electrochemical mobility:

$$D = U_e R'T/ZF, \tag{25}$$

where U_e is the electrochemical mobility, R' the gas constant, F the Faraday constant, T the absolute temperature, D the diffusion coefficient, and Z the electrochemical valence.

Some authors (Helferich, 1962; Illschner, 1958; Spiegler, 1963) have suggested that this relationship can be applied as an approximation to ion-exchange resins and membranes. From the equation it can be concluded that the diffusion of an ion through a membrane is related to the conductivity of the membrane.

C. TYPE AND CONCENTRATION OF ELECTROLYTES ABSORBED BY THE MEMBRANE

The type and concentration of the electrolytes absorbed in the ion-exchange polymer influence the diffusion constant of the membrane. Generally, the diffusion constant increases with an increase in the quantity of electrolytes inside the membrane. According to some authors (Soldano,

1953; Soldano and Boyd, 1953), this is because the co-ions in the ion-exchange polymer control the diffusion of the electrolyte (see Table II). As the concentration of coions in the ion-exchanger polymer increases with an increase in swelling, a higher diffusion rate is obtained with highly swollen resins (Korngold, 1976).

D. INFLUENCE OF TEMPERATURE ON DIFFUSION CONSTANT

The diffusion coefficient of a membrane increases with a rise in temperature. The rate of increase of the diffusion coefficient is approximately the same as the increase of diffusion in solution, and is equal to 2–2.5% for every degree Centigrade (Tobiano, 1972).

VI. Water Transport through Permselective Membranes

Water transport through membranes from the diluate to the brine cell can occur in one of the two ways:

(1) by the transport of water molecules together with ions through the membranes—the flux is proportional to the electric current and is called electrosmosis, or
(2) by osmosis caused by a difference in concentration.

A. ELECTROOSMOSIS

Water transport by electrosmosis is characterized by the following properties (Demarthy, 1973; Despics and Mills, 1956; Lakshminarayanaiah, 1967; Spiegler, 1958):

(1) For a number of cations it increases in the order of $Li^+ > Na^+ > K^+ > H^+$; this is also the order of water hydration of the ions;
(2) it is higher at lower concentrations;
(3) little change occurs with a rise in temperature; and
(4) it appears to increase with an increase in the swelling of the membrane (Demarthy, 1973).

Measurements of water transport rates are given in Table III (Demarthy, 1973), which shows the influence of the type of membrane and cation on the rate of water transport by electrosmosis.

In practice, water transport values are small in an ED process with solutions of low concentration and do not significantly influence the total efficiency of the process. At high concentrations, however, the amount of water transport is more important and must be taken into account in the calculation of efficiency. A simple calculation shows that when four molecules of water are considered for the electrotransport of Cl^- and eight molecules of water for Na^+, it is possible to obtain up to a 4.5N concentration of NaCl.

B. OSMOSIS

Water transport by osmosis through permselective membranes is slow at low osmotic pressures but begins to increase considerably at high osmotic pressures (200–400 atm). Experiments that were performed with a number of cation permselective membranes placed between solutions of 1M NaCl and 4M $MgCl_2$ are presented in Table IV (Korngold, 1972).

It is clear from Table IV that when commercial permselective membranes are used for desalination, water flux is low and insignificant. It is

TABLE III

Rate of Water Transport by Electroosmosis[a]

Cation	Concentration (g eq/liter)	Neginst HD homogeneous membrane as water molecules/Faraday (swelling 42%)	Ionac MC-3470 heterogeneous membrane as water molecules/Faraday (swelling 25%)
Li^+	0.1	28.3	13.1
	1	14.5	10.8
Na^+	0.1	18.3	8.2
	1	8.4	8.1
K^+	0.1	12.2	6.6
	1	5.0	6.7
H^+	0.1	—	2.3
	1	—	2.1

[a] M. Demarthy, Thesis, University Rouen, France (1973).

TABLE IV

Flux of Water Transport by Osmosis[a]

Membrane	Thickness of membrane (mm)	Water flux (liters/m^{2}day)
Ionac MC-3470	0.4	5.6
Neginst HC	0.3	3.0
Cellulose acetate membrane (E-400-25)	0.2	296

[a] E. Korngold, Ann. Meet. Isr. Chem. Soc., *178* (1972).

only important with highly concentrated solutions, especially with multivalent ions.

VII. Electrotransport of Large Ions through Permselective Membranes

The use of ED is generally restricted to small ions, such as chloride, sulfate, sodium, and calcium, because the electrochemical properties of permselective membranes are changed upon their association with large ions. When these ions are present in the solution, the electrical conductivity and permselectivity of the membrane decreases. If large ions are propelled by an electrical current into permselective membranes, they may become blocked there, poisoning the membrane.

A. Influence of the Molecular Weight of Ions on the Electrochemical Properties of the Membrane

Investigation of the permeability of anion-exchange membranes for carboxylate anions showed (Dohno *et al.*, 1975; Lightfoot and Friedman, 1954) that the permselectivity decreases gradually from 98 to 30% when the number of carbon atoms in the molecule is increased from two (MW = 59) to nine (MW = 171). The electrical resistance of the membrane increases with the rise in MW. The ED process can be used with solutes ranging in MW up to about 100 without significant changes in the electrochemical properties of the membrane.

B. FOULING AND POISONING BY LARGE IONS

Materials that cause fouling and poisoning can be classified into three categories (Kobias and Heertjes, 1972; Korngold, 1973, 1976; Korngold *et al.,* 1970; Kusomoto *et al.,* 1973; Tamamushi and Tamaki, 1959; van Duin, 1970, 1973).

(1) Organic anions that are too large to penetrate the membrane and accumulate on its surface: Mechanical cleaning can restore the original electrical resistance of the membrane. Anions such as humates and algenates can precipitate on the anion-exchange membrane under polarization conditions in the form of humic acid and alignic acid, causing a sharp increase in electrical resistance. These precipitates can be dissolved with a dilute base, $0.1N$ NaOH, and the original values of electrical resistance can be restored (Korngold *et al.,* 1970).

(2) Organic anions that are small enough to penetrate the membranes but whose electromobility is so low that they remain inside the membrane, causing considerable increase of resistance: Different kinds of detergents can cause this type of poisoning, and it is difficult to restore the original electrical resistance of membranes poisoned in this way (Korngold, 1976).

(3) Organic anions that are smaller than those of category (2), but still cause a certain increase in electrical resistance of the membrane: These anions can be eluted by electroelution with sodium chloride, and the original properties of the membrane can be obtained.

Experiments conducted with dodecylbenzenesulfonate (DBS, MW = 347.45), methyl orange (MW = 327.34), and congo red (MW = 696.67) with different commercial anion-exchange membranes illustrate the various types of behavior of permselective membranes with large ions (Korngold, 1976). DBS passed through all membranes. Methyl orange did not cross the Ionac and Neosepta membranes at all, while a small amount was transported through the Neginst HD-(I) membrane and a much larger amount through the highly swollen Neginst HD-(II) membrane. The electrical efficiency was low (less than 4%), and most of the current caused co-ion passage and water splitting rather than electrotransport of counterions. The congo red indicator did not pass through any of the membranes tested and did not cause significant changes in electrical resistance. At the end of the experiment with this indicator, slight precipitation was observed on the surface of the membrane and the precipitate was easily removed by mechanical cleaning (Korngold, 1976).

VIII. Energy and Membrane Area Requirements

A. ENERGY REQUIREMENT FOR DESALINATION BY ELECTRODIALYSIS

The energy needed for reducing the salinity of water is the sum of the energy requirements for ED separation and pumping. The energy required for ED separation can be expressed by the equation

$$E = I^2Rt, \tag{26}$$

where E is the energy requirement, I the current intensity, R the electrical resistance, and t the time.

The current intensity required to decrease the concentration of 1 m^3 of solution from C_1 to C_2 can be expressed by the equation

$$I = F(C_1 - C_2)/tn\eta, \tag{27}$$

where F is the Faraday number, C_1, C_2 the initial and final concentrations, respectively, n the number of cell pairs, t the time, and η the efficiency.

If t is expressed in hours and the value of the Faraday number is introduced into Eq. 27, then

$$I = 26.8(C_1 - C_2)/nt\eta. \tag{28}$$

Equations (26) and (28) give

$$E = 26.8(C_1 - C_2)IR/n\eta. \tag{29}$$

If large numbers of cell pairs are introduced between the electrodes, the voltage drop of the electrode can be disregarded, and it is possible to express the electrical resistance of the unit (R) by the equation

$$R = nr/A, \tag{30}$$

where r is the specific resistance of a cell pair and A the membrane surface area.

Equations (29) and (30) then give

$$E = 26.8(C_1 - C_2)ir/\eta 1000, \tag{31}$$

where i is the current density (I/A).

Equation (31) shows that the energy required for the ED process is proportional to the amount of salt removed, the electrical specific resistance, and the current density.

The electrical resistance of a cell pair is a function of several parameters, as can be seen from equation (32).

$$r = rc/S_1 + ra/S_2 + 850d_1/Cav_D P_1 + 1100d_2/Cav_B P_2 + 2r_\delta/S_1), \tag{32}$$

where rc, ra are the specific electrical resistances of cation and anion exchange membranes, d_1, d_2 the thickness of diluate and brine cells, respectively, S_1, S_2 the shadowing coefficients of diluate and brine spacers, respectively, P_1, P_2 the porosity factors of diluate and brine spacers, respectively, Cav_D, Cav_B the logarithmic average concentration of diluate and brine cells, respectively, and r_δ the electrical resistance of boundary layer.

Calculation of the electrical resistance of the diluate cell and brine cells [the third and fourth terms in Eq. (32)] respectively, was based on the electrical resistance of 0.02 N and 0.5 N sodium chloride solution at 25°C.

It is possible to reduce the specific electrical resistance of a cell pair by decreasing the membrane's electrical resistance by decreasing the distance between membranes and operating the unit under conditions at which the resistance of the boundary layer will approach to zero. Introduction of an ion-exchange spacer will further decrease the resistance (Kedem, 1975). Operation of the unit at high temperature will decrease the electrical resistance of the membranes and the solutions at a rate of about 2% per degree Centigrade and the energy requirement will decrease concomitantly.

Instead of the value ir, the voltage drop for one cell pair (V) can be introduced into Eq. (32) as:

$$E = 26.8(C_1 - C_2)V/\eta. \tag{33}$$

Equations (31) and (33) must be corrected to account for the energy lost in the electrode cells through the decomposition of water (1.7 V), ohmic resistance, and membrane potential.

Electrical Efficiency

The Faraday efficiency of an ED unit can be calculated according to the equation

$$\eta = 26.8(C_1 - C_2)Q/nI, \tag{34}$$

where I is the current density, n the number of cell pairs, C_1, C_2 the initial and final concentrations, and Q the flow rate.

The efficiency is affected by membrane permselectivity, current leakage, back diffusion, diluate and brine leakage between membranes, and water transport by electroosmosis and osmosis.

B. Pumping Energy

The pumping energy is the sum of the energies required for passing diluate, brine, and electrode rinse solution through the ED unit. The equation for the calculation of these energies is

$$E = 2.72\ \Delta pQ/\eta_p, \tag{35}$$

where E is the energy consumption, Δp the pressure drop, η_p the pump efficiency, and Q the flow rate.

The total pumping E_T will be

$$E_T = E_D + E_B + E_R, \tag{36}$$

where E_D is the energy requirement for diluate pumping, E_B the energy requirement for brine pumping, and E_R the energy requirement for rinse pumping.

The pressure drop in the cells is a function of three parameters: (1) geometry of the spacers, (2) cell thickness, and (3) geometry of cell distribution.

C. Membrane Area Requirement

The specific surface area of anionic and cationic membranes required for desalination to reduce the concentration from C_1 to C_2 can be expressed by the equation

$$A = 0.224(C_1 - C_2)/i\eta_e, \tag{37}$$

where i is the current density, C_1, C_2 the initial and final concentrations, A the specific membrane surface, and η_e the electrical efficiency.

The maximum current density is given by the equation [derived from Eqs. (9) and (10)]

$$i_{\max} = (C_1 - C_2)KU^b/1000 \cdot 2.3 \log(C_1/C_2). \tag{38}$$

Equations (37) and (38) give the minimum area for desalination:

$$A_{\min} = [516 \log(C_1/C_2)]/\eta_e KU^b. \tag{39}$$

For operation with $i/C = 350$ ($=KU^b$), the minimum effective area obtained will be

$$A_{\min} = (1.47/\eta) \log(C_1/C_2). \tag{40}$$

IX. Resin-Filled Cells

Several possibilities exist for the operation of ED systems with ion-exchange resins filling the cells between the permselective membranes. Generally, such systems are complicated because two unit operations, ED and ion-exchange, are incorporated. The development of this system showed that it is advantageous to use this to obtain water of low salinity and perform continuous separation (Korngold, 1967, 1975; Selegny and Korngold, 1968, 1972; Walters *et al.*, 1955).

A. MASS TRANSFER IN ELECTRODIALYSIS WITH ION-EXCHANGE RESIN*

In resin-filled cells, electromigration of ions occurs in the phase of the grains, in the solution phase or in certain combinations of solution–resin pathways (Kedem, 1975; Korngold, 1967; Spiegler *et al.*, 1956). The electrotransport of ions near the surface of the membrane is achieved in one of two ways:

(1) by direct electrotransport from the solution to the membrane i_s, or

(2) by electrotransport of ions through the resin to the membrane of the same sign i_r.

The limiting current density for solution–membrane electrotransport i_s will be similar to that in Eq. (6), taking into account lower surface contact (between solution and membrane):

* Korngold (1975).

$$i_s^{\lim} = FCDA_s/\delta(\bar{t}_m - t_s)A_t, \tag{41}$$

where A_s is the surface in direct contact with the solution and A_t the total surface of the membrane.

The limiting current density for resin–membrane electrotransport i_r appears to be infinite, as there is no inhibition of diffusion rate on these surfaces. Polarization, however, can take place on the surface of every ion-exchange grain in the cell in the same manner that it occurs on the membrane. The surface of the resins, which is generally much larger than the surface of the membrane, must be taken into account in estimating polarization.

It is obvious that the sum of i_s and i_r will represent the total current density:

$$i = i_s + i_r. \tag{42}$$

The relationship between i_s and i_r depends on the specific resin and solution conductivities and the contact between the ion exchanger and the membrane surface. The latter parameter is usually constant and depends on a number of geometric parameters. If the specific conductivity of the resin ρ_r is higher by orders of magnitude than the specific conductivity of the solution (ρ_s) (i.e., $\rho_r \gg \rho_s$) and a resin of the same sign of charge is located near the membrane, then the total limiting current density near the membrane will be high and the limiting current density on the surface of the ion-exchange resin beads will determine the limiting current density of the unit.

In the opposite case, where the specific resin conductivity is much lower than the specific solution conductivity (i.e., $\rho_s \gg \rho_r$), the limiting current density will be according to Eq. (40). It is obvious that only resins with high specific conductivity should be introduced into the cells.

If the resin and membrane are of opposite charge signs, polarization will be faster than in the absense of resin, for two reasons:

(1) There is less free membrane surface available; and

(2) the solution between the membrane and the ion-exchange grains is desalinized faster, and thus polarization is accelerated.

It can be concluded that to minimize polarization in the diluate cells, the cation-exchange resin must be near the cation-exchange membrane, and the anion-exchange resin near the anion-exchange membrane.

In this situation, when the stack operates under polarizing conditions, decomposition of water will take place at the contact interface between the two resins. However, the H^+ and OH^- ions formed will be reexchanged with anions and cations located in the beads, and thus, higher current efficiency will be obtained. If the diluate cell is filled only with cation-exchange resin, i^{lim} of polarization will be high near the cation-exchange membrane and low near the anion-exchange membrane. Conversely, if anion-exchange resin alone is introduced into the cell, the opposite situation will occur.

The type of resin introduced into the brine cell determines the location of the layer of concentrated salts under the polarizing conditions caused by ion electrotransport. If anion-exchange resin alone is placed in these cells, the concentrated layer will be near the cation-exchange membrane, and vice-versa for cation-exchange resin. If cation-exchange resin is placed near the cation-exchange membrane and anion-exchange resin near the anion-exchange membrane, the concentrated salt layer will be in the contact interface of the two resins. Anion exchanger in the brine compartment will decrease the danger of scaling near the anion-exchange membrane as a concentrated layer will not exist in this place.

B. MIXED-BED RESIN BETWEEN MEMBRANES

If mixed cation- and anion-exchange resins in the exhausted form are introduced between the cation-exchange and anion-exchange membranes and an electric field is applied, the ion-exchange resins will undergo electrical regeneration according to the equation (Glueckauf, 1959; Walters *et al.,* 1955).

$$R^1Cl + R^2Na + HOH \rightleftarrows HR^2 + R^1OH + NaCl$$

The NaCl is removed by the electrical current and the resins are regenerated. The electrical efficiency of such a reaction decreases as the resins are regenerated, and electrotransport of H^+ and OH^- competes with the removal of Cl^- and Na^+ ions.

C. ELECTRICAL REGENERATION OF ION-EXCHANGE RESINS

When water is electrolyzed, H^+ ions migrate to the cathode and OH^- ions to the anode. It is possible to use these OH^- and H^+ ions to regenerate *in situ* anion- and cation-exchange resins, respectively. It is thus possible to use ion-exchange techniques for salt removal or for continuous separation without the addition of other chemicals (Korngold, 1967; Selegny and Korngold, 1968, 1972).

In such a system the cation-exchange resin is introduced between two cation-exchange membranes, and the anion-exchange resin is introduced between two anion-exchange membranes.

References

Asahi Glass Co. Ltd. (1970). "Selemion-Ion Exchange Membrane." Tokyo, Japan, Commercial Prospectus.

Belfort, G., and Gutter, G. A. (1968). *Desalination* **5,** 267.

Belfort, G., and Gutter, G. (1972). *Desalination* **10,** 221.

Cooke, B. A. (1965). *First Int. Symp. Water Desalination Oct.* 3–9.

Cowan, D. A., and Brown, T. H. (1959). *Ind. Eng. Chem.* **51**(12), 1445.

Davis, A., and Lacey, E. (1970). "Forced-flow electrodesalination" (Progress Report No. 557). Webster, South Dakota, O.S.W. Research and Development.

de Körösy, F., and Shorr, Y. (1963). Israel Patent No. 14,720; U.S.A. Patent No. 3,388,080; British Patent No. 891,562.

Demarthy, M. (1973). Ph.D Thesis, University of Rouen, France.

Despics, A., and Mills, G. J. (1956). *Discuss Faraday Soc.* **21,** 150–162.

Dohno, R., Azumi, T., and Takashima, S. (1975). *Desalination* **16,** 55–64.

Donnan, F. G. (1934). *Z. Physik. Chem.* **A168,** 369.

Forgacs, C., Tshibashi, N., Leibovitz, J., Senkovic, T., and Spiegler, K. S. (1972). *Desalination* **10,** 181–144.

Glueckauf, E. (1959). *Br. Chem. Eng.* **4,** 646.

Grossman, G., and Sonin, A. A. (1972). *Desalination* **10,** 157.

Kedem, O. (1975). *Desalination* **16,** 105–118.

Kobus, E. J. M., and Heertjes, D. M. (1972). *Desalination* **10,** 383.

Helferich, F. (1962). "Ion-Exchange". McGraw-Hill, London.

Hodgen, R. B., Witt, E., and Alexander, S. S. (1973). *Desalination* **13,** 105–127.

Illschner, B. Z. (1958). *Elektrochem* **62,** 989.

Korngold, E. (1967). Etude chimique et électro-chimique des résines echangeuses d'ions application à des séparations d'ions voisins, en continue, par chromato-électrophorèse en colonne. Thèse, Faculté des Sciences de Rouen.

Korngold, E. (1970). "Present state of technological development of permselective polyethylene membranes at the Negev Institute for Arid Zone Research." *Water Desalination Symposium,* Beersheva, Israel.

Korngold, E. (1972). *Ann. Meet. Isr. Chem. Soc.* 178.

Korngold, E. (1973). The development of new membranes for use in desalination. *Proc. Fourth Int. Symp. Fresh Water Sea,* Vol. 3, 99–109.

Korngold, E. (1974). *Desalination* **14,** 359–367.

Korngold, E. (1975). *Desalination* **16,** 225–233.

Korngold, E. (1976). *Proc. Fifth Int. Symp. Fresh Water Sea,* Vol. 3.

Korngold, E. (1972). Novel permselective membranes. British Pat. Application No. 59,988/72; U.S. Pat Application No. N319,684; French Pat. No. 2166382.

Korngold, E., and de Körösy, F. "Reinforced cast polyethylene sulfochloride-based permselective membranes" (Report No. 131). Israel National Council for Research & Development.

Korngold, E., de Körösy, F., Rahav, R., and Taboch, M. F. (1970). *Desalination* **8,** 195–220.
Kusomoto, K., Sata, T., and Mizutani, Y. (1973). New anion exchange membrane resistant to organic fouling. *Proc. Fourth Int. Symp. Fresh Water Sea,* Vol. 3, 111–118.
Kusomoto, E., Sata, T., and Mizutani, Y. (1973). *Proc. Fourth Int. Symp. Fresh Water Sea,* Vol. 3, pp. 111–118.
Lakshminarayanaiah, I. (1967). *Desalination* **3,** 97–105.
Lightfoot, E. N., and Friedman, J. Y. (1954). *Ind. Eng. Chem.* **46,** 1579.
Mason-Rust (1970). "Webster test facilities and electrodialysis test bed plant" (Progress Report No. 568). Webster, South Dakota, O.S.W. Research & Development.
Maurel, A. (1972). "Dessalement des Eeaux Par Electrodialysis" (Techniques de l'engenieur). C.E.N., Cadarach.
Nishiwaki, Tadashi, and Itoi, Shigeru (1969). *Jpn. Chem. Q.* **5**(1), 36–40.
Prigent, Y. (1968). Ph.D. Thesis, University of Rouen, France.
Selegny, E., and Korngold, E. (1968). U.S. Patent No. 3,686,089; British Patent No. 1,240,710; French Patent No. 116,620 and 150,621.
Solan, A., Winograde, Y., and Katz, U. (1971). *Desalination* **9,** 89.
Soldano, B. A. (1953). *Ann. N.Y. Acad. Sci.* **57,** 116.
Soldano, B. A., and Boyd, G. E., *J. Am. Chem. Soc.* **75,** 6107.
Sonin, A. A., and Probstein, R. F. (1968). *Desalination* **5,** 293.
Spiegler, K. S. (1958). *Trans. Faraday Soc.* **54,** 1408–1428.
Spiegler, K. S. (1963). *J. Electrochem. Soc.* **100,** C-315.
Spiegler, K. S. (1971). *Desalination* **9,** 367.
Spiegler, K. S., Yoest, R. L., and Wyllie, M. R. J. (1956). *Discuss Faraday Soc.* **21,** 174.
Sybron Corporation (1972). U.S.A. Patent No. 1,270,621.
Tamamushi, B., and Tamaki, K. (1959). *Trans. Faraday Soc.* **55,** 1013.
Tobiano, V. (1972). Ph.D. Thesis, University of Rouen, France.
van Duin, P. J. (1970). *Proc. Third Int. Symp. Fresh Water Sea,* Vol. 2, p. 141.
van Duin, P. J. (1973). *Proc. Fourth Int. Symp. Fresh Water Sea,* Vol. 3, pp. 253–259.
Walters, W. R., Weisner, D. M., and Marek, L. Y. (1955). *Ind. Eng. Chem.* **47,** 61.
Winograde, Y., Solan, A., and Toren, M. (1972). Mixing efficiency of net-type turbulence promoters (Report). Technion-Israel Institute of Technology.

7

Desalting Experience by Hyperfiltration (Reverse Osmosis) in the United States

GEORGES BELFORT
Rensselar Polytechnic Institute
Troy, New York

I. Introduction

Although it has been known for some time that, using pressure as a driving force, synthetic membranes such as those made from cellulosic

ISBN 0-12-085480-5

derivatives are able to pass water in preference to salt, it was only in the early 1960s that a practical membrane with reasonably high water fluxes and excellent salt rejection was developed (Loeb and Sourirajan, 1962). This was the major technological breakthrough that established reverse osmosis (RO) as a viable, attractive process with a great many potential applications. Details describing the historical and theoretical development and the engineering applications of this process are available in several texts (Dresner and Johnson, 1981; Harris *et al.,* 1976; Madsen, 1977; Mears, 1976; Merten, 1966; Sourirajan, 1970, 1977). Also the proceedings of several international conferences have appeared in the literature and are useful in providing an overview of the general activities in the hyperfiltration (HF) (and membrane) field (Proceedings, 1980, 1981).

The main objective of this chapter is to cover the desalting experience by HF in the United States, and the following chapter will cover HF experience in Europe and Japan. The subjects that will be covered in this chapter include developments in new membrane-polymer characteristics, some elementary theoretical transport considerations, and plant equipment including membrane permeator design. The main feature of this chapter is a review of the RO process applications. Applications to water desalination, industrial and municipal wastewater applications, and the treatment of polluted rivers will be covered. Because membrane fouling is the single most important limiting process in pressure-driven membrane processes, attention to the control of product flux decline will be given. In this regard, the important and complementary role of pretreatment and membrane cleaning methods will be discussed.

Because of their prime significance in the HF process, several other subjects such as membrane stability and life (Chapter 3), polarization phenomena (Chapter 4), and the dynamics of solute and solvent during the process (Chapter 5) are covered elsewhere in this volume.

II. The Concept

If two reservoirs, one filled with fresh water and the other filled with salt water (brine), are separated by a semipermeable membrane that can pass water in preference to salt, water will spontaneously pass from the freshwater reservoir into the brine reservoir. This process will continue until osmotic equilibrium is approached, at which time water transport will terminate and an equilibrium osmotic pressure π° will be set up across the semipermeable membrane. If one now imposes an applied pressure greater than the osmotic pressure at osmotic equilibrium on the brine solution, water will pass through the membrane from the brine to the freshwater reservoir. This process is called RO (or HF). This discussion is schematically presented in Fig. 1.

To attain reasonable water fluxes ($J \approx 10$ gal/ft^2 day), the brine solution must be pressurized well above the equilibrium osmotic pressure as estimated from the brine concentration. In practice, RO systems are usually operated from about 4 to 20 times the equilibrium osmotic pressure π° for sea and brackish water, respectively. For seawater, the operating pressure is of the order of 100 atm,* whereas for brackish water and wastewater it is about 40 atm. Typical osmotic pressures can be obtained for dilute solutions from van't Hoff's equation

$$\pi^\circ = nRT/v_\mathrm{m}, \tag{1}$$

where n is the number of moles of solute, v_m the molar volume of water, R the universal gas constant, and T the absolute temperature. For more concentrated solutions, the osmotic pressure coefficient ϕ is used to modify Eq. (1):

$$\pi^\circ = \phi nRT/v_\mathrm{m}. \tag{2}$$

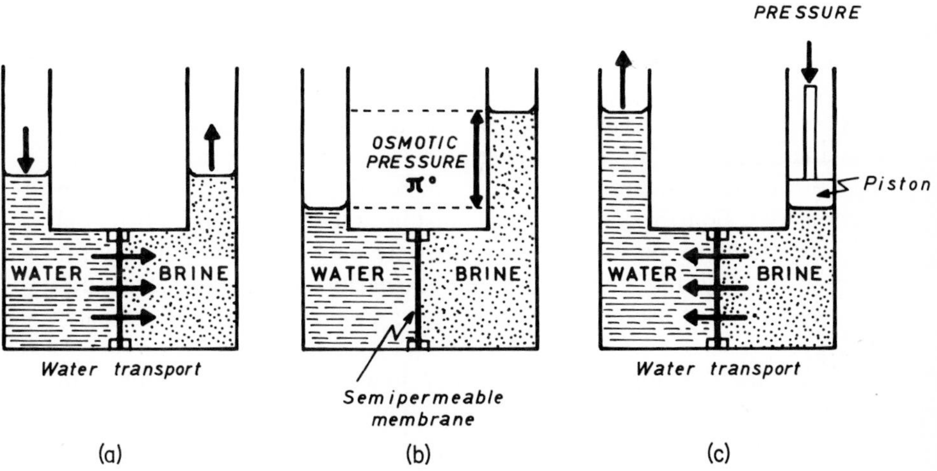

Fig. 1 Principles of normal and reverse osmosis.

* The most commonly used units in the United States for flux and pressure are gallons per square foot per day and atmospheres, respectively. Multiplication by 4.7158×10^{-7} and 1.0133×10^5 converts to meters per second and Newtons per centimeters squared, respectively.

The osmotic pressure coefficients of many pure solutions are tabulated and available in the literature (Sourirajan, 1970). A useful rule of thumb for estimating the osmotic pressure of a natural water is 1 psi (pounds per square inch) per 100 milligrams per liter. Thus for municipal wastewater containing about 1000 mg/liter dissolved salts, if we ignore for the moment the osmotic pressure exerted by the dissolved organics, we can estimate osmotic pressure of 10 psi and a corresponding applied pressure (multiply by 20) of about 200 psi (or 13.6 atm). In Section VII, we shall see that experimental RO wastewater systems have been operated at even higher pressures, from 400 to 600 psi (27–41 atm).

III. Reverse Osmosis Membranes

A practical RO membrane for water applications should possess several characteristics. First and foremost, it should be permeable to water in preference to all other components in the feed solution. Second, the rate of permeation of water per unit surface area (water flux) must be high

enough to produce reasonable product volumes per unit time, which can only be defined with respect to an economic analysis of the process. Third, the membrane must be durable, physically, chemically, and biologically and have a reasonably extended life. Lifetimes for commercial RO membranes using a brackish water feed are of the order of 1 to 3 yr. Wastewater feeds with a high level of dissolved organics may naturally reduce the life of membranes. This will be discussed in more detail in Section VIII. Fourth, the membrane must be able to withstand substantial

pressure gradients on its own or with some porous backing or support material. Finally, the membrane should be easily cast into the configuration necessary for use and preferably have an asymmetric skinned morphology.

Since the development in the early 1960s by Loeb and Sourirajan (1962) of a method for casting membrane films with high water fluxes and excellent salt rejections, the "anisotropic" cellulose acetate (CA) membrane has been considered the leading commercial membrane. This "anisotropic" membrane has about 2.5 acetyl groups per monomer molecule and consists of a very thin, dense skin, 0.15–0.25 μm in thickness, on top of a highly porous (>50% void space), thick (>100 μm) substructure. The asymmetric nature of a flat CA membrane is illustrated by the electromicrograph in Fig. 2a. The desalination properties of the membrane are determined solely by the characteristics of the thin, dense skin. Essentially, all the salt rejection takes place at the skin, and water flux rates through the membrane are limited by the permeability of the skin. The properties of the skin can be varied during the membrane preparation procedure, which has as its last phase an annealing step in water from 70 to 90°C. By accurately controlling the temperature during annealing, a range of different salt-rejecting membranes can be produced from a loose (70°C) to a tight (90°C) membrane.* The water permeability of the membrane is inversely (although not linearly) related to the salt rejection. At present, typical average water fluxes are about 2.5 gal/ft^2 day per 100 psi of applied pressure for salt rejections of greater than 95%. It has also been shown that the thin, dense film (or skin) and the porous substructure need not be made from one material. By optimizing each layer and sandwiching them together, a membrane with superior performance characteristics can be made (Riley *et al.*, 1971; Rozelle *et al.*, 1973). Thus, the permeability presumably can be increased without a decrease in rejection by reducing the dense film thickness. Of course, with very thin films (~1000 Å), mechanical strength of the membrane may suffer. These composite membranes are able to desalinate seawater (salt rejection ~99.5%) with excellent water fluxes of 30 gal/ft^2 day at 1000 psi applied pressure (Cadotte *et al.*, 1980; Rozelle *et al.*, 1977).

Thin film composite RO membranes consist of three layers: (1) a backing material often made from a woven or nonwoven fabric or polyester; (2) a microporous support cast directly on the fabric with a thickness of about 50 μm having an asymmetric pore structure of coarse pores (fraction of micrometer size) facing the backing material and very fine pores (about 300 Å) facing the feed solution; and (3) a thin, dense polymer coating (barrier) bonded to the 300Å face surface of the microporous support film. Polysulfone is the most popular polymer used for the microporous support film. The barrier is usually only a few thousand angstroms thick. Barriers were first made from cellulosic polymers but later, because of their temperature, pressure, and pH stability, noncellulosic polymers have been used (Cadotte *et al.*, 1980). The first type used was produced by an interfacial condensation reaction of polyethylenimine and toluene diisocyanate on the surface of a polysulfone support fiber (Cadotte and Rozelle, 1972). By choosing the barrier propitiously, Cadotte and co-workers (1980) reported excellent results with their FT-30 membrane. As shown in Fig. 3, they exhibit excellent flux and rejection values as a function of brine (concentration, pressure, tempera-

* In contrast to the Loeb and Sourirajan (1962) *wet* method for casting asymmetric RO membranes, a *dry* method has also been developed by Kesting (1973).

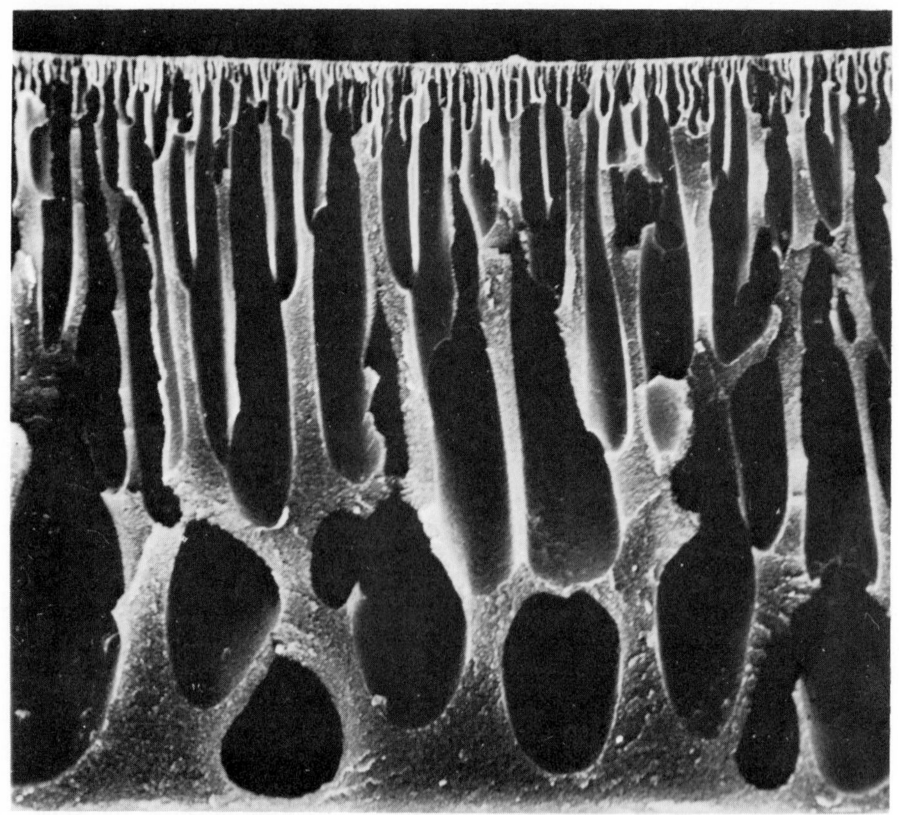

Fig. 2a Scanning electron micrographs of the substructure of asymmetric membranes showing dense skin and porous substructure. Flat cellulose–acetate membrane (total thickness ~100 μm). (ID ~500 μm).

ture, and pH. In addition, this membrane exhibits excellent nonbiodegradability and can withstand 50 ppm chlorine oxidation in immersion tests for at least 20 days at 800 psi, 25°C at pH <6. Also, methyl-, ethyl-, and isopropyl alcohol at 1000 ppm feed concentrations show 29, 72, and 92% rejections, respectively (Cadotte, private communication). Cellulose acetate membranes are commercially available in several configurations, for instance, as a tube with the dense film on the inside, as a flat sheet, or as a fiber with the dense film on either the inside or outside.

Another successful asymmetric membrane, developed primarily for the hollow-fiber configuration, has been made from aromatic polyamide formulations or Nomex. See Fig. 2b for a scanning electron micrograph (SEM) of a hollow-fiber membrane with the skin on the inside surface.

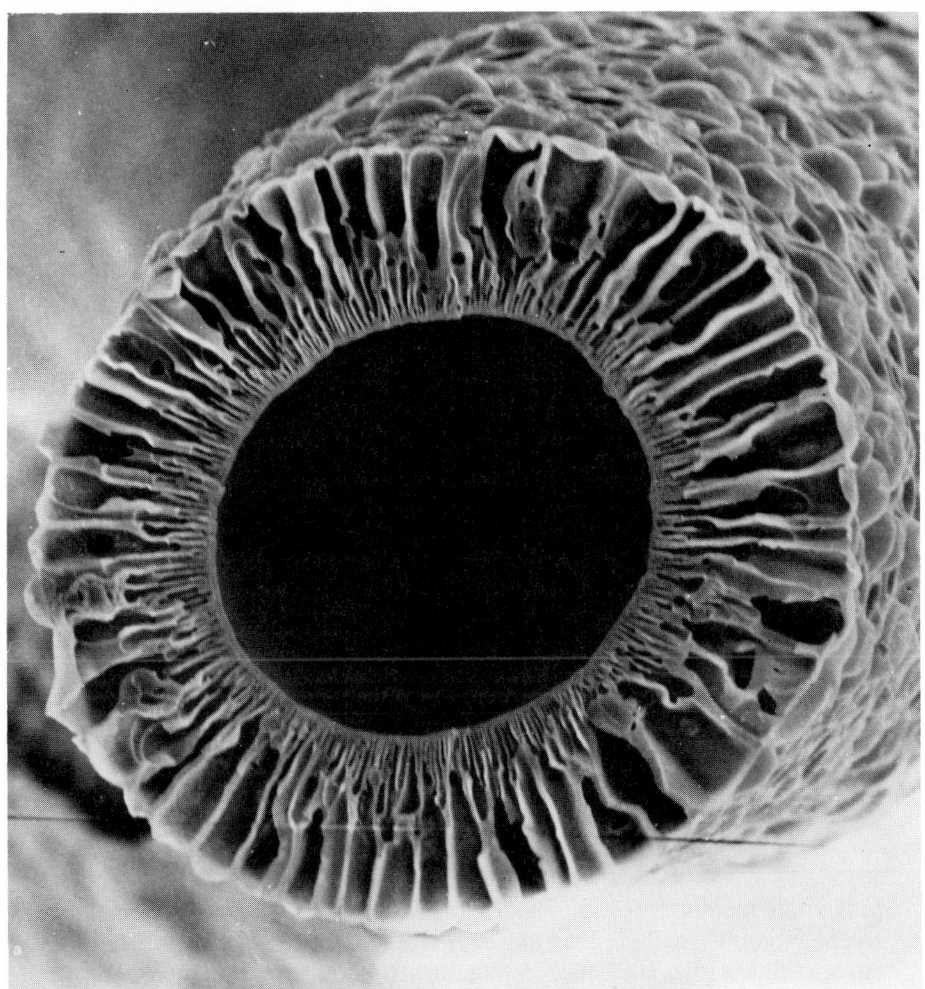

Fig. 2b Scanning electron micrographs of the substructure of asymmetric membranes showing dense skin and porous substructure. Hollow fiber polyamide membrane (ID ~500 μm). (Courtesy of H. Strathmann, Forschungsinstitute, Berghof Gmb, West Germany.)

Although the water permeabilities through these membranes are about an order of magnitude less than through the CA membranes, their packing density (square feet of membrane surface per unit volume of permeator) is about an order of magnitude higher.

A nonpolysaccaride membrane was shown to reject several model compounds found in a typical sewage effluent better than the CA or flat

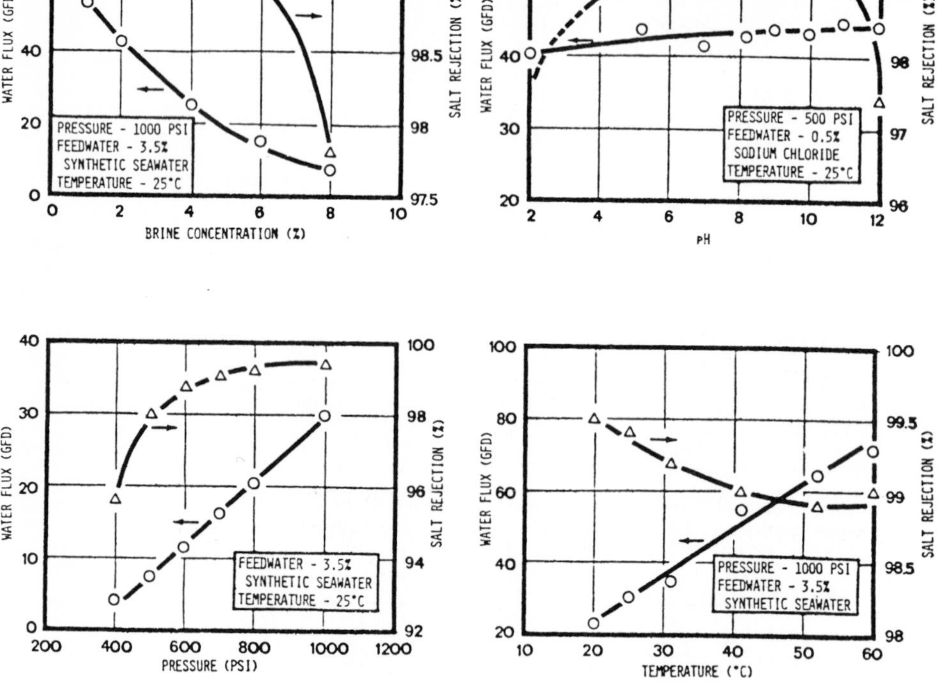

Fig. 3 Salt rejection and water flux for FT-30 membrane.

polyamide membranes (Chian and Fang, 1973).* This membrane (referred to as "NS-100" by Rozelle *et al.*, 1973 and 1977) consists of a polysulfone support film with a 6000-Å-thick coating of polyethyleimine reacted with diisocyanate (TDI).

The NS-100 membrane has exhibited excellent performance in rejecting salt from synthetic seawater (18 gal/ft^2 day at 99.4% R after 1200 hr), cyanide from simulated zinc cyanide electroplating rinse water (14 gal/ft^2 day at 94.5% R after 350 hr at pH = 12.9), and copper ions from a simulated acid copper rinse water (9 gal/ft^2 day at 99.8% R after passing 550 l at pH = 0.5) (Rozelle *et al.*, 1977). The excellent rejection of various organic solutes (alcohols, aldehydes, ketones, acids, amines, and

* The model compounds were from the following groups: aromatics, acids, alcohols, aldehydes, ketones, amines, ethers, and esters.

acetates) especially in comparison with CA strongly suggests that the NS-100 membrane could have wide applicability in many industrial and wastewater applications.

In Table I we present a comparison of the CA, aromatic polyamide, NS-100, and FT-30 membranes. The salt rejection is slightly better for CA NS-100 and FT-30 membranes than for the aromatic polyamide membranes. Aside from degradation resulting from free chlorine, the AP and NS-100 membranes are chemically inert in water and are also less liable to biodegradation. The FT-30 membrane is also resistant to Cl_2 oxidation at pH < 6, an important point when treating wastewaters. Note also that some materials frequently present in wastewaters are poorly rejected by the CA and aromatic polyamide membranes. These include phenols, aldehydes, urea, methanol, and methyl acetate (Chian and Fang, 1973; Duvel and Helfgott, 1975). As will be seen in Section VIII, an important part of RO operation is membrane cleaning. Thus the membranes must possess high stability and not degrade with time. Cellulose acetate is known to hydrolyze slowly with time (Chapter 3). The rate of hydrolysis is dependent on the feed constituents and pH (Sachs and Zisner, 1972)

A wide variety of other membranes have also been developed (Lonsdale and Podall, 1972). These include the cellulose acetate–butyrate and cellulose acetate–methacrylate blends, polybenzimidazole (PBI) (Model *et al.*, 1977), and sulfonated polyphenylene oxide (SPPO) (LaConti, 1977). The blend membranes have slightly different and, only under specific conditions, slightly better performance characteristics than CA membranes. PBI membranes were developed because of their high stability to extreme chemical, physical, and biological attack and because of their high water absorbability (twice that of CA). The PBI membranes are cast into flat sheets (spiral wound) and hollow fibers and have been tested on synthetic seawater and space vehicle washwater at elevated temperatures (75°C) and have performed well over periods as long as 3 months (Model *et al.*, 1977). Although the SPPO membranes do not compete with CA for brackish and seawater desalting, they are expected to do well on caustic and acid waste and organic effluents at relatively high temperatures (165°F) (LaConti, 1977).

Thus for some uses these particular membranes have more desirable properties than the standard asymmetric CA membranes, but in most cases overall performance has not yet warranted their replacing CA membranes completely.

Another interesting development, especially with a view to wastewater applications, is the ability of certain materials, when deposited dynamically on a porous substructure or support, to reject salt at high water permeabilities (Johnson, 1972; Johnson *et al.*, 1973; Tanny, 1980;

TABLE I

Comparison of Membrane Characteristics

Membrane material	pH range (continuous exposure)	Temperature limits (°F)	Operating pressure (ψ)	Materials causing membrane dissolution	Materials poorly rejected	Salt rejection (%)	Flux (gal/ft^2 day)
Cellulose acetate (tubes or flat sheets)	3–8	65–85	To 1500	Strong oxidizing agents, solvents, bacteria	Boric acid, phenols, detergents (ABS), carbon, chloroform extract, ammonia, urea, methyl acetate	95–99.5	8[a]–15
Aromatic polymides (fibers or flat sheets)	4–11	32–95	400 (fiber)	Strong oxidizing agents, particularly free Cl_2	Aldehydes, phenols, methanol, methyl acetate	85–95	2[a]–5
NS-100 (flat sheets or hollow)	2–12	50–165	To 1500	Strong oxidizing agents, particularly free Cl_2	Methanol	95–99.5	4–12
FT-30 (flat sheets)	3–11	50–165	To 1500	Not affected by Cl_2 at $pH < 6$	Methanol	96–99.5	5–60

[a] Lower fluxes were obtained using a laboratory model sewage (Chian and Fang, 1973).

Thomas, 1977). A wide variety of additives have been found to form salt rejecting dynamic membranes. These include materials such as humic and fulvic acid, Zr(IV) oxide, tannins, and polyelectrolytes (Sephadex CM-C-25) (Kraus, 1970). Fluxes of the order of 100 gal/ft^2 day with salt rejections greater than 50% have been reported for these dynamic membranes (Thomas *et al.*, 1973). It should be remembered that these low salt rejections may not be a disadvantage in municipal wastewater treatment because secondary effluents are usually in the range 750–1500 ppm TDS (Sachs and Zisner, 1972). The two major disadvantages of dynamic membranes are their inherent instability and, because of the high permeation fluxes, the need for high pumping rates of the feed stream to reduce concentration polarization or salt buildup at the membrane–solution interface.

In conclusion, commercially available membranes, such as the asymmetric CA and the aromatic polyamide hollow fibers have proved to be satisfactory for brackish and sea water applications. However, they are not perfectly suited to wastewater applications. The CA membrane has problems with leakage of certain feed constituents into the product and is susceptible to biological attack. The aromatic polyamide hollow-fiber membranes are chemically and biologically stable toward typical municipal wastewaters (except for highly chlorinated water), but because of their fiber design, the poor hydrodynamic condition of the feed solution results in a high potential for plugging and membrane fouling. This will be discussed in detail later. The NS-100 (Rozelle *et al.*, 1977), FT-30 (Cadotte *et al.*, 1980), PBI (Model *et al.*, 1977), SPPO (LaConti, 1977), and cynamic polyelectrolyte membranes have not yet proven themselves in large-capacity commercial plant operation, although all seem to be promising. See Strathman (1980) for an excellent review of RO membrane advances.

IV. Theoretical Considerations

A. TRANSPORT

Several theories exist that describe and predict transport behavior of RO membranes. The viscous flow theory assumes all flow occurs through pores in the membrane, flow rate and permselectivity being governed by porosity, pore size distribution, and interactions (chemical and electrical) with the surface of the pores. This approach is widely used to describe transport through UF membranes (Michaels, 1968), although it has also been invoked for special kinds of RO membranes such as the porous glass system (Belfort, 1972).

With respect to the asymmetric Loeb–Sourirajan-type membrane, the solution–diffusion theory has been successfully used (Merten, 1966). This theory explains the rejection phenomenon in terms of two steps. In the first step, the salt and water dissolve in the membrane film, whereas in the second step each molecular species is thought to move through the membrane by independent diffusion.

Returning now to the thermodynamics of irreversible processes (Section II.E of Chapter 1) and neglecting the last term (grad $\phi = 0$) for the RO process, Eq. (5) of Chapter 1 must be integrated across the thickness of the membrane. Using the subscript 1 to designate the solvent (water) and the subscript 2 to designate the solute in a two-component system, we obtain for the solvent:

$$\begin{aligned}\Delta\mu_1 &= \int (\partial\mu_1/\partial C_1)_{\mathrm{P,T}}\, dC_1 + \int v_1\, dP \\ &= \int (\partial\mu_1/\partial C_2)_{\mathrm{P,T}}\, dC_2 + \int v_1\, dP. \end{aligned} \tag{3}$$

We know that, when $\Delta\mu_1 = 0$, we are left with the osmotic pressure difference $\Delta\pi$. Thus for constant v_1,

$$v_1\, \Delta\pi = -\int (\partial\mu_1/\partial C_2)_{\mathrm{P,T}}\, dC_2 \tag{4}$$

and

$$\Delta\mu_1 = v_1(\Delta P - \Delta\pi). \tag{5}$$

For the solute

$$\Delta\mu_2 = \int (\partial\mu_2/\partial C_2)_{\mathrm{P,T}}\, dC_2 + \int v_2\, dP, \tag{6}$$

and for dilute solutions $\mu_2 = \mu_2^{\circ} + RT \ln C_2$, and constant v_2, we obtain

$$\Delta\mu_2 = RT\, \Delta \ln C_2 + v_2\, \Delta P, \tag{7}$$

where the second term on the right-hand side of Eq. (7) is negligible with respect to the first for the RO process (Merten, 1966). Thus we obtain

$$\Delta\mu_2 = RT\, \Delta \ln C_2 \approx (RT/C_2)\, \Delta C_2 \tag{8}$$

and can incorporate $\Delta\mu_i$ (or X_i) into Eq. (1) of Chapter 1 by neglecting the cross coefficients, which are relatively small for the asymmetric CA membrane (Bennion and Rhee 1969):
For water,

$$J_1 = K_1(\Delta P - \Delta\pi), \tag{9}$$

and for salt,

$$J_2 = K_3 \, \Delta C_2 = K_3 C_2' R, \tag{10}$$

where K_1 and K_3 are the water and salt permeability coefficients, respectively, and are related to the phenomenological coefficients, ΔC_2 is the difference in salt concentration between bulk streams of product and feed, C_2' the salt concentration of the bulk feed stream, and R the observed coefficient of salt rejection defined by

$$R = 1 - C_2''/C_2'. \tag{11}$$

To achieve greater accuracy, the concentrations at the membrane surface interfaces (C_{2m}' and C_{2m}'') should be used in Eqs. (10) and (11) instead of the bulk stream concentrations. Bulk stream values are, however, easier to measure and are usually used.

Thus using equilibrium and irreversible thermodynamics and ignoring coupled flows, we have derived the flux equations. The resultant equations, assuming constant coefficients [Eqs. (9) and (10)], are formulations of Ficks law of diffusion. Thus K_1 has been described in terms of a diffusion coefficient, water concentration, partial molar volume of water, absolute temperature, and effective membrane thickness. The coefficient K_3 has been described in terms of a diffusion coefficient, distribution coefficient, and effective membrane thickness (Merten, 1966). For further discussion the reader is referred to Soltanieh and Gill (1981) for an excellent review of RO membranes and transport models.

B. FOULING ANALYSIS

Because of its complexity and enormous variability, fouling of membranes has usually been analyzed empirically on a case to case basis. Several theoretical models have been proposed to describe the fouling process, although only a few fundamental studies have been reported in the literature (Doshi and Trettin 1981; Nakao and Kimura, 1981; Probstein *et al.*, 1981). The reader is referred to the detailed review of membrane fouling models presented in Chapter 5, Section II.B.3 and to Mathiasson and Sivik's review (1980). Here, we present additional analysis of the gel polarization model incorporating lateral migration and the resistance layers-in-series model, originally developed and refined for membrane fouling by Belfort and co-workers (Belfort and Marx, 1979; Green and Belfort, 1980; and Reed and Belfort, 1982) and adapted by Schippers and co-workers (Schippers and Verdouw, 1980; Schippers *et al.*, 1981) for colloidal fouling.

1. Gel Polarization Model Incorporating Lateral Migration

Consider the case where the permeation wall flux has minimal effect on the axial velocity profile, where the average axial velocity is much greater than the radial velocity at the membrane–solution interface. This condition is easily met for RO and may be met for some UF modules.

From a mass balance of particles at steady state, we obtain (Belfort and Chin, 1982)

$$C(V_r - V_L) = +D\, dC/dr, \tag{12}$$

with boundary conditions

$$C = C_w \quad \text{at } s = R_p \quad \text{or} \quad r = R_T - R_p \approx R_T, \tag{13}$$
$$C = C_b \quad \text{at } s = \delta \quad \text{or} \quad r = R_T - \delta, \tag{14}$$
$$V_r = V_w \quad \text{at } s = R_p \quad \text{or} \quad r = R_T - R_p, \tag{15}$$
$$V_L = V_L\,|_w \ \text{at } s = R_p \quad \text{or} \quad r = R_T - R_p. \tag{16}$$

In Eq. (12), V_r can be rewritten with the dimensional radial coordinate r as

$$V_r = V_w[2(r/R_T) - (r/R_T)^3]. \tag{17}$$

Assuming that the suction at the walls has negligible effect in the velocity profile, and hence the lift force, the Segre and Silberberg (1962) equation can be used to describe the lift velocity†

$$V_L = -0.17\langle U_f\rangle \text{Re}_T (R_p/R_T)^{2.84} r/R_T[1 - (r/r^*)], \tag{18}$$

where r^* is the dimensional equilibrium position. Substituting Eqs. (17) and (18) into Eq. (12) and rearranging for integration we obtain.

$$\int_{R-\delta}^{R} \left\{ V_w \left[\left(2\left(\frac{r'}{R_T}\right) - \left(\frac{r'}{R_T}\right)^3 \right] \right. \right.$$
$$\left. \left. + 0.17\langle U_f\rangle \text{Re}_T \left(\frac{R_p}{R_T}\right)^{2.84} \frac{r'}{R_T} \left[1 - \left(\frac{r'}{r^*}\right)\right] \right\} dr'$$
$$= \mathcal{D} \int_{C_b}^{C_w} \frac{dC'}{C'} = \mathcal{D} \ln \left(\frac{C_w}{C_b}\right). \tag{19}$$

After integrating across δ according to the boundary conditions stated in Eqs. (13)–(16), the following is obtained with $\eta \equiv \delta/R_T$:

$$V_w K(\eta) + 0.17\langle U_f\rangle \text{Re}_T (R_p/R_T)^{2.84} L(\eta) = k \ln(C_w/C_b), \tag{20}$$

† This equation is used here for simplicity. More complicated theoretically derived equations from Ho and Leal (1974) and Vasseur and Cox (1976) could also be used.

which gives the flux on rearrangement

$$V_w = [k/K(\eta)] \ln(C_w/C_b) - 0.17\langle U_f\rangle \mathrm{Re}_T(R_p/R_T)^{2.84}[L(\eta)/K(\eta)], \quad (21)$$

where $k = \mathcal{D}/\delta$, the mass-transfer coefficient given for laminar flow by the Graetz or Levêque solutions as

$$k = 0.816(\dot{\gamma}\, \mathcal{D}^2/L)^{1/3} \quad (22)$$

and

$$K(\eta) = (1 + \tfrac{1}{2}\eta - \eta^2 + \tfrac{1}{4}\eta^3), \quad (23)$$

with $K(0) \to 1$ as $\eta \to 0$, $K(1) \to \frac{3}{4}$ as $\eta \to 1$ and

$$L(\eta) = [(1 - \tfrac{1}{2}\eta) - (R_T/r^*)(1 - \eta + \tfrac{1}{3}\eta^3)], \quad (24)$$

with $L(0) \to 1 - (R_T/r^*)$ as $\eta \to 0$, $L(0) \to 0$ as $r^* \to R_T$.

In the limit $\eta \to 0$ when $\delta \ll R_T$ or $\eta \ll 1$, and substituting for k from Eq. (22) into Eq. (21),

$$V_w(\eta = 0) = \{[1.295(\mathcal{D}^2/R_T L)^{1/3} \ln(C_w/C_b)]\langle U_f\rangle^{1/3}\} + \{[0.34(R_T/\nu)(R_p/R_T)^{2.84}(R_T/r^*) - 1)]\langle U_f\rangle^2\}, \quad (25)$$

where the first set of braces pertains to convection and the second to lateral migration. The first term on the right-hand side of Eq. (25) is identical to the gel polarization model of Blatt *et al.* (1970), and the second term incorporates the lateral migration phenomenon. Clearly the power dependence between V_w and $\langle U_f\rangle$ could only increase above 0.333 depending on the relative importance of the lateral migration or second term on the right-hand side of Eq. (25).

2. *Resistance-in-Series Model*

Another phenomenological approach uses the standard filtration equation and adapts it to pressure-driven membrane processes (Belfort and Marx, 1979):

$$\left(\frac{1}{s}\frac{dv}{d\theta}\right)^{-1} = \left(\frac{\mu\alpha\omega}{\Delta PS}\right) V + \left(\frac{\mu r'}{\Delta P}\right)\left(1 + \beta V^n\right) + \cdots, \quad (26)$$

where $n = 1$ during the initial transient period and 0 during the steady-state period; β is a constant, characteristic of the membrane; $\omega = \omega(\theta)$ for variable suspended feed concentration with θ and ω_o for constant suspended feed concentration with $\theta = 0$. Note that $\Delta P = \Delta P_c + (\Delta P_m - \sigma\, \Delta\Pi)$ is the total pressure drop across the cake and RO or UF membrane, serially. For the latter case with negligible osmotic effects, $\Delta\Pi = 0$.

For the constant suspended feed concentration case and the initial period ($\omega(\theta) = \omega_0$, $n = 1$), Belfort and Marx obtained on integration:

$$\theta/V = \tfrac{1}{2}k_3V + k_2 + k_4N, \quad (27)$$

for $0 < V < V_{crit}$ and $0 < \theta < \theta_{crit}$, where

$$k_1 = \mu\alpha\omega_0/\Delta PS^2, \quad (28)$$

$$k_2 = \mu r'/\Delta PS, \quad (29)$$

$$k_3 = k_1 + \beta k_2, \quad (30)$$

$$k_4 = \text{integration constant}, \quad (31)$$

with a minimum at $d(\theta/V)/dV = 0$ or at $V_{min} = \pm(2k_4/k_3)^{1/2}$. As $V \rightarrow \infty$, Eq. (27) reduces to

$$\theta/V = \tfrac{1}{2}k_3V + k_2. \quad (32)$$

For the steady-state period ($n = 0$ and $\omega(\theta) = \omega_0$), one obtains after integration

$$\theta/V = \tfrac{1}{2}k_1V + k_5 + (k_6/V), \quad (33)$$

for $V > V_{unit}$ and $\theta > \theta_{unit}$, where $k_5 = k_2(1 + \beta)$ and k_6 is the integration constant. Because of continuity at the critical point (V_{crit}, θ_{crit}), the following necessary condition results:

$$k_4 - k_6 = \tfrac{1}{2}k_2 \,|_{V=V_{crit}}. \quad (34)$$

Once again, as $V \rightarrow \infty$, Eq. (33) reduces to

$$\theta/V = \tfrac{1}{2}k_1 + k_2(1 + \beta) \quad (35)$$

for $V \gg V_{crit}$ and $\theta \gg \theta_{crit}$, which is analogous to the standard filtration equation.

Schippers *et al.* (1981) have also proposed a transient or initial period during which "blocking filtration" or concentration polarization dominates the fouling process. They define a "modified fouling index" (MFI) at steady state as

$$\text{MFI} = \mu I/2\ \Delta PS^2 = (I/\alpha\omega_\theta)k_1, \quad (36)$$

where I is a measure of the fouling potential of a water, $\alpha\omega_\theta$ the product of the average specific cake resistance α (centimeter per gram), and the concentration of fouling solute (e.g., colloids) in the water.

V. Plant Equipment

A pressure-driven membrane separation plant consists of components shown in the simplified flow diagram in Fig. 4. Typically, the feed solution is first filtered and the pH adjusted between 5 and 6 for the asymmetric CA membranes. The feed is then pressurized (to say 40 atm for 3000 ppm TDS) and passed through the various membrane permeators or modules. Because the volume of the feed solution decreases with path length, fewer modules are needed for each successive stage, as seen in Fig. 4. The permeate of each stage is combined and stored for use. A back pressure regulator is used to reduce the pressure of the brine stream after the last stage. It has been suggested that this waste pressure be used to drive a turbine to produce electricity that could be used to run the motors of the pumps. Figure 4 is a once-through flow scheme. Many other alternatives are also used, including partial or total brine recycle to improve the system's recovery ratio. The recovery ratio is defined as the total volume of product obtained divided by the initial volume of feed. The recovery ratio is an economically significant performance parameter often neglected or minimized by manufacturers and researchers.

The filters, tanks, pumps, back pressure regulators, dampeners, and piping employed in these plants are conventional items commonly available to the chemical processing industry. The only special component is the membrane permeator or RO module. A photograph of a commercial plan of the spiral-wrap design is presented in Fig. 5.

VI. Membrane Permeators

The main requirement of a RO membrane permeator is that it house the membranes in such a way that the feed stream is sealed from the product steam. All other requirements are concerned with:

(1) mechanical stability, i.e., supporting a fragile membrane under high differential pressures (200–1500 psi), preventing pressure leaks between the feed and product streams, between the feed stream and its surroundings (air), and avoiding large pressure drops in the feed or product streams;

(2) hydrodynamic considerations, i.e., minimizing the buildup of both salt and fouling layers on the membrane surface, which might impede membrane performance; and

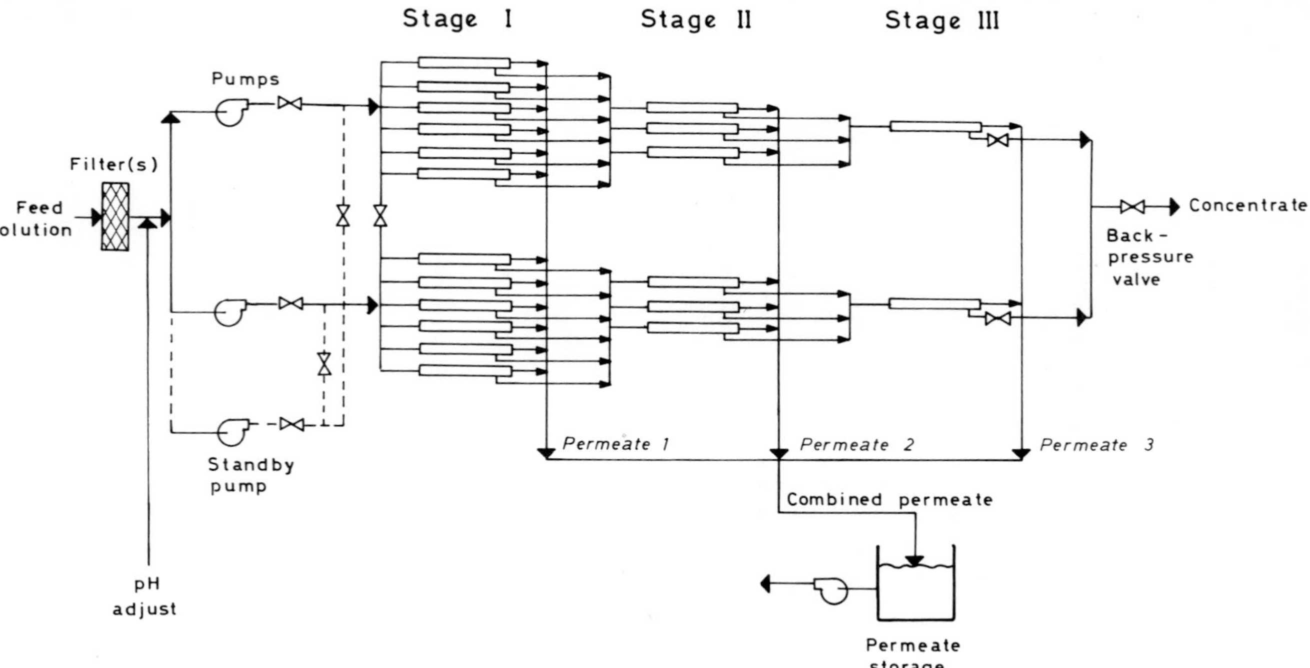

Fig. 4 Flow diagram of a pressure-driven membrane separator plant (after Lacey, 1972a.)

Fig. 5 Commercial spiral-wrap RO plant. (Courtesy of R. L. Riley Fluid Systems Div., UOP. San Diego, California.)

(3) economic considerations, i.e., obtaining high membrane packing density to reduce capital costs on the pressure vessels and designing the unit for ease of membrane replacement.

Several types of RO membrane permeators that meet these requirements are commercially available (Golomb and Besik, 1970). Based on the geometry of the membrane, they can be classified into five broad design categories: Tubular, spiral wrap, hollow fiber, flat plate, and dynamic. Subclasses within each category are described in Table II, along with the manufacturer's address for each item.

A sketch of each major commercial RO membrane permeator is shown in Figs. 6 and 7. Several performance and structural characteristics for the different permeators are also presented in Table III. A brief explanation of the module designs is presented subsequently with reference to Table III.

Notice first in Table III that the permeator with the lowest water output per unit volume (tubular with inside flow) is most easily cleaned, whereas the permeator with the highest water output per unit volume (fibers with brine flow on the outside) is the most difficult to clean. We are especially concerned with ease of cleaning when we have a turbid feed

TABLE II

Reverse Osmosis Membrane Permeators

Class	Designation	Description of available designs	Manufacturer[a]
Tubular	1a	Brine flow inside straight rigid support tube	A,C,E,L,M,N,P,Q
	1b	Brine flow inside helical support tube	
	1c	Brine flow inside straight squashed tube	Q
	1d	Brine flow outside straight rigid support tube	R
	1e	Brine flow outside flexible rigid support tube	Q
Spiral wrap	2a	Brine flow between alternate leaves of a spiral wrap	C,E,J,P,S,U
Fiber	3a	Brine flow outside flexible hollow-fiber membranes	B,G,I,T
	3b	Brine flow inside flexible hollow-fiber membranes	—
Flat plate	4a	Horizontal filterpress design with brine flow radially between leaves	F,K
	4b	Same as 4a with whole unit spinning	H
Dynamic membrane	5a	A dynamic precoat membrane is laid down on a porous support	D,O

[a] The following letters are used to designate major manufacturers: A=Abcor, Inc., Wilmington, Massachusetts; B=Asahi Chemical Co., Tokyo, Japan; C=Calgon Havens Systems, San Diego, California; D=Carre, Inc., Seneca, SC; E=Culligan International Co., Northbrook, Illinois; F=DeDanske Sukkerfabrikker, DK-1001, Copenhagen, Denmark; G=Dow Chemical Co., Walnut Creek, California; H=Dresser, Advanced Technology Center, Burroughs, Irvine, California; I=E.I. DuPont De Nemours & Co., Wilmington, Delaware; J=Envirogenics Co., El Monte, California; K=GKSS-Forschungszentrum, Germany; L=Israel Desalination Engineering, Tel Baruch, Israel; M=Kalle, West Germany; N=Nitto Electric Co., Osaka, Japan; O=Oak Ridge National Laboratory, Oak Ridge, Tennessee; P=Osmonics, Inc., Hopkins, Minnesota; Q=Paterson Candy Int., Laverstoke Mills, Whitchurch, Hamps, U.K.; R=Raypak, Inc., Westlake Village, California; S=Toray Industries, Inc., Otsu, Shiga, Japan; T=Toyobo Co., Ltd., Katata Research Center, Otsu, Shiga, Japan; U=UOP Fluid Systems Div., San Diego, California.

such as typical wastewater. Unfortunately, it is the permeators, which are virtually untried commercially, that offer the best combination: (1) tubular design with brine flow on the outside, (2) fiber design with brine flow on the inside, and (3) the dynamic membrane concept. The flow channel size, in the penultimate column of Table III, is presented as a measure of the cross-sectional area available for brine flow. In a crude calculation, a large value in the flow channel size should indicate little chance of fluid holdup owing to blockage from floatables or suspended solids. A small value should suggest that a high degree of prefiltering is necessary.

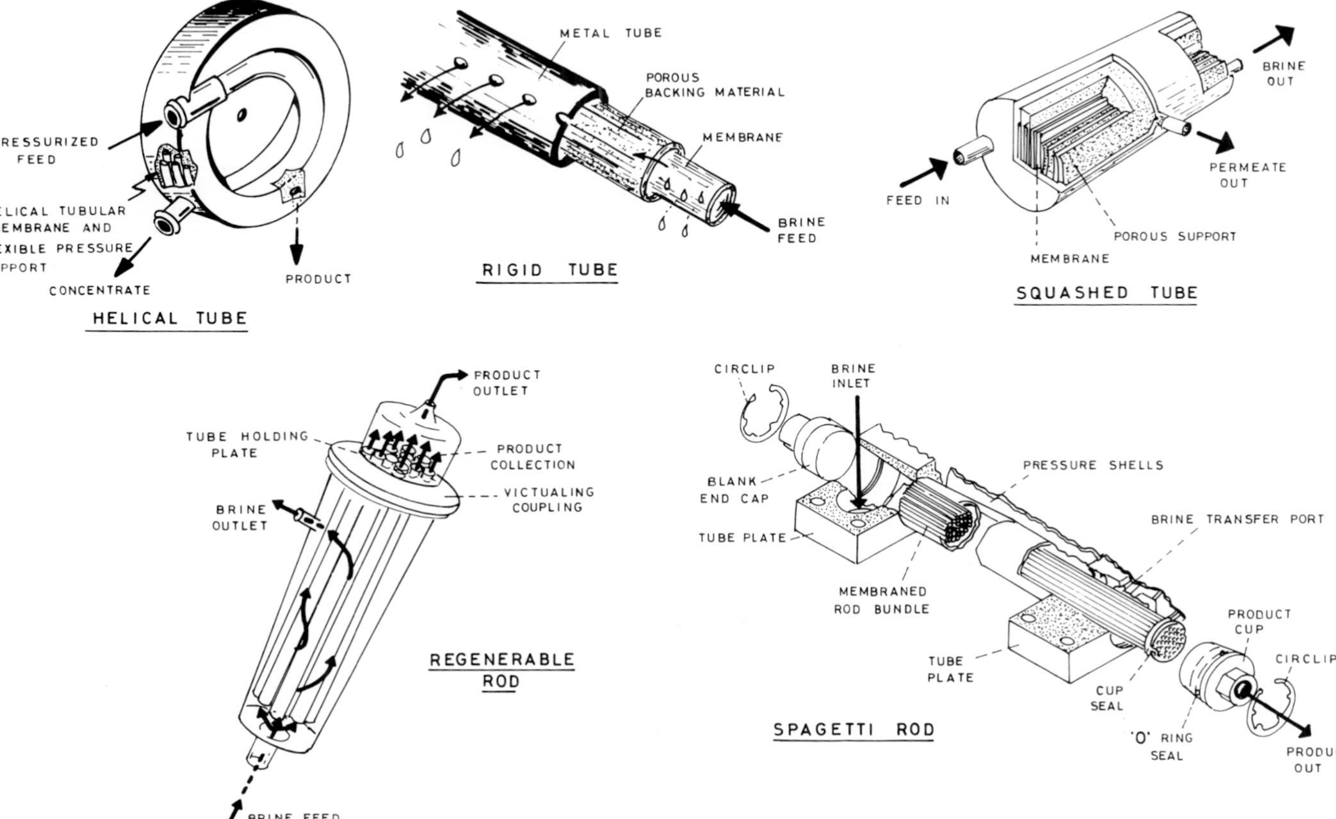

PRESSURIZED FEED
HELICAL TUBULAR MEMBRANE AND FLEXIBLE PRESSURE SUPPORT
CONCENTRATE
PRODUCT
HELICAL TUBE
METAL TUBE
POROUS BACKING MATERIAL
MEMBRANE
BRINE FEED
RIGID TUBE
BRINE OUT
PERMEATE OUT
FEED IN
POROUS SUPPORT
MEMBRANE
SQUASHED TUBE
PRODUCT OUTLET
TUBE HOLDING PLATE
PRODUCT COLLECTION
VICTUALING COUPLING
BRINE OUTLET
REGENERABLE ROD
BRINE FEED
CIRCLIP
BRINE INLET
BLANK END CAP
TUBE PLATE
PRESSURE SHELLS
BRINE TRANSFER PORT
MEMBRANED ROD BUNDLE
PRODUCT CUP
CIRCLIP
TUBE PLATE
CUP SEAL
'O' RING SEAL
PRODUCT OUT
SPAGETTI ROD

Fig. 6 Reverse osmosis membrane permeator designs.

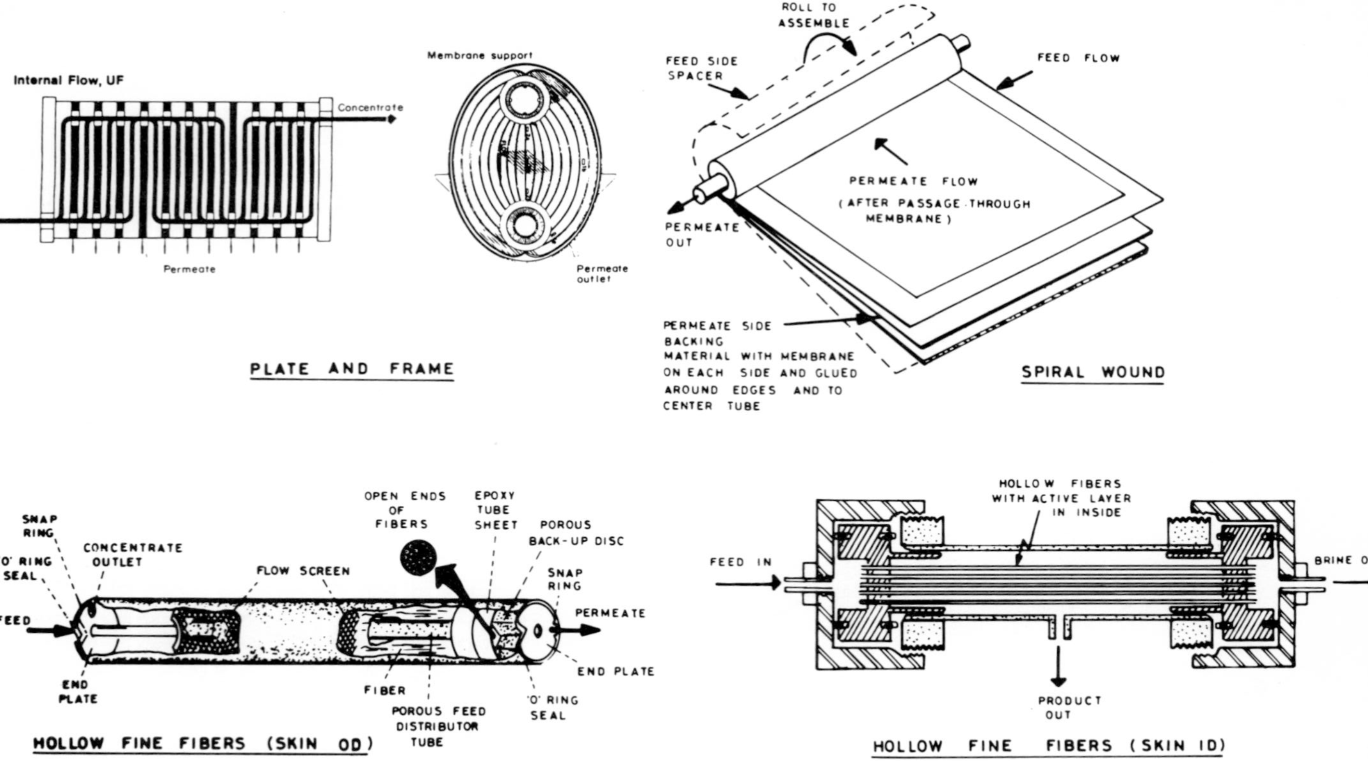

Fig. 7 Reverse osmosis membrane permeator designs.

TABLE III

Comparison of Reverse Osmosis Membrane Permeators

Module design	Packing density (ft^2/ft^3)	Water flux at 600 psi (gal/ft^2 day)	Salt rejection	Water output per unit volume (gal/ft^3 day)	Flow channel size (in.)	Ease of cleaning
Tubular						
1a Brine flow inside tube	30–50	10	Good	300–500	0.5–1.0	Very good
1b Brine flow outside tube[a]	140	10	Good	1400	0.0–0.125[a]	Good
Spiral wrap[b]	250	10	Good	2500	0.1	Fair
Fiber						
3a Brine flow inside fiber[c]	1000	5	Fair	5000	0.254	Fair
3b Brine flow outside fiber	5000–2500	1–3	Fair	5000–7500	0.002	Poor
Flat plate[d]	35	10	Good	350	0.01–0.02	Good
Dynamic membrane[e]	50	100	Poor	5000	~0.25	Good

[a] Data for spaghetti permeator obtained from Grover *et al.* (1973). The flow channel dimension can vary from zero (tubes touching) to about 0.125 in.

[b] Two different spiral-wound designs are commercially available. In the one case (UOP, CA) the permeate spirals to the center manifold, whereas for the other design (Toray, Japan) the brine stream spirals to the center manifold. Their performances are essentially equivalent.

[c] Data for fiber with brine flow inside obtained from Strathman (1973). Maximum internal pressure for this unit is 28 atm (410 psi).

[d] Data for flat-plate design obtained from Nielsen (1972).

[e] Data for dynamic membrane design estimated from Thomas *et al.* (1973).

A. HELICAL AND RIGID TUBE PERMEATORS

The membrane is cast as a tube on the inside of a porous support tube (e.g., paper or cloth), which is then placed inside a pressure vessel (see Fig. 6). The brine stream flows through the tube, and the product permeates the membrane radially. The pressure vessel may be a steel pipe with perforated holes (rigid design) or, if the support tube can withstand the pressure differential, a plastic or low-pressure housing (helical design) can be used to collect the product. Rods or spheres sometimes placed inside the tube along the center lines act to increase fluid velocity and axial shear at the membrane–solution interface. Sanderson (private communication, 1980) reports the production of a relatively inexpensive tubular vessel made from cast plastic elements held together with a rod. By slightly misaligning the elements, the cast membrane tube will have protrusions desirable for fluid mixing. Reynold's numbers as high as 130,000 have been used in tubular systems (Sachs *et al.*, 1975). Tubular units are easily cleaned and much operating data exists for them. Their disadvantages include low water production per unit volume, high water holdup per unit area of membrane, and relatively expensive membranes (about \$10–20/ft^2).

B. SPIRAL-WRAP PERMEATORS

Several flat or planar membranes are sandwiched between porous plastic screen supports and then rolled up, as shown in Fig. 7. The edges of the membranes are sealed to each other and to the central perforated tube. The resultant spiral-wrap module is fitted into a tubular steel pressure vessel, such as a 4-in. nominal pipe. The pressurized feed solution is fed into the pipe so it flows through the plastic mesh screens along the surface of the membranes. The product, which permeates the membranes, flows into the closed alternate compartments and spirals radially toward the weep holes in the central tube, where it is removed. Advantages of this design include relatively high water output per unit membrane area, and vast amount of operating data. The spiral-wrap design has probably been exposed to more hours of municipal effluent than any other design, aside from the rigid tubular design. However, it can only be cleaned chemically, not mechanically like the tubular design, and because of the small dimensions of the flow channel (Table III), it has a high probability of plugging. This design is one of the chosen designs for large-scale municipal treatment (Ajax International Corp., 1973). A similar but alternative design is also commercially available (footnote *b* in Table III).

C. Fiber Permeators

Several million hollow fibers almost as fine as human hair (100–200 μm outside diameter) are bundled together in a U-shaped configuration for brine flow on the outside or in a straight configuration for brine flow on the inside (see Fig. 7). The end of the fibers are epoxied into a tube sheet while making sure each fiber is not blocked. Thus when the brine flows at high pressure on the outside of the hollow fibers, the product permeates radially inward through the unsupported fiber. The product then moves inside the hollow-fiber bore to the product collection chamber. In the other case, when the design is similar to a typical heat exchanger, the brine flows into the bore of the hollow fibers at one end, and, after moving along the inside of the fiber, flows out of the other end of the unit. The product continually permeates radially outward through the fiber walls.

The shell-side feed hollow-fiber design (i.e., brine on the outside) is compact, low cost, has a low water holdup, and because of the compressive strength of the small diameter fibers, can withstand relatively high differential pressures (400 psi). It unfortunately plugs easily and is hard to clean.

The inside-feed hollow-fiber design has the advantages of the shell-side feed design plus the added advantage of well-controlled hydrodynamics of the feed, which improves the possibility of cleaning, but meager operating data are available.

D. Plate-and-Frame Permeators

The original plate-and-frame unit was similar in principle to the filter press (Aerojet General Corp., 1964, 1966). This design became defunct in the late 1960s because of many problems, the most important of which was the extreme difficulty and high expense of changing degraded membranes. Another unit using the flat-plate design (see Fig. 7) has been made commercially available (Madsen *et al.*, 1973; Nielsen, 1972). The original design was similar to a stack of phonograph records. Resulting from several years of intensive development, an improved design is available (Madsen, 1977). Alternate oval membranes and separating frames, used also for manifolding and sealing, are placed together and arranged for automatic internal staging in series. End plates are used to compress the stack similar to a conventional filter press. This design does not need a cylindrical pressure vessel because each membrane is individually sealed by its neighboring separating frame, one of the main advances over the original unwieldy plate and frame design. General advantages of this design include the low brine holdup per unit membrane area and the ability to desalt highly viscous solutions because of the thin channel height (0.01–0.02 in.). Its disadvantages include susceptibility to channel plugging and difficulties in cleaning, but membrane replacement is extremely easy.

The other designs shown in Figs. 6 and 7 and presented in Tables II and III are either not commercially available, such as the regenerable membrane designs, or have not made any impact on the market, such as the squashed membrane or the flat-plate spinning unit. The concept of a membrane regenerated *in situ* has been successfully proven and reported in the literature (Belfort *et al.*, 1973), but has not been commercialized.

VII. Applications of Reverse Osmosis to Water Desalination

The original purpose for supporting and developing the RO process in the 1960s as outlined by the Office of Saline Water and the U.S. Department of the Interior was to provide potable water from brackish water for isolated communities. Desalination of seawater and demineralization of wastewater by RO followed shortly thereafter. Early potential locations for brackish water membrane desalination were in the arid west (i.e., Buckeye, Arizona), and seawater desalting using distillation technology was in operation in Baja California, Mexico, and the Carribean. After the 1973 oil price hikes, RO became attractive for desalting seawater because of its lower energy (from one-half to one-third) consumption when compared with evaporative processes (Null, 1980). Other areas with needs for desalting plants include Florida and Texas.

For example, the largest seawater single-pass RO plant (3 mgd) was built in the Florida Keys. The plant uses polyamide hollow-fiber membranes and reduces the 38,000 ppm TDS feed to less than 500 ppm TDS in one stage (Boesch, 1981). Total costs including fixed and variable costs, are estimated to be \$3.83/kgal, excluding plant armortization and membrane replacement costs.

Because of radioactivity infiltration (Radium-226) into the drinking well waters of Sarasota County, Florida, a study to evaluate the effectiveness of several RO systems (hollow fiber and spiral wound) to remove Radium-226 was initiated (Sorg *et al.*, 1980). The largest operating plant eventually built in Sarasota County was 1 mgd, and several smaller plants were also installed at mobile homes and trailer parks. The membrane process successfully removed radioactivity (96%) from the water, however, the operating costs were estimated at about $1.00/kgal, which were considered high when compared with lime softening (75–96% removal) and ion exchange (81–97%). The added advantage of RO is that it removes excessive TDS (hardness, sulfates, and chlorides) efficiently.

VIII. Applications of Reverse Osmosis to Wastewater Renovation

One of the principal uses of RO is the desalination of brackish waters containing less than 10,000 ppm TDS. As an outgrowth of this application, many other applications have been pursued. Developments in the 1970s made one-pass desalination of seawater (35,000 ppm) to drinking water (500 ppm) a reality (Johnson and McCutchan, 1973; Riley *et al.*, 1973). Reverse osmosis, together with the ion exchange process, has been successfully used to produce ultrapure water for the electronics industry (deBussy and Whitmore, 1972; Haight, 1971; Riedinger and Nusbaum, 1972). Evidently, water from the combined process is less expensive than from either process separately. In addition, the treatment of feedwater for boiler units has been attempted by RO (Kosarek, 1979; Leitner, 1973; Riedinger and Nusbaum, 1972). The application of RO to industrial processes with a high value product is being tested. Examples of this include food processing, such as the concentration of whey (Goldsmith *et al.*, 1974; Madsen *et al.*, 1973; McDonough and Mattingly, 1970), orange juice (Merson and Morgan, 1968), maple sap (Willits *et al.*, 1967), coffee (Underwood and Willits, 1969), and apple and grape juices (Baxter *et al.*, 1979).

The remainder of this section will be devoted to the application of RO to wastewater renovation and will be divided into three areas: Industrial applications, municipal applications, and treatment of polluted rivers.

A. INDUSTRIAL APPLICATIONS

We shall be concerned primarily with the concentration of a given wastewater stream and the simultaneous production of improved water quality for reuse (Okey, 1972). The concentrate stream in the liquid phase may contain valuable constituents that can also be recycled and reused in the process or contain sludge or brine to be disposed of. Little attention has been given to disposal or its potential costs, although deep-well disposal, landfilling, evaporation methods, and transporting to the ocean have been suggested. Several relevant monographs describing various applications of membrane processes are available (Dresner and Johnson, 1981; Flynn, 1970; Lacey and Loeb, 1972; Mears, 1976; Sourirajan, 1977).

1. Pulp and Paper Industry

Considerable research has been conducted to test the applicability of RO to concentrate dilute pulp wastewaters (Anonymous, 1975; Wiley *et al.,* 1972a). The objective is to concentrate the dilute streams of 0.5 to 1% dissolved solids to about 10% solids using RO, and then to further concentrate the solids to say 50% by evaporation with eventual by-product recovery or disposal (Leitner, 1972). The overall objective is also to reduce direct discharge into rivers or lakes and to ensure that the biological oxygen demand (BOD), TDS, and odor and color are acceptable to the receiving waters. The definition of acceptability is defined by governmental authorities.

Average flux rates, using a rigid tubular RO system of about 2.5 to 5 gal/ft^2 day, with 80% water recovery, and rejections of more than 95% were reported for a wastewater effluent from a high-yield, sodium sulfite, chemicomechanical pulping process (Wiley *et al.,* 1972b). Problems included high osmotic pressures because of the high salt (Na_2SO_4) content and membrane fouling because of hydration of polysaccharides. Operating costs for various capacity plants (125,000–1,000,000 gal/day) were estimated to vary from $0.82–1.48/1000 gal of product water. Based on these encouraging results, plans were made for a 300,000–400,000 gal/day RO plant to be installed in Wisconsin.

Concentrates of high MW lignosulfonates at 30% solids and reducing sugars at 20% total solids have been fractionated and concentrated from spent sulfite liquors using both UF and RO (Bansal and Wiley, 1975).

Other studies by Wiley and his colleagues have evaluated a combination RO and freezing process at three bleached pulp mills. The brine was concentrated from 5 to 50 g/liter by RO and then to 200 g/liter by freezing resulting in a 50-fold concentration. Costs were high and disposal of the concentrates posed a problem (Josephson, 1978).

In another study, two dynamic (Zr–Si and Zr–PAA) and spiral-wound polysulfone membranes were evaluated on three paper mill process wastewaters (Porter and Edwards, 1977). Excellent color and fair-to-poor conductivity rejection was reported for short-term laboratory experiments.

2. *Mine-Drainage Pollution Control*

Present methods for treating acid mine drainage, such as neutralization and aeration, are wasteful in their inability to produce reusable water for industrial or municipal use. The residual water is high in dissolved solids, hardness, and sulfate concentration. The application of RO to the removal of nearly all dissolved solids and the production of a concentrate from which valuable heavy metals can be recovered has been conducted and reviewed by Wilmoth and Hill (1972). The spiral-wrap, hollow fine fiber, and tubular configurations have been tested and compared. Table IV summarizes typical rejection by these configurations. Unfortunately, because of the maximum allowable standards on iron and manganese concentrations of 0.3 and 0.05 mg/liter, respectively, the product water in Table IV is not potable. Field tests disclosed that two major causes of chemical fouling of RO membranes are calcium sulfate and iron (Mason and Gupta, 1972; Sleigh and Kremen, 1971). Bacteria have been shown to concentrate on the membrane and to oxidize ferrous ion, thereby fouling the membrane. The limiting factor in this high recovery RO operation was calcium sulfate precipitation. Various flushing techniques were used to remove the fouling layers, including high-velocity flushes to dislodge the precipitate, acidified flush to render the salts soluble, and a BIZ flush

TABLE IV

Typical Rejections by Reverse Osmosis Systems on Mine Drainage Effluents[a,b]

Systems	pH	Conditions[c]	Acidity	Ca	Mg	Total Fe	Fe^{2+}	Al	SO_4	Mn
Spiral wrap[d]										
Feed water	3.1	2070	460	260	170	77	64	12	1340	43
Product	4.4	17	38	0.4	0.3	0.4	0.3	0.2	0.9	0.5
Rejection (%)[e]	—	99.2	91.7	99.8	99.8	99.8	99.8	99.2	99.9	98.8
Tubular[d]										
Feed water	3.4	1050	250	125	92	78	61	12	660	14
Product	4.2	46	46	2.2	1.4	0.9	1.0	1.0	4.4	0.3
Rejection (%)[e]	—	95.6	81.6	98.2	98.5	98.8	98.4	91.7	99.3	97.8
Hollow fiber[d]										
Feed water	3.4	1020	210	150	115	110	71	15	940	14
Product	4.3	32	32	1.2	1.4	1.2	0.8	0.8	4.6	0.1
Rejection (%)[e]	—	96.9	84.8	99.2	98.8	98.9	98.9	94.7	99.5	99.1

[a] Source: Wilmoth and Hill, 1972.

[b] All units are mg/liter except pH and conductivity.

[c] Cond.—Specific conductane ($\mu\Omega$/cm).

[d] Recovery, 75%.

[e] Rejection = 100 (feed concentration − product concentration)/feed concentration.

(enzyme-active laundry presoak) to remove the organics. Ion exchange of cations was suggested as a pretreatment to reduce precipitation and allow for higher water recoveries (Pappano *et al.*, 1975).

An extended period of operation reported for mine-drainage feed waters on a single set of membranes was on the spiral modules that accumulated 4400 hr on approximately a −0.015 log–log slope* (Wilmoth and Hill, 1972). The longest period of operation on the hollow-fiber modules was 2670 hr on a −0.037 log–log slope. For the tubular unit, the longest successful run was 807 hr of operation on a −0.04 slope. No loss of solute rejection was noted in any of these long-term tests. Estimates of cost per 1000 gal of product vary between $0.35 and $1.50, depending on such variables as plant size, water quality, and disposal techniques.

In a study to determine the efficiency of removing selected trace elements from acid mine waters, Wilmoth *et al.* (1979) compared lime neutralization with RO and IE. Although the former process was slightly less efficient in removing trace elements, it was chosen as the favored process because of its power cost. Reverse osmosis is also being considered to treat oil shale wastewaters. Using laboratory test data Hicks *et al.* (1980) indicated the efficacy of this treatment. Besides phenol (>50%), all other major solutes such as TDS, ammonia, boron, and fluoride exhibit >80% rejections. They suggest a 2-yr membrane life, provided the system is operated properly without debilitating fouling and scaling. Together with resin adsorption, biological oxidation and steam stripping, they estimate treatment costs from $2.60 to $8.50/m^3 ($0.10–1.60/barrel) of air produced.

3. *Plating and Metal Finishing Operations*

Here again is the possibility of using RO to treat rinsewaters and wastes from plating and metal-finishing operations (Anonymous, 1975; Rozelle, 1971). The objective is to recover reusable water and to salvage valuable metals. Intensive laboratory-scale work has been conducted in Canada on electroplating effluents, especially on nickel-plating wastes (Golomb, 1972). Based on a small capacity plant of 2400 gal/day, processing costs for a nickel-plating rinse of about $3.00/1000 gal have been estimated (Leitner, 1973).

For treatment of cyanide bath wastes, which are generally at pH

* If one plots the log of the permeate flux versus the log of time, after a short initial nonlinear period, a linear negatively sloping plot is usually observed. This log–log slope designated as *b* has been used as an indicator of the severity of membrane fouling and a rough predictor of membrane life.

levels higher than 8, membranes other than CA are considered (Allegrezza *et al.*, 1975). In addition, rejections for the cyanide ion from cyanide or cyanide-complex solutions have been poor for all membranes tried (Hauk and Sourirajan, 1972; Leitner, 1973).

Hexavalent chromate rejection, though has been found to be highly dependent on pH. At pH ~7, the chromate rejection is greater than 99%, whereas for pH ~2.5, the chromate rejection is less than 94% (Riedinger and Nusbaum, 1972). Increasing the pH may cause other compounds such as aluminum hydrate from aluminum anodyzing to precipitate at pH ~4, causing RO membrane fouling problems.

Laboratory experiments on nine different plating bath rinsewaters were examined by Donnelly *et al.* (1976) using tubular and spiral wound, both with cellulose acetate membranes and hollow nylon fiber modules. The efficacy of using RO was confirmed, although the spiral wound and hollow fiber modules were chosen over the tubular module because of cost. Also, the new FT-30 membrane with wide pH tolerances would probably surpass the cellulose membranes in this application.

The RO process was also evaluated by treating wastewater for a steel container facility. The objective was to recycle as much of the wastewater as possible while treating the concentrate prior to disposal to the existing sewage plant (Marino *et al.*, 1978). Because of severe fouling by precipitation onto the membrane surface (zinc, iron, and calcium phosphates), lower than expected TDS rejections (average 88.2%) and product water recovery (83%) resulted. The authors conclude that neither IE, because of its high operating costs, nor RO because of the less than expected rejection and recovery, are suitable for this application.

Field and bench scale RO experiments in wastewater derived from automotive electrocoat paint operations have been reported (Anderson *et al.*, 1981). Spiral-wound CA membranes operating at 3100 kPa (450 psi) and 24–27°C were used. Some low MW solutes (ethyl, butyl, and hexyl glycol ethers) were purposely allowed to permeate the membrane for recycling to the paint operations. Weakly cleaning with dilute lactic acid to remove colloidal lead [$Pb_3(PO_4)_2$] was successful in maintaining respectable fluxes.

4. *Photographic Processing Industries*

Chemical recovery and rinsewater renovation have been studied using the tubular RO process in photographic processing industries. One study reports that 80–90% of the final washwater can be renovated for reuse (Cohen, 1972). In addition, ferro- and ferricyanides can be concentrated and reused as a bleach replenisher, and electrolytic recovery of the concentrated silver is possible (Mahoney *et al.,* 1970; reported by Cohen, 1972).

5. Food Processing Plant Effluent

In the United States, two laws governing the operation of food processing plants present obstacles to the use of the RO process. The first prohibits the reuse of water for food manufacturing, although some secondary applications may be considered, such as boiler feed and cleanup. The second requires sterilization temperatures that CA membranes (and probably most others) are not able to withstand.

Despite these restrictions, RO, ED, and UF have all been applied to food industry wastes. Ultrafiltration and RO have been used in series to concentrate whey from a dairy plant resulting in two saleable by-products: Protein concentrate and lactose (Horton *et al.,* 1972). Pilot plant testing in a dairy research center using a tubular unit, was also conducted in Ireland (reported by Cohen, 1972). The new flat-plate thin-channel configuration was tested on a whey effluent from a dairy plant (Madsen *et al.,* 1973; Nielsen, 1972). They point out the advantages of their thin-channel design, emphasizing the hydrodynamic advantage over other units for highly viscous solutions.

Tubular modules were used to concentrate the effluents from a potato starch factory (Porter *et al.,* 1970). The permeate is good for reuse, and the concentrate with some further treatment can produce soluble proteins and free amino acids. Reverse osmosis costs, however, will restrict the commercial feasibility of such treatment until low-cost membranes become available.

Because of legal pressure on industries to prevent their effluents from polluting the natural surface and groundwaters, the financing arrangements of municipalities (Chapter 1 in this volume), and disparity between municipal volume needs and the present capacities of membrane technology, we expect to see industry lead the way in using membrane processes. Other unmentioned applications include the treatment of cooling tower make-up water, laundry wastes, petrochemical effluents, detergents removal from nuclear wastewater, treatment of hospital wastes, removal of pesticides, reuse of textile dyeing wastes, and waste-cutting oils (Brandon *et al.,* 1975; Chian *et al.,* 1975; Cohen, 1972; Cohen and Loeb, 1973; Fang and Chian, 1974; Gaddis *et al.,* 1979; Gollan *et al.,* 1975; Johnson *et al.,* 1973; Markind *et al.,* 1973, 1974; Minard *et al.,* 1975; Riley *et al.,* 1980; Sammon, 1976; Schmitt, 1974; Sonksen *et al.,* 1979).

B. MUNICIPAL APPLICATIONS

Because of the stringent requirements concerning the quality of municipal effluents and because of the general need to consider wastewater reuse, advanced methods such as RO have been extensively tested to determine their future role in various municipal treatment trains. Thus the RO process has been tested not only on raw sewage (Conn, 1971; Sprague, private communication), but also on secondary effluents with different types of pretreatment. At least two intensive studies comparing the performance of several RO configurations on municipal wastes have been conducted in the United States (Boen and Johannsen, 1974; Smith *et al.*, 1970). In addition, many studies on municipal wastewaters have been reported and are discussed in the following sections.

1. *Initial Studies*

That CA membranes could indeed concentrate organic as well as inorganic solutes from municipal wastewaters was established in the early 1960s (Aerojet General Corp., 1965). Thereafter, a series of field tests was conducted to evaluate spiral-wound modules with diatomaceous earth-filtered secondary effluent (Bray and Merten, 1966). These tests showed effective removal of many secondary effluent constituents, but revealed fast permeate flux decline (within hours). This, as will be evident later, is a major problem encountered in treating sewage with membrane processes. Considerable effort has gone into attempts to minimize this flux decline phenomenon.

A series of pilot plant investigations of the spiral-wrap and the tubular configurations, as well as of the flat-plate unit (Smith *et al.*, 1970), led to the following conclusions.

(1) Cellulose acetate membranes were capable of rejecting more than 90% of the TDS, phosphates, particulate matter, total organic carbon, and chemical oxygen demand (COD). Ammonia nitrogen and nitrate nitrogen are only rejected from 80 to 90% and 60 to 70%, respectively. The rejection performance exhibited by the three types of modules were similar.

(2) The pH control of the feed was successful in minimizing inorganic precipitation at recoveries as high as 80%.

(3) The principal cause of flux decline was attributed to organic fouling of the membranes. A correct identification of the class of material responsible for flux decline was not determined, but soluble, colloidal, and suspended species are all fundamentally involved in fouling.

(4) Organic fouling was most successfully controlled by routinely depressurizing the unit and flushing it with an enzyme-active detergent.

(5) The tubular unit, because of its well-defined hydrodynamics, showed less tendency to plug.

2. *Later Studies*

Several comprehensive investigations have been conducted on individual configurations at various sewage plants (Bishop, 1970; Cruver and Nusbaum, 1974; Cruver *et al.*, 1972; Eden *et al.*, 1970; Feuerstein and Bursztynsky, 1969; Fisher and Lowell, 1970; Grover and Delve, 1972; Hardwick, 1970; Loeb *et al.*, 1974) or on several configurations tested simultaneously under similar conditions of feed, pretreatment, and pressure (Currie, 1972). Results from a few of the most important studies are presented in Tables V and VI and Fig. 8.

In Table V, the results of three different studies are summarized, each using a different membrane configuration or permeator design. Within each study, the type of effluent (Cruver *et al.*, 1972) or the membrane formulation was varied (Feuerstein and Bursztynsky, 1969; Fisher and Lowell, 1970). The purpose of the first two studies (spiral wrap and tubular) reported in Table V was to determine reliable flux decline slopes for long-term tests on different types of municipal effluents. The purpose of the last study (flat plate) was to evaluate the effectiveness of the blended CA membranes for wastewater treatment.

Figure 8 includes details of three typical lifetime tests (i.e., of 50- , 56- , and 77-day duration, see Table V) for product water flux versus time of operation. The following conclusions are drawn from the results shown in Table V and Fig. 8.

(1) If necessary pretreatment (sand filtering) and periodic membrane cleaning are rigorously adhered to, CA membranes perform satisfactorily for extended periods at reasonable flux decline rates ($b \approx -0.10$) in the spiral-wrap and tubular configurations.

(2) Activated carbon pretreatment has been shown to be unnecessary for successful RO operation on secondary effluent (see tests on 50- and 65-day duration in Table V and Fig. 8), although sand-filtered secondary effluent is necessary to maintain reasonable product flux.

(3) Flux decline slopes for adequately treated and cleaned secondary effluents are in the range -0.025–-0.052, whereas for primary effluents a correspondingly higher range -0.040–-0.074 is to be expected.

(4) Although not obvious from Table V, it was found experimentally that the higher the initial flux of the membrane, the worse the flux

TABLE V

Performance of Cellulose Acetate Membranes with Sewage Effluents[a]

Configuration and membrane	Type of[b] effluent	Cleaning during operation	Initial[c] flux (gal/ft^2 day)	Flux[d] decline slope	Extra-polated[e] average flux after 1 yr (%)	Recovery (%)	Feed pressure lb/in.^{2}	Duration of test days
Spiral wrap and cellulose diacetate[f]	CFSE	Daily 15 min tap water flush and 2% BIZ flush/week	10.7	−0.025	79	80	320	56
	SFSE	Daily 15 min tap water flush and twice 2% BIZ flush p/week	10.9	−0.037	71	80	320	50
	SFSE[g]	Same	10.0	−0.052	62	80	320–690	104
	CCPE	Same	15.8	−0.074	51	75–80	400	146
	CFCCPE	Same	15.1	−0.049	64	75–80	400	66
Tubular and cellulose diacetate[h]	CCSFPE	None	13.0	−0.056	60	80–95	700	77
Tubular, cellulose acetate, and cellulose triacetate blend[h]	CCSFPE	None	25.0	−0.040	69	80	700	11
Flat plate and ds 2.55 blend (type 35C)[i]	Secondary effluent	None	46.0	−0.032	75	—	600	17

(*continues*)

TABLE V (*continued*)

Configuration and membrane	Type of[b] effluent	Cleaning during operation	Initial[c] flux (gal/ft^2 day)	Flux[d] decline slope	Extra-polated[e] average flux after 1 yr (%)	Recovery (%)	Feed pressure lb/in.^{2}	Duration of test days
Flat plate and E-383-40 cellulose[i]	Secondary effluent	2 g/liter BIZ solution at 600 psi for 30 min/day	50.0	−0.054	61	—	600	6
Flat plate and CAM-360 70°C[i]	Secondary effluent	2 g/liter BIZ solution at 600 psi for 30 min/day	58.0	−0.049	64	—	600	6

[a] In general, all parameters had rejections above 85%, except for the NO_3^- ion which was rejected between 50 and 70%.

[b] The following abbreviations are used: CFSE, carbon filtered secondary effluent; SFSE, sand filtered secondary effluent; CCPE, chemically coagulated (alum) primary effluent; CFCCPE, carbon filtered chemically coagulated primary effluent; CCSFPC, chemically coagulated sand-filtered primary effluent.

[c] Multiply gal/ft^2 day by 0.0407 to get m^3/m^2 day.

[d] Flux decline slope is defined by $b = \Delta \log(\text{flux})/\Delta \log(\text{time})$, dimensionless.

[e] Expressed as a percent of the initial flux.

[f] Source: Cruver, Beckman, and Bevage (1972).

[g] Constant product experiment at 10 gal/ft^2 day.

[h] Source: Feuerstein and Bursztynsky (1969).

[i] Source: Fisher and Lowell (1970).

TABLE VI

Rejection of Various Constituents from Municipal Sewage Plant Secondary Effluents by Cellulose–Acetate Membranes

	Feuerstein and Bursztynsky (1969)[a]	Eden *et al.* (1970)	Cruver *et al.*, (1972)[a]		Eastern Municipal Water District, Hemet, California[b] Boen and Johannsen (1974)						
Configuration pretreatment conditions	Tubular alum	Tubular sand filtered	Spiral sand filtered	Spiral carbon filtered	Tubular sand filtered	Fiber sand filtered	Spiral sand filtered	Rod sand filtered	Tubular none	Fiber none	Rod none
Duration of test (days)	—	83	50	56	17	17	17	17	14	14	14
Pressure (psi)	700	—	320	320	400	400	400	400	400	400	400
Recovery (%)	80	56	80	80	44	56	37	9	37	37	6
Rejections (%)											
(a) COD	96	98	96	91	>99	91	96	99	97	83	96
(b) Conductivity	93	92	95	—	—	—	—	80	—	—	92
(c) Sulfate	—	97	>99	>99	>99	80	97	97	>99	77	99
(d) Chloride	—	93	95	93	—	—	—	—	—	—	—
(e) Nitrate nitrogen	41	81	—	—	80	71	55	77	82	63	73
(f) Phosphate	98	98	>99	>99	98	80	90	96	>99	90	98
(g) $Ca^{2+} + Mg^{2+}$	—	95	>99	>99	99	86	98	94	99	88	96
(h) Ammonia nitrogen	98	—	95	97	—	—	—	—	—	—	—
(i) Turbidity (JTU)	—	62	—	—	92	88	94	89	—	—	—

[a] Product flux results appear in Table V.

[b] For 17 days all the RO units received a feed of sand-filtered, postchlorinated secondary effluent; thereafter for 14 days the sand filtration was bypassed. The tubular, fiber, spiral, and rod (membrane on the outside of a porous stick) permeators were supplied by Universal, Du Pont, Gulf and Raypack, respectively, and the percent reduction for the product flux per unit pressure difference, as a result of discontinuing sand filtration, was 8.01, 43.03, 22.38, and 26.16, respectively.

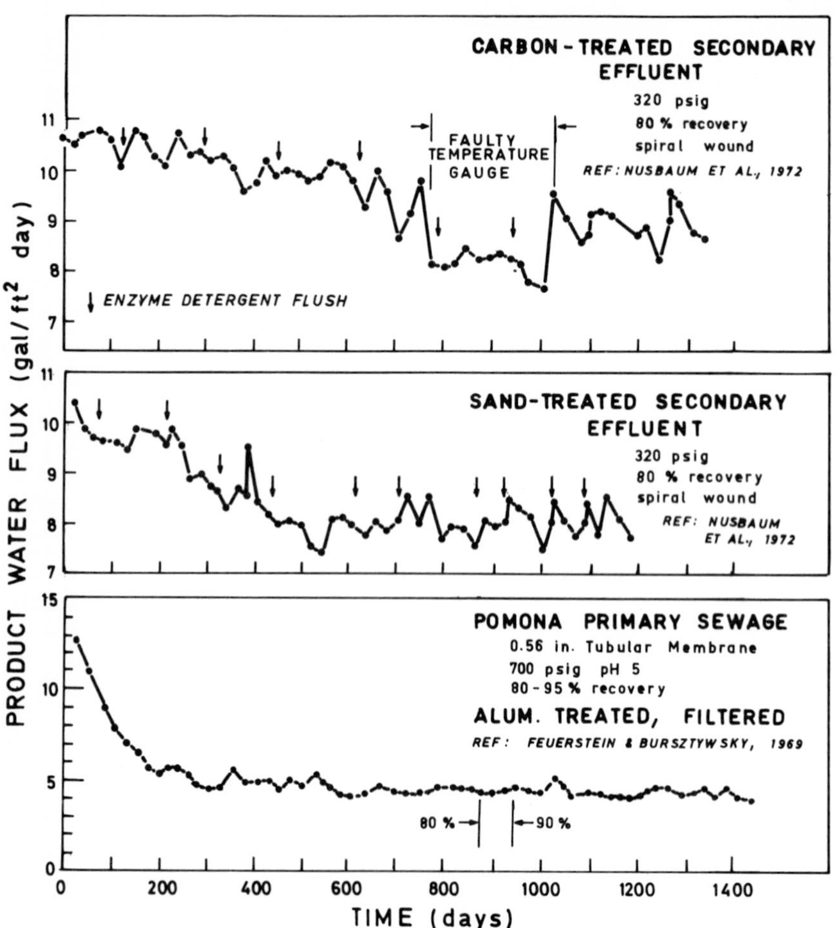

Fig. 8 Product water flux versus time for different pretreated municipal effluent feed waters.

decline slopes became (Eden *et al.*, 1970; Nusbaum *et al.*, 1972a). This phenomenon is probably related to membrane fouling.

(5) Rejections to various constituents for the tests in Table V were all above 85%, except for NO_3^- ions.

Details from four studies on the rejection of various constituents from municipal sewage plant secondary effluents by cellulose acetate membranes are presented in Table VI. Only nitrate nitrogen consistently showed poor rejections (between 41 and 82%). All other compounds tested exhibited rejections of over 85%. For the spiral unit, sand-filtered secondary effluent showed the same rejection percentages as did carbon-filtered secondary effluent (Cruver *et al.*, 1972). Results from a comparative study (Boen and Johannsen, 1974) indicate that with or without pretreatment of secondary effluent feed, the tubular and spiral units displayed the best performance. The fiber unit had slightly lower rejections, whereas the rod unit (membrane cast on the outside of a porous ceramic stick) had very low percent recoveries. The tubular, fiber, spiral, and rod designs displayed K_1 values [see Eq. (9)] of 1.4077, 0.1097, 0.5439, and 0.7922×10^{-5} gm/cm^2 sec atm, respectively.

Hamoda *et al.* (1973) and Matsuura and Sourirajan (1972) tested a porous cellulose acetate membrane (designated Batch 316) for low-pressure applications. Hamoda *et al.* (1973) evaluated both water and solute permeability of a series of organic compounds (glucose, sucrose, soluble starch, beef extract, glutamic acid, sodium stearate, and detergents) as a function of solute concentration. For these compounds, they obtained rejections greater than 88.5% (with most rejections above 95%) and fluxes between 35.6 and 42.7 gal/ft^2 day.

Other investigators have used RO to concentrate trace organic contaminants in drinking water (Deinzer *et al.*, 1974, 1975) and to evaluate the removal efficiencies and product fluxes of selected organic chemical species (sometimes termed organic refractories) found in abundance in most treated sewage and other waste effluents (Bennet *et al.*, 1968; Edwards and Schubert, 1974).

3. *Further Studies*

Because of commercial availability, advanced development, and fair stability to oxidants such as chlorine, cellulose acetate membranes as indicated previously, were predominantly used in laboratory and pilot plant studies for treating wastewater from the mid-1960s to the mid-1970s. Most researchers were aware of the potential biodegradation of cellulosic material in municipal wastewater environments and anticipated the development of more suitable membranes. The new thin-film composite RO membranes discussed previously have excellent resistance to biodegradation and wide pH and temperature tolerance (see Table I). In addition, the FT-30 membrane is stable with respect to chlorine. A concerted effort to develop the knowledge to reduce the mineral content of municipal and other effluents has been expended in California (Blanton, 1977). A desalting program to evaluate RO on tertiary treated water in both pilot and field scale has been conducted at the advanced wastewater treatment plant in Orange County. Because the design, operation, and general experience obtained at this plant, named "Factory 21," is the subject of Chapter 10 in this volume, it will not be reviewed here.

Another project was between the City of San Diego and the State of California. A comprehensive evaluation of tubular RO membranes to treat raw sewage with minimum pretreatment was conducted. Periodic 8-hr addition of precoat (carbon, diatomaceous earth, and hexametaphosphate) are added to the feed and deposited on the membrane surface to protect it (Sprague *et al.*, 1975). The project was eventually terminated because of lack of funds, but the feasibility of treating raw sewage with membrane precoats was clearly proven.

Hansen and Bailey (private communication, 1978) developed and tested a treatment train consisting of primary clarification, extended aeration, activated sludge treatment, clarification, and filtration followed by RO and carbon adsorption for organic polishing wastewater effluent at a Sierra ski resort. An interesting feature of the spiral-wound design was the convertibility of a two-stage series RO section to a less efficient one-stage parallel section in the case of high wastewater flow rates. Problems with membrane fouling during the first year led to the installation of a secondary effluent polishing unit (physical–chemical treatment process) prior to the membrane section. Thereafter, excellent performance over four seasons was reported. Total unit treatment costs to produce a final effluent closely matching commercially available water were $22/kgal. This price, although high, is considerably less than the other possible alternative—tank trucking the effluent out of the Kirkwood Creek Basin.

Another feasibility study funded by EPA consisted of treating 150,000 gpd of secondary effluent from the Escondido facility in California (Beckman, 1979). Although the study was plagued by pump problems, a known problem for RO, the pretreatment system included a polishing clarifier, multimedia filter, gravity filter, and the RO system. Cost estimates from this system were similar to that of the Factory 21 application (see Chapter 10).

C. TREATMENT OF POLLUTED RIVERS

1. *Pilot Plants*

Several feasibility studies were conducted in the 1970s to determine whether RO is capable of treating polluted river water (Kuiper *et al.*, 1973, 1974; Melbourne, private communication, 1973; Miller, private communication, 1973). All tests were carried out with asymmetric cellulose acetate membranes in tubular configurations. The purpose of the studies was to determine the effect of polluted river water on process variables such as flux decline rate, required pretreatment, effectiveness of various membrane cleaning techniques, recovery ratios, and percent rejections.

One of the earliest studies focused on the effect of tangential brine velocities on the arrest of product flux decline for a feed of untreated river water (Sheppard and Thomas, 1970). This study was the first to show that, for a particular feed from the Tennessee River, the absolute flux and the flux decline rate were directly related to the tangential brine velocity. It suggested that there exists a threshold velocity above which the product flux J remains relatively constant and also that the tighter the membrane, the lower the thresold velocity. This study also showed that at high tangential velocities such as 24 ft/sec, the flux decline parameter b was consistently low (−0.02 to −0.03) irrespective of feed composition. These and other studies have highlighted the relationship between the hydrodynamics of the brine stream and the reduction of the membrane flux decline parameter b (Sheppard and Thomas, 1971; Sheppard *et al.*, 1972). The next question is one of economics: What pumping power is required to maintain these high velocities, and how does this relate to the optimum economic condition for plant operation?

Other studies using RO on sand-filtered Rhine River water in the Netherlands, and Thames and Trent river waters in the United Kingdom have been conducted. Detailed operation over 19 months for a 16 m^3/day RO pilot plant using Rhine River water has been reported (Kuiper *et al.*, 1973). Aside from some anomalous behavior at the end of the study, the main problem of membrane fouling and associated flux decline was adequately controlled by using chlorination, coagulation with iron, rapid sand filtration as a pretreatment, and membrane-cleaning procedures comprising daily depressurization, washing with acid, flushing, and mechanical cleaning. The average applied pressure was 40 atm and the average recovery about 70%.

A 7 m^3/day (1850 gpd) pilot plant was operated on sand-filtered Thames River water for 18 months at 27–34 atm (400–600 psi) (Miller, private communication). This study sought to establish the pretreatment techniques and/or membrane-cleaning methods required for the treatment of river waters. They showed that membrane fouling strongly correlated with the turbidity of the feed waters and the percent recovery of feed water. The study used detergent membrane cleaning and foam ball swabbing and showed that both methods adequately arrested product flux decline. During the period of operation (8000 hr), they were able to maintain permeabilities from 0.22 to 0.20 m^3/atm day.

Another pilot plant using Trent River water as a feed was able to reduce TDS from between 650 to 700 mg/liter with adequate BOD removal (equivalent to 30 min through an activated carbon stage) (Melbourne, private communication, 1973). This tubular plant was operated with rough upflow sand filtration, pH adjustment to 6, and membrane cleaning by a soap solution and foam swab at regular intervals.

2. *Large-Scale Commercial Plants*

When the 96 mgd Yuma, Arizona, desalting plant began operation in 1982, it dwarfed all other desalting plants in operation (Unknown, 1980). The quantity of membranes needed for such a plant probably surpasses the total world manufacturing capacity for manufacturing membranes. The plant was commissioned by the U.S. Government to reduce the TDS concentration of irrigation return flows from 3200 to 283 mg/liter as part of the U.S. commitments to supply Mexico with sufficient quantity of water at a high enough quality level for irrigation.

Because of the extremely large capacity of the desalting plant and the need for high reliability, a series of pilot plant studies was initiated in 1974 to provide the necessary design and procurement data. As a result of these preliminary studies, RO using spiral-wound modules was chosen as the most efficient from a physical cost point of view. One manufacturer was awarded $20.6 million for a 73.3 mgd plant, and another was awarded $7.2 million for a 22.4 mgd plant. Two pretreatment systems were tested, and partial line softening followed by multimedia filtration was chosen for the final design. According to Lopez (1979) and on the basis of January 1979 prices, the total cost of the desalting complex was estimated at $190 million (including interest during construction). He estimated the operation and maintenance cost at $0.77/kgal with a $2/kgal investment per daily gallon installed capacity.

IX. Control of Product Flux Decline

When any conventional municipal effluent is treated with a pressure-driven membrane process, the product flux will decline with time, eventually rendering the process uneconomical. Three phenomena were shown to be responsible for product flux decline in long-term performance of RO membranes (Belfort, 1974b). They are (a) membrane *hydrolysis* resulting in an increase in both water and salt flux; (b) membrane *compaction* resulting in an initial flux decline with little effect on rejection (Bennion and Rhee, 1969); and (c) membrane *fouling* resulting in a decrease of water flux and possibly salt flux, or a relatively constant water flux with a decrease in rejection (Kuiper *et al.*, 1973, 1974).

As we shall see, feed pretreatment and membrane-cleaning methods have been developed to curtail product flux decline. These two processes are closely linked and a determination of the economic optimum for the control of product flux decline must include a consideration of each.

A. PRETREATMENT

Most researchers, using municipal effluents as a feed, have assumed that fewer problems might be encountered if effluents of the highest available quality were treated. However, some feel this assumption may not be entirely valid (Belfort, 1974; Sachs, private communication, 1973).

Several different pretreatment methods are shown in the flow diagram in Fig. 9 and Table VII. The most frequently used pretreatment methods for municipal feeds include clarification with chemical coagulants, such as alum, to remove large suspended solids and some dissolved organics and activated carbon filtration to remove small suspended solids and a large proportion of the dissolve organics.

Conventional water treatment technologies are used to pretreat the particular feed water so the membrane process can operate with acceptable performance (flux, flux decline, and retention characteristics). Table VII, taken from Pepper's (1981) excellent review of pretreatment methods indicates methods to remove and/or stabilize various fouling problems. In some cases a particular method would not be suitable; explanations to this effect appear in the column "Counter-Indicator." To eliminate a particular treatment train, long-term variations of feed (or source) water quality should be known. In addition to that discussed in Table VII, temperature of the feed water is an important characteristic. The temperature range at which the membranes are stable is clearly an important limiting factor. Solubility limits for various inorganic precipitates ($CaSO_4$) are also temperature dependent.

Acidification reduces the rate of product flux decline by increasing the solubility of inorganic precipitates such as $CaSO_4$, $CaCO_3$, or $Mg(OH)_2$ and is essential for minimizing the rate of hydrolysis of the CA

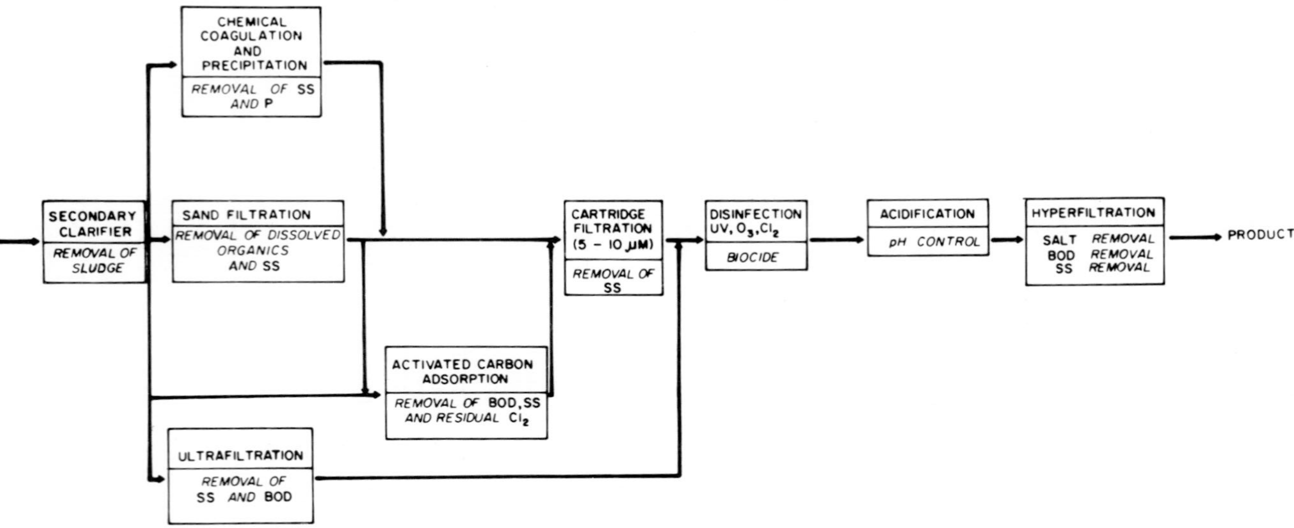

Fig. 9 Reverse osmosis pretreatment for a secondary effluent feed.

TABLE VII

Summary of Pretreatment Methods[a]

Problem	Primary (secondary) pretreatment	Purpose	Counter indications
(1) Ca-Mg bicarbonate	(a) Base exchange softening	Removes Replaces Ca/Mg by Na, which has a soluble bicarbonate	High TDS causes slip, max. 2000 mg/liter. Too expensive at greater than 115 m^3/day
	(Sequestrant)	(Delays any tendency for slip to cause precipitation)	
	(b) Lime softening	Removes Precipitates Ca and bicarbonate as calcium carbonate and Mg as magnesium hydroxide	Not suitable for less than 2000 m^3/day
	(Sequestrant) or acid dose)	(Prevents post precipitation)	
	(c) Acid dose	Removes Replaces bicarbonate with the more soluble chloride or sulfate	Difficulty and cost of obtaining acid
	(Sequestrant)	(Back up if acid dosing fails)	
	(d) Add sequestrant	Stabilizes Delays the formation and modifies the nature of the precipitate	
(2) Ca sulfate scale	(a) Base exchange softening	See (1)a	
	(b) Add sequestrant	See (1)d	
(3) Silica scale	(a) Raise temperature	Stabilizes Increases solubility	Cost of heat
	(b) Lime softening	Removes Brings down some silica with $CaCO_3$ and $Mg(OH)_2$	Not suitable for less than 2000 m^3/day
(4) Iron Precipitation	(a) Oxidize (aerate) and filter	Removes Precipitates iron as ferric hydroxide, easily removed by filter	Presence of other oxidizable material such as H_2S

(continues)

TABLE VII (*continued*)

Problem	Primary (secondary) pretreatment	Purpose	Counter indications
	(Acid dose)	(Prevents further precipitation)	
	(b) Exclude oxidizing agents e.g., air or Cl_2	Stabilizes Keeps the iron in the soluble ferrous state	Not good for intermittent use or where operating and maintenance staff are of poor quality
	(Acid dose)	(Increases solubility)	
	(c) Acid dose	Stabilizes Keeps iron in solution	pH to 5, acid expensive
(5) Colloids	(a) Coagulate and filter	Removes Causes the colloids to form larger particles which can be filtered out.	
	(Acid dose for some coagulants)	(Prevents post precipitation of coagulant	
	(b) Base exchange softening	Stabilizes Discourages coagulation, the solution is concentrated and colloids go through the RO plant into the concentrate	Not suitable for high TDS water where slip through the softener causes more than 5 ppm total hardness in the softened water
(6) Bacteria	Sterilize (Cl_2 and filter	Removes	
	(a) Dose sodium metabisulphate	Removes Cl_2 that would otherwise damage membranes	
	(b) Carbon filter		
(7) H_2S	(a) Degas and add Cl_2	Removes Most of the H_2S comes off as gas, the remainder is oxidized to sulfate	Hard waters require acid dosing before degasser to prevent scaling of degasser packing
	(See (6)a and b)		

TABLE VII (*continued*)

Problem	Primary (secondary) pretreatment	Purpose	Counter indications
	(b) Exclude oxidizing agents e.g., air & Cl_2 (Permeate must be degassed)	Stabilizes Stays in solution	Presence of bacteria, or poor operating/maintenance staff
(8) Cl_2	(a) Dose sodium bisulphite (Back up with second dose)	Removes Chemically destroys the Cl_2	Disastrous if dosing fails
	(b) Carbon filter	Removes	Filter often becomes a breeding ground for bacteria

[a] After Pepper, 1981. There is evidence that pretreating with activated carbon filtration is in the long run (>50 days) no better than sand filtration (Nusbaum *et al.*, 1972a, b) and fouling also occurs even after the UF pretreatment (Bailey *et al.*, 1973). What does this evidence lead us to conclude?

membranes. The pH is usually kept between 5 and 6 by using a mineral acid.

Cartridge filters are often used for final polishing of the feed solution, with their removal size dependent on the permeator configuration.

Increasing the temperature of the feed stream has improved the efficiency of the ED process (Forgacs, 1967). It was thought one might gain by increasing the temperature of the secondary effluent feed for the RO process. Unfortunately, it was found that increased fouling at higher temperatures offset any potential advantage (Bailey *et al.*, 1973). Results by the author have confirmed such an effect with ammonia-based spent sulfite liquor separation by UF.

Ultrafiltration has been used with some success as a pretreatment process for RO (Bailey *et al.*, 1973). It appears that some fouling material for the RO membranes is dissolved or in fine colloidal suspension, the particles being smaller than the average pore size of the UF, sand, and activated carbon prefilters. O'Melia (1980) and Yao *et al.* (1971) showed that particles in the size range 0.2–2 μm exhibit relatively small Brownian motion and are not easily removed by other mechanisms such as gravity, inertia, etc.

The more treatment a raw sewage is subjected to, the more likely it will contain very small particles, i.e., the spread of the particle size distribution will widen with increased treatment. Thus, if minute particles ($\ll 1$ μm) contribute to fouling, the assumption stated at the beginning of this section, namely, fewer problems will be encountered with an effluent that has a higher degree of treatment, may not necessarily hold. In fact, raw sewage with virtually no pretreatment (minor pulverization of the feed to protect the centrifugal pump) has been successfully treated by RO (Conn, 1971). Radically different membrane-cleaning techniques are used for this system. They will be discussed in the next section.

Before discussing cleaning methods, it is appropriate to examine the most hazardous fouling constituents of domestic sewage. They include biological activity, dissolved organics (as measured by BOD, COD, or TOC), and suspended solids (SS). The exact contribution to fouling of each of these constituents is not known. As mentioned previously, model compounds found in sewage have been tested for their rejections. In addition, a similar study to evaluate potential fouling constituents has been conducted (Belfort *et al.,* 1975). Biological activity is, however, a particular hazard, because the CA membrane is susceptible to biological degradation, and the sewage feed source is an ideal environment for microbial growth. Chlorination of the feed water prior to RO treatment has been used to prevent biological attack. Residual chlorine, however, is dangerous to the membranes, especially to the aromatic polyamide membranes. Results from a study of the causes of biodegradation on tubular CA membranes by agricultural runoff wastewater suggest a two-step mechanism (Richard and Cooper, 1975). The first step is the metabolism of the acetyl groups and then the cellulutic microorganisms are able to act on the remaining modified material. Quoting Richard and Cooper (1975, pp. 97 and 103): "Acidification to pH 4–6 with sulfuric acid or removal of dissolved oxygen with catalyzed sodium sulfite (best method) or chlorination at 0.1–0.2 ppm total residual prevented bacterial attack of the membranes. Spongeball cleaning appeared not to appreciably reduce colonization of the membrane surface by bacteria nor subsequent biodeterioration of the colonized membranes."

Bettinger (1981) reported an effective method to reduce the degradative effects of biological activity in a surface water RO plant containing aromatic polyamide hollow fibers sensitive to chlorination or other strong oxidation. They used shock iodine treatment administered daily for a 30 min period. As a result of this treatment, the plant did not require any cleaning during an 8-month period. The membrane was expected to be slowly iodinated, exhibiting a "somewhat faster rate of flux decline". The economics of this continuous operation, however, far outweigh the disadvantage.

B. Membrane Cleaning Techniques

Several approaches have been used in attempts to reduce the rate of product flux decline in systems treating municipal effluents (Belfort, 1974). See Table VIII for a detailed summary of these techniques. A summary follows of the membrane cleaning experience of several groups in RO and wastewater treatment.

The early studies with spiral-wrap and tubular modules were moderately successful in maintaining fluxes by daily air–water flushes at low pressure (Merten and Bray, 1967). Initial studies with citric acid and an anionic detergent solution proved unsuccessful in arresting flux decline in the flat-plate unit (Smith *et al.,* 1970). Periodic cleaning (approximately every 10 days) with an enzyme active presoak solution (BIZ) was successful in holding the flux relatively constant for tubular modules with carbon treated-secondary effluent (Belfort *et al.,* 1973). Other studies on a similarly treated feed solution with spiral-wrap modules, used daily air and water flushes and weekly flushes with a solution of 10,000 mg/liter of an enzyme presoak product (Nusbaum *et al.,* 1972a). These methods were able to maintain a relatively low product flux decline rate ($b = -0.05$). Cruver and associates (1972) showed sodium perborate, EDTA, and BIZ flushes each restored the flux to 80 to 85% of the initial value, each additive giving approximately equal results. One problem with the application of these flushes is that their pH is usually harmfully high (>9) for CA membrane. Raw sewage was treated successfully by RO in San Diego using the Havens tubular units. The product flux decline rate was arrested by precoating the membranes every 8 hr under pressure with diatomaceous earth, powdered activated carbon, and a surface-active agent (Conn, 1971). The precoat was probably protecting the membranes from fouling. One disadvantage of this method may be the abrasiveness of the precoat, which may reduce the lifetime of the membranes. Most systems operating with tubular modules have used the foam swab flushing technique for feeds with high suspended solids concentration. A disadvantage of this type of flushing is any abrasive material adhering to the swab could appreciably damage the membrane. Two RO tubular pilot plant studies with river water feed have successfully maintained fluxes by using daily depressurization, washing with HCl (pH = 3), foam ball flushing (Kuiper *et al.,* 1973) and by detergent flushing and foam swabbing (Miller, personal communication, 1973).

From the typical chemical analysis of acid mine-drainage feeds presented in Table IV, the two main causes of membrane fouling could be anticipated (Wilmoth and Hill, 1972). They are the bacterial conversion of

Fe^{2+} to Fe^{3+}, which is precipitated on the membrane, and $CaSO_4$ inorganic scale precipitation. Disinfection by uv light or sudden decreases in pH to 2.5 delay bacterial growth for about 100 hr. Careful pH control is also used to minimize the $CaSO_4$ scale. The most successful cleaning method reported involves flushing with acidified water (pH = 2.5) and leaving the unit idle for 1 week! Evidently, long periods of depressurization cause reverse product flow and the membrane to relax or destress, leading to improved fluxes. Specific flushes such as an enzyme-active presoak (BIZ), ammoniated citric acid, and sodium hydrosulfide were successful in the same study (Wilmoth and Hill, 1972) in removing organics, $CaSO_4$, and iron, respectively. Ultrasonic techniques have also been successful in cleaning the membranes (Smith and Grube, 1972; Wilmoth and Hill, 1972).

Another promising technique for cleaning membranes has been pursued by Thomas (1972) and co-workers (1973). By increasing the tangential velocity in a tubular permeator, they were able to define a threshold velocity, above which flux decline was markedly smaller than at lower velocities. Their results were surprisingly similar for river water and primary treated sewage. This method has also been successfully adapted to the treatment of dilute pulp and paper effluents (Wiley *et al.*, 1972a). Green and Belfort (1980) proposed a theoretical basis for the correlation obtained by Thomas and co-workers (1973) and by Shen and Hoffman (1980). These ideas were reviewed theoretically in Section IV.

From Table VIII we see that the most common membrane-cleaning techniques for RO are foam ball swabbing and flushing with additives at low pressure. Several new approaches to membrane cleaning are listed in the same table, but from a practical point of view they need further development. See Belfort (1974) for further discussion of membrane-cleaning methods.

At present, membrane-cleaning methods are able to reduce the flux decline rate to a relatively small value ($b < 0.10$) for secondary effluents with some pretreatment. These methods can be costly in downtime, expense of chemicals, and their degradative effect on membranes. Cheaper and more effective membrane-cleaning methods would reduce the operating cost of RO and render it more competitive.

X. Economics of Reverse Osmosis

Although many cost estimates for brackish water treatment have been made (Currie, 1972; Dresner and Johnson, 1981; Harris *et al.*, 1969;

TABLE VIII

Membrane Cleaning Techniques for Hyperfiltration

Technique	Method	Description	Investigator
Physical	Mechanical	Foam ball swabbing	Loeb and Selover (1967)
			Kuiper *et al.* (1973)
	Hydrodynamical	Tangential velocity variation	Sheppard and Thomas (1970)
		Turbulence promoters	Thomas (1972)
	Backwashing	Depressure and forced or osmotic reverse flow of product	—
	Air–Water flushing	Daily 15 min depressurized flush	Merten and Bray (1967)
	Sonication	Regular ultrasonic cleaning with wetting agent	Smith and Grube (1972)
			Wilmoth and Hill (1972)
	Reverse flow	Reverse flow direction of feed	Goel and McCutchan (1976)
Chemical	Additives to feed	pH control to reduce hydrolysis and scale deposit 5 ml/gal of 5% sodium hyperchlorite at pH 5	Fisher and Lowell (1970)
		Friction reducing additives (polyethyleneglycol) soil dispersants (sodium silicate)	Bailey *et al.* (1973)
	Flushing with additives at low pressure	Complexing agents (EDTA, Sodium hexametaphosphate)	Cruver *et al.* (1972)
		Oxidizing agents (citric acid)	Grover and Delve (1972)
			Bailey *et al.* (1973)
		Detergents (1% BIZ)	Miller (private communication, 1973)
		High concentration of NaCl (18%)	Lacey and Huffman (1971)
Other	Membrane replacement	*in situ* membrane replacement	Belfort *et al.* (1973)
	Inorganic membranes	Encourage biogrowth to consume fouling film	Belfort (1974)
	Active insoluble enzymes attached to membrane	Degradation of fouling film	Fisher and Lowell (1970)
	Polyelectrolyte membranes	Composite membranes or dynamic layer technique	Marcinkowsky *et al.* (1966)
	Precoat protection	Precoat (diatomaceous earth, activated carbon, and surface-active agent)	Conn (1971)
		Deposit a porous diatomaceous earth coating	Belfort (1980)
			Belfort and Marx (1978)

Lacey, 1972b; LeGros *et al.*, 1970), cost estimates for large seawater desalination and wastewater treatment processes are less reliable because insufficient large-scale data are available. The reader is referred to Chapters 12 and 13 for additional information on the economics of membrane processes.

XI. Conclusions

We have marshaled the advantages and disadvantages of each commercial membrane and module and have reviewed desalination and treatment of industrial, brackish, sea, and various waste waters. Finally, it is inferred from the many applications and the increasing interest and activity in pressure-driven membrane processes that they are being taken seriously for water desalination and wastewater treatment, and should play an expanding role in industrial applications.

Given this bright picture, we are prompted to ask why these membrane processes have not received wider acceptance in either industrial or municipal applications? One important reason is that they are structurally and operationally different from conventional municipal wastewater treatment processes, such as sedimentation, filtration, and even biological treatment (Belfort, 1974). Second, the membrane processes have higher operating (and sometimes total) costs than the latter processes. In addition, trained operators are needed for these pressure-driven membrane processes. Understandably, there exists a certain hesitancy in accepting these new processes. Yet, with respect to wastewater reclamation for potable use, RO, with its capacity to reject viruses, bacteria, and dangerous organic and inorganic compounds, is certainly an attractive, viable process.

As for industrial applications, especially for pollution abatement, it is a lamentable, but undeniable, fact that most commercial polluters need to be "legally encouraged" to act on their plant effluents. It is also true that, although laws exist in most developed countries, they are not adequately enforced. With increased environmental awareness and a realization that humanity must save its surface and groundwaters from industrial and agricultural degradation, prosecutions for illegal polluting are becoming more numerous, especially in the advanced industrial countries. Because of this, UF and RO will surely find wide acceptance for industrial uses.

With energy costs rising, many traditional industrial separation processes, especially those involving phase changes such as distillation will be carefully reevaluated. Membrane separation processes, with their relatively low energy requirements, will undoubtedly play an increasingly active role within industry. Applications in new industries such as the gene splicing industry will probably also evolve.

Clearly, with the construction and successful operation of the large Yuma desalting plant (96 mgd), RO or HF will have come of age.

Acknowledgment

This chapter is dedicated to David, Gabriel, Jonathan, Marlene, Sophie, and Grete.

Thanks are due to Professor K. Sam Spiegler for his continual encouragement and advice, and to all the students with whom the author has shared his research. Their enthusiasm, creativity, and sheer cleverness have made the voyage that much more enjoyable.

References

Aerojet General Corp. (1964). U.S. Office Saline Water, Research Development Progress Report No. 86.

Aerojet General Corp. (1965). "Reverse osmosis as a treatment for wastewater" (Contract No. 86-63-277). U.S. Pub. Health Serv., Publ. No. 2962.

Aerojet General Corp. (1966). U.S. Office Saline Water, Research Development Progress Report No. 213.

Ajax International Corp. (1973). "Sales New Flash 73-10-1", p. 1.

Allegrezza, A. E., Jr., Charpentier, J. M., Davis, R. B., and Coplan, M. J. (1975). "Hollow Fiber Reverse Osmosis Membranes" (Paper No. 34b). Presented at 68th Annual AIChE Meeting, Los Angeles.

Anderson, J. E., Springer, W. S., and Strosberg, G. G. (1981). *Desalination* **36,** 179–188.

Anonymous (1975). *Chem. Week*. September, 31–32.

Bailey, D. A., Jones, K., and Mitchell, C. (1973). "The Reclamation of Water from Sewage Effluents by Reverse Osmosis." Presented to the Joint Meeting of the Scottish Branch of the IWPC, IPHE, and IWE. Department of Environment, Water Pollution Research Laboratory, Stevenage, United Kingdom.

Bansal, I. K., and Wiley, A. J. (1975). *Tappi* **58,** 125–130.

Baxter, A. G., Bednas, M. E., Matsuura, T., and Sourirajan, S. (1979). "Reverse Osmosis Concentration of Flavor Components in Apple Juice—and Grape Juice—Waters (Paper No. 17e)." Presented at 87th AIChE National Meeting, Boston, Massachusetts.

Beckman, J. E. (1979). Reverse Osmosis Renovation of Secondary Effluent, NTIS PB–293–761. U.S. Gov. Report, Washington, D.C.

Belfort, G. (1972). The role of water in porous glass desalination membranes. Ph.D. Thesis, University of California, Irvine.

Belfort, G. (1974). "Cleaning of Reverse Osmosis Membranes in Wastewater Renovation." Presented at Joint AIChE Meeting, Germany.

Belfort, G. (1980). *Desalination* **34** 159–169.

Belfort, G., and Marx. (1978). Artificial Particulate Fouling of Hyperfiltration Membranes III Mechanism of Membrane Protection. *Proc. 6th Int. Symp. Fresh Water Sea* **4,** 183–192.

Belfort, G., and Marx, B. (1979). *Desalination* **28,** 13–30.

Belfort, G., and Chin, P. C., and Dziewulski, D. M. (1982). "A New Gel-Polarization Model Incorporating Lateral Migration for Membrane Fouling." Proceedings of World Filtration. Congress III, Vol 2, pp. 548–555. The Filtration Society, England.

Belfort, G., Alexandrowicz, G., and Marx, B. (1975). "A Study of the Mechanism and Prevention of Membrane Fouling in the Application of Hyperfiltration (Reverse Osmosis) to Wastewater Treatment." National Council Research Development, Prime Minister's Office, Jerusalem.

Belfort, G., Littman, F., and Bishop, H. K. (1973). *Water Res.* **7,** 1547–1559.

Bennet, P. J., Narayarian, S., and Hindin, E. (1968). *Eng. Bull. Purdue Univ. Eng. Ext. Ser.* **132,** 1000–1017.

Bennion, D. N., and Rhee, B. W. (1969). *Ind. Eng. Chem. Fundam.* **8,** 36.

Bettinger, G. E. (1981). *Desalination* **38,** 419–424.

Bishop, H. K. (1970). Use of improved membranes in testiary treatment by reverse osmosis. Water Pollution Research Series 17020 DHR 12/70. U.S. Environ. Protect. Agency, Washington, D.C.

Blanton, M. (1977). *Water Sewage Works* 60–62.

Blatt, W. F., Dravid, A., Michaels, A. S., and Nelson, L. (1970). "Solute Polarization and Cake Formation in Membrane Ultrafiltration: Causes, Consequences and Control Techniques." *In* Membrane Science and Technology (J. E. Flinn, ed.), pp. 47–97. Plenum, New York.

Boen, D. F., and Johannsen, G. L. (1974). Reverse osmosis of treated and untreated secondary sewage effluent. Environ. Protect. Technol. Ser. EPA 670/2-74-077.

Boesch, W. W. (1981). *Desalination* **38,** 485–496.

Brandon, D. A., El-Nashar, A., and Porter, J. J. (1975). "Reuse of Wastewater Renovated by Reverse Osmosis in Textile Dyeing" Presented at 2nd National Conference on Complete Water Uses, Chicago.

Bray, D. T., and Merten, U. (1966). *J. Water Pollut. Control Fed.* **100**(3), 315.

Cadotte, J. E., and Rozelle, L. T. (1972). In Situ-formed Condensation Polymers for Reverse Osmosis Membranes (Report. No. PB 229337). NTIS, Springfield, Virginia.

Cadotte, J. R., Petersen, R. J., Larson, R. E., and Erickson, E. E. (1980). "A new thin-film sea water reverse osmosis membrane." Presented at 5th Seminar on Membrane Separation Technology, Clemson University, Clemson, South Carolina.

Chian, E. S. K., and Fang, H. H. P. (1973). "Evaluation of New Reverse Osmosis Membranes for Separation of Toxic Compounds from Water." Presented at 75th National AIChE Meeting, Detroit.

Chian, E. S. K., Bruce, W. N., and Fang, H. H. P. (1975). *Environ. Sci. Technol.* **9,** 52–59.

Cohen, H. (1972). "The use of pressure-driven membranes as a unit operation in the treatment of industrial waste streams." *In* Utilization of Brackish Water (G. A. Levite, ed.), pp. 63–69. National Council Research Development, Prime Minister's Office, Jerusalem.

Cohen, H., and Loeb, S. (1973). Industrial Wastewater Treatment in Israel using Membrane Process" (Rep. No. 132). Negev Institute for Arid Zone Research, Beer-Sheva, Israel.

Conn, W. M. (1971). "Raw Sewage Reverse Osmosis." Presented at 69th Annual AIChE

Meeting, Cincinnati. City of San Diego, Pt. Loma Sewage Treatment Plant, San Diego, California.

Cruver, J. E., and Nusbaum, I. (1974). *J. Water Pollut. Control Fed.* **46**(2), 301–311.

Cruver, J. E., Beckman, J. E., and Bevage, E. (1972). "Water Renovation of Municipal Effluents by Reverse Osmosis" (EPA Proj. No. EPA 17040EOR). Gulf Environmental Systems Co., San Diego, California.

Currie, R. J. (1972). "Study of Reutilization of Wastewater Recycle Through Groundwater" (preliminary copy). Eastern Municipal Water District, Hemet California (also cited in final project report of Boen and Johannsen, 1974).

deBussy, R. P., and Whitmore, H. B. (1972). Nat. Eng. (February issue).

Deinzer, M., Melton, R., Mitchell, D., and Kopfler, E. (1974). "Trace Organic Contaminants in Drinking Water: Their Concentration by Reverse Osmosis". Presented at Am. Chem. Soc., Los Angeles.

Deinzer, M., Melton, R., and Mitchell, D. (1975). *Water Res.* **9,** 799–805.

Donnelly, R. G., Goldsmith, R. L., McNulty, K. J., Grant, D. C., and Tan, M. (1976). Treatment of Electroplating Waters by Reverse Osmosis. Environ. Prot. Techn. Series EPA-600/2-76-261. U.S. Environ. Protect. Agency, Washington, D.C.

Doshi, M. R., and Trettin, D. R. (1981), *I&EC Fundam* **20**(3), 221.

Dresner, L., and Johnson, J. S., Jr. (1981). "Hyperfiltration (reverse osmosis)." *In* Principles of Desalination (K. S. Spiegler and A. D. K. Lairds, eds.). Academic Press, New York.

Duvel, W. A., Jr., and Helfgott, T. (1975). *J. Water Pollut. Control. Fed.* **47,** 57–65.

Eden, G. E., Jones, K., and Hodgson, T. D. (1970). Recent development in water reclamation. *Chem. Eng. London* Jan./Feb. Issue, CE24-CE29.

Edwards, V. H., and Schubert, P. E. (1974). *J. Am. Water Works Assoc.* October, 610–616.

Fang, H. H. P., and Chian, E. S. K. (1974). "RO Treatment of Power Cooling Tower Blowdown for Reuse" (Paper No. 40C). Presented at 67th Annual AIChE Meeting, Washington, D.C.

Feuerstein, D. L., and Bursztynsky, T. A. (1969). Reverse osmosis renovation of municipal wastewater. Water Pollution Control Research Series ORD-17040 FFO12/69. U.S. Environ. Protect. Agency, Washington, D.C.

Fisher, B. S., and Lowell, J. R., Jr. (1970). New technology for treating wastewater by reverse osmosis. Water Pollution Control Research Series 17020 DUDO9/70. U.S. Environ. Protect. Agency, Washington, D.C.

Flinn, F. (ed.) (1970). "Membrane Science and Technology." Plenum, New York.

Forgacs, C. (1967). Proc. Int. Symp. Water Desalination, 1st 1965 paper No. SWD/83. U.S. Office Saline Water, U.S. Dept. of the Interior, Washington, D.C.

Gaddis, J. L., Spencer, H. G. and Wilson, S. C. (1979). Separation of Materials in Dye Manufacturing Process Effluent by HF, Water 78. *AIChE Symp. Series* **75**(190), 156–161.

Goldsmith, R. L., deFilippi, R. P., and Hossain, S. (1974). *AIChE Symp. Ser.* **120,** 7–14.

Goel, V., and McCutchan, J. W. (1976). Colorado River Desalting by Reverse Osmosis. *Proc. Fifth Int. Symp. on Fresh Water from the Sea, May 16–20,* Alghero, 385–395.

Gollan, A., Goldsmith, R., and Kleper, M. (1975). "Advanced Treatment of MUST Hospital Wastewaters." Presented at 5th Intersociety Conference on Environmental Systems, San Francisco.

Golomb, A. (1972). An example of economic plating waste treatment. Proc. Int. Conf. Water Pollut. Res., 6th, 1972 Paper 15/2/31.

Golomb, A., and Besik, F. (1970). RO- a review of the applications to waste treatment. *Ind. Water Eng.* **7,** 16.

Green, G., and Belfort, G. (1980). *Desalination* **35,** 129–147.

Grover, J. R., and Delve, M. H. (1972). Operating experience with a $23m^3$/day reverse osmosis pilot plant. *Chem. Eng. London* January Issue, pp. 24–29.

Grover, J. R., Gaylor, R., and Delve, M. H. (1973). *Proc. Int. Symp. Fresh Water Sea, 4th* **4** 159–169.

Haight, A. G. (1971). "Demineralized Water through Reverse Osmosis and Ion Exchange." Presented at American Association for Contamination Control meeting, Washington, D.C.

Hamoda, M. F., Brodersen, K. T., and Sourirajan, S. (1973). *J. Water Pollut. Control Fed.* **45,** 2146–2154.

Hardwick, W. H. (1970). Water renovation by reverse osmosis. *Chem. Ind. London* February Issue, pp. 297–301.

Harris, F. L., Humphreys, G. B., Isakari, H., and Reynolds, G. (1969). Engineering and economic evaluation study of reverse osmosis. U.S. Off. Saline Water, Research Development Progress No. 509.

Harris, F. L., Humphreys, G. B., and Spiegler, K. S. (1976). "Reverse osmosis (Hyperfiltration) in Water Desalination," Chapter 4, pp. 127–186. *In* Membrane Separation Process (P. Mears, ed.). Elsevier, Amsterdam.

Hauk, A. R., and Sourirajan, S. (1972). *J. Water Pollut. Control* Fed. **44,** 1372–1382.

Hicks, E., Probstein, R. F., and Wei, I. (1980). "Water Management in Oil Shale Production." Proc. Indust. Wastes Symp. 53rd Ann. WPCF Conf. 1–23.

Ho, B. P., and Leal, L. G. (1974). *J. Fluid Mech.* **65**(2) 365–400.

Horton, B. S., Goldsmith, R. L., and Zall, R. R. (1972). *Food Technol.* **26,** 30–35.

Johnson, J. S., Jr. (1972). *In* "Reverse Osmosis Membranes Research" (H. K. Lonsdale and H. E. Podall, eds.)., pp. 379–403. Plenum, New York.

Johnson, J. S., Jr., and McCutchan, J. W. (1973). "Desalination of Sea Water by Reverse Osmosis." Presented at AIChE Meeting, Dallas, Texas. (Obtainable from J. W. McCutchan, University of California, Los Angeles.)

Johnson, J. S., Jr., Minturn, R. E., Westmoreland, C. G., Csurny, J., Harrison, N., Noore, G. E., and Shor, A. J. (1973). "Filtration Techniques for Treatment of Aqueous Solutions" (Annu. Progr. Rep., ORNL-4891). Chem. Div., Oak Ridge Nat. Lab., Oak Ridge, Tennessee.

Josephson, J. (1978). *Envir. Sci. Techn.* **12**(6) 629–632.

Kesting, R. E. (1973). *J. Appl. Polym. Sci.* **17,** 1771–1784.

Kosarek, L. J. (1979). *AIChE Symp. Ser. 190,* **75,** 148–155.

Kraus, J. A. (1970). Application of hyperfiltration to treatment of municipal sewage effluents. Water Pollution Control Research Series ORD 17030 EOHO1/70. U.S. Environ. Protect. Agency, Washington, D.C.

Kuiper, D., Born, C. A., van Hezel, J. L., and Verdouw, J. (1973). *Proc. Int. Symp. Fresh Water Sea,* 4th Proc. No. 4.205.

Kuiper, D., van Hezel, J. L., and Bom, C. A. (1974). *Desalination* **15,** 193–212.

LaConti, A. (1977). "Advances in Development of Sulphonated PPO and Modified PPO Membrane Systems for Some Unique Reverse Osmosis Applications, Chapt. 10, pp. 211–230" *In* Reverse Osmosis and Synthetic Membranes (S. Sourirajan, ed.). National Research Council, Canada.

Lacey, R. E. (1972a). Membrane separation process. *Chem Eng*. London September 4, pp. 56–74.

Lacey, R. E. (1972b). "The costs of reverse osmosis." *In* Industrial Processing with Membranes (R. E. Lacey and S. Loeb, eds.), Chapter 9, p. 179. Wiley (Interscience), New York.

Lacey, R. E., and Huffman, E. L. (1971). Demineralization of wastewater by the transport depletion process. Water Pollution Control Research Series 17040 EUN02/71.

Lacey, R. E., and Loeb, S., eds. (1972). "Industrial Processing with Membrane." Wiley (Interscience), New York.

Larson, R. E., Cadotte, J. E., and Petersen, R. J. (1981). *Desalination* **38,** 473–483.

LeGros, P. G., Gustafson, C. E. Sheppard, B. P., and McIlhenny, W. F. (1970). U.S. Off. Saline Water, Research Development Progress Rep. No. 587.

Leitner, G. F. (1972). Tappi **55,** 258–261.

Leitner, G. F. (1973). *Chem. Eng. Prog.* **69,** 83–85.

Loeb, S., and Sourirajan, S. (1962). *Adv. Chem. Ser.* **38,** 117.

Loeb, S., and Selover, E. (1967). *Desalination* **2,** 63–68.

Loeb, S., Levy, D., and Melamed, A. (1974). "Reclamation of Municipal Wastewater for Reuse" (Final Report, NEG-ES-73-1/2). Presented to the Israel National Council for Research and Development, Jerusalem, Israel, Ben Gurion University Research and Development Authority.

Lonsdale, H. K., and Podall, H. E. (eds.) (1972). "Reverse Osmosis Membrane Research." Plenum, New York.

Lopez, M. (1979). *Desalination* **30,** 15–21.

Madsen, R. F. (1977). "Hyperfiltration and Ultrafiltration in Plate-and-Frame Systems." Elsevier, Amsterdam.

Madsen, R. F., Olsen, O. J., Nielsen, I. K., and Nielson, W. K. (1973). "Use of hyperfiltration and ultrafiltration with chemical and biochemical industries." *In* Environmental Engineering, A Chemical Engineering Discipline (G. Linder and K. Nyberg, eds.), pp. 320–330. Reidel, Dordrecht, Netherlands.

Mahoney, J. G., Rowley, M. E., and West, L. E. (1970). *In* "Membrane Science and Technology" (J. E. Flinn, ed.), pp. 196–208. Plenum, New York.

Marcinskowsky, A. E., Kraus, K. A., Phillips, H. O., and Shor, A. J. (1966).*J. Amer. Chem. Soc.* **88,** 5744.

Marino, M., Terril, M. E., Burke, B., and Simon, A. (1978). Application of the Reverse Osmosis Process for Achieving Industrial Wastewater Reuse. Presented at the WPLF 51st Annual Conf., Anaheim, California.

Markind, J., Minard, P. G., Neri, J. S., and Stana, R. R. (1973). Use of Reverse Osmosis for Concentrating Waste Curring Oils. Proc. Am. Inst. Chem. Eng.—Canod. Soc. Chem. Engr., 4th Joint.

Markind, J., Neri, J. S., and Stana, R. R. (1974). "Use of Reverse Osmosis for Concentrating Oil Coolants." Presented at 78th National AIChE Meeting, Salt Lake City, Utah.

Mason, D. G., and Gupta, M. K. (1972). Ameanability of reverse osmosis concentration to activated sludge treatment. Water Pollution Control Research Series 14010 FOR03/72.

Matsuura, T., and Sourirajan, S. (1972). *Water Res.* **6,** 1073–1086.

McDonough, F. E., and Mattingly, W. A. (1970). *Food Technol.* **24,** 88.

Mattiasson, E., and Sivik, B. (1980). Desalination **35,** 59–103.

Mears, P. (1976). "The Physical Chemistry of Transport and Separation by Membranes" Chapter 1, pp. 1–38. *In* Membrane Separation Processes (P. Mears, ed.), Elsevier, Amsterdam.

Merson, R. L., and Morgan, A. I., Jr. (1968). *Food Technol.* **22,** 631.

Merten, U. (1966). "Desalination by Reverse Osmosis." MIT Press, Cambridge, Massachusetts.

Merten, U., and Bray, D. T. (1967). Reverse osmosis for water reclamation. Adv. Water Pollution Research, Proc. Int. Conf., 3rd, Vol. 3, p. 000.

Michaels, A. S. (1968). Ultrafiltration in "Progress in Separation and Purification" (E. S. Perry, ed.), Vol. 1, pp. 297–334. Wiley (Interscience), New York.

Minard, P. G., Stana, R. R., and DeMeritt, E. (1975). "Two years experience with a reverse osmosis radioactive laundry water concentrator." Presented at 2nd National Conference Complete Water-reuse, Chicago.

Model, F. S., Davis, H. J., and Poist, J. E. (1977). "PBI Membranes for Reverse Osmosis," Chapter 11, pp. 231–248. *In* Reverse Osmosis and Synthetic Membranes (S. Sourirajan, ed.). National Research Council, Canada.

Nakao, S., and Kimura, S. (1981). In *Synthetic Membranes: Hyper- and ultrafiltration Uses* (Albien F. Turbak, ed.), Vol. 2, pp. 119–132. American Chemical Society, Washington, D.C.

Nielsen, W. K. (1972). "The Use of Ultrafiltration and Reverse Osmosis in the Food Industry and for Wastewaters from the Food Industry" (Paper No. WKN/1h). (Obtainable from DDS, Nakshov, Denmark.)

Null, H. R. (1980). *CEP* **76**(8) 42–49.

Nusbaum, I., Cruver, J. E., Sr., and Kremen, S. S. (1972a). "Recent Progress in Reverse Osmosis Treatment of Municipal Wastewaters" (Rep. No. Gulf-EN-A10994). Fluid System Division UOP, San Diego, California.

Nusbaum, I., Cruver, J. E., and Sleigh, J. H., Jr. (1972b). *Chem. Eng. Prog.* **68,** 69–70.

Okey, R. W. (1972). "The treatment of industrial wastes by pressure driven membrane processes." *In* Industrial Processing with Membranes (R. E. Lacey and S. Loeb, eds.), Chapter 12, p. 249. Wiley (Interscience), New York.

O'Melia, C. R. (1980). *ES&T* **14**(9) 1052–1060.

Pappano, A. W., Blackshaw, G. L., and Chang, S. Y. (1975). "Coupled Ion Exchange Reverse Osmosis Treatment of Acid Mine Drainage" (Paper No. 44d). Presented at 80th National AIChE Meeting, Boston, Massachusetts.

Pepper, D. (1981). *Desalination* **38,** 403–417.

Porter, J. J., and Edwards, J. L., Jr. (1977). *South. Pulp Pap. Manuf.* **12,** 24–31.

Porter, W. L., Siciliano, J., Krulik, S., and Heisler, E. G. (1970). "Reverse osmosis: Application to potato-starch factory waste effluents." *In* Membrane Science and Technology (J. E. Flinn, ed.), pp. 220–230.

Probstein, R. F. Chan, K. K., Cohen, R., and Rubenstein, I. (1981). *In* "Synthetic Membranes: Desalination" (Albin F. Turbak, ed.), Vol. 1, pp. 131–145. American Chemical Society, Washington, D.C.

Proceedings of the International Congress on Desalination and Water Reuse (1981). *Manama,* Nov. 29–Dec. 3, 1981; *Desalination* **39** (1/2/3).

Proceedings of the Symposium on Membrane Technology in the 80's (1980). *Desalination* **35** (1/2/3).

Reed, R. H., and Belfort, G. (1982). "Characterization of Fouling Potential for Pressure-Driven Membrane Processes: A New Simulation Flow Cell." Wat. Sci. Tech. **14,** 499–522.

Richard, M. G., and Cooper, R. C. (1975). "Prevention of Biodegradation and Slime Formation in Tubular Reverse Osmosis Units." Presented at the Annual Conference of the National Water Supply Improvement Association, Key Largo, Florida, 1975, and "Prevention of Biodegradation and Slime Formation in Reverse Osmosis Units Operated at Firebaugh, California." Final report for Calif. Dept. of Water Resources, School of Public Health, Univ. of Calif. Berkeley, CA, June 1975.

Riedinger, A. B., and Nusbaum, I. (1972). Reverse osmosis applied to wastewater reuse. Amer. Soc. Mech. Eng. Publ. No. 72-PID-8.

Riley, R. L., Lonsdale, H. K., and Lyons, C. R. (1971). *J. Appl. Polym. Sci.* **15,** 1267–1276.

Riley, R. L., Hightower, G. R., Lyons, C. R., and Tagami, M. (1973). Thin film composite membranes for single stage seawater desalination by reverse osmosis. *Proc. Symp. Fresh Water Sea 4th,* **4,** 333–347.

Riley, R. L., Case, P. A., Lloyd, A. L., Milstead, C. E., and Tagami, M. (1980). "Recent Developments in Thin-Film Composite Reverse Osmosis Membrane Systems." Presented at Joint Symp. on Water Filtration and Purification, AIChE and the Filtration Society, Philadelphia, Pennsylvania.

Rozelle, L. T. (1971). Water Pollution Control Research Series 12010 DRH 11/71. U.S. Environ. Protect. Agency, Washington, D.C.

Rozelle, L. T., Cadotte, J. E., Nelson, B. R., and Kopp, C. U. (1973). *Polym. Symp.* **22,** 223–239.

Rozelle, L. T., Cadotte, J. E., Cobian, K. E., and Kopp, C. V., Jr. (1977). "Nonpolysacharide Membranes for Reverse Osmosis: NS-100 Membranes," Chapter 12, pp. 249–262. *In* Reverse Osmosis and Synthetic Membranes (S. Sourirajan, ed.). Natural Research Council, Canada.

Sachs, B., and Zisner, E. (1972). "Reverse osmosis for wastewater reclamation." *In* Utilization of Brackish Water (G. A. Levite, ed.), pp. 70–80. National Council Research Development, Prime Minister's Office, Jerusalem, Israel.

Sachs, B., Shelef, G., and Ronen, M. (1975). "Renovation of Municipal Effluents by Sewage Ultrafiltration." Department of Membrane Processes, Israel Desalination Engineering, Tel Aviv, Israel.

Sammon, D. C. (1976). "The Treatment of Aqueous Waters and Foods by Membrane Processes," Chapter 13, pp. 499–527. *In* Membrane Separation Processes (P. Mears, ed.). Elsevier, Amsterdam.

Schippers, J. C., and Verdouw, J. (1980). *Desalination* **32,** 137–148.

Schippers, J. C., Hanemaayer, J. H., Smolders, C. A., and Kostense, A. (1981). *Desalination* **38,** 339–348.

Schmitt, R. P. (1974). Reverse osmosis and future army water supply. Am. Soc. Mech. Eng. Publ. No. 74-ENAs-6, New York.

Segre, G., and Silberberg, A. (1962). *J. Fluid Mech.* **14,** 136.

Shen, J. J. S., and Hoffman, C. R. (1980). "A comparison of Ultrafiltration of Latex Emulsions and Macromolecules Solutions." Presented at 5th Membrane Seminar, Clemson University, Clemson, South Carolina, May 12–14.

Sheppard, J. D., and Thomas, D. G. (1970). *Desalination* **8,** 1–12.

Sheppard, J. D., and Thomas, D. G. (1971). *AIChE J.* **17,** 910–915.

Sheppard, J. D., Thomas, D. G., and Channabasappa, K. C. (1972). *Desalination* **11,** 385–398.

Sleigh, J. H., and Kremen, S. S. (1971). Acid mine waste treatment using reverse osmosis. Water Pollution Control Research Series 14010 DYG08/71.

Smith, R., and Grube, W. (1972). *In* Wilmoth and Hill (1972).

Smith, J. M., Maase, A. N., and Miele, R. P. (1970). Renovation of municipal wastewater by reverse osmosis. *Water Pollut. Contr. Res. Ser.* **17040** 05/70.

Soltanieh, M., and Gill, W. N. (1981). *Chem. Eng. Commun.* **12,** 279–363.

Sonksen, M. K., Sittig, F. M., and Maziarz, E. F. (1979). Treatment of Oily Wastes by Ultrafiltration and Reverse Osmosis. Presented at 33rd Annual Purdue Industrial Waste Conference, Purdue University, West Lafayette, Indiana, May 1978.

Sorg, T. J., Forbes, R. W., and Chambers, D. S. (1980). *J. AWWA* **72,** 230–237, April 1980.

Sourirajan, S. (1970). "Reverse Osmosis." Academic Press, New York.

Sourirajan, S. (ed.) (1977). "Reverse Osmosis and Synthetic Membranes—Theory, Technology and Engineering." National Research Council Publication, Ottawa, Canada.

Sprague, W. H., Konopka, W. F., Jr., and Pearson, E. S. (1975). Reverse Osmosis Reclamation of Wastewater at Point Loma Contract W. Conn, City of San Diego, Water Quality Lab., Pt. Loma, California.

Strathman, H. (1973). In "International Symposium on Membranes and Wastewater Treatment" (G. Belfort, organizer). Hebrew University, Jerusalem, Israel.

Strathman, H. (1981). *J. Membrane Sci.* **9**, 121–189.

Tanny, G. (1980). "Recent Progress in the Theory and Application of Dynamically Formed Membranes." Presented at 5th Seminar on Membrane Separation Technology, Clemson University, Clemson, South Carolina.

Thomas, D. G. (1972). *Membrane Dig.* **1**, 71–201.

Thomas, D. G. (1977). "Dynamic Membranes—Their Technological and Engineering Aspects, Chapter 14, pp. 295–312." *In* Reverse Osmosis and Synthetic Membranes (S. Sourirajan, ed.). National Research Council, Canada.

Thomas, D. G., Gallaher, R. B., and Johnson, J. S., Jr. (1973). Hydrodynamic flux control for wastewater application of hyperfiltration system. Environ. Protect. Technol. Ser. EPA-R2-73-228.

Underwood, J. C., and Willits, C. O. (1969). *Food Technol.* **23**, 787.

Unknown (1980). *C&EN* **58**(5) 26–30.

Vasseur, P., and Cox, R. G. (1976). *J. Fluid Mech.* **78**(2) 385–413.

Wiley, A. J., Dubrey, G. A., and Bansul, I. K. (1972a). Reverse osmosis concentration of dilute in pulp and paper effluents. Water Pollution Control Research Series 12040 EEL02/72.

Wiley, A. J., Scharpf, K., Bansul, I., and Arps, D. (1972b). *Tappi* **55**, 1671–1675.

Willits, C. O., Underwood, J. C., and Merten, U. (1967). *Food Technol.* **21**, 24.

Wilmoth, R. C., and Hill, R. D. (1972). "Mine Drainage Pollution Control by Reverse Osmosis." Presented at American Institute of Mining, Metallurgical and Petroleum Engineers. (Obtainable from R. C. Wilmoth, EPA, Box 555, Riversville, West Virginia.)

Wilmoth, Roger C., Baught, T. L., and Decker, D. W. (1979). "Removal of Selected Trace Elements from Acid Mine Drainage Using Existing Technology." Proceedings of the 33rd Industrial Waste Conference, Purdue University, Lafayette, Indiana, pp. 886–894. Ann Arbor Science, Ann Arbor, Michigan.

Yao, Kuar Mu, Habibian, M. T., and O'Melia, C. R. (1971). Water and Wastewater Filter, Conceptive Application ES&T 1105–1112.

8

Desalting Experience Using Hyperfiltration in Europe and Japan

EBERHARD STAUDE
Institüt für Technische Chemie, Universitat Essen
Bundesrepublik Deutschland

ISBN 0-12-085480-5

I. Introduction

Hyperfiltration (HF) has been included in many papers that describe the transport phenomena and a mathematical description of the process. It is considered a separation process. The module that is the central apparatus for the HF process consists of membranes and a pressure-retaining element. Standard membrane performance with respect to sodium chloride solution at an operating pressure of 42 bars is usually reported by manufacturers, thus facilitating the possibility of comparing different studies' results. Unfortunately, these data often lack extended operating time and product recovery information, and the tendency to give information on membrane characteristics is noticeable. It is assumed, however, that the preparation of membranes using cellulose 2.5-acetate (CA) as a membrane polymer is well known, and membrane casting procedures using other synthetic polymers are also sometimes given.

In addition to traditional desalting, other applications using HF membranes will be discussed here. Thus, results will be presented on the treatment of effluents from municipal and industrial applications, as well as streams within specific industries, such as the food industry. Also, the treatment of streams containing organics is considered.

Frequently, results presented here are obtained on a laboratory scale using small test cells as opposed to some larger scale and even industrial scale studies also included. This distinction should be kept in mind. From the reference list at the end of the chapter it appears that activities in Japan are rising, and may even represent a larger effort in HF research and development than in Europe.

II. Seawater Desalination

A. Introduction

In Europe, the typical high density population areas often experiencing a lack of potable water supply are the coastal regions relatively far from rivers and islands off the Atlantic and Mediterranean coasts. Rain collection reservoirs and ground water pumping help reduce this potable water need. The ever present danger of seawater intrusion usually accompanies overpumping and must be avoided at all costs. Because of its proximity, seawater desalination represents a tangible and (possibly) economic water supply.

Seawater desalination by distillation has, in the past, been the choice method for large capacity needs, whereas membrane processes were usually chosen for small capacities using brackish water as a feed. However, because of the high relative operating costs of distillation (see Chapter 12 of this text), and the development of new reverse osmosis (RO) modules for seawater desalination, the preference for the latter methods is slowly becoming reality. Pretreatment and membrane fabrication costs still remain appreciable and are being worked on extensively. Extension of membrane lifetime and the development of suitable energy recovery techniques are also major development goals.

Research and development efforts in Japan have centered around the Japanese Water Reuse Promotion Center, which has initiated a six-year program to solve some of the main problems associated with HF (Ohya, 1976). No comparable center exists in Europe, although several governments have developed extensive funding programs for desalination methods. For example, in the German Federal Republic, the Bundesministerium für Forschung und Technologie sponsored a 9×10^6 DM desalting program in 1977. Of this funding, at least one-half was for membrane related projects (Hauser, 1977).

The major water-shortage areas in Italy are in the arid south, islands of Sardinia and Sicily, and remote locations where new tourist hotels and industry are being established. Distillation plants built before the development of a viable RO process account for most of the present total seawater desalting capacity of 150,000 m^3/day. The development of new membrane materials and module testing is being pursued in Italy (Di Pinto and Santori, 1977). Desalination of brackish water sources by electrodialysis (ED) has also been used extensively.

Because of the commercial success with cellulosic membranes (i.e., 2.5-CA) on brackish water desalination, it was first widely used to desalt seawater. However, because of hydrolysis and susceptibility to biological degradation, polymers other than CA have been sought (see Chapters 3 and 5). Applications of several new membranes to seawater desalination have been attempted and will be discussed later.

B. Seawater Desalination in Europe

One question often asked concerns economic preferability of the one- or two-stage seawater desalination process. A fresh impetus was given to this question after the development of new polymers suitable for membranes that can physically operate with adequate efficiency in a one-stage process. Thus, cross-linked polyether–amide can be used as a membrane polymer that apparently meets all necessary specifications (Pusch and Riley, 1977). Recently developed membranes are not commercially available on a large scale, leaving CA membranes to suffice for this purpose. Using tubular modules fitted with CA membranes, calculations were performed to establish optimum arrangement in a HF plant (Pasternack *et al.*, 1973). Tubes with turbulent brine flow inside the tube are combined into multitubular mass exchangers, similar to a tube bundle, in heat exchangers with a tube- and brine-side compartment. Possible arrangements are shown in Fig. 1 as a function of cost and salt concentration. The graph indicates a two-stage process is preferred to a one-stage process using commercially available CA membranes for seawater salt concentration above 1.25 wt%. The membranes in this (paper reinforced)

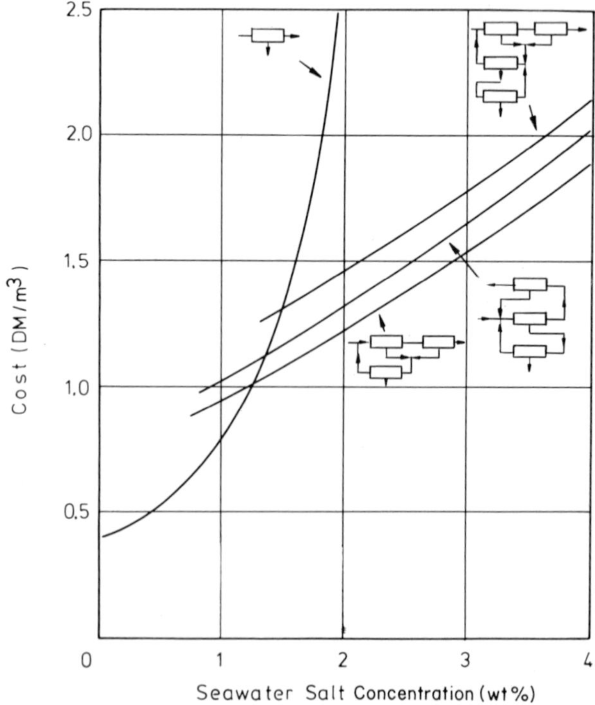

Fig. 1 Product cost–price (DM/m^3) as function of seawater salt concentration (weight %) using different module arrangements. Tube length 7500 mm, diameter 15 mm (from Pasternack *et al.*, 1973).

were prepared by Kalle (West Germany). The cost–price for fresh water from this analysis is relatively high using larger units. Therefore, the suggested optimum application of this arrangement leads to plants with a capacity of about 1000 m^3/day.

One of the factors controlling the cost–price is membrane lifetime. For the just-mentioned calculations, an approximate lifetime of 2 yr was assumed. Various factors requiring membrane replacement during operation include: hydrolysis, resulting in an increase in water and salt permeability; compaction, causing an initial flow decline; and fouling, resulting in the deposition of water constituents such as suspended and dissolved organics and inorganics. Thus, the search for more resistant membrane materials is understandable, though fouling is not only a matter of membrane stability itself, but of feed pretreatment and periodic cleaning.

Inorganic materials such as porous glass usually do not noticeably compact. For two-stage seawater desalination, hollow-fiber membranes have been prepared from phase-separated $Na_2O–SiO_2–B_2O_3$–glass. The hollow fibers had an outer diameter in the range of 20 to 200 μm and wall thicknesses of 5 to 40 μm, respectively. Similar to phase inversion, which occurs during membrane preparation from polymer solutions, porous glass is produced by the function of two contiguous phases and a leaching of the borate-rich phase by dilute acid. The rate of this separation effect is temperature dependent and the range for heat treatment is generally between 450–600°C. Hollow fibers constructed by this method have a salt rejection of 83% and a product flow of about 1.0 m^3/day at 120 bars using 3.5% NaCl (Crozier *et al.*, 1973). The rejection was increased to 99.6% by means of surface chemical modifiers using 0.5% NaCl solution in the feed at 100 bars. The volume flow was about 0.1 m^3/day. The hollow fibers had an outer diameter of 80 μm, an inner diameter of 60 μm and a length of 200 mm. In the HF experiment with 3.5% NaCl solution, rejection decreases to 98% (Schnabel, 1976).

There are, however, other candidate membrane materials suitable for seawater desalination with better stability than CA and porous glass. In the form of composite membranes (Riley *et al.*, 1970), cellulose triacetate (CTA) is most suitable for one-stage seawater desalination, for this polymer is endowed with high intrinsic salt rejection, high resistance to hydrolytic attack, and good mechanical stability. The production of ultrathin membranes required for composite membrane development necessitates a highly sophisticated manufacturing technique. Because of their excellent performance characteristics, the efforts to fabricate integral CTA membranes are understandable. The limited solubility of CTA in suitable solvent systems, however, hinders this aim. By choosing and testing efficient solvents, promising casting solutions were obtained that could be used at ambiant temperature or temperatures below 0°C. The membrane preparation procedure includes the same steps as described by Loeb, with the exception that modifying agents such as acetic acid were added to the annealing bath (Nussbaumer *et al.,* 1976). With this method, integral membranes can be prepared for one-stage seawater desalination. The results from long-term tests are unpublished, so *m* values are not yet available.* The stability of these new types of membranes in real seawater desalination has not been established.

The use of sulfonated polysulfone to produce asymmetric membranes by dissolving the polymer in a solvent mixture, including a volatile solvent, has been reported (Chapurlat, 1973). After casting a thin film using a short evaporation period, and coagulation and thermal treatment in an aqueous solution, the resulting membrane is suitable for desalination. Using a 35 g/liter sodium chloride solution under a pressure of 60 bars, the rejection was about 92–95% at a corresponding volume flux of 400 to 500 liter/m^2 day. These membranes exhibit excellent resistance to chemical and bacteriological agents, but for single-stage seawater desalination, an improvement in performance is still needed. It has not always been straightforward to go from laboratory trials using synthetic seawater or even a 35,000 ppm NaCl solution, to field tests using actual seawater.

To reduce complications, however, the HF pilot plants working under realistic conditions are usually equipped with commercially available modules. Seawater pilot plants have been installed by Degremont and situated on the Island of Cavallo (Corsica) (Treille, 1970) and the Island of Houat on the Atlantic coast (Treille and Rovel, 1973). The tubular modules at Cavallo had a membrane area of 2.4 m^2, whereas the plant at Houat is fitted with modules of the MP 36–18 type manufactured by Rhone–Poulenc, with an effective surface of 2.8 m^2. Module Polytubulaire (MP) means each module consisted of multiple plastic tubes as a membrane support and drainage device enclosed in a plastic envelope. The feed was outside the tube, and the membranes were CA type. The average characteristics of the modules before installation were 35 g/liter sodium chloride solution under 60 bars at 20°C. Rejection was 97.3%, flux 235 liter/m^2 day. Additional data are given in Table I.

The product water from the Houat plant with chloride salinity of less than 250 mg/liter was injected with carbon dioxide and passed through a neutralite filter. The final step prior to being pumped to the water tower on the island was an injection of dilute sodium hypochlorite solution for sterilization. In contrast to the two-stage process, the product of the one-stage unit at Cavallo resulted in a chloride content of 880 mg/liter. The

* *m* values are usually defined as the slope of the log(flux) versus log(time) curve.

TABLE I

French Pilot Plants

Site	Year of construction	Stages	Capacity (m^3/day)	Recovery (%)
Cavallo	1969	1	8	—
Houat	1971	2	50	17

average lifetime of the membranes at Houat exceeded 1 yr, and the efficiency of the membranes at Cavallo did not vary much from this. These results are supported by continuous experiments with the same MP modules at Carro near Marseille. After 4800 hr of operation, the performance of the membranes diminished from 96.9 to 96.2% in rejection, using real seawater at 60 bars (Brun *et al.*, 1970). These units were equipped with intake and pretreatment devices that ensured the required quality of seawater for proper operation and membrane stability. Normally, these operations include filtration and chlorine injection. Under more severe conditions in regions with a strong tide, supplementary measures must be taken into account. Thus at Houat, the seawater is pumped only at high tide and then stored in an intermediary tank. This permits easy removal of the algae and sand. For installation of the desalination facility onboard ship, a different treatment system is used. A plate-and-frame module (Bödekker *et al.*, 1976a) was placed on two decks of the nuclear research vessel *NS Otto Hahn* (Bödekker *et al.*, 1976b) to evaluate seawater membranes and test the newly designated module with an effective membrane area of 1.03 m^2. Under these actual conditions during a round-trip journey from Rotterdam, The Netherlands, to Durban, South Africa, it was found only prefiltration was necessary. Figure 2 shows the results obtained at 70 bars using experimental CTA membranes with the plate-and-frame module. As seen in the figure, the flux (isolated squares) follows the variations in seawater temperature with time. The dotted–dashed line is corrected to a temperature of 20°C. The temperature does not noticeably influence the salt rejection.

Similar results were obtained using DuPont's hollow-fiber B–10 modules, installed by Permo–Degremont on sailing boats or motor yachts. The boats cruised in the Mediterranean or in the Atlantic Ocean. This made it possible to purify seawater directly by HF in its raw state because the numerous forms of pollution in coastal regions disappear at a distance varying from 3 to 10 nautical miles from the polluting shores. The feed for the HF installation onboard ship needed only to be pretreated by physical

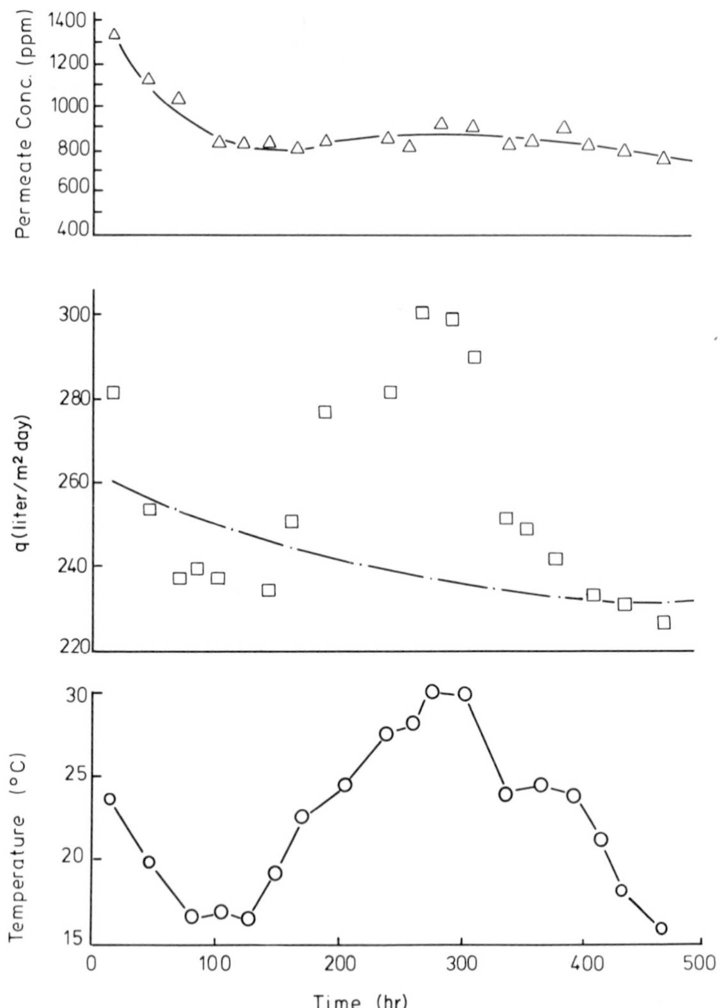

Fig. 2 Total salt content of permeate (ppm), flux (liter/m^2 day) and seawater temperature (°C) versus time. Feed flow rate 170 liter/hr, the total operation time lasted about 1100 hr with a shutdown period of about 200 hr (from Böddeker *et al.*, 1976b).

means. Cleaning procedures of the modules did not need be undertaken until at least 6 months after operation. The recovery was below 30%, the results for an operating pressure of 55 bars, as well as the feed concentrations, are illustrated in Table II for the Atlantic Ocean and the Mediterranean, respectively. Thus potable water was produced for the crews on a daily average demand of 3.8 to 4.5 m^3 (Lerat, 1976). These 25 units are in

TABLE II

Hyperfiltration of Seawater with Hollow-Fiber Module[a]

Composition (mg/liter)	Atlantic Ocean			Mediterranean Sea (near Sicily)		
	Feed	Product	Rejection (%)	Feed	Product	Rejection (%)
Ca^{2+}	371	3.0	99.2	350	3.2	99.1
Mg^{2+}	1132	11.2	99.0	1330	15.1	98.8
Na^{+}	9300	87.5	99.0	10300	91.0	99.1
Fe^{2+}	0			0		
HCO_3^-	142	3.0	97.9	170	6.2	96.3
Cl^-	17129	167.0	99.0	20400	241.2	98.8
SO_4^{2-}	2310	5.4	99.7	2850	7.0	99.7
TDS	32700	290.0	99.1	36600	370.0	99.0

[a] The shell diameter of the hollow-fiber module was 4 in.

service and demonstrate the versatility of this system (Shields, 1978). The rejection decreases only slightly, for instance, for Cl ion decreases from 98.8% in June 1975 to 97.7% in April 1978, and the same holds true for the Ca, Mg, and bicarbonate ions (Lerat, 1978).

The same author (Lerat, 1980) has also reported on the good performance of 40 desalting units installed on boats and platforms using seawater as a feed from different locations such as the Indian Ocean, the Red Sea, and the Caribbean Sea.

A plate-and-frame module equipped with DDS membranes type 999 was erected aboard a prawn fishing ship for freshwater supply. For pretreatment, a sandfilter with an activated carbon was used (Nielsen *et al.*, 1980). A report is available on the production of freshwater using seawater along the Swedish coast at the Baltic and North seas by Lindner and Nilsson (1980). They operated with several types of modules, e.g., hollow fiber, spiral wound, plate-and-frame, and tubular. The latter was used for the most part with good results. Prefiltration and aeration for iron removal proved to be necessary.

Shortage of potable water is a well-known problem in the Middle East. In the Sinai, for example, along the Red Sea shore, a serious lack of potable water exists. Therefore, near Eilat, field trials have been conducted to evaluate the economics of HF seawater desalination as an alternative to thermal desalination. This investigation was stimulated by the rapidly increasing cost of thermal energy. The commercially available Chemical Systems Inc. spiral-wound dry RO module, as well as DuPont's B-10 were tested (Glueckstern *et al.*, 1978b). Preliminary results were reported after a run of 2300 hr for the hollow-fiber modules put into operation in September 1977 (Glueckstern *et al.*, 1978c). Seawater (42,180 TDS) and seawater blended with reject brine from a nearby brackish water HF plant (26,500 TDS) were used as alternative feedwater sources. The pretreatment consisted of alum dosing before sand filtration, followed by pH regulation. After passing through an activated carbon filter and UV sterilizer, the feed was pumped into the modules at an operating pressure of 56 bars. Though the productivity was only 10–20% below the nominal value, the results indicate the reliability of the unit in maintaining rejection between 98.0 and 99.4%. Every 400–700 hr of operation, a module cleaning was practiced with PTB.

C. SEAWATER DESALINATION IN JAPAN

The water supply problem in Japan is more severe than in Europe. Consequently, attempts to obtain information about the advantages of either the one- or two-stage HF process and the performance of newly developed membranes have been more intense. It is advisable to operate in a two-stage process if CA membranes are used (Ohya *et al.*, 1975). The advantage of this polymer lies in its simple and inexpensive conversion into membranes. Thus it is well suited for the production of various membrane configurations when applied to modular development. Hence the spiral-wound modules developed by Toray are provided with CA membranes (Ohya, 1976). A two-stage demonstration plant was constructed; the flow-sheet depicted in Fig. 3 shows both the proposed salt concentra-

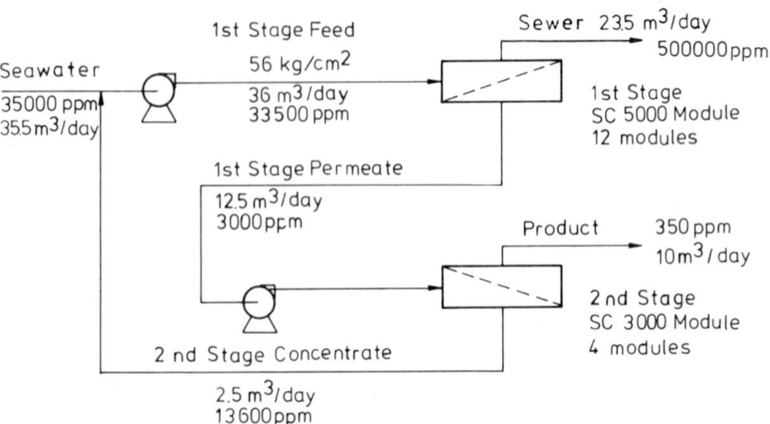

Fig. 3 Two-stage HF system (from Ohya, 1976).

tions and the output of the two stages. The plant is designed for an overall recovery of 30 to 35%, the daily product rate is about 10 m^3. In addition to CA, other synthetic polymers suitable for membrane materials are under development. Table III presents the performance characteristics of these membranes for spiral-wound modules. The values clearly indicate that the plant has to operate in two stages, for the salt rejections are not adequate for a one-stage process ($R \geq 98.5\%$). Clearly, using higher operating pressures, a slight but insufficient increase in desalination performance might be possible.

In addition to the development of new modules, another problem in seawater desalination is the development of cheap and productive membranes that exhibit good resistance to heat, chlorine and other membrane deteriorating agents. Normally, chlorine is used for disinfection purposes. As is well known, the stability of CA membranes is somewhat limited by these demands. A polybenzimidazolone derivative (PBIL) seems to be a potential candidate polymer (Hara *et al.*, 1976). In investigating the resistance of different membranes to chlorine, a 1% sodium chloride solution containing 0.05% high-test bleaching powder with 60% effective chlorine was used as testing solution. The polyamide hydrazide membrane (PAH) was eliminated just after beginning the test runs because of chlorine attack. The effect of chlorine on the CA and PBIL membranes is shown in Fig. 4. Hence, PBIL could be a promising successor to CA for chlorinated feed waters.

The performance in a seawater test run carried out for up to 720 hr at 80 bars using 3.5% NaCl solution indicates the applicability of this new material to one-stage seawater desalination. The different annealing temperatures result in different membrane characteristics, PBIL (I) was treated with hot water for 10 min at 98°C, and PBIL (II) at 80°C, for the

TABLE III

Performance of New Spiral-Wound Modules[a]

Module No.	Membrane materials	Productivity (m^3/day)	Rejection (%)
SC-1000	CA standard type	5.6–6.2	93–95
SC-2000	CA high-flux type	7.0–8.5	90–92
SC-3000	CA high-rejection type	3.2–4.0	96–98
SA-1500	Aromatic PA standard type	4.5–5.5	94–96
SA-3000	Aromatic PA high-rejection type	3.0–3.5	98–99

[a] Test conditions: 25°C, 30 bars, 1500 ppm NaCl solution. Size of modules: od 0.1 m, length 1 m.

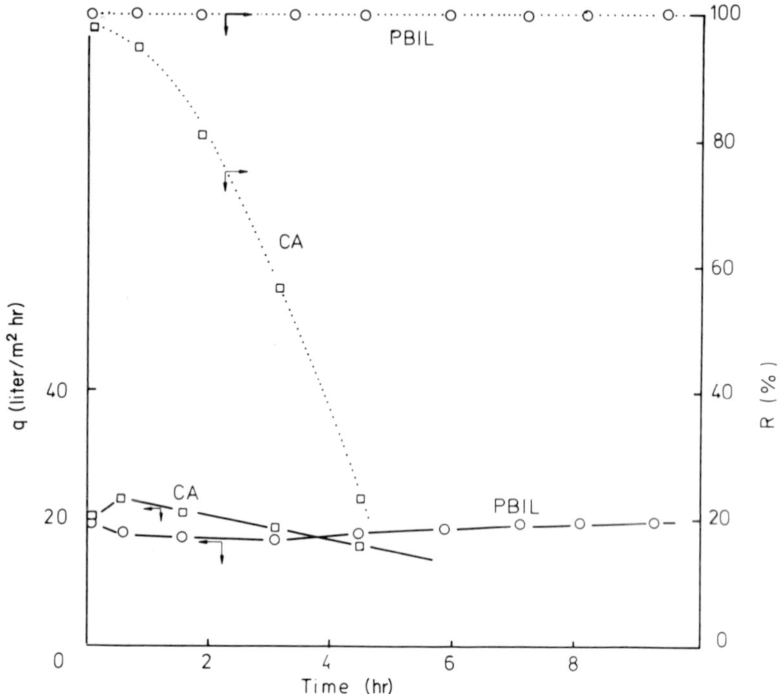

Fig. 4 Chlorine resistance of membranes. Operation conditions: Pressure 80 bars, temperature 36 ± 2°C, feed flow rate 100 liter/hr (from Hara *et al.*, 1976).

same period. As to the flux decline of the membranes used in these experiments, the best values are shown by the PAH membrane. The m values, obtained from the log–log flux–time relationships were as follows: PBIL (I): −0.04, PBIL (II): −0.09 and PAH: −0.002. The CA value obtained at 40 bars, was about −0.03–−0.04 (Kunst *et al.*, 1973).

In spite of the insufficient resistance of polyamide and even PAH membranes to chlorine, the good desalting properties of membranes made from such or similar polymers continually induce scientists to examine these substances (Endoh *et al.*, 1976). Aromatic polyamides with carboxylic groups (I) prepared by low temperature solution polymerization from the combination of aromatic diamine and aromatic acid derivatives exhibit better HF properties than those membrane materials without carboxylic groups (II). This can be seen in Table IV. These new aromatic polyamides were fabricated into asymmetric membranes using the Loeb–Sourirajan technique. The performance at 70 bars using a feed of

TABLE IV

Life Test of Carboxylic Aromatic Polyamide Membranes[a]

Polymer	Initial rejection (%)	Volume flux (m/day)	*m* value[b]
With COOH groups	98.2	0.73	−0.015
Without COOH groups	97.7	0.40	−0.120

[a] Test conditions: 5000 ppm NaCl, p = 40 bars, elapsed.
[b] Slope of the log(flux) versus log(time) plot.

35,000 ppm NaCl solution was for polymer (II) as follows: R = 98.6%, flux = 0.30 m/day. This is borderline and does not seem to be sufficient for a single-stage process. Results for these membranes in real seawater have not been published. Therefore, their suitability remains to be demonstrated. Practical and cheap production processes for forming a polymer and converting the polymer into a membrane play a significant role in the choice of such a polymer. To what extent the membranes examined so far meet these expectations cannot be deduced from the papers cited. The need for such research on new membrane polymers, however, is clearly unquestionable if seawater desalination by HF is to be economically competitive for fresh water production.

The field tests of the Water Reuse Promotion Center in Japan are devoted to performance evaluation under realistic conditions. For the purpose of establishing operational techniques for seawater desalination by HF, including pretreatment systems, commercially available U.S. modules as well as modules developed in Japan were installed in small units at the Chigasaki Laboratory, a test facility sponsored by the Japan MITI. Some of these modules were operated for about 6000 hr. As indicated in Table V, the commercial modules operate in a one-stage process, whereas those under development are planned for a two-stage process. The module of Toyobo Co, Ltd. is provided with a new type of cellulose diacetate hollow fiber, designed to minimize the depositing of fouling materials and to be easily cleaned. DuPont's B-10 hollow-fiber module may be cleaned effectively by PTB or citric acid treatment.

Data are reported for the single-stage desalination modules (Ohya, 1976) and especially for DuPont's B-10 module (Murayama *et al.*, 1976). The seawater intake for the test facility is positioned at a depth of 7 and 700 m off shore. This necessitates a thorough feed pretreatment prior to HF. Depending on the module's construction and membrane materials, the pretreatment systems will differ somewhat. The overall operations

TABLE V

Different Hyperfiltration Modules in the Chigasaki Test Plant[a]

Module type	One-stage desalination	Two-stage desalination
Hollow fiber	DuPont B-10 Dow Chemical XFS-4167.08	Toyobo
Spiral wound	UOP CTA composite membrane UOP P-300 composite membrane	Toray
Tubular	—	Kobe Steel[b] Nitto Electric Industries

[a] Common pretreatment plant: 300 m^3/day.
[b] Kunisada (1977, private communication).

include chlorination of the raw water, flocculation combined with pH adjustment, followed by multimedia filtration. This suggests that seawater in coastal regions requires almost the same pretreatment as surface water. After a 2500-hr test run, the B-10 hollow fiber module showed no significant product flow decline, the product water quality after 2300 hr was 613 μmho/cm, or equivalent to a rejection of 98.7%. To maintain the flux at 280–300 liters/hr at 25°C during the overall operation, the inlet pressure was set at 51.7 bars. After 450 hr it was elevated to 52.2 bars, and after 1000 hr the pressure was adjusted to the maximum value of 56 bars. By this means the recovery varied between 28 and 30%. The module characterization may be expressed by the parameters proposed by Kimura and Sourirajan (1967). Using their analytical method for large HF plants (Ohya and Sourirajan, 1969), a decrease in productivity for the B-10 module of one-third and for the spiral-wound module of one-sixth after 10,000 hr of operation can be predicted. Kunısada *et al.* (1977) published additional results concerning the two-stage processes. The Toray spiral-wound modules operated for about 4500 hr, the recovery was 35 and 80% for the first and the second stage, respectively. The TDS value of the second-stage product was 120 ppm.

The recovery of the tubular module unit was 35% for both the first and second stage. The operating pressures ranged between 40 and 50 bars. The performance of these modules decreased only slightly during an elapsed operation time of 600 to 1600 hr. The experiments are in progress to determine the superiority of the two designs for HF seawater desalination practice. A contribution to this question is given by Obiya (1977), who tested the performance of DuPont's B-10 hollow-fiber module. For the first run of about 900 hr, the recovery was 20–30% at feed temperatures of 16 to 26°C, whereas at higher feed temperatures (30–35°C) the recovery was 30–35%. The relatively simple pretreatment consisted of a clarifier and one fine filter. To prevent bacterial growth, NaOCl was added and residual chlorine was reduced by $NaHSO_3$. Plugging of modules was avoided by a 10 μm cartridge filter. Thus for an overall operation time of 1300 hr, the effectiveness of the plant was highly satisfactory and data for scale-up could be obtained. In addition, the newly developed CTA hollow-fiber modules "Hollosep high rejection type" have been installed at the Laboratory of MITI. They have been run from June 1978 to March 1980 (53 bars, recovery ratio 27%) in a single-stage seawater desalination mode. A small change in salt rejection was observed, and the value of the compaction factor $m = -0.015$. The feed seawater was pretreated with dual mediafilter and a 10 μm cartridge filter (Matsumoto *et al.*, 1980). Normally, such simplified pretreatment schemes are not the rule. As may be seen from the previous examples, the type and intensity of pretreatment depend partially on the site of the intake system as well as on the degree of pollution of the coastal sea. Possibly a two-stage HF process using tubular modules is the appropriate method for reducing pretreatment costs. Thus advantages of the well-defined flow characteristics inside the tubes with their minimal fouling tendency can be utilized. The investigations by Tsuge *et al.* (1977) were performed with this aspect in mind. The behavior of tubular modules listed in Table V containing cellulose diacetate membranes and cellulose diacetate and triacetate blended membranes, respectively, with a rated 98% salt rejection at 42 bars using a 5000 ppm NaCl solution, was investigated with seawater at 50 bars. The seawater concentration was adjusted to about 4% to operate under practical conditions with expected mean concentrations within the modules. The CA membrane modules exhibited a decrease in salt rejection from 93 to 91% and a m value of -0.076 after an elapsed time of 1000 hr. The pH of the feed was adjusted to about 6 and filtered by a 50 μm cartridge filter. The pretreatment for the modules containing blended membranes consisted of 100 μm filter. Nevertheless the flux decline was small. The main metallic compounds from the brown deposit on the membranes observed after 500 hr of operation were Na, Si, K, Fe, and Cr. The fouling layer dissolving behavior was tested with different cleaning solutions; oxalic and citric acid proved to be most effective. Realistic comparison between different systems is not yet possible because of the different fouling characteristics of the various types of polluted seawater.

D. CONCLUSION

Because research and development of seawater desalination by HF is supported in several advanced countries by national authorities, it demonstrates the interest of this objective. Moreover, the need for such public support of these activities indicates the difficulties connected with seawater HF. If this were not the case, more private companies would be engaged in the field of producing fresh water. Compared with HF of brackish water, the number of seawater desalination plants by pressure-driven membrane separation processes is small, as can be seen from Table XII in Section III.C. From the efforts of Kurita, the largest private Japanese company that distributes HF plants, it is possible to compare the brackish water desalination capacity with that of seawater. This comparison shows that seawater capacity is dwarfed by brackish water capacity. From this it may be concluded that either many problems for seawater desalination by HF have not yet been solved, or that this method is not effective or economical. It should be remembered, however, that where abundant brackish water is available, it has usually been tapped in preference to seawater. Also, the first major applications of HF were for brackish water desalination. Thus pretreatment and cleaning procedures now need to be developed and optimized for seawater applications.

III. Hyperfiltration of Brackish Water

A. INTRODUCTION

In the past, the main application of HF was the production of fresh water from brackish water. Today, this is still the dominant application of the process. In this section, brackish water will be considered to be not only well water or ground water with a total dissolved solids (TDS) content of 1500 to 5000 ppm, but also surface waters such as rivers, lakes, and even tapwater. This wide range of water quality implies that HF yields different water quality products. Thus it is used to produce drinking water for households and hotels, water for general process applications, boiler feed water in industries, and for use in hospitals and renal units. Also ultrapure water production for rinsing electronic components uses HF.

The different water sources accordingly have an influence on the pretreatment needs, on the choice of the appropriate module, as well as on the membrane lifetime. Large-scale pretreatment of well water is usually unnecessary, if only soluble inorganic solutes are present. For the most part, chlorination and filtration is usually sufficient. The prefilter mesh size depends on the modules used. For hollow-fiber modules a 5 μm filter is necessary and coarser filters (20 μm) are used for the spiral-wound modules. Even coarser filters (100 μm) have been used for tubular modules. High levels in sulfate and carbonate require a careful examination to determine whether their saturation limits may be reached in the course of concentration that may consequently result in scaling. To avoid precipitation, chemical pretreatment using complexing agents to sequester the cations is often required. Similarly, basic well waters must be treated with chemicals to reduce the pH. Macromolecular organic substances such as humic and fulvic acids cause rapid fouling. Often in these cases the usual pretreatment is not sufficient. To ensure adequate membrane performance, frequent cleaning is unavoidable. Organic materials are normally found in surface water, but tap water is not free from such fouling substances. Taking all this into account, the best all-purpose module for handling brackish water seems to be the tubular design. Excellent results, however, have also been obtained using the plate-and-frame modules and even the hollow-fiber modules. This means that each type will find its appropriate application if the corresponding pretreatment is selected. A survey of commercially available HF modules with their recommended application is given by McBain (1976).

B. EUROPEAN ACTIVITIES

1. *Groundwater*

With the "spaghetti" module system developed in England by the UKEAE at Harwell in collaboration with Paterson Candy International (PCI), field trials were undertaken at two boreholes near Appleby Parva in Leicestershire and at Harwell. These experiments were carried out in the course of a test program to determine the behavior of membranes cast on the outside of supporting tubes. This led to a new type of module. The water analysis of the boreholes and values of the product water after HF are shown in Table VI. The rejection for the TDS was about 95.8% for the plant at Appleby Parva, the recovery was 66.6%. For the plant at Harwell, a rejection of 97.5% was obtained. The operating pressures and characteristics of the CA membranes are not given (Grover *et al.*, 1973). Using well water with a higher calcium content, the applicability of the

TABLE VI

Water Analysis from Hyperfiltration Tests on Several Sites

	Appleby Parva		Harwell		Corfu		Cyprus		
								Prod.	
Analysis[a]	Feed	Prod.	Feed	Prod.	Feed	Prod.	Feed	PCI	B-9
pH	7.0	6.3	7.3	5.9	6.7	6.15	7.0	6.5	6.8
EC	1950	110	560	16.4	2150	249	4200	430	260
TDS	1682	70	410	10	2533	120	3080	275	175
Hardness	1215	36	204	2	1282[b]	49	580	40	5
Ca^{2+}	334	8	32	1	562	30	79	10	1
Mg^{2+}	—	—	—	—	37.6	1.2	94	3	1
Na^{+}	—	—	—	—	15.2	2.5	820	84	54
SO_4^{2-}	950	12.9	42.4	0	1251	43.6	1040	10	18
CO_3^{2-}	155	15	—	—	—	—	—	—	—
HCO_3^{-}	—	—	—	—	332	35.4	210	42	39
Cl^{-}	28	8	20	1	35.5	7.8	815	121	57

[a] EC is the electrical conductivity (μmho cm^{-1}), the concentrations are expressed as parts per million, hardness as parts per million $CaCO_3$.

[b] Noncarbonate hardness as parts per million $CaCO_3$.

spaghetti module was tested. The test site was on the east coast of Corfu (Pepper *et al.*, 1973). The aim was to demonstrate that HF is suitable for treating water rich in calcium sulfate to generate potable water if the potable water supply at remote places is limited. In this particular case the permeate was to be used for a hotel. The feed water analysis (column 6, Table VI) shows that the water was close to saturation. It was shown experimentally during the operating time of 1225 hr that it was possible to chemically clean the modules and restore their performance by periodical chemical flushing. In addition, the feed was dosed with 6 ppm Calgon S (sodium hexametaphosphate) to prevent scaling on the membranes. To maintain a constant volume flow the plant started with clean membranes at 26.7 bars. The pressure was increased to 40 bars because of the increasing precipitation of calcium sulfate on the membranes. This could be observed by the decreasing permeation flux. At this point the membranes were cleaned, the operating pressure returned to 26.7 bars, and the procedure began anew. The rejection based on TDS was about 95.2%.

A similar project was initiated in the test program started in 1974 in Cyprus (Lloyd *et al.*, 1976). The HF units were mounted on trailers; one type was fitted with two spaghetti modules of different CA membrane performance, the other with a spiral-wound module and hollow-fiber DuPont B-9. The trailers were also equipped for automatic operation. The selection of the five boreholes where the tests were carried out was based on different criteria: (1) The salinity of the brackish water should be within the range of 1500 to 6000 ppm TDS, (2) a sufficient flow rate should be available to supply the units, (3) a suitable possibility for the disposal of the concentrate should be available, and (4) a low concentration of iron (<1 ppm) and a low concentration of suspended solids should be in the feed water. In Table VI, the water analysis of one borehole and the water analysis produced with spaghetti CA and hollow-fiber polyamide membranes are shown. With regard to TDS, the rejection of the CA membranes was 91.4%, whereas for the polyamide hollow fibers a rejection of 94.3% was measured at an inlet pressure of about 25 bars. At this well the operation time was about 1000 hr, the water recovery 60%. The pretreatment procedures are not described in detail, but sand filters and pH adjustment were provided. The flux decline was 9.09% per 1000 hr. The start of the project was delayed 14 months by political events, and the modules were exposed to two hot summers in Cyprus. The membranes, however, withstood these conditions that were beyond the allowable temperature limits.

The same authors present in another paper (Lloyd *et al.*, 1978) detailed data obtained from operations at all the boreholes. Generally, the process performance was satisfactory. A definitive statement on membrane lifetime cannot be given, but for the cost estimation the membrane lifetime was assumed to be 4 yr.

An HF unit consisting of DuPont B-9 modules was installed in Tarragona, Spain for real production of feed water (Dangeon *et al.*, 1978). The raw water was obtained from a borehole with seawater intrusion resulting in increasing salt content. The start-up of the HF plant was in February 1978, the capacity was 2640 m^3/day. The rejection 1 month after the start-up was Na 92.3, Cl 97, and SiO_2 96.5%, respectively.

On the island of Heligoland four Dow hollow-fiber modules RO-4 K, three in parallel followed by the fourth in series, were installed by Krupp (Weise, 1977, private communication). The feed water had 1300 ppm chloride and 250 ppm sulfate. The pretreatment consisted of coarse filtration, chlorine dosing, sand filtration and polishing cartridge filtration (<5 μm). The pH was adjusted to 5.3 with sulfuric acid. After 6 months of operation the rejection was about 92% at a recovery of 68%, and the pressure was 40 bars.

2. *Surface Water*

A detailed study on HF plant performance on surface water was made during a period of 19 months. The difference between surface water and ground water is primarily because of a change in the water quality.

The trials were carried out with water from the Rhine River. The major difficulties for the HF process were caused by the fluctuating feed water quality. Maximum and minimum values of various water quality parameters for the Rhine water during a 2-yr period are presented in Table VII (Kuiper *et al.*, 1974a). The HF pilot plant was located at a pumping station of the municipal water works of The Hague. The daily capacity was 15–20 m^3, and it was provided with 12 Paterson Candy tubular modules. Each contained 18 tubular CA membranes in series, having a length of 8 ft and an inside diameter of 0.5 in.

The Rhine water was pretreated with chlorination, coagulated with ferric chloride addition, and followed by rapid sand filtration. The chlorine residual was reduced by sulfite to about 0.05–0.15 ppm. Four different module arrangements were used. In each arrangement, three modules were initially operated in parallel and provided with different membranes annealed at 75°, 80°, and 86°C. These three modules were followed by either two modules in parallel and one in series, or all in series. In the latter case a different arrangement is ensured by different types of membranes. The average operating pressure was 40 bars, the overall pressure drop amounted to 3 to 5 bars.

The average rejection of various cations and anions for the different types of membranes during 6 months is shown in Table VIII. From the experimental results, it was concluded that to avoid fouling, tight membranes and a high transverse feed velocity should be used. Consequently,

TABLE VII

Rhine River Water Quality as Feed to Hyperfiltration Process[a]

	1971		1972	
Analysis	maximum	minimum	maximum	minimum
pH	7.9	7.2	8.0	7.4
EC (μmho/cm)	1210	660	1260	490
Temp. (°C)	23	0.7	24	1
Ca^{2+} (ppm)	103	72	115	73
Mg^{2+} (ppm)	20	10.6	17.2	9.6
Fe^{2+} (ppm)	5.7	0.7	6.5	0.5
Cl^- (ppm)	286	127	301	66
SO_4^{2-} (ppm)	127	72	134	76
HCO_3^- (ppm)	198	143	192	134
NO_3^- (ppm)	22	10	26	9.6
Organics[b]	43	23	44	18

[a] After Kuiper *et al.* (1974a).
[b] Expressed as $KMnO_4$ number.

TABLE VIII

Rejection of Rhine Water Ingredients Using Different Cured Membranes[a]

	Percent rejection at curing temperature		
Ions	75°C	80°C	86°C
Cl^-	63	77	90
NO_3^-	35	44	73
SO_4^{2-}	92	96	97
Ca^{2+}	83	92	97
Mg^{2+}	79	87	94
Na^+	60	69	85
NH_4^+	57	66	81

[a] After Kuiper *et al.* (1974a).

membranes cured at 86°C and operated at feed velocities of about 1.5 m/sec were hardly effected by fouling. Using velocities of about 2 m/sec the fouling effect decreased remarkably, even for the membranes cured at lower temperatures. It is well known, however, that fouling is membrane specific, and extension of these findings to other types of membranes may not be valid.

To limit the fouling effect significantly, several cleaning procedures were applied. These procedures are an important result of this study, for during the 19-month period it was shown that daily mechanical cleaning with foam balls, followed by a washing with hydrochloric acid (pH = 3) and flushing was the best membrane cleaning method.

Other experiments were also conducted with small-scale pilot plants that allowed the control of any desired process parameter so as to optimize the HF performance. This installation consisted of several units provided with the same modules and membranes as used in the previously mentioned experiments (Kuiper *et al.*, 1974b). The crucial process parameters, such as pretreatment, membrane characteristics, feed velocities, recovery, and applied pressure were measured. Because the feed pretreatment greatly affects membrane fouling and, consequently, the lifetime of the membranes, five different feed water qualities were produced, as illustrated in Fig. 5.

During the transportation through the 45 km pipe from the Rhine River to the water works, chlorination with 3 ppm Cl_2 was applied at the intake works, especially if the water temperature was higher than 8°C. As may be seen from the flow sheet, the quality of the feed water ranges from

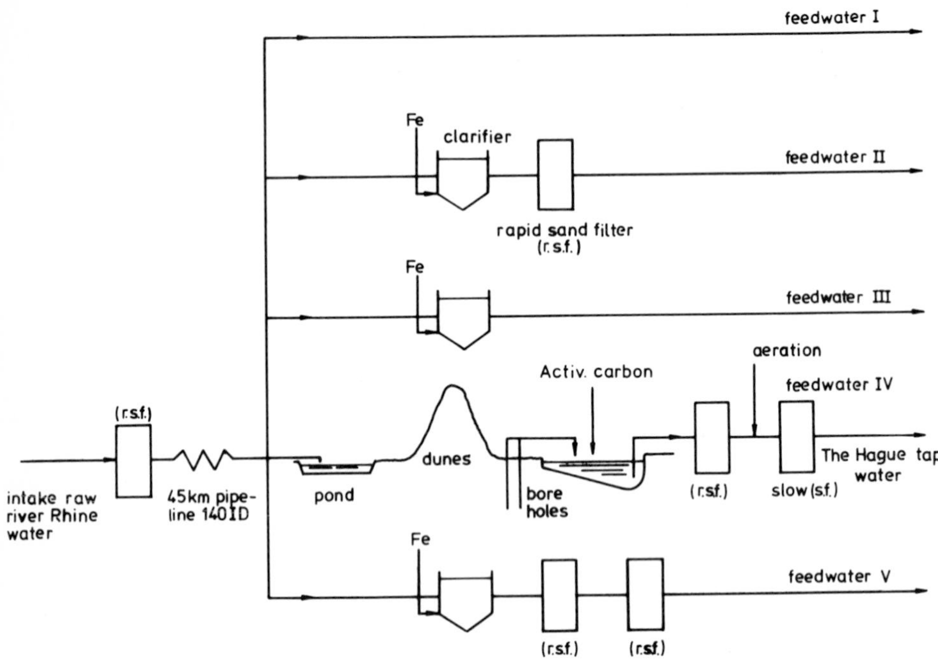

Fig. 5 Pretreatment flowsheet for five different feedwaters (from Kuiper *et al.*, 1974b).

type IV, The Hague tapwater, to type I, rapid sand-filtered Rhine water only partially free of suspended matter. In chemical clarification, the water is flocculated with iron and the resulting flocs removed by sedimentation and sand-filtration.

The tests conditions were as follows: Pressures were 20, 30, and 40 bars; feed velocities were 0.75, 1.25, and 1.5 m/sec; and recoveries were 50, 75, and 90%. The results show the large effects of each of the process variables on membrane fouling and flux decline. The worst case is for membranes annealed at the lowest temperature. Both water type V and tap water hardly caused a noticeable decrease in product flow rate. In addition, high feed velocities and low applied pressures reduced membrane fouling. Considering these results, the most promising conditions are to pretreat raw surface water by chemical clarification and sand filtration, then operate the HF at 20 bars with a feed velocity of about 0.75 m/sec. The recovery was at 90%. These experiments were undertaken primarily to study membrane fouling and lifetime. To optimize for rejection or volume flow, the experimental conditions must be changed. On this basis, it could be demonstrated that the extensive pretreatment of tapwater may be unnecessary. The results using the feed type V show that even this water required membrane cleaning after only several days or even a few weeks. It may be concluded from this investigation that Rhine water can be successfully desalted by HF.

The continuation of these trials was directed toward the removal of inorganic and organic micropollutants as well as to develop an additional pretreatment scheme for the spiral-wound and hollow-fiber modules. Schippers *et al.* (1978) report their experience in continuing the Rhine water studies. A pretreatment scheme consisting of coagulation, neutralization, sedimentation, and rapid sand filtration resulted in considerable reduction of the metal content. The HF tubular module unit that follows this pretreatment scheme contained highly annealed CA membranes and produced a product water having a low concentration of most metals. The removal by HF or organic compounds was found to be sufficiently effective using total organic carbon as a measure of this removal. The removal of low MW organics, however, was rather poor, as expected. Also, chlorination before removing the humic acids should be avoided because this results in halogenated organics that are poorly rejected by HF. Besides the tubular modules, the spiral-wound and hollow-fiber modules are under investigation. These latter two module types require a more complicated pretreatment scheme. In-line coagulation with cationic polyelectrolytes should be promising.

Although not all rivers in Europe are as severely polluted as the Rhine, it seems preferable to pretreat any raw river water by chemical clarification and sand filtration. Water from the river Ebre serves as feed for the production of demineralized boiler water. To prolong the time between the regeneration cycles of the ion exchanger, a two-stage HF unit fitted with DuPont B-9 modules was installed. The overall conversion was about 70% with a daily production of 288 m^3. With HF the cycle could be extended by a factor of about 25, but the real production depended on the wide fluctuation of the river water quality, between 310 and 810 ppm (expressed as $CaCO_3$) during June 1977 and February 1978. An effective pretreatment scheme was used to reduce module fouling caused by colloids in the raw water (Dangeon *et al.* 1978).

According to Di Pinto and Santori (1977) there are three hollow-fiber module units installed to supply potable water for hotels in Italy. The feed is brackish water, but more precise details are not available. An extensive study, however, was carried out by Boari *et al.* (1978), using five different commercially available HF modules to evaluate their performance. Membrane lifetime, different kinds of cleaning techniques, and operational parameters such as applied pressure, feed flow rate, and recovery were investigated. The specification for each of the modules is presented in Table IX. Brackish water was used as the feed with a salinity of 3000 ppm

TABLE IX

Module Specifications for Brackish Water Hyperfiltration Study[a]

	Envirogenics	Westing-house	UOP (Havens)	UOP	Du Pont
Type of membrane	Blend membrane heat treated 14 ff.	CA 090 series	CA type 620	CA model 4004	PA
Configuration of membrane	tubular	tubular	tubular	spiral wound	hollow fiber
Max. operating temperature (°C)	30	28	30	30	30
Max. operating pressure (bars)	70	28	42	42	28
Membrane area (m^2)	3.83	0.86	1.48	2.37	164
Packing density	19.7	18.8	73.3	147	6919

[a] After Boari *et al.* (1978).

expressed as NaCl. It was pretreated as follows: The pH was adjusted by HCl dosing to a value of 5.5 (±0.5); sodium hypochlorite was added (10–12% active chlorine) for a residual chlorine of 0.5 mg/liter; and suspended solids were filtered using two filters. The feed for the hollow-fiber modules was not chlorinated to prevent membrane degradation. Washing was carried out whenever the pressure drop increased and module performance decreased significantly, indicating an increase in membrane fouling. To recover the original module performance 0.2% ammonium citrate solution (for removing iron-containing precipitates), 0.02% alcozyme solutions (a synthetic detergent for organic foulants), or hydrochloric acid solution at a pH of 3.5 to 4.0 (for dissolving carbonates and hydroxides of bivalent and trivalent metals) were used alternatively.

Figure 6 shows the volume flux as function of the mean effective pressure for each module. The results were obtained at a maximum feed velocity to minimize the effects resulting from concentration polarization, thus the Reynolds number for the tubular modules varied between 30,000 and 50,000. For the total volume production per day of a particular module the results shown in Fig. 6 should be weighted by the packing density in the bottom row in Table IX. Figure 7 shows the mean observed rejection as a function of operating time. Some results from the long-term tests are also listed in Table X. From the results it may be concluded that the modules were tested under well-defined field conditions. The DuPont

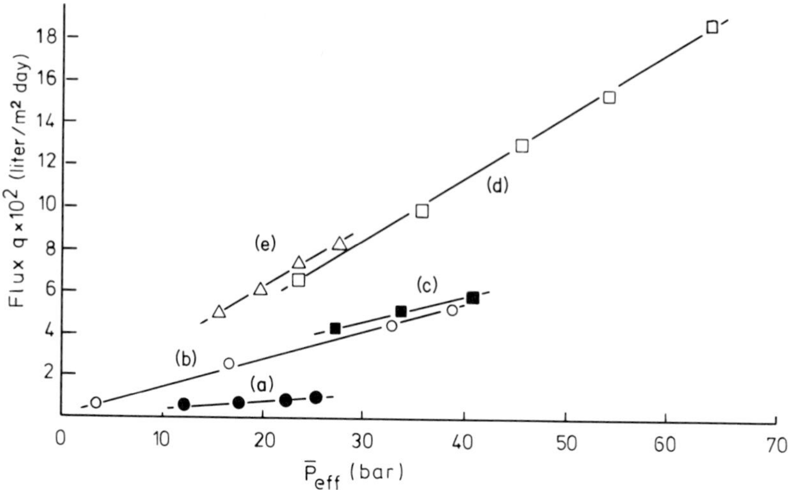

Fig. 6 Volume flux (q) as function of the mean effective pressure ($\bar{p}_{eff}$) for different modules (see Table IX). (a) DuPont, (b) UOP tubular, (c) UOP spiral, (d) Envirogenics, and (e) Westinghouse. The effective pressure $\bar{p}_{eff} = (p_f + p_b)/2 - \{(\pi_f - \pi_b)/2 - \pi_p\}$, f = feed, b = brine, p = permeate (from Boari *et al.*, 1978).

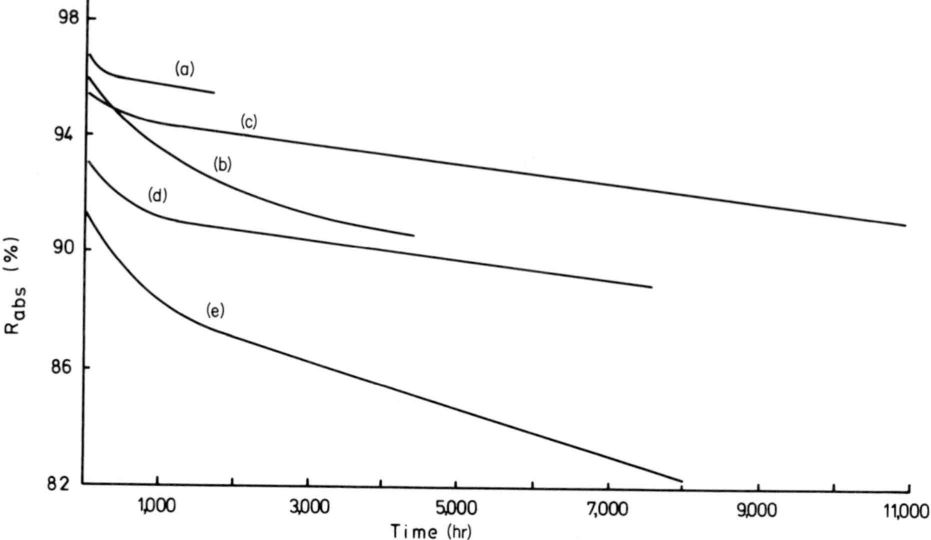

Fig. 7 Mean observed rejection $\bar{R}_{obs}$ as function of operation time h for different modules (see Table X (a) DuPont, (b) UOP tubular, (c) UOP spiral wound, (d) Envirogenics, and (e) Westinghouse. $\bar{R}_{obs} = (\bar{c} - c_p)/\bar{c}$; $\bar{c} = (\bar{c}_f + c_b)/2$ (from Boari *et al.*, 1978).

TABLE X

Long-Term Results for Brackish Water Hyperfiltration Study[a]

Plant	Operating pressure (bars)	Recovery (%)	Rejection (%)	Flux (%)	Decrease in rejection (%)	Operating time (hr)
Envirogenics	40	40	93	18	4.5	7700
Westinghouse	28	50	91	10	8.9	8000
UOP (Havens)	42	75	87.5	12	7	4500
UOP (spiral wrap)	42	75	97	18	3.5–4.5	11000
DuPont	28	50	98	3	1.2	1700

[a] After Boari *et al.* (1978).

hollow-fiber module and the UOP spiral-wrap module appeared to give the best performance of the whole experiment.

To obtain potable water and to achieve a concentrate for conversion into table salt and gypsum by evaporation to dryness, brackish coal mine water was treated by HF. Several pretreatment procedures were examined to obtain a suitable quality brackish water feed for the process (Kepinski *et al.*, 1976). The product water contained less than 500 mg/liter TDS, and the concentrate has 23 g/liter TDS. No flux values were presented in the paper.

In addition to the HF of seawater (see Section II.B), a program for the HF of brackish water was initiated to compare it with the present expensive distillation process to produce tapwater for the city of Eilat. From January 1973 to December 1974, four different HF modules were examined in Eilat. They included the DuPont hollow-fiber B-9 module, the Eastman Kodak spiral-wound MK 96 module, the Ajax spiral-wound Roga 4100 module and the Israel Desalination Engineering tubular module. Except for the B-9, which had polymide hollow fibers, all the other modules were fitted with membranes made from CA. Feedwater with a salinity of 2500 ppm TDS was supplied from a well about 30–40 km north of Eilat (Glueckstern and Greenberger, 1976). With the aid of a sand filter, a 20 μm cartridge filter, additional pretreatment from a polishing filter and doses of acids and complexing agents, the performance of the modules could be maintained with predefined limits. The $-m$ value for the hollow-fiber modules was 0.047, the corresponding values for the Roga and MK-96 were 0.030 and 0.038, respectively. In 1973, because of military action, the modules were exposed to untreated feedwater for a 2-week period. This resulted in the destruction of the CA membranes by biological attack. Nevertheless, on the basis of this experience, a second site was selected to obtain more results on tests under field conditions using well water high in salinity. In October 1975 at Sabha, about 3 km north of Eilat, the operation commenced, using DuPont B-9 modules, Roga 4100 modules, and Dow triacetate hollow-fiber Dowex RO-4 K modules. The recovery range was from 17 to 66%, and the rejection of all the membranes could be maintained above 90% by monthly chemical cleaning. Using the operating parameters of the HF pilot plant, a commercial plant was built to meet the annual increase rate of about 10% in water supply in the Eilat region (Glueckstern *et al.*, 1978a). The first HF unit was fitted with Dowex RO-20 K spiral-wound modules and had an initial capacity of 700 m^3/day. An expansion program is scheduled to increase production up to 8000 m^3/day in 1981.

3. *Tap Water*

Hyperfiltration has been recommended even for municipal tap water supplies when used for rinsing electronic components during manufacture. A 6 m^3/hr HF plant was installed in such a factory at Tours, France, which is supplied with slightly mineralized (200 ppm TDS) surface water that is normally used for drinking water (Allard *et al.*, 1976). The excellent water quality necessitated a simple pretreatment in the form of ion exchange to prevent calcium carbonate scaling inside the modules. No reference was made to the kind of modules employed in this HF plant although hollow fibers are suspected. The plant operated satisfactorily for several months until the concentration of iron and manganese increased to several mg/liter. This demanded cleaning at 2-week intervals. Comprehensive chlorination and flocculation as well as filtration and dechlorination reduced the cleaning frequency to 20-week intervals. The investment costs of the added pretreatment were entirely compensated for by the savings in cleaning, membrane replacement and filter lifetime.

Rovel (1977) reports on a 4 yr experience using an HF plant in Poland fitted with hollow-fiber B-9 modules to produce 1500 m^3/day. The permeate was again used as feed for an ion exchange process. The demineralized water was used for rinsing electronic components. During the 4-yr operating time, the data showed that the plant operated without any serious trouble. Some problems were encountered in the pretreatment scheme. The TDS value was reduced from 710 to 70 ppm, and SiO_2 from 21 to 2.5 ppm, respectively, the origin of the raw water was not stated. But in some cases the measures were not necessarily predicted from laboratory experiments. In a German dairy an HF plant was fitted with tubular modules having an overall installed area of 226 m^2. After several months, the pretreatment was reduced to only prefiltration. As may be seen from Table XI, the tap water feed permitted this procedure with a recovery of about 70%. The tubular CA membranes (paper-reinforced Kalle type) had the following performance characteristics using a 5000 ppm NaCl solution at 40 bars: Solute rejection was 93%, and volume flux was 1100 $1/m^2$ day. The membrane tubes were connected (fastened) only at the ends of the 3 m long tubes (24.4 mm ID). Therefore, the units shutdown had to be carried out under full operating pressure to prevent the collapse of the membranes. This unit has been operating for 18 months continuously producing flushing water without changing the membranes (Dobias, 1977, private communication). Often the production of ultrapure water using tap water feed is accomplished using hollow-fiber modules (Marquardt, 1977), because the floor space area required is small compared to tubular modules. A combined pretreatment of phosphate and sulphuric acid (Marquardt, 1975) has proved to be especially effective. Tap water may also be treated with plate-and-frame modules, DDS (Nakskov, Denmark) has installed several plants for the generation of dialysis water in hospitals and of boiler feed water, respectively (Poulsen, 1977, Private communication).

C. JAPANESE ACTIVITIES

Although it is not always easy to obtain capacity data from European companies engaged in brackish water desalination by HF, a survey of Japanese activities in HF is available (Ohya, 1976, 1977) and is shown in Table XII. The capacities for 1976, the year of the survey, will probably be larger by 10–20% than shown in Table XII. The reason is that the sales

TABLE XI

Quality of the Tapwater Feed for Dairy Hyperfiltration Plant

Analysis	Feed	Product	Rejection (%)
Temperature (°C)	13	—	—
pH	7	—	—
Cl^- (ppm)	245	50	79.6
Hardness[a]	428	53.6	87.5

[a] Expressed as ppm $CaCO_3$.

TABLE XII

Capacity of Various Hyperfiltration Plants in Japan (m^3/day)[a]

	1976[a]	1977[b]
Brackish water desalination	41,595	54,813
Ultrapure water generation	8,934	11,078
electronic industry	1,500	—
medical and pharmaceutical industries	554	—
Wastewater treatment and recovery	5,197	9,326
Food and enzyme	315	—
Seawater desalination	185	138.4
Miscellaneous	—	4,754
Total (m^3/day)	56,226	80,114.4

[a] From Ohya (1976).
[b] From Ohya (1977).

volume of Daicel Co.—the largest HF producer—was not available when the data for Table XII were prepared. Nevertheless, these data will demonstrate the trends in Japan. The largest HF distributor, Kurita Ltd., has installed approximately 40,000 m^3/day total capacity since 1971 for brackish water desalination, an additional 5,100 m^3/day for high purity water generation, and 230 m^3/day for medical application (Taniguchi, 1977, private communication). The daily production of desalted water using seawater as a feed amounts to about 63 m^3. These figures indicate that the most significant application of HF until the survey was the generation of industrial feed water from brackish water, e.g., boiler feed water. For this purpose the HF plants are often combined with an ion-exchange installation. Ultrapure water production for rinsing components of electronic industries or medical and pharmaceutical applications is the next most frequent application. Detailed engineering data on long-term studies with large capacity HF installations are given by Kurita (Taniguchi, 1977). One unit at the Kashima Steel Works has a daily output of 13,400 m^3. In 1980, Horio reported that the firm had gained valuable experiences in pretreatment, membrane lifetime, and operation conditions since start-up of HF equipment.

The HF is followed by ion-exchange demineralization to produce high-purity boiler feed water. The feed for the HF plant is surface water taken from the Kitaura Lake, which is sometimes supplied with inflowing seawater. Because of this, there is a remarkable change in salinity of the feed and, because of the yearly temperature change, an increase and decrease in algae growth is observed. This may be seen from Fig. 8. In contrast to the pretreatment scheme chosen for Rhine water (Kuiper, *et*

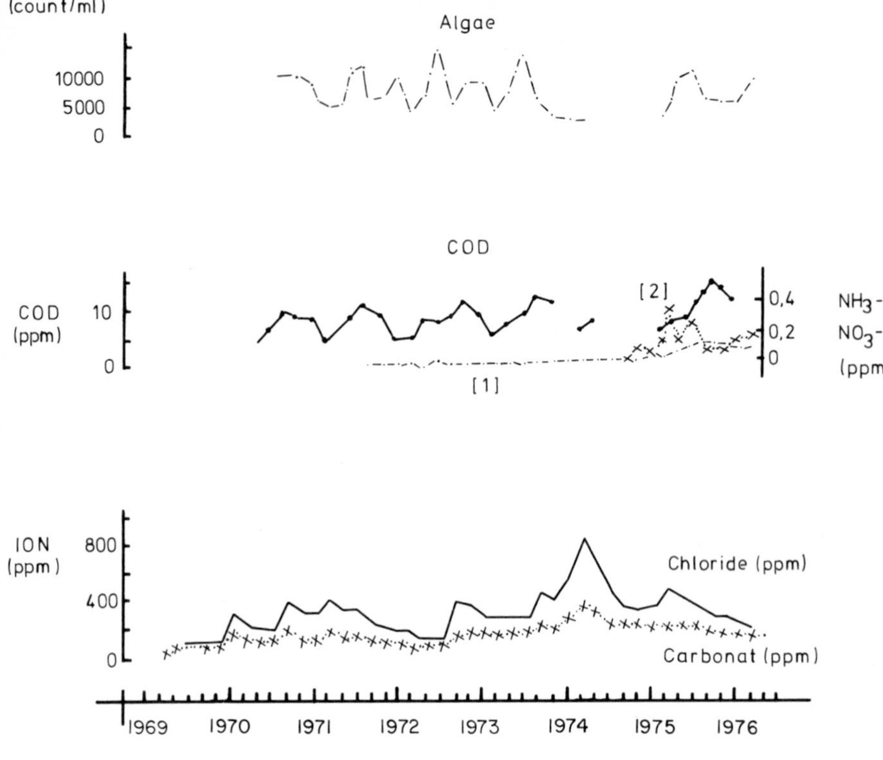

Fig. 8 Water quality of Kitaura Lake as function of time (from Taniguchi, 1977).

al., 1974a), more intensive pretreatment was installed for this case. This more sophisticated pretreatment resulted in less frequent membrane cleaning operations than that used for the Rhine water study where daily cleaning was used. The pretreatment consisted of a heat exchanger, settling basin, dual media filter followed by a polishing filter, acid injection and a safety filter. Using these processes, membrane cleaning required a two-hour water flush every 20 days, and an 8-hr chemical cleaning every 6 months. The plant availability could be maintained at 98%. The start-up of the plant was in late 1971, and completed in March 1975. The m value was about -0.02 at an operating pressure of 25 to 30 bars, and the module exchange rate was between 5 and 6% per year. Some of the HF results shown in Table XIII demonstrate the performance of the installed spiral-wound Roga 4160 HR modules.

TABLE XIII

Hyperfiltration Results Using Intensively Pretreated Surface Water and Spiral-Wound Modules[a]

Analysis	Expressed as	Feed	Product	Concentrate	Rejection (%)
EC	(μmho/cm)	2445	73.2	7000	98.45
Total hardness	ppm $CaCO_3$	265	4	880	99.3
Ca hardness	ppm $CaCO_3$	85	2	250	98.8
Mg hardness	ppm $CaCO_3$	180	2	630	99.5
Cl^-	ppm	650	20	1900	98.4
SO_4^{2-}	ppm	134	1	380	100
SiO_2	ppm	12	1	33.6	95.6

[a] After Taniguchi (1977).

D. CONCLUSION

According to Table XII, HF of brackish water represents the largest application for membrane separation processes. The feed concentration expressed as TDS ranges between 1000 and 5000 ppm. This is the preferred range of operation for ED, which thus competes with HF.

All types of HF modules are practical, depending on the intensity and type of feed pretreatment. Everything being equal, in the long run, those modules with the higher packing density will probably prevail. Whereas much work has been done on pretreatment and cleaning procedures, little effort is being invested into the problem of the concentrate disposal.

IV. Hyperfiltration as Unit Operation in Industry

A. INTRODUCTION

In contrast to the considerable amount of published papers dealing with HF of seawater and brackish water, there is relatively little information on the use of membranes in industry for the purpose of the recovering reuseable substances or of water purification. Several factors may be responsible for this. Unlike seawater HF research, which is in most cases partly or entirely supported by government funds, resulting in open publication of the results, companies involved in industrial applications are reticent to release information that may give an undue advantage to their competition. Other limitations include excessively large feed volumes and unusually high feed salt concentrations.

B. Small-Scale Trials

With a view to the removal and recovery of harmful substances, plating rinse effluents are often treated by HF. The classic work by Golomb (1970) gives experimental data for this type of practical application. Fundamental research concerning chromic plating rinse were carried out by Masuda *et al.* (1976). They used synthetically prepared plating rinse. The experiments were performed using different annealed CA membranes at 40 bars. The influence of chromic acid solution pH (0.01–0.2 *M*) on the transport behavior of the membranes was quantitatively studied. With increasing pH values, the rejection of the chromic species increased according to the increasing concentration of the divalent chromate ion, CrO_4^{2-}, formed by dissociation. Concentrating chromic acid solution at pH 7.1, the rejection remained constant up to a concentration of 15,000 ppm, when the volume flow decreased with increasing concentration, as expected. Similar findings using CrO_4^{2-} and CrO_7^{2-} solutions were reported by Stenger *et al.* (1975). In Table XIV, the rejection of the sodium and potassium salts are compared with the rejection of NaCl. The experiments were carried out with stirred batch cells at 40 bars. The rejection values are initial values at a recovery of about 10%.

TABLE XIV

Rejection of Sodium and Potassium Chromic Salts by Cellulose Acetate Hyperfiltration Membranes[a]

Salt	Concentration (M)	pH	Rejection (%)
NaCl	0.085	—	84.0
Na_2CrO_4	0.031	8.67	98.5
K_2CrO_4	0.026	8.07	97.0
$Na_2Cr_2O_7$	0.019	4.02	90.5
$K_2Cr_2O_7$	0.017	4.20	87.0

[a] After Stenger *et al.* (1975).

C. LARGE-SCALE APPLICATIONS

The Austrian firm of Bran & Lübbe has installed a HF unit to treat nickel-containing rinse effluents. The rinse water is concentrated from 5 to 10 g/liter nickel salts up to 50 g/liter, and the concentrate is then recycled to the plating bath. The nickel rejection is about 99% at 28 bars operating pressure (Dobias, 1977, private communication). The modules are of the spiral-wound type produced by Chemical Systems Inc., and are fitted with CA membranes that are stable after a repeated wet–dry cycle without any decrease in membrane performance.

An excellent example of the use of HF in industry is the "Alclose RO" process developed by Shinko–Pfaudler (Koga and Ushikoshi, 1977). Instead of only using UF in a closed system for the recovery of electrocoatings and the production of a permeate free of resins and dyes to rinse electrocoated fabrics, DuPont's B-9 hollow fiber HF modules were added to the flow sheet. Because electrocoating paints for aluminum products contain low MW resins that permeate UF membranes but not the HF membrane, a better quality product was obtained. An additional advantage was the good performance of the polyamide membranes at pH 8.0–9.5 of the electrocoating recovery system. Figure 9 shows that the UF modules serve as a pretreatment device to prevent plugging of HF modules by airborne dust particles that may contaminate the open electrocoating bath. The HF permeate contains less than 0.03% nonvolatile matter,

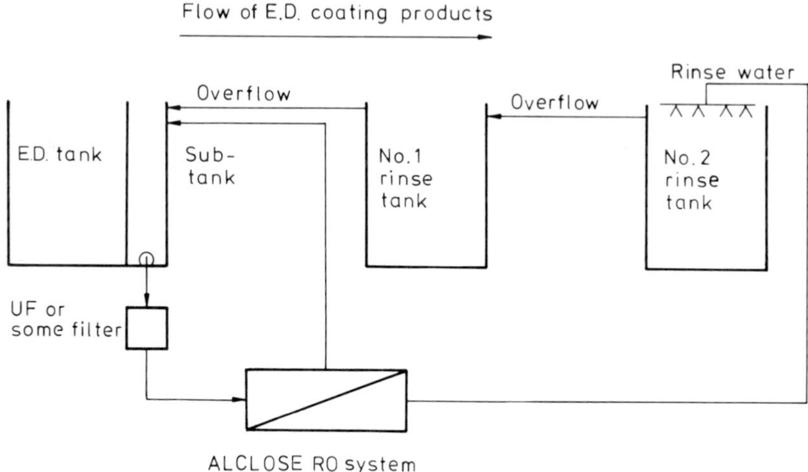

Fig. 9 Flow diagram of ALCLOSE RO system (from Koga and Ushikoshi, 1977).

whereas the concentration in the bath ranges between 8 and 15%. The first plant was erected in October 1975 and until now no replacement of modules has been necessary. Largely similar to the UF of electrocoating, this process operates extremely well and saves paint costs.

V. Hyperfiltration of Liquid Foods

A. INTRODUCTION

Food technology comprises various treatment processes that are used to reduce transportation costs by concentration, or to ameliorate storage capability by water removal. Various techniques are used to obtain specific ingredients from natural products, but even the transformation of natural foods via some intermediate process stages to new synthetic foods should be included under the aegis of food technology. Furthermore, cleaning processes are important where unwanted natural substances are partially removed from the foods. Examples include: The concentration of apple juice, the production of sugar from sugar beets, and the brewing of beer. An example of the cleaning procedures is the reduction of a particular food stream's salt content.

For these fluid-phase processes, traditional well-proven unit operations exist. However, the search for better treatment methods continues undiminished. The problem is either to reduce the cost of the product or to preserve typical aromatic ingredients in the product and intensify their effect. In addition, the nutritive quality of foods often decreases during the manufacturing process. Therefore, new processes are sought in which the substrate quality is preserved. A nondegradable process for fluid food treatment is separation with membranes. With this technique, some disadvantages associated with conventional unit operations may possibly be overcome. Many large-scale applications of membrane processes in the food industry attest to this fact (Merson and Ginnette, 1972; Roosmain *et al.,* 1974).

The osmotic pressure of apple, grapefruit, and pineapple juice is about 21 bars, which is close to that of seawater. To attain suitable concentration ratios (e.g., 1 : 4), extremely high operating pressures are required (Leightell, 1972). Under these circumstances, the relatively low permeation rates make the process economics questionable. Moreover, the variety of ingredients creates additional problems. Fruit juices are multicomponent systems composed of sugar, acids, esters, alcohols, aldehydes, and pectines as macromolecules. The latter hardly contributes to the osmotic pressure of the solution, but they increase the viscosity and consequently the pumping costs. Besides this, concentration of macromolecules such as pectines, sugars, and acids pose few problems for membrane treatment. Suitable membranes to reject these solutes already exist. The rejection and transport of flavoring substances such as esters, alcohols, and aldehydes, however, is crucial (Harrison, 1970). Whereas oil-soluble substances found in orange juice are rejected by membranes, water soluble substances permeate CA membranes (Merson and Morgan, 1968). Hence, the juice reconstituted from hyperfiltration concentrate frequently lacks essential taste components. The taste of the new product does not correspond to the original one. The same taste substances, by contrast, are volatilized from the original concentrate and are to some extent absent anyway. This occurs using conventional methods such as evaporation. For these reasons, the transfer from a laboratory scale to a technical scale appears to be long in coming. Some examples of using HF in food processing are presented subsequently.

B. HYPERFILTRATION OF ALCOHOL-CONTAINING BEVERAGES

The fact of low molecular weight alcohols permeating CA membranes is used during beer treatment to purposely reduce the alcohol content. One reason for this, is the stipulation of a definite maximum alcohol level in the blood of motorists. Other reasons are health-related, such as for dieting when the consumption of low-alcohol drinks are desired by those with metabolic diseases. Prohibition was also an impetus in this direction (Hürlimann, 1919).

Conventional methods for the production of beer with low alcohol content are based on the intervention of the fermentation process or dilution of the end product. These products do not have an attractive taste. An alternative method is the careful removal of alcohol from the end product. Hyperfiltration now competes with the currently used boiling process and vacuum distillation (Kieninger *et al.*, 1975).

The HF of beer was demonstrated using two different operational procedures. In one case the full beer was concentrated and a water–alcohol mixture permeated the membranes. A corresponding volume of pure water was added to the concentrate after completing the operation. The other procedure was to dilute the full beer with pure water and concentrate it by HF to its original volume. Patents and reports exist for both variants.

In the first procedure different types of beer in quantities of 30 to 200 liters were pumped in a closed-loop apparatus fitted with a plate-and-frame module. The concentration ratio at 60 bars was 0.5–0.75. The results are shown in Table XV. The membranes were CA type 990 fabricated by DDS, (Nakskov, Denmark), the rejection of alcohol using a 5 volume % alcohol-in-water solution was about 5% at a recovery ratio of 0.5 (Wucherpfennig and Neubert, 1976). Footnote *b* in the table shows that the reconstitution was performed using the dealcoholized permeate. Thus components of the permeate such as acids and wort could be used for improving the taste of the alcohol-lowered beer. The volume flow rate was extremely low, and with increasing concentration the value decreased from 15 to 6 liter/m^2 day at a recovery of 0.6. Economically, these results are not promising. The other procedure should operate more efficiently, because diluting the full beer with a known volume of deionized water ensures a more economical permeation rate. Careful treatment of beer has suggested operation at a preliminary pressure of 1 to 5 bars at temperatures of 0 to 10°C. The concentrate is collected at a pressure of 1 to 4 bars under a CO_2 atmosphere in prestressed tanks. Using these measures, the generation of foam and the absorption of oxygen is prevented (Adler, 1974).

With CA membranes from PCI (England, UK) (T1/12 and T2/15, which had a rejection of 97 and 93%, respectively, using NaCl solution at 40 bars), trials were carried out to show the influence of different dilution ratios (Kieninger *et al.*, 1975). With regard to bitter substances such as tannin, anthocyanogenes, and "bitter values," nearly no change in bitter concentration was detectable in the concentrate as compared with the original solution. Color and pH value were also not noticeably altered. With increasing operating pressure, a certain amount of nitrogeneous

TABLE XV

Alcohol Reduction by Hyperfiltration Using Different Types of Beer[a]

	Alcohol (g/liter)				
Type	Beer	Concentrate	Permeate	Reconstituted beer	Recovery (%)
Export	40.1	45.3	38.0	17.0	0.5
Diät	51.7	59.4	48.8	13.1	0.75
Diät[b]	51.3	59.5	47.1	19.4	0.75[b]

[a] After Wucherpfennig and Neubert (1976).
[b] See text.

substances and polyphenols were lost. As is to be expected, a certain quantity of aroma compounds were lost along with the removal of ethanol, which caused a reduction in taste in the alcohol-lowered product. This is more pronounced as more ethanol is removed. Table XVI shows the content of aroma compounds in alcohol-lowered beer as a function of dilution. The first column in Table XVI shows the original quality of the full beer. According to these experimental findings, esters permeate the membranes more easily than alcohols.

There is no published data for alcohol-reduced beer on a plant-production scale, though some experts regard the separation with membranes as the most promising method. The market for this low-alcohol beer is limited, and the taste retention problem must still be solved. Again, the latter problem is always coupled with the alcohol content. The development of appropriate membranes with selective permeability to only alcohol seems difficult.

With regard to aroma compounds in apple cider, the concentration trials with HF have led to the same results (Schobinger *et al.*, 1974). These experiments were carried out to compare HF with the conventional thermal process of evaporation. The tests were performed with a DDS module equipped with CA membranes type 995 at 42 bars. The maximum recovery amounted to about 0.88. Whereas the extract compounds were completely rejected, the acids were partially retained by the membranes. The aroma compounds behaved similarly to those in the case of beer concentration. There was no noticeable difference in concentrate quality from the conventional process.

Although must does not contain alcohol, it is involved in the first step in wine production. This section deals with concentration experiments of this precursor of wine. When the grapes contain insufficient sugar, wine

TABLE XVI

Aromatic Compounds in Alcohol-Lowered Beer[a]

	Alcohol content (%)			
Components in beer	3.95	2.50	1.40	0.50
Ethyl acetate (mg/liter)	18.4	8.6	3.6	1.0
Isoamyl acetate (mg/liter)	1.65	0.75	0.4	0.25
n-Propyl alcohol (mg/liter)	12.6	8.4	6.0	3.6
Isobutyl alcohol (mg/liter)	6.4	4.8	4.0	2.5
Isoamyl alcohol (mg/liter)	53.0	39.0	31.0	18.0

[a] After Kieninger *et al.* (1975).

with a low alcohol content will result. This reduces its sales potential and storage stability. Because it is forbidden by law in some countries to add alcohol to the natural wine, the desired concentration of must should be obtained before fermentation, by either vacuum evaporation or freezing. Concentration by HF, however, could also be competitive. The feasibility of this operation using a tubular module plant was tested with 5.3 m^3 grape must (Peynaud and Allard, 1970). At a recovery of about 0.2 it was found that only small amounts of potassium (2%) and maleic acid (6%) permeated the membranes. The loss in sugar was negligible (0.2%), and even tartaric acid could not be detected in the permeate, but its content was lower in the concentrate compared with the original values. Tartaric acid was probably precipitated in the form of tartar. More detailed results on grape must concentration were presented by Peri and Pompei (1975). They used flat membrane laboratory modules and CA membranes types DDS 990, 995, and 999. The trials were carried out at pressures from 30 to 70 bars and at a temperature of 6°C (±1°C). The results showed that an economic concentration of must was limited to a maximum amount of 20% in sugar. This corresponded to an osmotic pressure difference of more than 30 bars across the membrane. The flow rate was below 200 liter/m^2 day using the least annealed membrane (990) and about 50 liter/m^2 day using the tightest membrane (999). In any case, the rejection of sugar was greater than 99%, whereas the rejection of acids (expressed as total acid) ranged from 70 to nearly 90%. These values correspond to the findings of Peynaud and Allard (1970).

There is no commercial plant to concentrate grape must, though the results are qualitatively superior compared with those attained by evaporation. One reason for this is the seasonally limited usage of the plant capacity, making the economics less attractive.

The precipitation of tartar in the course of concentrating tartar-containing solutions is, however, used for the stabilization of wine. This usually means the reduction of potassium hydrogen tartrate in wine to prevent tartar precipitation in bottles filled with wine or champagne. These beverages are normally supersaturated in tartar. The conventional methods applied are cooling or ion-exchange. Electrodialysis has been utilized successfully (Wucherpfennig and Krueger, 1975). By means of HF the wine is concentrated to a certain ratio, ensuring the recrystallization of tartar in the concentrate. This concentrate is filtered after expansion to normal pressure using a normal filter press. After clarification, the concentrate and permeate are recombined.

A pilot plant fitted with tubular modules containing a membrane area of 58 m^2 was installed for producing champagne (Rhein, 1975). The Kalle-type tubular reinforced CA membranes had a rejection of 85–90% for sodium chloride, using a 0.085 *m* solution at 40 bars. They did not exhibit any noticeable deterioration or scaling formation after 18 months of operation. According to the fluctuating concentration of the ingredients in the feed from one year to the next, the technical feasibility of plant operation was problematic. The influence of most essential wine compounds on recrystallization tendency by HF was examined using a synthetic solution and wine (Staude *et al.*, 1976). In addition to polyphenols, the concentration of ethanol and glycerol affected the tartar precipitation. In Fig. 10 the range from the beginning of recrystallization to the formation of visible crystals using different ethanol–water compositions is shown. The content of tartar in the feed was 3 g/liter. During each trial the ethanol–water ratio was constant and the membranes (Kalle-type reinforced CA membranes) did not reject ethanol. In Fig. 11 the temperature dependence of the beginning of recrystallization as a function of recovery ratio is depicted. In addition to the synthetic solutions used in Fig. 10, the results for natural wine are also given. They show the same slope as those for synthetic solutions. Using the temperature dependency of the volume flow, an experimental activation energy of about 27.1 kJ/m was calculated. The tartar content in the wine obtained from recombination of the permeate and filtered concentrate was dependent of the concentration ratio. The best conditions for the concentration range was from 0.4 to 0.6. The product's taste was not altered when compared with the original

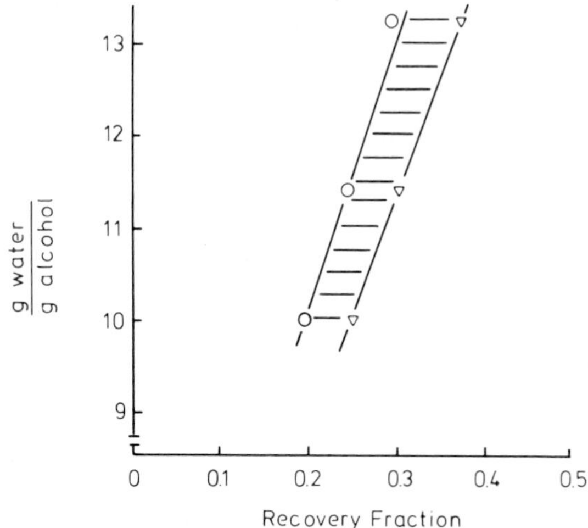

Fig. 10 Influence of alcohol concentration on tartar crystallization as function of recovery. Formation of microcrystallites in solution ○, precipitation of visible tartar crystals ▽ (from Staude *et al.*, 1976).

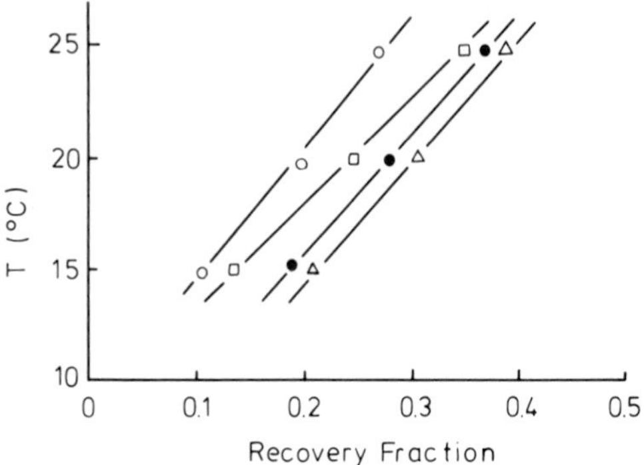

Fig. 11 Influence of temperature on the beginning of tartar crystallization as function of recovery. Each solution contained 3 g/liter tartar; 8 g/liter glycerol, 90 g/liter ethanol (○); 6 g/liter glycerol, 80 g/liter ethanol (□); 4 g/liter glycerol, 70 g/liter ethanol (△); natural wine (●) (from Staude *et al.*, 1976).

wine. This technical application is, however, for the future. One reason for this delay may be because of legislative problems associated with the wine industry.

C. CONCENTRATION OF FRUIT JUICES

Hyperfiltration of grape must is aimed at enriching the sugar content. The main aroma compounds are, however, developed during fermentation. The concentration of fruit juices tends to remove water while retaining all essential compounds. As mentioned previously, all experiments with fruit juices show that water-soluble higher organic molecules like sugars or even organic acids are rejected by selecting appropriate membranes. The flavor substances, which are primarily C_2–C_6 alcohols, C_4–C_8 esters, and C_2–C_6 aldehydes permeate the membranes. This is one of the problems in concentrating fruit juices by HF. This can probably be solved by using membranes prepared from polymers other than CA (Chian and Fang, 1973). Another problem is the high osmotic pressure of the juice. To ensure flow rates at the necessary fourfold concentration, high operating pressures of 105 bars are required. In addition, the solid matter in the juices often causes fouling requiring frequent cleaning operations. This may explain why there has been no previous large-scale operation.

To prevent membrane coating, Pompei and Rho (1974) centrifuged passion-fruit juice before concentration by HF. With different CA membranes of the same specification used in must concentration experiments (Peri and Pompei, 1975), and different pressures, it was shown that this juice behaves similarly to other juices. The higher MW substances are rejected well, whereas the volatile aroma constituents permeate the membranes to a certain degree. Thus sugar is rejected by more than 99.9% over a pressure range from 40 to 70 bars using membrane type DDS 999, whereas the rejection of the polyphenols, expressed as gallic acid, decreases from 96.8 to 95.9% under the same conditions. The rejection of the four most essential flavor substances was calculated to be 70%. The corresponding pressure was not noted. Organic compounds with a high distribution coefficient in the membrane phase such as phenol (Lonsdale *et al.*, 1967) normally exhibit a decrease in rejection for increasing pressure (Pusch *et al.*, 1976). Whereas the reconstituted juice obtained by diluting the concentrate and by adding the solids did not show any taste difference compared with the original juice, the technical application of HF appeared to be restricted by the low-volume flow rate (permeation flux). Additional work was conducted to study the concentration behavior of fruit juices such as apple, orange, and pineapple juice by HF. Information on the pressure–permeation relationship of volatile aroma compounds was not provided (Gherardi *et al.*, 1972), though detailed experimental results are presented that correspond to other experimental findings using fruit juices. The trials were performed with an Abcor tubular module plant (membrane area 1 m^2) equipped with CA membranes, and the operating pressure for each application was about 50 bars.

D. HYPERFILTRATION IN DAIRY INDUSTRY

Initially, membrane separation processes were used in the dairy industry to reduce pollution problems resulting from the discharge of high BOD whey. A partial solution was found using UF in which the protein is rejected. This protein-containing concentrate with high nutritive value serves as provender. It is also suitable for human consumption, supplying the daily needs of amino acids with 14.5 g of whey proteins. In contrast, HF plays a minor role in the dairy industry. This is not because the osmotic pressure of cheese whey is particularly high, because it is only about 7 bars. Fouling as a result of proteins is the most significant problem. The larger protein molecules are transported to the membrane surface by convection, but because of their low diffusion coefficients, their back diffusion from the membrane toward the bulk solution is small. This results in severe concentration polarization and fouling. A recommended counter measure is to increase the tangential flow velocity within the modules. This is limited, however, by the increasing viscosity of the concentrated whey and appears to be the main problem in cheese whey HF. Because of viscosity limitations, a concentration from 6 to 7% solids, up to about 20%, appears to be the practical limit (Leightell, 1972). Like UF, two important aspects in using HF for processing cheese–whey present themselves. These are the separation of the feed into its useful components, and the waste disposal problem. Hyperfiltration of whey can be carried out in a one-stage process or in a combined two-stage process. The latter combines UF and HF in one operation.

In the one-stage process the nutritious protein and lactose are concentrated, whereas the lactic acid and monovalent salts permeate the membrane and are removed. The mass balance for the HF of cheese–whey is shown in Fig. 12. These results are obtained using a 28 m^2 plate-and-frame module plant of the Danish firm DDS, fitted with CA membranes type 985. The membranes exhibit a salt rejection of 87% at 42 bars using a 2500 ppm NaCl solution. The operating pressure was about 50 bars (Nielsen *et al.*, 1972). The permeate with 0.1% dry matter has a BOD of less than 300 mg O_2/liter, which can be discharged to the sewer. The advantage of this one-stage whey treatment process lies in minimizing the pollution problem because the concentrate can usually find a ready outlet after being dried or evaporated. This concentrate corresponds to ordinary evaporated whey with a lower salt content.

The two-stage process, however, produces different quality streams. In the first stage the whey is ultrafiltered. In the second stage the protein-free permeate is concentrated about four times by HF. In this manner a product is obtained with slightly more than 20% dry matter and 80%

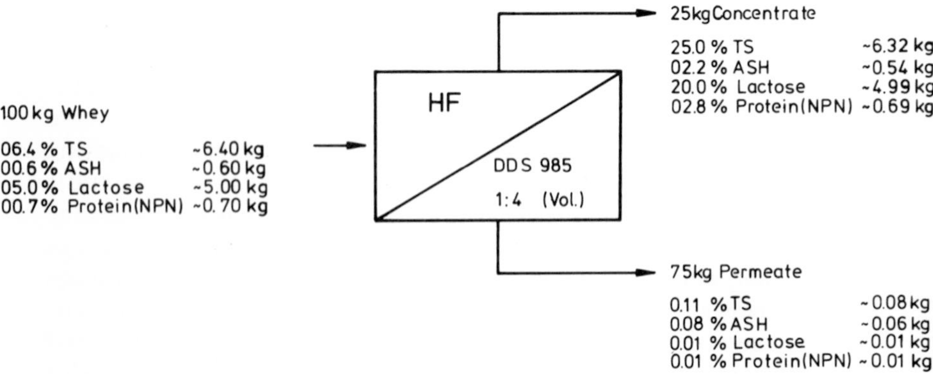

Fig. 12 Mass balance for sweet whey HF. (From Nielsen *et al.*, 1972).

lactose. This lactose concentrate is used as a raw material for the production of pure lactose because it contains no contaminant proteins. The demand for lactose, however, does not meet the supply of lactose derived from whey. Thus using the additional second step of HF depends on market conditions and not the viability of the process itself.

In the Netherlands DDS has delivered an HF plant for the treatment of 180 t/day of cheese whey. The plant consists of 16 plate-and-frame modules with a membrane area of 448 m^2. The modules are fitted with CA membranes type 990 (Poulsen, 1977, private communication). Like whey, skim milk may also be concentrated using HF. The usual concentration ratio lies between 1 : 2 and 1 : 3. The concentrate is used for making dairy products such as ice cream or powdered skim milk.

The most suitable module designs for whey treatment by HF are the plate-and-frame and tubular modules. In Europe the current membranes used in these modules are CA types. Besides the poor market conditions for the products, the sole use of CA membranes is a further restriction on the application of HF. To avoid bacterial growth, whey should be treated at temperatures below 15°C or above 35°C. In the latter case the membrane (CA) stability limitation that is recommended by membrane producers is exceeded. Moreover, the daily cleaning procedures in the dairy industry are often carried out using acidic solutions. This could also effect the membranes' performance if the acidic cleaning solution contacts the modules for a period longer than a few minutes. The modules could be cleaned instead with enzyme-active solutions. Because of these limitations, more stable membranes are actively being sought. Polysulfone is one such possible membrane polymer.

E. HYPERFILTRATION OF CARBOHYDRATES

The components in the permeate of ultrafiltrated whey can be separated into different fractions by HF. The carbohydrate lactose remains in the concentrate. By selecting a suitable membrane, the carbohydrate solution may be fractionated by HF. Glucose syrups are classified as carbohydrate solutions and used as raw materials for food manufacture. They are ideal substances for processing by HF because they are clear aqueous solutions of maltooligosaccharides and maltodextrins. The specification index is the dextrose equivalent (DE) that can be obtained at any level by fractionating and combining products by this process (Birch and Kearsley, 1974). For these experiments, PCI (UK) modules with different types of CA membranes (see characteristics in Table XVII) were used.

TABLE XVII

Cellulose Acetate Membrane Characteristics for Carbohydrate Fractionation Using Substances with Different Molecular Weights[a]

MW	Membrane rejection (%) (25°C, 2.1 bars)		
	T2/40	T2/A	T4/A
10,000[b]	100	100.0	100.0
1,000[b]	100	97.0	70.0
500	100	75.0	25.0
100	95	0	0
maximum pressure (bars)	40(80)	27.0	13.5

[a] After Birch and Kearsley (1974).
[b] Dextran.

The HF system was composed of a holding tank with cooling coils, filter, pump and the modules. Each membrane was operated at maximum pressure. The feed material was a 43% DE acid converted glucose syrup at solids concentration of 15–20%, which was continuously recirculated until the desired DE was achieved. The feed rate to the membranes was about 16 liters/min. Using the tightest membrane of the three types, with an overall rejection of 87%, the T2/40 yielded a permeate concentrate of up to 80–85% DE, showing a slight permeation from glucose up to about maltopenose. The T4/A membrane (overall rejection of about 83%) is, however, the most suitable membrane to produce syrups in the range of 15–43% DE. The permeate of this membrane had DE's only up to about 70%. With the loosest membrane the feed was simply divided into similar fractions of nearly identical composition.

Syrups with higher DE values of, say, more than 19% glucose are hygroscopic, they cannot be spray dried. Alternately, high MW components can contribute to turbidity or unwanted starch flavors. Experimental findings show that by using HF with suitable membranes and module combinations, carbohydrate syrups can be obtained free of both low and high MW components.

Additional detailed fractionation studies were presented by Kearsley (1976). It was demonstrated that the glucose fractions consisted of a narrower sugar spectrum compared with the commercially available products of the same DE values. By using HF and UF, (the membrane Type T4/A is also used for UF), the glucose syrups users in the sugar industry may be supplied with more specialized sugars.

F. CONCLUSION

According to Leightell (1972), the situation of commercial HF of foods was similar to that of nuclear power in the early 1960s. The commercial situation still appears to be at the same level as in 1972. Hyperfiltration is only used extensively in the dairy industry. In 1977, DDS (Denmark) and PCI (UK) delivered 13 HF plants for a total membrane area of about 3100 m^2 for whey concentration (Eriksson, 1977). It should be stated, however, that the demand for desalted protein powder is as low as that for lactose. By contrast, the demand for concentrated fruit juices seems to be sufficiently high. Thus the small loss of aroma compounds in the reconstituted product may be tolerated. Obviously, however, economic realities cannot tolerate the solution of HF problems during industrial operation. As discussed previously, many experimental results obtained on a laboratory scale since 1972 have increased our knowledge in treating foods by HF. By selecting appropriate membranes and modules and optimizing process parameters, the necessary results can be obtained for feasibility evaluations. Hence, it should be possible, even despite the restrictions stated previously, to realize the commercialization of HF in this industry.

VI. Wastewater Renovation by Hyperfiltration

A. INTRODUCTION

The growth in population, coupled with higher standards of hygiene, as well as a higher level of industrialization, has transformed society's approach in dealing with domestic and industrial sewage. The destination of this raw wastewater should not be by direct disposal into rivers and sewers. More or less complete purification is now required by law. Effluents often contain valuable substances worth recovering. Dairy effluents are examples of this, because after treatment they can be used as raw materials to produce proteins and lactose. Also, heavy metals found in rinsing waters from the galvanic industry could be recycled, especially in view of the growing shortage of workable ores and the toxicity of these metals. These systems are not treated in this section.

Although the danger of converting rivers into waste streams has not been fully averted, river water should not be seen as wastewater but as a polluted surface water. The application of HF to wastewater reclamation to produce water for reuse will be divided into three topics: (1) municipal sewage, (2) effluents from industry and the food industry, and (3) wastewater from steam power stations. The goal of most wastewater reclamation projects is the production and disposal of relatively high-quality water into the sewer. Normally, rivers that often act as sewers also represent an important source of water supply in highly industrialized countries. To improve the quality of river water, it is necessary to use advanced waste treatment processes to improve the quality of effluents from sewage treatment plants before they enter such surface streams. This means that high-quality water produced from wastewater renovation may be suitable for direct reuse and the need for river water may be reduced.

If HF is used as an advanced unit process, the problem is to find the appropriate place in the sequence of treatment processes where the total performance may be optimized. For municipal wastewater, HF is often installed to treat secondary effluents, whereas for wastewater with high COD, HF is used to reduce this value before the permeate is fed into the biological sewage treatment plant. Wherever it is placed, however, the brine produced by HF gives rise to a disposal problem. High recovery rates may reduce this problem to a certain degree, but this decreases the product flow rate. All types of HF modular designs are applicable to wastewater reclamation, but the tubular configuration seems to be preferable to the others because of its well-defined flow conditions.

B. MUNICIPAL WASTEWATER

There is no doubt that product water of high quality can be obtained from sewage effluents by HF. Detailed data on the rejection of specific substances are given in the literature (Hindin and Bennett, 1969). The problem in sewage HF is membrane fouling. Therefore, research deals with flux decline restoration and methods to maintain an economic flow rate either by prevention of fouling or by efficient removal of the deposits. A pilot plant of tubular modules (PCI B-type with CA membranes inside the 12-mm diameter tubes) was used in the studies at Rye Meads, England (Bailey *et al.*, 1974). Sand-filtered and carbon-treated effluents were used as feed (Table XVIII). By sulfuric acid dosing the pH was adjusted to a value below 6.5. Other feed additives such as sodium silicate, carboxymethylcellulose, or polyoxyethylene glycol did not prove effective. The best results were obtained using a 1% solution of biologically active laundry presoak (BIZ), which ensured a nearly constant volume flow of about

TABLE XVIII

Quality of Municipal Effluents for Hyperfiltration Treatment

Characteristic	Rye Meads		Tokyo
	Sand filtered	Sand and carbon filtered	
BOD (ppm)	3.0	2.0	5.0
TDS (ppm)	754.0	720.0	406.0
EC (μmho/cm)	1252.0	1124.0	600.0
Color (Hazen units)	24.4	1.8	15.0
Turbidity (ATU)	5.6	1.7	2.0

0.43 m/day during a period of 400 days. There was no noticeable difference after 160 days of operation in the volume flow, using either a sand- and carbon-filtered feed or only sand-filtered feed. This is the most noteworthy result of this investigation because by additional activated-carbon treatment, the COD value was reduced about 65%. The product rate was maintained between 0.4 and 0.5 m/day at a pressure of 34.4 bars. The rejection with regard to TDS was 92.7% (sand filtered) and 89.3% (sand and carbon filtered), respectively. Pretreatment with UF membranes of the sewage effluents did not prove effective as a measure to prevent flux decline, although the quality of the UF permeate was suitable for many industrial applications. All results were obtained with a minimum recovery of about 2.4%.

The influence of recovery on product water quality using secondary effluents from the municipal sewage installations of the city of Leiden, The Netherlands, was investigated by Wechsler (1977). The HF pretreatment train consisted of flocculation (after addition of 550 ppm alum and agitation), and filtration (dual media filter followed by a 20 μm cartridge filter). Paterson Candy International (UK) tubular modules with CA membranes annealed at 80° and 86°C, respectively, were installed in the closed-loop plant, which operated at a pressure of 40 bars and a temperature of 25°C. As can be seen from the experimental findings in Fig. 13, the chloride concentration increased with rising recovery, whereas the level of organic materials remained constant up to a recovery factor of 0.93. The organic concentration was measured by the permanganate number. The results indicate the normal behavior of certain organic species soluble in the CA membrane phase that were not rejected by the membrane. In this case, 12% of the permanganate oxidizable organics in the feed were permeable through the HF plant. Thus, even at high recovery, the installation costs of activated carbon posttreatment do not increase signifi-

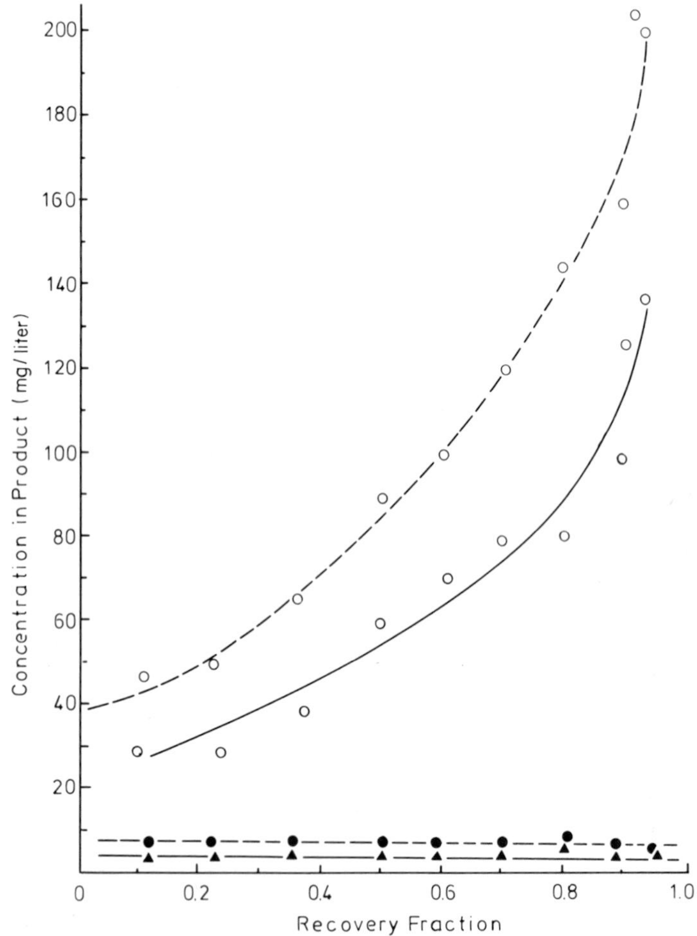

Fig. 13 Product solute concentration as function of recovery. Dashed lines are for membranes annealed at 80°C, open symbols the chloride concentration, filled symbols the permanganate oxygen demand (from Wechsler, 1977).

cantly, provided all organics show the same behavior as the permanganate oxidizable substances with respect to their membrane permeation. At a recovery of 80%, the *m* value for the flux decline was about −0.03 for the 86°C annealed membrane and about −0.04 for the 80°C annealed membrane. At higher recovery values the precipitation of lime must be prevented. If this is maintained, membrane fouling could be controlled by depressurization, followed by product water flushing. Foamball swabbing and chemical cleaning were also used. Obviously, the deposits on the membranes are normally not removed with that kind of ease. A continuous sponge ball cleaning system to prevent scaling was developed by the Japanese firm Hitachi for tubular module systems (Shimozato *et al.*, 1976). The balls remove soft scale by mechanically cleaning the membranes and also promote turbulent flow. In the case of hard scale caused by metal hydroxides or calcium salts, a combined treatment using chemicals and automatic ball cleaning removes the precipitates. By cleaning the module every 0.5 hr using the sponge ball procedure, the volume flux could be maintained at 0.65–0.75 m^3/m^2 day, at 25°C. Secondary effluents from the terminal plant at Osaka, Japan were used as feed (Yanagi and Mori, 1980).

Secondary effluents are suitable as a source of industrial water supply. This fact is made use of by the Tokyo Metropolitan Authority. This beneficial use is even more necessary as the level of the groundwater decreases with increased pumping because of a rising demand for industrial water. The quality of this effluent water is relatively good compared with other effluents (Table XVIII). Applying this for feedwater, Kimura and Nakao (1975) investigated the fouling phenomena to obtain data transferable to other effluents. The tests were performed at 40 bars using CA membranes annealed at different temperatures. The membranes were characterized by pure water permeability and the solute transport parameter according to Kimura and Sourirajan (1967) using a 1000 ppm NaCl solution. The compaction of the membranes with pure water was determined using the equation $(J/J_\circ) = (\tau/\tau_\circ)^{-m}$, where J is the volume flow, τ the elapsed time, and the index $\circ$ designates the initial value. For these experiments the feed supplied by the Tokyo Authority was used without any pretreatment. The results show in Fig. 14 that (1) the flux decline does not obey the power law but tends to merge into a single line, (2) fouling does not proceed after having reached a critical value, and (3) axial velocity influences the deposit accumulation. With respect to electrical conductivity rejection, it was found that it became constant after 30 hr. As for the 85°C annealed membrane, the value increased from 96 to 97.5%. This agrees with the observation using Rhine water (Kuiper *et al.*, 1974b). The rejection ability of the deposit layer could be demonstrated and is discussed using a dynamic membrane model. It was, however, difficult to obtain unambiguous results.

To study the effects of different pretreatment processes on the performance of HF, several pretreatment techniques such as coagulation and sedimentation (CS), sand filtration (SF), and activated carbon adsorption (AC) were evaluated. Tubular modules were used because of their lower sensitivity to fouling materials as compared to other modules. The modules were developed by Kobe Steel and Nitto Electric Industrial. The

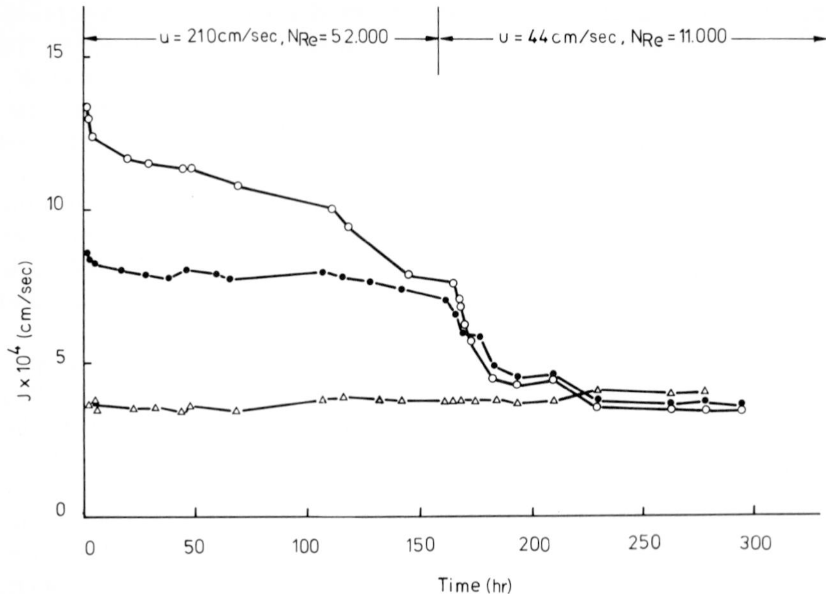

Fig. 14 Volume flow *J* as function of operation time *h* for the HF treatment of secondary effluent for different cured CA membranes (from Kimura and Nakao, 1975).

experiments were carried out at 42 bars and the membrane rejection with a 5000 ppm NaCl aqueous solution was 90% (Tsuge and Mori, 1977). Secondary treated municipal effluents (SE) were used as feed. The pretreatment trains for the different systems were as follows: (I) (SE–CS–SF–AC–HF), (II) (SE–CS–SF–HF), (III) (SE–SF–HF) and (IV) (SE–HF). The rejection related to electrical conductivity was 93.4 and 94.3% for system (I) and (IV), respectively. The flux decline of system (IV) is shown in Fig. 15. It could be demonstrated that a daily-simple flushing is not satisfactory and that additional cleaning by chemicals ($NaBO_3$ and Triton X100 or detergents with enzyme) twice a week ensured a relatively stable volume flow. Because of the high product quality, it seems advisable to blend the HF permeate with lower quality water when the reclaimed water is to be used for industrial purposes.

C. Industrial Wastewaters

The possibility of reusing industrial effluents after treatment by HF was investigated with respect to cost. In Japan, it appears to be an eco-

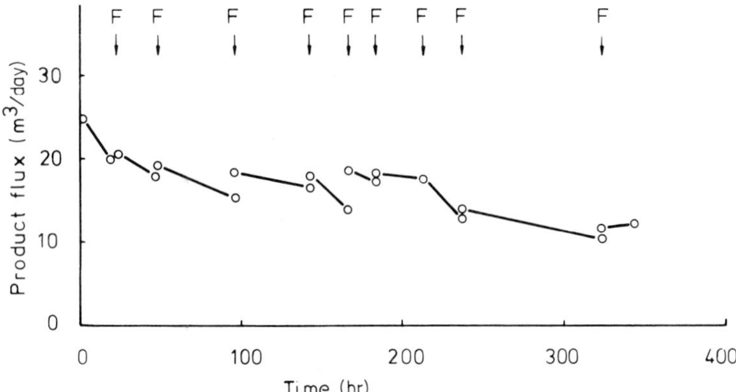

Fig. 15 Volume flow as function of operation time for the HF treatment of secondary effluent for pretreatment scheme No. IV: secondary effluent, coagulation–sedimentation, sand filtration. Flushing is represented by F. Operation pressure at 42 bars, feed flow 24–28 liter/min (from Tsuge and Mori, 1977).

nomic necessity, as cities have raised the water supply cost, including sewage costs. Experiments were performed with the Toray spiral-wound modules type SC-3000 (see Table III) at 30 bars. The feed was a mixture of chemical factory wastewater, secondary treatment effluent, and raw wastewater. The product was of a high quality. The pilot plant ran for 2 yr without any membrane degradation; the m value was about -0.02, and cleaning with oxalic acid was effective for improving the performance (Kojima *et al.*, 1977). The salt rejection was about 95% (from Fig. 4 in the cited paper), but the salt type was not specified. Likewise, Kojima and Tatsumi (1977) installed Toray's modules to reclaim wastewater from an air conditioning unit. The plant consisted of a dual-media filter, pH adjustment, modules in a two-bank array, and neutralization. At a recovery rate of 90% and a total capacity of 300 m^3/20 hr, the electrical conductivity rejection was 97.6%. Because of the good raw water quality (TDS = 250 ppm, BOD = 5 ppm), no trouble occurred during operation. The product was used as plant water in the factory. Roga-type spiral-wound modules were installed for the reclamation of plating wastewater containing chrome and cyanide, respectively. After passing through primary treatment systems consisting of conventional oxidation-reduction processes, sedimentation unit, dual-media filter, and polishing sand filter, the effluents were fed into the HF plant with the modules in a five-bank array and a total capacity of 370 m^3/16 hr (Sato *et al.*, 1977). At pressures of 25 to 28 bars, temperatures around 20–30°C, and a recovery of 92%, the plant operated for 18 months with an almost negligible flux decline (m value: -0.015 for 30 bars and 25°C). The electric conductivity rejection was

96.7%. The product was reused after further demineralizing by ion exchange and the heavy metals in the concentrate were exchanged by ion-exchange resins and discharged.

Effluents from food industries often contain constituents with high MW, these substances are frequently intended for animal or human consumption. Most of these materials, however, have high BOD values, so that discharging into rivers or sewers requires an efficient removal of this BOD. Normally, this may be performed by UF, which is well known for treating cheese whey. But often the constituents suffer chemical breakdown because of bond cleavage with increasing waste age. The smaller molecules formed from this process are not removed as well by UF. This may be shown for two different effluents of the starch industry, where the rejection ranged from 85.1 to 99.1% using UF (Pepper, 1975). Contrary to these findings, olive oil effluent water does not appear to alter after 3 months' storage. Using CA membranes DDS type 999 at 70 bars, the rejection of COD was about 91.3% for fresh water and about 94.7% for stored water (Pompei and Codovilli, 1974). In general, a better quality effluent can be ensured by a two-step process working with HF in the second stage. Nevertheless, the application depends on the economic feasibility. Using a pilot plant fitted with tubular modules developed by the Israel Desalting Engineering Ltd., Matz *et al.*, (1978) tested a combined UF–HF process to reduce the high COD of citrus washwater to a COD value that does not need further treatment when discharged into the sewer. The membranes used for the UF modules were based on the aromatic polyamide polymer designated HUF 33040 (Sachs *et al.*, 1976), whereas the polymer for the HF membranes was CA. The CA's membrane characteristics were as follows: Rejection R = 96–98% and volume flow q = 500–800 liters/m^2 d using a 5000 ppm NaCl solution at 55 bars and 25°C. The citrus washwater is a mixture of pulp, sugar, and other ingredients. In the UF stage, the process operates within the pressure range of 3.5 to 3.9 bars at 30°C with a final volume flow of 400 liter/m^2 day at a conversion of 58%. Because of the increasing viscosity (from 130 to 800 cp) of the concentrate within the tubes, the linear velocity decreased from 329 to 79 cm/sec. The UF permeate contains sugar and other substances. The pulp-containing concentrate may be used as an animal-feed supplement. The HF trials were performed at 40–50 bars at 35°C using the UF permeate that is clear and free of suspended solids as feed. This feed was concentrated fourfold by HF. The efficiency of the HF process was controlled by measuring the sugar concentration. The volume-flow declined from 500 to 180 liters/m^2 day and depended on the degree of concentration. The permeate was nearly free of sugar, the conductivity was around 100–300 μmho/cm. This product water may be reused for washing purposes. Otherwise, it may be discharged after neutralization (having a pH of 4) into the sewer without further treatment because its COD is about 800 mgO_2/liter. In comparison, the HF feed had a COD of 34,000 mgO_2/liter, and the original washwater a value of 105,000 mgO_2/liter. In addition, the authors present a detailed evaluation of the economics of this combined UF–HF process, but because the values are from 1976, they require updating.

The combination of UF with HF is also common in the dairy industry if the BOD of the UF permeate is to be reduced. The most polluting substances in the UF permeate are lactic acid and nonprotein nitrogen compounds, so their removal can be accomplished by HF from 40,000 mgO_2/liter to less than 300 mgO_2/liter (Bundgaard *et al.*, 1972). The concentrate from this process may be used for calf feed.

In some branches of industry, most of the water is used for cooling and other purposes and, therefore, it is usually low in pollutants, as reported for a brewery (Pompei, 1971). Compared with this, flushing effluents from filters and fermentation tanks with a volume of about 3% of the total water volume used contribute most to pollution with regard to BOD (>90%). Therefore, it appears economically efficient to treat these volume streams separately, especially if renovation with membranes is contemplated. With the less-polluted streams a high recovery rate is possible and the permeate is fit for reuse. With the highly polluted streams, a small volume is treated to reduce the BOD values in the effluents that are fed to the sewage treatment plant or the sewer. The concentrate can be used as provender. Much work has been done to reject the lower MW substances that are responsible for the high BOD value in such effluents. As is well known, rejection depends on the solubility of these substances in the CA membranes. In addition, the water content in these membranes decreases with increasing concentration of the organics, resulting in a reduced volume flow (Pusch *et al.*, 1976; Staude, 1976). Contrary to these findings, the rejection of low MW substances is often considerably improved using mixed solutes. For phenol, a rejection of 18% is reported in the presence of chlorophenols (Hindin and Bennett, 1969), these results were obtained with synthetic solutions. Synthetically prepared mixtures of chlorobenzene in water were also used for HF experiments. To remove chlorobenzene, several types of membranes were tested, with the best rejection obtained from a "D/f GIAP" membrane. Cellulose acetate membranes exhibited deterioration. In addition, wastewater was also used (Korneva *et al.*, 1976).

Detailed studies show that in practice the situation is more complicated, especially if the solution contains organics and inorganics. With dye-containing effluents, the removal of thio-urea and zinc salts was investigated (Staude, 1973). These results are given in Table XIX, where the feed solutions were flushing effluents from dyeing machines (No. 1)

TABLE XIX

Hyperfiltration of a Dye Solution[a]

Feed solution[b]	Zinc concentration (mg/liter)	Zinc rejection (%)	Thiourea concentration (mg/liter)	Thiourea rejection (%)	Recovery
1	311.0	99.5	1052	74.0	0.74
2	71.5	99.2	327	53.5	0.64
3	233.0	98.9	501	11.0	0.20

[a] After Staude (1973).

[b] See text for explanation of each feed solution.

and effluents withdrawn from a settling basin (No. 2). The rejection for the diazosalts was more than 97%. Contrary to these experimental findings, the rejection of thiourea was low when using a solution without dyes and other ingredients necessary for dye production (No. 3). The tests were performed in test cells at 40 bars and at a temperature of 20°C, the CA membranes (Kalle, West Germany) had a rejection of 93% and a volume flow of 0.85 m/day using 0.085 *m* NaCl solution.

The effluents from an acrylic fiber manufacturing plant were treated with high recovery ratios to ascertain economically feasible conditions for incineration of the concentrate. Mappelli *et al.* (1978) used for their experiments a mobile unit fitted with three HF modules (UOP's tubular module, spiral-wound module, and DuPont's B-9 module) and Westinghouse's tubular UF modules. The feed pretreatment consisted of oxidation by different oxidizing agents to transform compounds that easily permeate the membranes to higher rejecting solutes. Moreover, the feed for the hollow-fiber module was ultrafiltered to avoid plugging by suspended matter. Operating with a tubular module (1.42 m^2 membrane area) at 42 bars, a maximum recovery of 97% was achieved by recycling the concentrate. The volume flow decreased from 83 liter/hr at 80% recovery to 21 liter/hr at 97%. Table XX shows some of the data. This treatment of effluent from a fiber-finishing plant with the previously mentioned HF modules did not present any problems. The successful application of this process depends solely on the estimated costs.

The investigations of Steiner and Zibinski (1973) were directed to the reduction of the COD in the effluent from the textile and fermentation industries. The trials were performed with a DDS laboratory HF module using CA membranes at 50 bars. Using the textile waste, the COD in the product could be decreased drastically, whereas HF of the fermentation waste was less effective. Similar results were obtained using fermentation

TABLE XX

Hyperfiltration of Effluents from an Acrylic Fiber Manufacturing Plant[a]

	Original effluent	After air treatment	Product (permeate)
COD (ppm)	1069.0	1061.0	64.0
EC (μmho/cm)	2320.0	2058.0	23.0
Residue at 105°C (ppm)	2066.0	[a]	45.0
Residue at 600°C (ppm)	1163.0	[a]	23.0
ASB complex (as N,ppm)	97.0	[a]	3.0
Sulfites and thiosulfates (ppm)	8.0	1.0	[a]
Acetaldehyde (ppm)	63.0	109.0	20.0
Monomer (ppm)	11.0	2.3	0.6

[a] UOP tubular module with CA membranes, 42 bars, 20°C, feed flow rate 406 liter/hr. Membrane performance: R = 98% (NaCl solution 3000 ppm), permeability 12.6 liter/m^2 day bar eff. Mean rejection (obs.) $1 - 2c_p/(c_f + c_b)$; p = product, f = feed, b = brine. After Mappelli *et al.* (1978).

wastes from a pharmaceutical factory (Stenger *et al.*, 1975). The experiments were conducted at 40 bars with different annealed Kalle- (West Germany) type CA membranes. Figure 16 shows the TOD rejection and volume flow as function of recovery. The rejection of TOD is independent of the membranes used, which agrees with the usual findings for the rejection of low MW organic compounds. In the pulp industry, HF is used to concentrate spent sulfite liquor. The DDS has equipped a Norwegian cellulose factory with a HF plant having a total membrane area of 392 m^2. At pressures of 40 bars, the sulfite liquor is concentrated from 6 to 12% TDS. This is further concentrated by evaporation. The plant has a capacity of 7 m^3/hr of reclaimed water (Poulsen, 1977, private communication). In Japan, the efforts of the Water Reuse Promotion Center (Ogasawara *et al.*, 1977) are directed toward the renovation of effluents generated when kraft pulp is bleached. Highly effective pretreatment procedures reduce the COD value from 360 to 8 ppm in the waste before it is fed into the HF tubular modules operating at 30 bars. Slime generation on the membranes was suppressed by adding copper sulfate. The goal of these trials was to upgrade the quality of treated wastewater and make it fit for reuse. Experiments using effluents from the soda carbonation neutralizer of an ethylene manufacturing plant containing large amounts of inorganics were also conducted with this objective in mind. The pretreated feed was desalted by hollow-fiber modules and spiral-wound modules at recovery ratios of 50–75 and 30%, respectively. The electrical conductivity rejection was 97.7%. The permeate quality was quite satisfactory, but the pretreatment

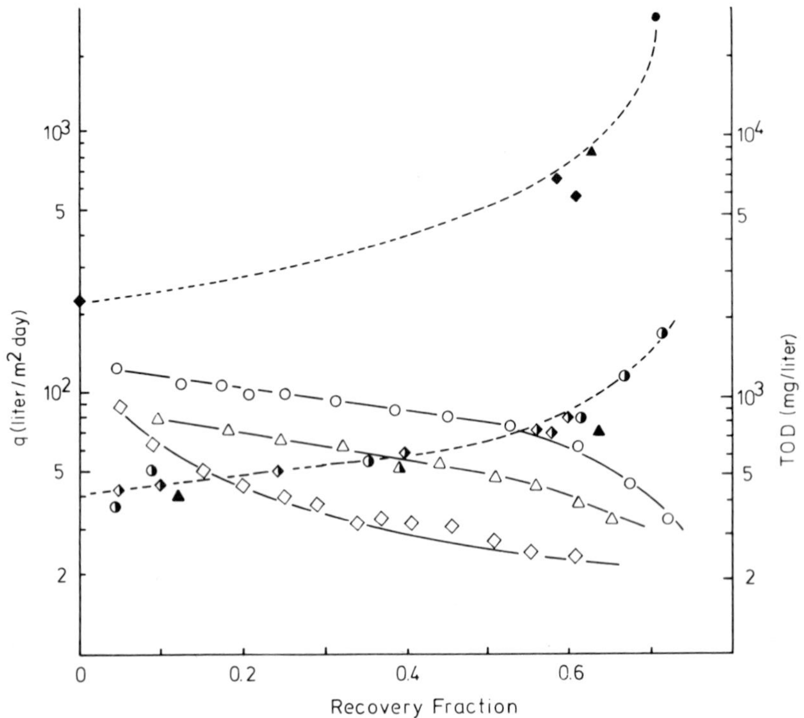

Fig. 16 Volume flow q and TOD as function of recovery using different annealed CA membranes, where R (5000 ppm NaCl solution): ◇ = 40%, △ = 70%, ○ = 80%. Open symbols indicate volume flow, semifilled symbols are for TOD in product, and filled symbols are for TOD in concentrate (from Stenger *et al.*, 1975).

systems proved in both cases to be expensive. Efforts should thus be directed toward lowering the pretreatment costs.

The washwater resulting from the acid treatment of crude kerosene fraction obtained during the petroleum oil refining process contains after neutralization sulfate and sulfonates, which may act as tensides. The TDS (at 180°C in mg/liter) may amount to 5200, whereas the COD lies in the range of 4700 mgO_2/liter. These composition values vary widely, however. With IDE tubular modules (Tel Baruch, Israel) and appropriate CA membranes (Matz *et al.*, 1978), the HF treatment resulted in considerable effluent cleaning. The rejection in COD and TDS was 85 and 94%, respectively. The concentration is not reported, but beginning from the TDS value of 14,000 ppm, a conversion of 44% is calculated. At that concentration the volume flow is reduced from 1030 liter/m^2 day to half its value.

D. POWER STATION WASTEWATER

In power stations, water is primarily used for cooling, washing purposes, and to produce demineralized water for boiler feed. To reduce the raw water demand of a thermal power station, an HF pilot plant was installed. It consisted of an up-flow depth filter and a precoat filter followed by the spiral-wound modules (Roga 4060 HR) fitted with CA membranes. In addition, an ammonium absorber was used for polishing the permeate. Two different types of wastewaters were examined (Kawabe *et al.*, 1975). One type of feedwater was chemically treated and the other type treated by the oily waste treatment system. The results indicate that pretreatment ensures a continuous operation of the HF process. Rejection ranged from 95 (monovalent ions) to more than 99% (divalent ions). For silica, the rejection varied from 85 to 91%. The operating pressure was 28 bars. The HF product water was fit for reuse. It is obvious from the study that better results can be obtained if the HF system is installed at the end of existing water and wastewater treatment plants. Evidently, this application seems to be important for reducing the fresh water demand. It is assumed that several HF plants have already been installed for this purpose. Detailed data on their efficiency are not available.

E. CONCLUSION

The applicability of HF for waste reclamation appears to be at the pilot plant stage. Many results have been reported in the rejection of various solutes, but little work has been published on concentrate disposal or reuse. This depends on the composition of these concentrates. Sammon and Stringer (1975) enumerate in their review some of the possible disposal methods such as multi-stage flash distillation, solar evaporation, discharge by pipes into the sea, or into cavities in suitable geological formations.

Membrane lifetime may also hinder the scale-up process, although substances that cause fouling and membrane deterioration seem to be well understood, and measures for chemical and mechanical cleaning are known. Information is also available on operating costs. If, in addition to these points, the separate treatment of streams differing in their intensity of pollution is taken into consideration, HF of wastewater will be feasible in the near future for the production of water, the most common solvent in the home and industry.

VII. Summary

Hyperfiltration is a low-energy-consuming separation process as compared with phase transition processes. This process promises to become a suitable method in the future because energy supply is likely to become limited and energy saving becomes a necessity. There is a general rule that the period between innovation and successful technical application is about 10 yr. Although the introductory period for HF has already exceeded 10 yr, this process has failed to reach the stage of large-scale technical application. This raises the question as to why.

Some possible answers may include the need for better membranes, higher permeation flux and rejection, and a longer lifetime. Economic reasons may also play a role because the investment costs for HF plants are relatively high. Although water of desired quality and certain chemicals that could be recycled are, at present, sufficiently available, it is necessary to plan for the future, when these limited resources may no longer be freely obtainable.

Seen as a whole, it is essential to ignore commercial considerations and furnish proof that HF systems are technically feasible. After commercial availability, it is expected that the superior system will dominate the market.

Acknowledgment

The author is obliged to all those people who sent him publications on request. Gratitude is especially due to several Japanese authors who sent original papers, some even partially translated into English. The author also thanks his wife, Magdalena, for her painstaking efforts in correcting and arranging the manuscript, and Barbara Steffens for her patience in typing it. He is also indebted to Richard Brunt for reading the manuscript and helping him avoid most of the pitfalls of the English language.

References

Adler, K.-W. (1974). *Brauwelt* **114,** 443–446.

Allard, J.-J., Rovel, J. M., and Treille, P. (1976). *Desalination* **19,** 169–174.

Bailey, D. A., Jones, K., and Mitchell, C. (1974). *Water Pollut. Control* **73,** 353–364.

Birch, B. B., and Kearsley, M. (1974). *Staerke* **26,** 220–224.
Boari, G., Carrieri, C., Mappelli, P., and Santori, M. (1978). *Desalination* **24,** 341–364.
Böddeker, K. W., Hilgendorff, W., and Kaschemekat, J. (1976a). *Chem. Ing. Tech.* **48,** 641.
Böddeker, K. W., Hilgendorff, W., and Kaschemekat, J. (1976b). *Desalination* **18,** 307–313.
Brun, R., Duriau, P., and Dussaussoy, P. (1970) *Proc. Int. Symp. Fresh Water Sea 3rd* **2,** 329–338.
Bundgaard, A. G., Olsen, O. J., and Madsen, R. F. (1972). *Dairy Ind.* **37,** 539–546.
Chapurlat, R. (1973). *Proc. Int. Symp. Fresh Water Sea 4th* **4,** 83–93.
Chian, E. S. K., and Fang, H. H. P. (1973). *AIChE Symp. Ser.* **69,** 497–507.
Crozier, D. S. McMillan, P. W., Phillips, S. V., and McC.Taylor, J. (1973). *Proc. Int. Symp. Fresh Water Sea 4th* **4,** 107–114.
Dangeon, F., Rovel, J. M., and Treille, P. (1978). *Proc. Int. Symp. Fresh Water Sea 6th,* **3,** 277–285.
Di Pinto, A. C., and Santori, M. (1977). *Desalination* **20,** 105–117.
Endoh, R., Tanaka, T., Kurihara, M., and Ikeda, K. (1976). *Proc. Int. Symp. Fresh Water Sea 5th* **4,** 31–39.
Eriksson, P. (1977). *Nord. Mejeri Tidisskr.* **7,** 238–246.
Gherardi, S., Porretta, A., and Dall'Aglio, G. (1972). *Ing. Conserve* **47,** 16–26.
Glueckstern, P., and Greenberger, M. (1976). *Proc. Int. Symp. Fresh Water Sea 5th* **4,** 301–313.
Glueckstern, P., Kantor, Y., and Mansdorf, Y. (1978a). *Proc. Int. Symp. Fresh Water Sea 6th* **3,** 297–305.
Glueckstern, P., Kantor, Y., and Wilf, M. (1978b). *Desalination* **24,** 365–376.
Glueckstern, P., Kantor, Y., and Wilf, M. (1978c). *Proc. Int. Symp. Fresh Water Sea 6th* **3,** 307–316.
Golomb, A. (1970). *Plating* **57,** 1001–1005.
Grover, J. R. Gayler, R., and Delve, M. H. (1973). *Proc. Int. Symp. Fresh Water Sea 4th* **4,** 159–169.
Hara, S., Mori, K., Taketani, Y., and Seno, M. (1976). *Proc. Int. Symp. Fresh Water Sea 5th* **4,** 53–62.
Harrison, P. S. (1970). *Chem. Ind.* 325–328.
Hauser, H. K. J. (1977). *Umsch. Wiss. Tech.* **77,** 267–273.
Hindin, E., and Bennett, P. J., (1969). *Water Sewage Works* **116,** 66–69.
Horio, K. (1980). *Desalination* **32,** 211–220.
Hürlimann, H. (1919). *Z. Gesamte Brauwes.* **42,** 323.
Kawabe, A., Kakimoto, Y., Karaso, H., Abe, O., Yabe, K., and Imaizumi, M. (1975). *Wastewater Reclamation by the Reverse Osmosis Process at a Power Station.* Presented at second National Conference on Complete Water Reuse, Chicago.
Kearsley, M. (1976). *Staerke* **28,** 138–145.
Kepinski, J., Lipinski, K., and Chlubek, N. (1976). *Proc. Int. Symp. Fresh Water Sea, 5th* **4,** 347–352.
Kieninger, H., Narziß, L., and Heil, G. (1975). *Sep. Processes Membr. Ion Exch. Freeze Conc. Food Ind. Int. Symp.* **A5,** 1–22.
Kimura, S., and Nakao, S.-I. (1975). *Desalination* **17,** 267–288.
Kimura, S., and Sourirajan, S. (1967). *AIChE J.* **13,** 497–503.
Koga, S., and Ushikoshi, K. (1977). *Desalination* **23,** 105–112.
Kojima, Y., and Tatsumi, M. (1977). *Desalination* **23,** 87–95.
Kojima, Y., Endoh, R., and Hirose, M. (1977). *Sangyo Kikai* **52,** 37–41.
Korneva, L. V., Avdonin, Y. A., and Olevskii, V. M. (1976). *Sov. Chem. Ind. Engl. Transl.* **8,** 21–23.

Kuiper, D., Bom, C. A., van Hezel, J. L., and Verdouw, J. (1974a). *Desalination* **14,** 163–172.
Kuiper, D., van Hezel, J. L., and Bom, C. A. (1974b). *Desalination* **14,** 193–212.
Kunisada, Y., Murayama, Y., and Hirai, M. (1977). *Desalination* **22,** 243–252.
Kunst, B., Basnec, A. M., and Arneri, G. (1973). *Proc. Int. Symp. Fresh Water Sea 4th* **4,** 217–225.
Leightell, B. (1972). *Process Biochem.* **7,** 40–42.
Lerat, H. (1976). *Desalination* **19,** 201–210.
Lerat, H. (1978). *Proc. Int. Symp. Fresh Water Sea 6th* **3,** 317–325.
Lerat, H. (1980). *Desalination* **32,** 201–210.
Lindner, G., and Nilsson, G. (1980). *Proc. Int. Symp. Fresh Water Sea 7th* **2,** 267–277.
Lloyd, A. I., Theodosiou, S. A., and Knibbs, R. H. (1976). *Proc. Int. Symp. Fresh Water Sea 5th* **4,** 353–364.
Lloyd, A. I., Theodosiou, S. A., and Knibbs, R. H. (1978). *Proc. Int. Symp. Fresh Water Sea 6th* **3,** 327–340.
Lonsdale, H. K., Merten, U., and Tagami, M. (1967). *J. Appl. Polym. Sci.* **11,** 1807–1820.
Mappelli, P., Santori, M., Chiolle, A., and Gianotti, G. (1978). *Desalination* **24,** 155–173.
Marquardt, K. (1975). *Vom Wasser* **45,** 129–158.
Marquardt, K. (1977). *Chem. Tech. Heidelberg* **6,** 127–133.
Masuda, H., Kamizawa, C., Matsuda, M., Nakane, T., and Ohira, T. (1976). *Bull. Chem. Soc. Jpn.* **49,** 675–678.
Matsumoto, H., Kawamura, C., Sekino, M., Hamada, K., Ukai, T., and Matusi, H. (1980). *Proc. Int. Symp. Fresh Water Sea 7th* **2,** 189–196.
Matz, R., Zisner, G., and Herscovici, G. (1978). *Desalination* **24,** 113–128.
McBain, D. (1976). *Chem. Br.* **12,** 281–284.
Merson, R. L., and Ginnette, L. F. (1972). In "Industrial Processing with Membranes" (R. E. Lacey and S. Loeb, eds.), pp. 191–221. Wiley, New York.
Merson, R. L., and Morgan, A. I., Jr. (1968). *J. Food Technol.* **22,** 631–635.
Murayama, T., Kasamatsu, T., and Gaydos, J. G. (1976). *Desalination* **19,** 439–446.
Nielsen, I. K., Bundgaard, A. G. Olsen, O. J., and Madsen, R. F. (1972). *Process Biochem.* **7,** 17–20.
Nielsen, W. K., Madsen, F. R., and Olsen, O. J. (1980). *Desalination* **32,** 309–326.
Nussbaumer, H., Perl, H., and Beer, H. (1976). *Proc. Int. Symp. Fresh Water Sea 5th* **4,** 135–141.
Obiya, N. (1977). *Desalination* **22,** 457–464.
Ogasawara, H., Kunisada, Y., Kaneda, H., and Hirai, M. (1977). *Desalination* **23,** 113–122.
Ohya, H. (1976). *Desalination* **19,** 411–419.
Ohya, H. (1977). *Desalination* **23,** 223–233.
Ohya, H., and Sourirajan, S. (1969). *AIChE J.* **15,** 780–782.
Ohya, H., Kasahara, S., and Sourirajan, S. (1975). *Desalination* **16,** 375–393.
Pasternack, K., Rauch, K., and Rautenbach, R. (1973). *Proc. Int. Symp. Fresh Water Sea 4th* **4,** 285–295.
Pepper, D. (1975). *Sep. Processes Membr. Ion Exch. Freeze Conc. Food Ind. Int. Symp.* **A7,** 1–11.
Pepper, D., Rogan, A. I., and Tanner, C. (1973). *Proc. Int. Symp. Fresh Water Sea 4th* **4,** 297–307.
Peri, C., and Pompei, C. (1975). *Vini Ital.* **17,** 179–185.
Peynaud, E., and Allard, J.-J. (1970). *C. R. Seances Acad. Agric. Fr.* **56,** 1454–1458.
Pompei, C., (1971). *Ind. Aliment. Agric.* **88,** 1585–1591.
Pompei, C., and Codovilli, F. (1974). *Sci. Technol. Alimenti* **4,** 363–364.

Pompei, C., and Rho, G. (1974). *Lebensm. Wiss. Technol.* **7,** 162–172.
Pusch, W., and Riley, R. L. (1977). *Desalination* **22,** 191–203.
Pusch, W., Burghoff, H. G., and Staude, E. (1976). *Proc. Int. Symp. Fresh Water Sea 5th* **4,** 143–156.
Rhein, O. (1975). *Weinwirtschaft Mainz* **111,** 134–138.
Riley, R. L., Lonsdale, H. K., and Lyon, C. R. (1970). *Proc. Int. Symp. Fresh Water Sea 3rd* **2,** 551–560.
Roosmain, A. B., Saroja, S., and Nanjundaswamy, A. M. (1974). *Indian Food Packer* **28,** 48–64.
Rovel, J. M. (1977). *Desalination* **22,** 485–493.
Sachs, S. B., Zisner, E., and Herscovici, G. (1976). *Desalination* **18,** 99–111.
Sammon, D. C., and Stringer, B. (1975). *Process Biochem.* **10,** 4–12.
Sato, T., Imaizumi, M., Kato, O., and Taniguchi, Y. (1977). *Desalination* **23,** 65–76.
Schippers, J. C., Bom, C. A., and Verdouw, J. (1978). *Proc. Int. Symp. Fresh Water Sea 6th* **3,** 363–375.
Schnabel, R. (1976). *Proc. Int. Symp. Fresh Water Sea 5th* **4,** 409–413.
Schobinger, U., Karnowska, K., and Grab, W. (1974). *Lebensm. Wiss. Technol.* **7,** 29–37.
Shields, C. P. (1978). *Proc. Int. Symp. Fresh Water Sea 6th* **3,** 395–414.
Shimozato, A., Takahashi, S., Koike, Y., Ebara, K., and Komori, S. (1976). *Hitachi Rev.* **25,** 147–152.
Staude, E. (1973). *Chemie Ing. Tech.* **45,** 1222–1225.
Staude, E. (1976). *Chemie Ing. Tech.* **48,** 711–712.
Staude, E., Stenger, K., and Wildhardt, J. (1976). *Dtsch. Lebensm. Rundsch.* **72,** 189–193.
Steiner, D., and Zibinski, E. (1973). *Verfahrenstechnik Mainz* **7,** 337–342.
Stenger, K., Wildhardt, J., and Staude, E. (1975). *Chem. Ztg.* **99,** 220–224.
Taniguchi, Y. (1977). *Desalination* **20,** 353–364.
Treille, P. (1970). *Proc. Int. Symp. Fresh Water Sea 3rd* **2,** 601–613.
Treille, P., and Rovel, J. M. (1973). *Proc. Int. Symp. Fresh Water Sea 4th* **4,** 425–434.
Tsuge, H., and Mori, K. (1977). *Desalination* **23,** 123–132.
Tsuge, H., Yanagi, C., and Mori, K. (1977). *Desalination* **23,** 235–243.
Wechsler, R. (1977). *Water Res.* **11,** 379–385.
Wucherpfennig, K., and Krueger, R. (1975). *Sep. Processes Membr. Ion Exch. Freeze. Conc. Food Ind. Int. Symp.* **A11,** 1–20.
Wucherpfennig, K., and Neubert, S. (1976). *Brauwelt* **116,** 1419–1420, 1422–1423.
Yanagi, C., and Mori, K. (1980). *Desalination* **32,** 391–398.

9

Water and Wastewater Treatment Experience in Europe and Japan Using Ultrafiltration

H. STRATHMANN
Fraunhofer-Institut Für Grenzflächen und Bisuer Fahren-stecknik
Stuttgart, Federal Republic of Germany

I. Introduction

The technical term ultrafiltration (UF) was first introduced by Bechhold (1907, 1908) in 1907 in Europe to describe the filtration of particles that

ISBN 0-12-085480-5

were so small they could be seen only by an "ultramicroscope." A series of synthetic membranes were developed in following years (Elford, 1930; Zsigmondy and Carius, 1927) and applied to the separation of small particles with diameters of less than a fraction of a micrometer. For more than 50 years, however, the process was used primarily by scientists as a tool to study basic mass transport and thermodynamic phenomena and did not have any commercial significance. Only in the early 1960s, when the asymmetric cellulose acetate (CA) membrane was developed (Loeb and Sourirajan, 1962) did UF become recognized as a new, efficient separation technique in the food, drug, and chemical process industries as well as for wastewater treatment and potable water purification systems. The basic principle and the potential of the process, problems related to membrane manufacturing, and the chemical engineering aspects of equipment design have been discussed extensively in the literature (Lonsdale *et al.*, 1965; Merten, 1966; Michaels, 1968b). A number of manufacturers provide large-scale UF equipment together with a series of membranes tailor-made for various separation problems.

Although the basic concept of UF was developed in Europe during the first decade of this century, its commercial and technical significance was first recognized in the United States. Because of this, the majority of the manufacturers of large-scale UF systems are located in the United States.

In UF and the closely related process of reverse osmosis (RO), a hydrostatic pressure driving force and a specially structured polymeric film are utilized to separate mixtures of molecules or particles on the basis of differences in their molecular size or shape. The difference between UF and ordinary filtration is rather arbitrary. The distinction is primarily determined by the size of the particles to be separated. The term UF is used today for the separation of molecules and particles with diameters varying from the submicrometer range, i.e., below the resolution of an optical microscope, down to molecules with an MW in the range of 2000 to 5000 (Strathmann, 1970).

Ultrafiltration is an economical mass separation technique that does not involve a phase change. The process can be carried out at ambient temperature, and the various chemical species to be separated are not chemically or physically altered. This is of special importance when temperature-sensitive biological materials must be processed. As of 1982, UF was largely confined to aqueous solutions. There is no intrinsic reason, however, why it should not be utilized in filtration of organic solvents, and in 1978 membranes resistant to most organic solvents have appeared in the literature (Strathmann, 1978).

II. Membrane Materials Used in Ultrafiltration Systems

Membranes used in UF should have high fluxes, a separation characteristic tailored to the required separation, a sharp molecular weight cut-off, (which means a narrow pore size distribution), and good mechanical, chemical, and thermal stability, which in general provides good flux stability and high life expectancy.

The first requirement of high fluxes is achieved by using asymmetric membranes first developed by Loeb and Sourirajan (1962). These membranes have a thin, dense skin layer supported on a much more open microporous sublayer. The membranes not only provide significantly higher fluxes than the symmetric structured porous membranes used before, but they also show significantly better flux stability. This is because they act as a true surface filter, retaining all particles and molecules at the surface where they can easily be removed by shear forces provided by the flow of the feed solution parallel to the membrane surface. The older, uniform-structured, ultrafine depth filters do not have this asymmetric structure and soon become plugged with particles lodged in their interstices.

Asymmetric UF membranes are manufactured from a variety of polymers by a phase inversion process (Kesting, 1971). In this process, a solution of 10 to 20 wt% polymer dissolved in a water-miscible solvent is cast onto a flat surface. A portion of the solvent is allowed to evaporate, after which the film is precipitated by immersing it into a water bath. The main criteria for polymer selection are chemical, mechanical, and thermal stability. In practice, the most widely used polymers are CA, various aromatic polyamides, polysulfones, and polyacrylonitrile–poly(vinyl chloride) copolymers. Each of these polymers have characteristic properties that lead to certain advantages and disadvantages in their use as UF membranes. The choice of the polymer depends largely on the separation process the membrane is to be used for. The properties of the principal UF membranes are summarized in Table I.

A. CELLULOSE ACETATE MEMBRANES

Asymmetric membranes were first made from CA by the phase inversion process in an attempt to make RO membranes for desalination of seawater (Kesting, 1971; Manjikian, 1967). Consequently, the first

TABLE I

Polymers Used for Manufacturing of Ultrafiltration Membranes

Polymer	Typical achievable molecular weight cut-off	pH Operating range	Maximum operating temperature (°C)	Chlorine resistance	Organic solvent resistance	Membrane configuration
Cellulose acetate	1,000–50,000	3.5–7	35	Good	Poor	Flat sheets, tubes
Polysulfone	5,000–50,000	0–14	100	Good	Fair	Flat sheets, tubes, capillaries
Aromatic polyamides	1,000–50,000	2–12	80	Poor	Fair	Flat sheets, tubes, capillaries
Polyacrylonitrile, poly(vinyl chloride), copolymer	30,000–100,000	2–12	50	Fair	Fair	Flat sheets, tubes, capillaries

membranes used in large-scale UF applications were made from CA. Cellulose acetate membranes with a wide range of pore sizes can be produced by changing the casting solution composition or the polymer precipitation procedure. Because of the hydrophilic nature of CA, relatively high membrane fluxes are generally obtained. It is possible to make CA membranes with low MW cut-offs that reject more than 99% sodium chloride from a 1% salt solution, and with MW cut-offs as high as 200,000. Cellulose acetate, however, is mechanically weak and more thermally and chemically unstable than most of the alternative polymers. For example, CA is attacked by most organic solvents even when they are present in low concentration mixed with water, and it hydrolyzes below pH 4 and above pH 8. The maximum operating temperature is also limited to about 35°C; higher temperatures lead to irreversible membrane shrinkage and lower fluxes. Finally, CA is susceptible to bacterial attack. In spite of all these short comings, CA membranes are still widely used today in UF, primarily because it is possible to produce uniform, high-quality membranes at a relatively low cost.

B. POLYSULFONE MEMBRANES

Soon after the development of the CA membrane, it was recognized that asymmetric skin-type membranes could be made by the phase inversion process from a series of other polymers. As a result, a variety of membranes were developed and initially sold for laboratory applications (Michaels, 1968a). Polysulfone in particular emerged as a useful membrane material. Polysulfone membranes have excellent chemical stability and are not attacked in strong acid or alkali solutions, even at elevated temperatures. Because of the hydrophobic nature of the polymer, however, polysulfone membranes cannot be produced in as wide a range of pore sizes as CA. Membranes with low MW cut-off characteristics are especially difficult to obtain when simultaneously high fluxes are desired. Also, because of the chemical nature of the polymer, certain materials are adsorbed at the membrane surface leading to severe flux decline and loss of material.

C. POLYAMIDE MEMBRANES

Aromatic polyamides are also widely used to make UF membranes. The properties of microporous membranes made from this polymer are similar to those made from polysulfone. Because of their higher water sorption of 12 to 15 wt%, compared to the water sorption of polysulfone of 4 or 5 wt%, membranes with low MW cut-off and good flux can be obtained (Saier, 1977). Other useful properties of the aromatic polyamide membranes are their excellent mechanical strength and thermal stability. A severe drawback of these membranes is their sensitivity to attack by free chlorine at concentrations as low as 5 to 10 ppm. The membranes are also hydrolyzed by alkaline solutions above pH 12, especially at elevated temperatures. Like the polysulfone membranes, polyamide membranes do adsorb certain materials at the surface, leading to fouling problems and eventually to a decline in the filtration rate.

D. POLYACRYLONITRILE MEMBRANES

Polyacrylonitrile is generally used as a copolymer with vinyl chloride for the preparation of UF membranes. A feature of these membranes is that they can be dried completely and rewetted without changing their filtration characteristics, unlike membranes made from CA, polysulfone, or polyamide, which must be kept wet or treated with glycerin, poly(vinyl alcohol), etc. before drying. This property offers some advantages in handling and shipping. Because of the relatively hydrophobic character of the polymer, however, only membranes with MW cut-offs exceeding 30,000 can be made. Membranes with lower MW cut-offs show generally poor water fluxes. In terms of their chemical and thermal stability, polyacrylonitrile membranes provide no advantage over polysulfone or polyamide membranes, and their mechanical strength is significantly lower.

Several other polymers have been used to make UF membranes as well as the polymers described above (So *et al.*, 1973; Walch, 1973). Most of these, however, are either experimental products or are applied only in the laboratory. As of 1982 these membranes were of minor significance in industrial UF systems.

Although there are significant differences between membranes made from CA, polysulfone, polyamide, or polyacrylonitrile as far as their intrinsic properties (chemical, thermal, and mechanical stability) are concerned, it is difficult to give a general evaluation of the usefulness of the four different polymers as UF membranes. This evaluation depends mainly on the separation problem to which the membrane is to be applied. For example, in many applications in the food industry, CA membranes have been used successfully (Bundgaard *et al.*, 1972; Merson and Ginnette, 1972) while in others, such as the recovery of poly(vinyl alcohol) from textile industry wastewaters at temperatures exceeding 70°C, CA membranes are completely useless, because of their low maximum operating temperature (Trauter and Reuss, 1977b). All four membrane materials have found special areas of application where they can be operated very successfully.

III. Ultrafiltration Modules and System Designs

For the successful application of UF as an efficient mass-separation process, the design of the module used to contain the membrane and the layout of the system in which the module is installed are as important as the selection of the proper membrane. This is because of concentration polarization that leads to gel- or cake-layer formation on the membrane surface. This gel layer, which can drastically alter the UF properties of the membrane, is sensitive to the module design. In addition to affecting the economics of the process by reducing the filtration rate, a gel-layer formation on the membrane surface can also change the separation characteristics of the membrane. This is of special importance where macromolecular materials are to be fractionated (Baker and Strathmann, 1970; Blatt *et al.*, 1970). One of the most significant parameters in designing a UF module is the flow control of the feed solution at the membrane surface (Shen and Probstein, 1977). As of 1982, several membrane module designs have been used having significant differences in the feed flow distribution, operating pressure, and capital and operating costs. The more important module concepts used in large-scale technical UF units are summarized in Table II.

TABLE II

Ultrafiltration Membrane Module Concepts

Module type	Membrane Surface per module (m^2/m^3)	Capital cost	Operating cost	Flow control	Ease of cleaning in place
Tubular	25–50	High	High	Good	Good
Plate and frame	400–600	High	Low	Fair	Poor
Spiral wound	800–1000	Very low	Low	Poor	Poor
Capillary	600–1200	Low	Low	Good	Fair
Grooved rods	200–300	Low	Low	Poor	Fair

A. THE TUBULAR MEMBRANE MODULE

The tubular membrane module was first developed for RO application. The system provides significant advantages in terms of concentration polarization control and membrane fouling. While RO tubular modules have been replaced to a large extent by hollow-fiber and spiral-wound systems because of high investment and operating costs, they are still widely used in UF. The main reason is that tubular systems are tolerant towards suspended solids in the feed solution. They can easily be mechanically cleaned by foam swabs without dismantling the equipment. The construction of a tubular membrane system is schematically shown in Fig. 1. In tubular membrane systems the membrane is placed on the inside of a porous stainless steel or fiberglass reinforced plastic tube. The pressurized feed solution flows down the tube bore, the product solution permeates the membrane and is collected in the outer shell. Tubes may be installed in series or parallel array. Tubular modules varying in tube diameter from 1.0 to 2.5 cm are produced by several manufacturers (Abcor, 1977; Paterson Candy, 1977; Kalle, 1976; Nitto, 1977; Büttner-Schilde-Haas, 1977). In some cases the membrane is directly cast on a porous tube (Nitto, 1977). In other modules the membrane is cast on a porous paper that is supported by an outer tube with outlet ports for the filtrate (Kalle, 1977). This second design allows for relatively easy membrane replacement, but sometimes implies operational difficulties (Kriegel *et al.*, 1976). The advantages of tubular systems are the control of concentration polarization effects, adjustment of the feed flow velocity over a

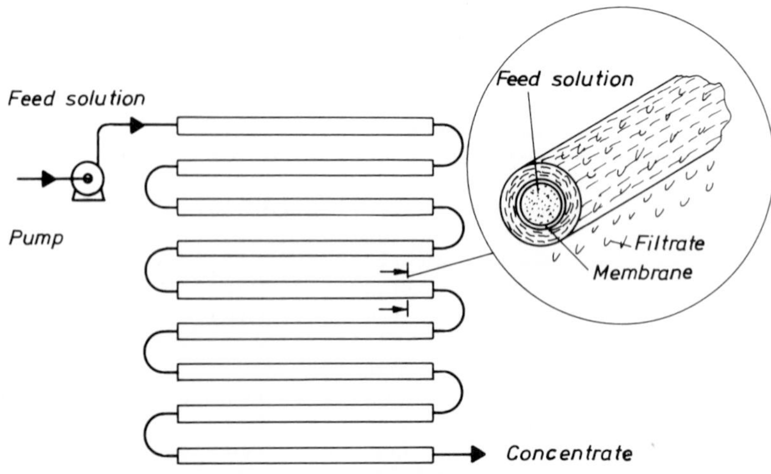

Fig. 1 Schematic diagram of a tubular UF module.

wide range, and mechanical cleansing of the system if excessive membrane fouling makes this necessary. The disadvantages of the tubular system are the relatively high investment, operating costs, and low ratio of membrane surface area to system volume.

B. THE PLATE-AND-FRAME MEMBRANE MODULE

Plate and frame membrane systems were among the first introduced in large-scale UF and RO units. The design has its origins in the conventional filter press concept. The membranes, porous membrane support materials, and spacers forming the feed flow channel are clamped together and stacked between two endplates. The feed solution is channelled across the surface of the membrane by the feed side spacers. The concept is shown in Fig. 2. There are a variety of plate-and-frame designs on the market differing primarily in the design of the feed flow channels (DDS, 1977; Rhone-Poulenc, 1977; Dorr Oliver, 1977). In some modules the membrane can be removed from the porous support plate (DDS, 1977), in others it is directly cast on a support structure (Dorr Oliver, 1977). All plate-and-frame systems provide a large membrane area per unit volume. In general, control of concentration polarization is more difficult with

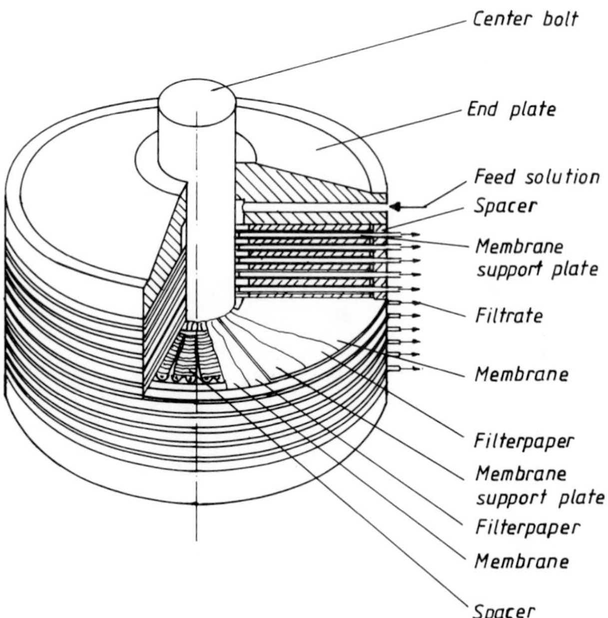

Fig. 2 Schematic diagram of a plate-and-frame UF module.

these units than with tubular systems, and plugging of the feed flow channel can be a problem especially with solutions containing larger quantities of suspended solid matter. It is usually possible to dismantle and mechanically clean the membranes in a plate-and-frame module, but this is considerately more time consuming than cleaning a tubular system. The capital costs of plate-and-frame units depend on the specific module design. In general, however, capital costs are somewhat lower than in tubular systems. Operating costs are also generally lower than in tubular systems (Madsen, 1977).

C. THE SPIRAL-WOUND MEMBRANE MODULE

The spiral-wound membrane module is widely used in RO. In principle, it is a plate-and-frame system that has been rolled up. The basic concept is shown schematically in Fig. 3.

The feed flow channel, the membrane, and the porous membrane support are rolled up and inserted in an outer tubular pressure shell. The filtrate is collected in a tube at the center of the roll. Several alternate spiral-wound module designs are possible, depending on the feed and filtrate flow paths. In all spiral-wound membrane modules, the membrane surface per unit volume is high and capital and operating costs are low. Unfortunately, it is difficult to control concentration polarization effects in these modules, and severe membrane fouling occurs even with solu-

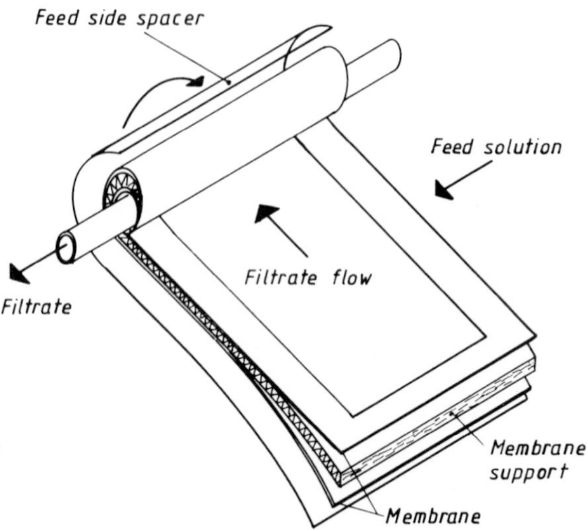

Fig. 3 Schematic diagram of a spiral-wound UF module.

tions containing only moderate concentrations of suspended solid materials. Therefore, the use of the spiral-wound membrane module in UF is limited.

D. THE CAPILLARY MEMBRANE MODULE

The capillary membrane module system consists of a large number of membrane capillaries with a diameter in the range of 0.5 to 1.5 mm. The feed solution is passed down the center of the capillary and the filtrate permeates the wall of the capillary. Because the membranes are produced by fiber-spinning technology, the capillaries are unsupported and capital costs are relatively low. The system also provides good feed flow control and a large membrane surface area per unit volume. The operating pressure, however, is limited and the system is relatively sensitive to operating errors. Plugging of the capillaries can be a problem when the inner diameter of the capillaries is too small. In all cases, an effective prefiltration of the feed solution must be provided. See Fig. 4.

E. THE ROD MEMBRANE MODULE

There are several other membrane module concepts described in the literature. One of these is the "rod system." This system is similar to the hollow-fiber membranes used in RO. The membrane, however, does not

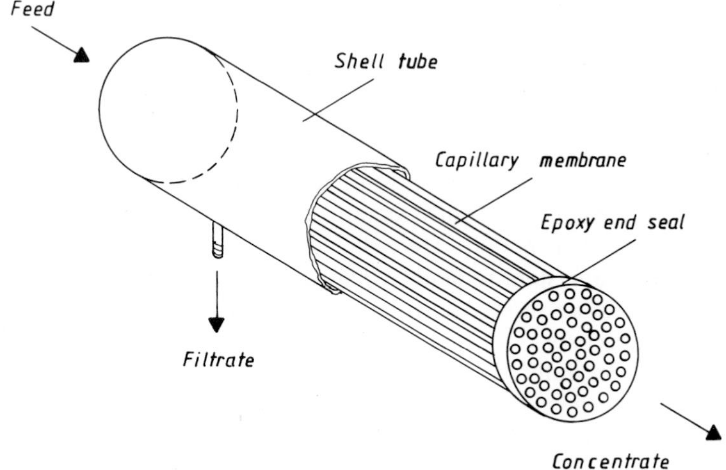

Fig. 4 Schematic diagram of a capillary membrane module.

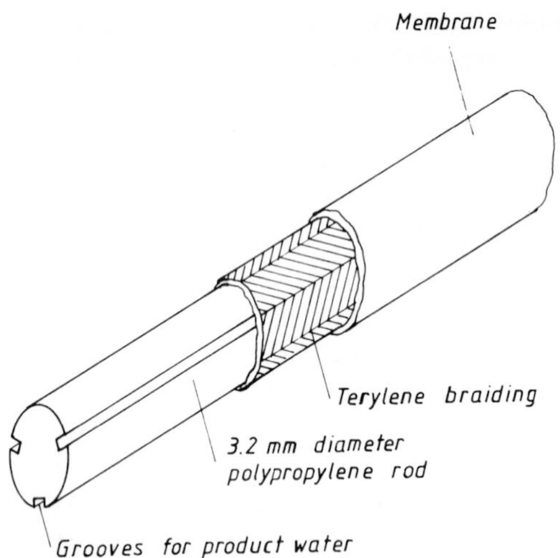

Fig. 5 Schematic diagram of a rod-type UF module.

consist of a thin, hollow fiber, but of grooved rods coated on the outside with an asymmetric membrane. A bundle of such rods is inserted in an outer tubular pressure shell. The feed solution is on the outside. The concept of this module is shown in Fig. 5. This system also provides a relatively high membrane surface area per unit volume, but the feed flow control is not as good as in the tubular or capillary modules.

F. COMPARISON OF ULTRAFILTRATION MODULES

Because available UF modules vary significantly in design and performance, a comparison of capital and operating costs is difficult. In general, the tubular module is the most expensive design in terms of capital and operating cost. But in some applications, as in the recovery of electrocoating paints from rinse water, tubular systems are more widely used than any other module. Here the control of concentration polarization, the proven system reliability, and the ability to dispense with prefiltration procedures more than compensates for the relatively high capital and operating costs (Goldsmith, 1975). Plate-and-frame systems have been used successfully in the dairy industry in spite of their relatively high capital costs because they offer significant advantages in terms of easy membrane replacement, cleaning procedures, and high operation reliability (Madsen, 1977). Capillary membranes have sucessfully been used for applications ranging from wastewater treatment to surface-water sterile filtration. Although a prefiltration procedure is generally required, their relatively low capital and operating costs make the system attractive (Wysocki, 1976). In summary, because of the diversified nature of UF applications, one module concept does not dominate the field and each system seems to have found its own area of application.

IV. Practical Experience in Ultrafiltration

Ultrafiltration made its first appearance on an industrial scale in Europe in about 1970. It is now applied on a commercial basis to various mass-separation problems, such as the advanced treatment of industrial effluents and process waters, the concentration, purification, and separation of macromolecular solutions in the chemical, the drug and food industry, and in analytical and medical laboratories. Although the use of UF in various areas has been explored in a large number of laboratory or pilot-plant studies, commercially operated installations are limited to a much smaller number of applications.

A. ULTRAFILTRATION IN ADVANCED WASTE AND PROCESS WATER TREATMENT SYSTEMS

One of the earliest and most popular commercial uses of UF was the treatment of industrial effluents, with the incentive to recover valuable constituents and recycle the water. Some of the industrial effluents that have been successfully treated by UF are summarized in Table III.

1. *Recovery of Electrophoretic Paints*

One of the most successful applications of UF is in the recovery of electrophoretic paints (Gäfgen, 1972). These paints are widely used in the automobile and household appliance industries for protective coatings. Metal parts are immersed in a water bath where the paint is deposited electrophoretically. Painted components emerging from the electropainting tank tend to drag out excess paint, which is removed by a rinsing cycle. The rinse water, which may contain 1–2% paint, is then passed back into the coating tank, diluting the original paint solution. A UF

TABLE III

Industrial Effluents Treated Successfully by Ultrafiltration

Industry	Effluent	Effluent constituents	Concentration of constituents (%)	Recovered materials	Discarded materials	Process Status
Automobile appliances	Rinse water from paint coating process	Electrocoat paint	0.5–2	Electrocoat paint, water	—	Commercial plant
Metal processing	Rinse water from processed metal	Oil emulsion	0.2–1	Oil emulsion water	—	Commercial plant
Metal processing	Rinse water from metal cleaning baths	Detergents, oil, et.	1	Water	Detergents, oil, etc.	Commercial plant
Textile	Rinse water from desizing process	Poly(vinyl alcohol)	0.2–2	Poly(vinyl alcohol water	Water	Pilot plant
Dairy	Rinse water	Proteins, lactose, etc.	0.5–1	Proteins	Water	Commercial plant
Meat	Rinse water	Proteins	0.5–1	Proteins	Water	Pilot plant
Starch	Process water	Starch	0.5–5	Starch	Water	Pilot plant
Yeast	Process water	Yeast	0.5–2	Yeast	Water	Pilot plant
Wool processing	Washwater	Lanolin	0.5–1	Lanolin	Water, detergent	Pilot plant
Paper pulp	Wastewater from bleaching process	Lignosulfonates	0.5–1	—	Lignosulfonates, water	Pilot plant

system is used for dewatering the paint by removing the same amount of water as is brought into the tank by the rinsing solution. The filtrate is directly reused in the washing cycle. Occasionally, a portion of the filtrate is discharged to remove unwanted salts or other impurities that may have been introduced into the tank. Considerably savings on paint and water have been achieved by the use of UF in this process, and several commercial plants are in operation in nearly all European automobile production plants. Most of these UF units use tubular module designs with CA membranes because no particular chemical or thermal stability is required from the material used. The average membrane lifetime generally exceeds 2 yr.

2. *Separation of Oil–Water Emulsions*

Another application of UF, leading to recycling of rinse waters and their constituents, is the removal of oil–water emulsions from rinsing waters of metal parts processed in a repetitive metal drawing or forming procedure. These rinse waters typically contain 0.8–1% oil and can be ultrafiltered to give a concentrated solution of approximately 10% oil, which can be reused directly in the metal drawing or forming process and in a oil-free permeate that can be reused as a rinsing solution. If necessary, UF can be used to concentrate the oil emulsion up to 50 to 60% mineral oil, after which it can be disposed of by incinerating, and the virtually mineral oil-free permeate is discharged into conventional wastewater treatment systems. The process is widely used in metal processing plants, especially in the automobile industry, and production lines of tin cans, etc. Commercial plants are available on the market using tubular, capillary, or plate-and-frame module designs with CA, polyamide, polysulfone, or other membranes.

3. *Processing of Rinse Waters from Alkaline Metal Cleaning Baths*

In processing rinse waters of alkaline baths used to clean greasy or dirty metal components by passing the solution through an ultrafilter, it is possible to concentrate impurities such as mineral oil, grease, etc. for discharge and recycle of the permeate to the rinsing cycle. Commercial plants are in operation using tubular, capillary, or other module concepts (Goldsmith, 1975). Thorough prefiltration is necessary where other than tubular modules are used. In addition, because the feed is a strong alkaline solution, the membranes must have good chemical stability at high pH values. Finally, because frequent washing cycles, often at elevated temperatures, are necessary to avoid membrane fouling, membranes with good temperature stability are required (Belfort, 1977). The membranes used predominantly in this particular application are made from polyamide or polysulfone.

4. *Cleaning of Textile Desizing Rinse Waters*

The recovery of water and poly(vinyl alcohol) from textile desizing effluents has been studied extensively in pilot-plant operations. The process usually is carried out at elevated temperatures to lower the viscosity of the solution. Membranes with good temperature stability are, therefore, required. There are no particular requirements as far as the module design is concerned and laboratory and pilot-plant tests have been carried out with several module concepts (Trauter and Ruess, 1977b).

5. *Recovery of Proteins and Starch from the Food Industry*

Food-industry effluents are an attractive candidate for UF because it is possible to simultaneously solve a pollution problem and recover valuable products. Dairy and meat processing-plant effluents have been studied in detail (Horten *et al.*, 1972; Porter and Michaels, 1971b; Rautenbach and Rauch, 1976), but effluents from the starch (Porter *et al.*, 1970), yeast (Rushton and Khoo, 1977), and wool processing (Wysocki, 1977) industries also contain valuable products that can be recovered and the filtrate reused.

In the dairy industry, a series of commercial plants are in operation recovering proteins from whey. In general, plate-and-frame or tubular systems are used (Güngerich, 1976), but capillary membranes have also been introduced for this process (Wysocki, 1977). The membranes used are either made from CA, polyamide, or polysulfone (Nielsen, 1976). The process has no particular difficulties, but frequent washing and sterilization cycles are required if severe fouling and bacterial growth are to be avoided. The recovery of proteins from meat processing effluents is still in the pilot-plant stage, as is the recovery of valuable constituents from starch and yeast processing effluents and lanolin from wool washwaters.

6. *Treatment of Pulp and Paper Industry Effluents*

Effluents from the pulp and paper industry are currently of critical concern as a major wastewater problem. The practical application of membrane processes for the treatment of these effluents has been the subject of extensive exploratory research and engineering studies. Data from these laboratory and pilot-plant evaluations have indicated the importance of a large-scale application of UF for concentrating dilute effluents, especially from the bleeding and pulp washing operations. The costs of UF processes are of particular concern because huge quantities of water have to be treated and the costs of the water and wastewater treatment procedure amounts to a significant part of the pulp or paper products. The recovery of valuable wastewater constituents and the reuse of the filtrate as process water can to some extent decrease the UF costs.

Membranes to be used in UF of pulp and paper mill effluents must show good chemical stability, especially against free chlorine and alkaline solutions (Ammerlaan, 1969). The membrane modules should have good flow control to avoid excess concentration polarization and membrane fouling because of the relatively high contents of suspended matter in the effluent. Most of the commercially available UF systems do not fulfill the requirements of chemical stability, flow control, and economics. This is probably the prime reason that UF, up to 1982, has not been introduced on a commercial basis in the pulp and paper industry despite the technical feasibility of the process and the huge market.

B. ULTRAFILTRATION IN FOOD PROCESSING AND CHEMICAL INDUSTRY

Although the treatment of industrial effluent is the largest market for UF, there are other areas of application that are very interesting from a commercial point of view. These are general mass-separation problems in the food and chemical process industry. In many of these applications the process costs are not as significant as in wastewater treatment because the products are relatively expensive and UF is often replacing technically less suitable or more costly mass-separation procedures. Typical applications of UF in the food and drug industry are summarized in Table IV.

1. *Ultrafiltration in the Food Industry*

The dehydration of various food products can often be carried out by UF at far lower costs than by conventional techniques such as evaporation or freeze-drying, and without loss of volatile flavor components or

TABLE IV

Ultrafiltration in the Food and Drug Industry

Industry	Feed Solution	Process	Product	Process status
Dairy	Milk, skim milk, whey	Concentration of milk or proteins	Concentrated milk, cheese, dietary products	Commercial plant
Wine	Wine	Sterile filtration deproteinization	Wine, most	Pilot plant
Vinegar	Vinegar	Sterial filtration deproteinization	Vinegar	Commercial plant
Brewery	Beer beer effluents	Recovery of yeast	Yeast	Pilot plant
Drug	Enzyme, protein and virus solutions	Concentration purification	Enzymes, viruses	Commercial plant
	Extracts of plants and animal matters	Concentration and purification of alkaloids, hormones, opiates, etc.	Alkaloids, hormones, opiates, etc.	Pilot plant

deterioration in flavor, which often accompanies evaporation concentration methods.

In the dairy industry, the use of UF appears particularly promising. Whole and skimmed milk can be concentrated by UF. Because of the permeability of UF membranes to salts and lactose, special dietary milks can be conveniently prepared. The recovery of proteins from whey not only solves a severe wastewater problem, but the concentrated proteins and the lactose and lactic acid-containing filtrate can be commercially utilized (Tomarelli and Bernhart, 1962; Webb and Whittier, 1948; Whittier and Webb, 1950). A series of commercially operated UF units are installed in the dairy industry.

The concentration of egg white is another application of UF that has been explored on laboratory and pilot plant base (Lowe *et al.*, 1969). It was demonstrated that dewatering of egg white by UF offers considerable advantages over other methods as far as the product quality is concerned.

The concentration of pectin solutions is generally accomplished by chemical precipitation followed by a rather complex extraction procedure (Tressler and Maynard, 1961). Using UF, pectins can be concentrated without degradation and the costly and complex chemical precipitation and extraction procedure is avoided.

The sterilization and clarification of beverages, especially wine, by UF has become a standard procedure. Wine as obtained from the fermentation process contains small amounts of proteins, polysaccharides, and other colloidal impurities that are generally removed by chemical precipitation with, e.g., bentonites. This treatment affects the quality of the wine and leads to a significant loss of wine with the precipitate and a considerable amount of sludge, causing a costly disposal problem (Millies, 1976). Ultrafiltration of the wine removes these impurities and sterilizes the wine.

Sterilization and clarification of liquid food products has become an attractive application of UF because of low process costs and the product quality.

2. *Ultrafiltration in the Chemical and Pharmaceutical Industry*

In the pharmaceutical industry a profitable application of UF is the isolation, concentration, and purification of biologically active substances such as enzymes, viruses, nucleic acids, specific proteins, etc. (Butterworth *et al.*, 1970; Fallick, 1969). Traditional methods of isolating these products include solvent extraction, dialysis, precipitation, and chromatographic procedures. These are often rather inefficient and costly, and lead to a considerable loss of product. By using UF isolation, product concentration can often be carried out in one step with considerable cost savings.

The fractionation of blood to isolate certain blood proteins on a large scale is an important aspect in modern medicine (Yamashita *et al.*, 1974). The combination of processes involving conventional techniques and UF have been carried out at the pilot-plant stage leading to lower costs and higher-quality products.

Pharmaceuticals produced by extraction from plants or animal matter such as alkaloids, hormones, opiates, etc. are frequently contaminated with macromolecular or particulate impurities. In many cases these impurities can be removed from the extract by UF, yielding a clear, high-quality product (Carlberg and Mannervik, 1975).

Another interesting application of UF that has the potential to become a substantial market is the sterile filtration of water, solutions used in intravenous transfusion, or in the preparation of pharmaceutical products. A similar application is the production of ultraclean water used in the semiconductor industry where microfiltration is more frequently replaced by UF because of better product quality and longer filter life (Marquardtand Dengler, 1977).

Finally, there are a series of other uses of UF in the chemical process industry and in medical, biomedical and chemical laboratories described in the literature (Porter and Michaels, 1971a). These, from a commercial point of view, are of great interest for membrane and equipment manufacturers and users of the process.

V. Ultrafiltration Plant Design and Process Costs

The design of UF plants and the process costs vary significantly with the specific application. In general, UF is an integral part of an overall separation procedure that may include several posttreatment and pretreatment procedures. Therefore, it is difficult to make any general statement about UF plant design and process costs. It is, however, possible to define the parameters that are significant for all applications of UF and, disregarding post- and pretreatment procedures, process costs can be predicted to some extent independent of the specific application.

A. GENERAL PROCESS DESIGN PARAMETERS AND COSTS

Because of various membrane module designs and the large variety of applications, UF has a great deal of flexibility with respect to operating modes. In large-scale industrial UF plants, continuous operation with maximum transmembrane flux at minimum operating pressure is desired. In UF, the transmembrane flux is generally determined by the resistance of a gel or cake layer formed on the membrane surface by the retained components because of concentration polarization effects (Strathmann, 1973). Concentration polarization and the formation of gel layers on the membrane surface cannot completely be avoided, its magnitude, however, can be controlled by the flow distribution at the membrane surface. The thickness of the gel layer, which directly effects the transmembrane flux, is determined by the concentration of the feed solution, the temperature, and the flow velocity of the feed solution parallel to the membrane surface. In particular, the feed flow velocity is of prime importance for a successful application of UF and has, therefore, been studied in great detail (Blatt *et al.*, 1970; Strathmann, 1973). In reality, it is not flow velocity parallel to the membrane surface, but the shear rate that controls the gel layer thickness. The shear rate is directly proportional to the pressure drop over the membrane and is largely independent of membrane geometry. The pressure drop in the UF system determines to some extent the energy cost of the process, and the investment costs are determined by transmembrane flux. A cost analysis of UF processes shows that additional energy costs resulting from pressure losses in the system are in most applications relatively low compared with savings in equipment costs resulting from higher membrane fluxes. By contrast, high shear rates at the membrane surfaces require, especially in tubular systems, large pumps and multiple feed-solution recirculation because only a small fraction of the feed solution is removed in each pass. The recirculation can also cause an increase in the feed-solution temperature. Therefore, heat exchangers are often necessary to control the feed-solution temperature. In other membrane modules the maximum operating pressure of the system limits the pressure drop that can be obtained. Capillary membranes generally cannot be operated at pressures in excess of 2 to 3 bars. The maximum shear rate obtainable in such a system is limited by the maximum pressure drop of 2 or 3 bars. Because of the different system designs, operating pressures and feed flow velocities can vary significantly. In tubular systems, operating pressures up to 10 bars and feed flow velocities of several meters per second are used because the operating pressures in capillary systems are generally less than 3 bars and feed flow velocities less than 5 m/sec.

Plate-and-frame membrane systems generally operate at even higher hydrostatic pressures, but at significantly lower feed flow velocities. A systematic flow diagram of a typical UF plant is shown in Fig. 6. In this

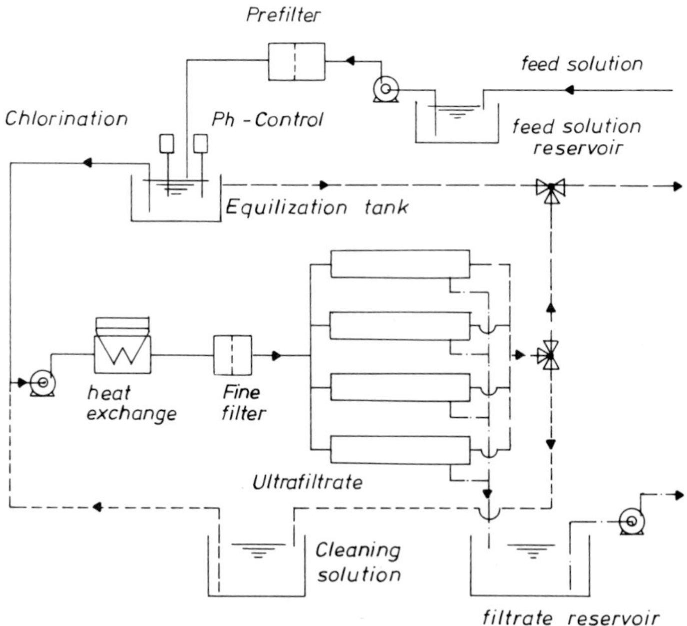

Fig. 6 Schematic diagram of a typical UF unit.

plant the feed solution is contained in a reservoir, the pH of the solution is adjusted to the desired value, and often chlorine is added for sterilization purposes. From the reservoir the feed solution is passed through a prefilter, which can be a simple sandbed or a paper filter, into an equalization tank. Before the feed solution enters the actual UF modules it generally passes through a cartridge filter and a heat exchanger. Part of the concentrated solution is fed back to the equalization tank, the rest of which is discharged for further treatment or other use. Frequent rinsing of the modules with the appropriate washing solutions is necessary to remove the gel layer if high filtration rates are to be maintained. Therefore, in most UF plants, a wash solution can be passed in a closed loop through the modules instead of the feed solution. Depending on the specific application and module type used, this general scheme might be varied to some extent. In some cases the pretreatment of feed solution can be omitted completely, while in others it is supplemented by carbon absorption, flocculation or other means. This is all reflected in the process cost.

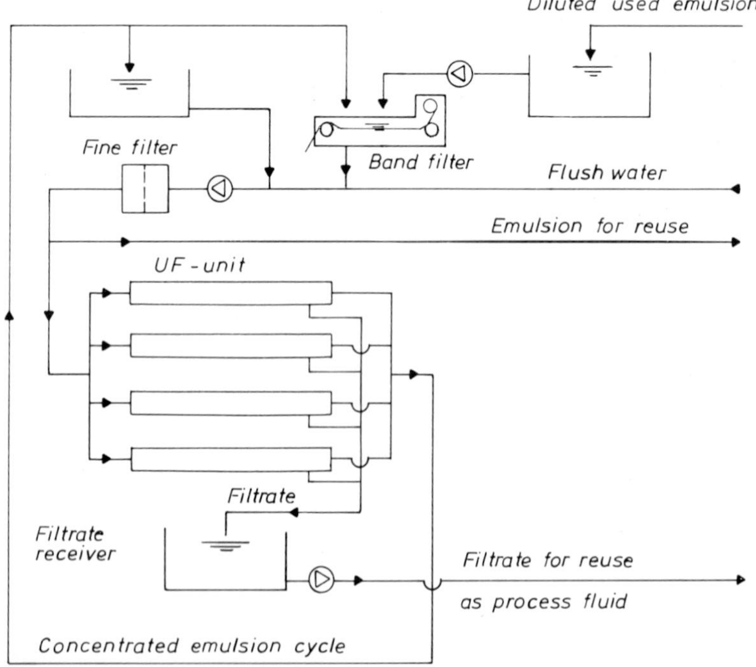

Fig. 7 Flow diagram of a UF unit used to concentrate a diluted oil emulsion (Wysocki, 1976b).

B. SPECIFIC APPLICATIONS AND PROCESS COSTS

From pilot-plant studies and commercial plants, the costs of UF have been analyzed quite thoroughly and compared with competitive processes for several applications. Reliable cost data are available (Beaton, 1977; Marquardt, 1976; Rautenbach and Rauch, 1977; Scheuermann, 1977; Wysocki, 1976a, 1976b, 1977), especially for the treatment of industrial effluents and in the food industry.

As an example of the costs of UF processes, the costs of treating a diluted oil in water emulsion system are analyzed below. This analysis is based on more than 18 months experience with a small commercial plant. The feed solution consists of an approximately 0.5% dilute oil emulsion. This emulsion is separated to an oil-free filtrate and a 10% oil concentrate, both of which are then recycled to the metal-machining process. The flow diagram of the UF unit is shown in Fig. 7. Pretreatment consists of a band and a cartridge filter. To maintain flux stability, the entire plant was flushed once a day with a cleaning solution for about 20 min before shutdown.

The specifications of the plant are given in the accompanying tabulation.

Operating Parameters	
Rated output of filtrate	400 liter/hr
Specific flux rate	33 liter/hr m^2
Total membrane area	12 m^2
Operating pressure	1.5 bar
Membrane system	Polyamide capillary membrane
M W cut-off	50000
Recirculation rate of the feed	2000 liter/hr
Total power input	4.0 kW
Total treatment costs (in U.S. dollars)	
Capital cost	30000.
Depreciation, 10% per year	3000.
Interest, 8% of half the purchase price	1200.
Operating costs	
Energy at $0.10/kW hr	1020.
Labor at $25.00 per hour	3750.
Membrane replacement at 8 m^2 per year	2620.
Chemicals and water for cleaning	1650.
Total costs per year	13690.
Costs per cubic meter of filtrate	7.1

Similar detailed cost analyses have been described in the literature for other applications, especially in the food industry (Marquardt, 1976; Wysocki, 1976a) and for the treatment of electrocoat paint baths. Prices per cubic meter of filtrate range between less than \$1 to more than \$10, depending on the specific application and plant size.

The total market for UF equipment in Europe and Japan is difficult to estimate. It was shared in 1977 by several companies with Abcor–Dürr probably being the market leader as far as the treatment of industrial effluents is concerned; DDS in the dairy industry, and Romicon, Rhone–Polenc and Berghof sharing a smaller part in the wastewater treatment field and the food industry, but they are well represented in sterile filtration and the production of ultraclean water for the semiconductor industry. The total UF market in Europe in 1977 is estimated from data made

Fig. 8 Ultrafiltration plant with a capacity of 10 m^3/hr, utilizing tubular membrane modules for the separation of in water emulsion (photograph provided by Abcor–Dürr GmbH).

available by the industry to be $8 millions. Data for Japan were not available, but the market is estimated to be about one-half that of Europe.

VI. Manufacturers and Distributors of Ultrafiltration Equipment in Japan and Europe

There are about 24 companies in Japan and Europe in one way or another active on the UF market. Many of them, however, are distributors or subsidiaries of U.S. companies. About one-half that number are manufacturers of original membranes and equipment. In Table V, manufacturers and distributors of UF membranes and equipment are summarized.

Table V shows that the tubular configuration is still the dominating membrane design. It is manufactured by Abcor, Patterson Candy, Kalle, Nitto, and others. The capillary system is manufactured by Amicon, Romicon, Berghof, and Asahi. The plate-and-frame system is manufac-

Fig. 9 Ultrafiltration plant with a capacity of 120 m^3/day, utilizing plate-and-frame modules for the concentration of whey (photograph provided by De Danske Sukkerfabrikker).

TABLE V

Suppliers of Ultrafiltration Membranes and Equipment in Europe and Japan

Suppliers	Type of activity	Module type	Membrane material	Area of activity	Status of development	Address
Abcor–Dürr	Equipment manufacturer	Tubular spiral	Cellulose acetate, polysulfone	Industrial effluents, food processing	Commercial plants	Abcor–Dürr GmbH Spitalwaldstr. 3 7 Stuttgart 40 Germany
Alfa–Laval	Distributor	Capillary (Romicon)	Polysulfone	Food processing	Commercial plants	Alfa–Laval GmbH Wilhelm–Bergner–Strasse 2056 Glinde Germany
Amicon	Membrane and equipment manufacturer	Capillary plate and frame	Polysulfone and others	Laboratory equipment		Amicon B.V. Mechelaastraat 11 Oosterhout, NB Holland
Berghof	Membrane and equipment manufacturer	Capillary	Polyamides	Industrial effluents, food and drug processing	Commercial plants	Forschungsinstitut Berghof GmbH P.O. Box 1523 74 Tübingen 1 Germany
BSH	Equipment manufacturer	Tubular	Cellulose acetate, polyamide	Industrial effluents	Commercial plants	Büttner–Schilde–Haas AG P.O. Box 4 415 Krefeld 11 Germany
DDS	Membrane and equipment manufacturer	Plate and frame	Cellulose acetate, polysulfone	Food and drug processing	Commercial plants	De Danske Sukker-fabrikker 6 Tietgensvej P.O. Box 149 4900 Nakskov Denmark

Dorr–Oliver	Membrane and equipment manufacturer	Plate-and-frame	Noncellulosic	Industrial effluents	Commercial plants	Dorr–Oliver GmbH Friedrich–Bergius–Strasse 62 Wiesbaden 12 Germany
Eisenmann	Distributor	Plate-and-frame (Rhone–Poulenc)	Noncellulosic	Industrial effluents	Pilot plant	Eisenmann KG Tübinger Str. 81 203 Böblingen Germany
Gütling	Distributor	Capillary (Romicon)	Polysulfone	Industrial effluents, food processing, ultraclean water	Commercial plants	Gütling GmbH Hofener Str. 47 7012 Fellbach Germany
Hager & Elsässer	Distributor	Tubular (PCI)	Cellulose acetate	Industrial effluents	Commercial plants	Hager & Elsässer GmbH Ruppmann Str. 26 7 Stuttgart 80 Germany
I.D. Engineering	Membrane and equipment manufacturer	Tubular	Cellulose acetate and others	Industrial effluents	Pilot plant	Israel Desalination Engineering Ltd. P.O. Box 18041 Tel-Aviv Israel
Kalle	Membrane manufacturer	Tubular	Cellulose acetate polyamide	Industrial effluents, food processing	Pilot plant	Kalle AG P.O. Box 3540 62 Wiesbaden 1 Germany
KIS	Equipment manufacturer	Capillary (Berghof)	Polyamide	Industrial effluents, food processing	Commercial plants	Fried. Krupp GmbH Krupp Ind. u. Stahlbau 42 Essen Germany
Millipore	Equipment manufacturer, distributor	Plant-and-frame	Cellulosic	Laboratory equipment	Commercial plants	Millipore GmbH Siemensstr. 20 6078 Neu–Isenburg Germany

(*continues*)

TABLE V (*continued*)

Suppliers	Type of activity	Module type	Membrane material	Area of activity	Status of development	Address
PCI	Membranes and equipment manufacturer	Tubular rod	Cellulose acetate	Industrial effluents,	Commercial plants	Patterson Candy Ltd. Laverstoke Mill Whitchurch Hampshire RG287NR England
Rhône–Poulenc	Membranes and equipment manufacturer	Plate-and-frame	Noncellulosic	Industrial effluents, food processing	Commercial plants	Rhône–Poulenc Chemie 21, rue Jean Goujan 25360 Paris France
Sartorius	Membranes and equipment	Plate-and-frame	Cellulosic	Laboratory equipment	Commercial plants	Sartorius–Membranfilter GmbH P.O. Box 142 34 Göttingen Germany
Asahi	Membranes and equipment manufacturer	Capillary	Cellulosic and noncellulosic	Industrial effluents, laboratory equipment	Commercial plants	Asahi Kasei 1-1-2 Hibiya Mitsui Bldg. Uhrokucho, Chiyoda-Ku Tokyo Japan
Daisel	Membranes and equipment manufacturer	Tubular spiral-wound	Cellulose acetate and others	Industrial effluents, laboratory	Commercial plants	Daisel 3-8-1 Toranomon Bldg. Kasumigasiki, Chiyoda-Ku Tokyo Japan
Hitachi	Distributor	Tubular (Nitto)	Cellulose acetate and others	Industrial effluents	Pilot plant	Hitachi Plant Co. 1-1-14 Kanda, Chiyoda-ku Tokyo Japan

IHI	Distributor	Rod (PCJ)	Cellulose	Industrial effluents	Pilot plant	Ishikawajma–Harima Ind. 2-2-1 Ohtemacki, Chiyoda-Ku Tokyo Japan
Mitsui	Equipment manufacturer, distributor	Tubular spiral (Daisel)	Cellulose acetate and others	Industrial effluents	Commercial plants	Mitsui Ship Constructing 5-6-4 Tsukigi, Chuo-Ku Tokyo Japan
Mitsubishi	Distributor	Tubular spiral hollow-fiber	Cellulose acetate and others	Industrial effluents	Pilot plant	Mitsubishi Rayon Co. 8, Kyobashi 2 Chome, Chuo-ku Tokyo Japan
Nitto	Membranes and equipment manufacturer	Tubular	Cellulose acetate and others	Industrial effluents	Commercial plants	Nitto Electric Industriai Co. 27 Umedacho, Kiba-ku Osaka Japan
Sumitomo	Equipment manufacturer	Tubular	Cellulose acetate	Industrial effluents	Pilot plant	Sumitomo–Juki 1-Mitoyo-cho, Kanda Chiyoda-Ku Tokyo Japan

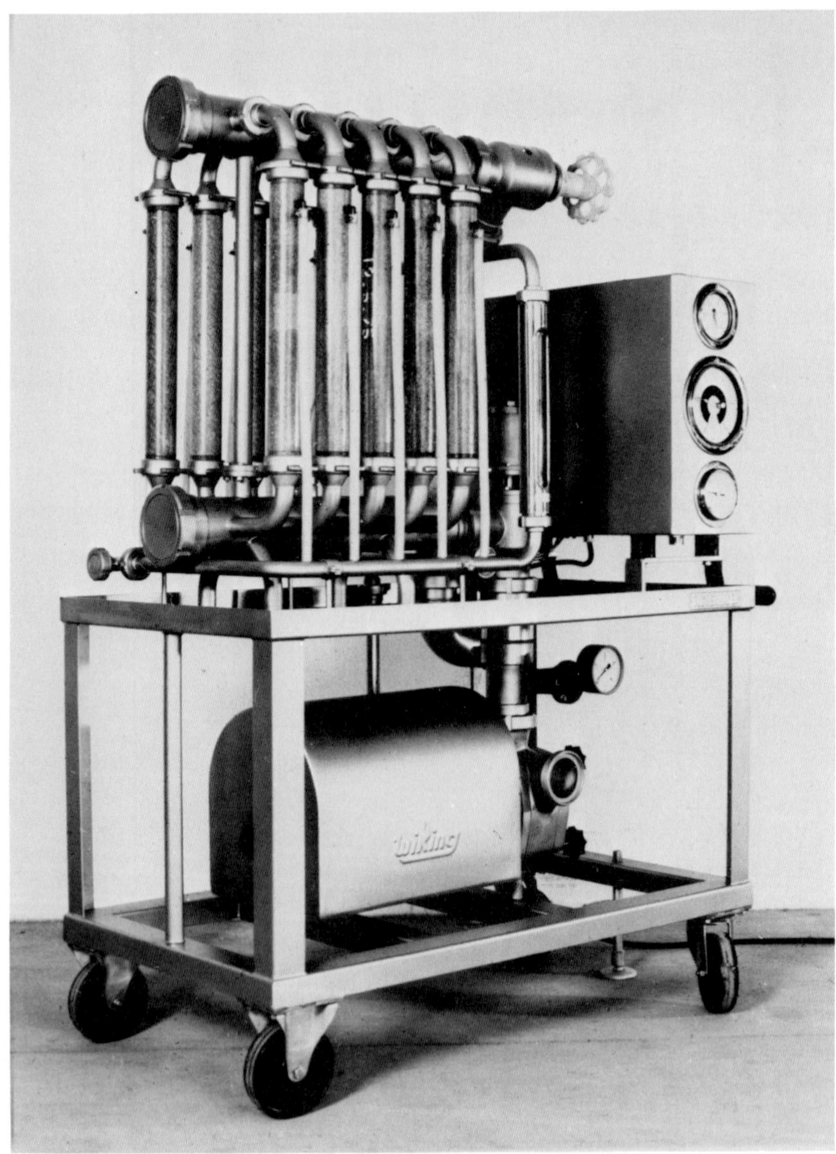

Fig. 10 Ultrafiltration pilot plant utilizing capillary membrane modules for sterilization and deproteinization of wine (photograph provided by Berghof GmbH).

tured by DDS and Rhône–Poulenc, although the Dorr–Oliver module is a plate-and-frame system only in the sense that the membranes are arranged in parallel as flat-sheet leaflets, but no spacers are used to build the feed flow channel. The rod module, as offered by PCI, has not yet achieved quite the relevance as the other module concepts. It is difficult to obtain reliable data for the market share of the various systems and manufacturers from the information that is available. However, there is a clear indication that certain systems and suppliers dominate in certain areas of application. The tubular system, as manufactured by Abcor–Dürr, clearly dominates the market for UF of electrocoat paint baths, but it is also well introduced in treating industrial effluents. The plate-and-frame system, as manufactured by DDS, is well established in the food and drug industry, while the capillary system, as manufactured by Romicon, Berghof, and Asahi, is used widely for the treatment of certain industrial wastewater, but it is also well introduced in the food and drug industry and for production of ultraclean water for the semiconductor industry.

The following three figures show photographs of commercial UF systems. Figure 8 shows a UF unit with tubular membrane modules. It is installed in an automobile plant for the separation of oil in water emulsions, and has a capacity of 10 m^3/hr. Figure 9 shows a plate-and-frame UF unit. It is used in the dairy industry for the concentration of whey. The total membrane area is 336 m^2 and the filtration capacity is 120 m^3/day. Figure 10 shows a UF unit with capillary membranes. It is used as pilot plant for sterilization and deprotonization of wine. The membrane area of the unit is 10 m^2 and the filtration capacity is 20 m^3/day.

References

Abcor-Dürr Product Catalogue (1977).

Ammerlaan, A. C. F., Lueck, B. F., and Wiley, A. J. (1969). *Tappi* **52,** 118.

Asahi Product Catalogue (1977).

Bailey, P. A. (1977). *Filtration Separation* **14,** 213.

Baker, R. W., and Strathmann, H. (1970). *J. Appl. Polym. Sci.* **14,** 1197.

Beaton, N. C. (1977). *Inst. Chem. Eng. Symp.* **51,** 59.

Bechhold, H. (1907). *Kolloidal Z.* **1,** 107.

Bechhold, H. (1908). *Biochem. Z.* **6,** 379.

Belfort, G. (1977). *Desalination,* **21,** 285.

Berghof Product Catalogue (1977).

Blatt, W. F., Dravid, A., Michaels, A. S., and Nelsen, L. M. (1970). *In* "Membrane Science and Technology" (J. E. Flinn, ed.). Plenum, New York.

Bundgaard, A. G., Olsen, O. J., and Madsen, R. F. (1972). *Dairy Ind.* **37,** 539.
Butterworth, T. A., Wang, D. I. C., Sinskey, A. J. (1970). *Biotechnol. Bioeng.* **12,** 615.
Büttner-Schilde-Haas Product Catalogue (1977).
Carlberg, I., and Mannervik, B. (1975). *J. Biol. Chem.* **250,** 5 475.
DDS Product Catalogue (1977).
Dorr Oliver Product Catalogue (1977).
Elford, W. (1930). *Proc. R. Soc.* **B106,** 216.
Fallick, G. J. (1969). *Process Biochem.* **4,** 9.
Gäfgen, K. (1972). "Ultrafiltration in EC-Anlagen" (Bulletin of Abcor-Dürr GmbH).
Goldsmith, R. L. (1975). *J. Eng. Ind.* **97,** 238.
Güngerich, Ch. (1976). *Deut. Milchwirtschaft* **27,** 322.
Horton, B. S., Goldsmith, R. L., and Zall, R. R. (1972). *Food Tech.* **26,** 30.
Kalle Product Catalogue (1976).
Kesting, R. E. (1971). "Synthetic Polymeric Membranes." McGraw-Hill, New York.
Kriegel, E., Effelsberg, H., Schaade, M., and Bockhorn, W. (1976). *Tech. Mitt. Krupp Werksber.* **34,** 21.
Loeb, S., and Sourirajan, S. (1962). *Adv. Chem. Ser.* **38,** 117.
Lonsdale, H. K., Merten, U., and Riley, R. L. (1965). *J. Appl. Polymer Sci.* **9,** 1341.
Lowe, E., Durkee, E. L., Merson, R. L., Ijichi, K., Cimino, S. L. (1969). *Food Technol.* **23,** 753.
Madsen, R. F. (1977). "Hyperfiltration and Ultrafiltration in Plate and Frame Systems," Elsevier, Amsterdam.
Manjikian, S. (1967). *Ind. Eng. Chem. Prod. Res. Dev.* **6,** 23.
Marquardt, K. (1976). *Chem.-Tech.* **5,** 411.
Marquardt, K., and Dengler, H. (1977). *Chem.-Tech.* **6,** 127.
Merson, R. L., and Ginnette, L. F. (1972). In "Industrial Processing with Membranes" (R. E. Lacey, and S. Loeb, eds.). Wiley (Interscience), New York.
Merten, U. (1966). In "Desalination by Reverse Osmosis" (U. Merten, ed.). M.I.T. Press, Cambridge, Massachusetts.
Michaels, A. S. (1968a). *Chem. Eng. Prog.* **64,** 31.
Michaels, A. S. (1968b). In "Advances in Separation and Purification." (E. S. Perry, ed.). Wiley, New York.
Millies, K. (1976). *Weinberg Keller* **33,** 141.
Nielsen, P. S. (1976). *Deut. Milchwirtschaft* **27,** 188.
Nitto Product Catalogue (1977).
Paterson Candy International Product Catalogue (1977).
Porter, M. C., and Michaels, A. S. (1971a). *Chem. Tech.* **1,** 56.
Porter, M. C., and Michaels, A. S. (1971b). *Chem. Tech.* **1,** 248.
Porter, W. L., Siciliano, J., Krulick, S., and Heisler, E. G. (1970). In "Membrane Science and Technology" (J. E. Flinn, ed.). Plenum, New York.
Rautenbach, R., and Rauch, R. (1976). *Verfahrenstechnik* **10,** 25.
Rautenbach, R., and Rauch, K. (1977). *Chem. Ing. Tech.* **49,** 223
Rhone-Poulenc Product Catalogue (1977).
Romicon Product Catalogue (1977).
Rushton, A., and Khoo, H. E. (1977). *J. Appl. Chem. Biotechnol.* **27,** 99.
Saier, H. D. (1977). *Chem. Labor Betr.* **28,** 6.
Scheuermann, E. A. (1977). *Lebensm. Technol.* **10,** 11 (1977).
Shen, J. S., and Probstein, R. F. (1977). *Ind. Eng. Chem. Fundam.* **16,** 459.
So, M. T., Eirich, F. R., Strathmann, H., and Baker, R. W. (1973). *Polymer Lett.* **11,** 201.
Strathmann, H. (1970). *Chem. Ing. Tech.* **42,** 1095.

Strathmann, H. (1973). *Chem. Ing. Tech.* **45,** 825.
Strathmann, H. (1976). *Pure Appl. Chem.* **46,** 213.
Strathmann, H. (1978). *Desalination* **26,** 728.
Trauter, J., and Ruess, B. (1977a). *Melliand Textilber.* **58,** 107.
Trauter, J., and Ruess, B. (1977b). *Melliand Textilber.* **58,** 711.
Tressler, D. K., and Maynard, J. (1961). "Fruit and Vegetable Juice Processing Technology." Avi, Westport, Connecticut.
Tomarelli, R. M., and Bernhart, F. W. (1962). *J. Nutr.* **78,** 44.
Walch, A. (1973). *CZ-Chem. Tech.* **2,** 7.
Webb, B. H., and Whittier, E. O. (1948). *J. Dairy Sci.* **31,** 139.
Whittier, E. O., and Webb, B. H. (1950). "By-Products from Milk." Reinhold, New York.
Wysocki, G. (1976a). *Deut. Molkerei-Zeitung* No. 29, July.
Wysocki, G. (1976b). *Wasser Luft Betr.* 76(4).
Wysocki, G. (1977). *Chem. Tech.* **6,** 285.
Yamashita, S., Yokota, K., and Ishikawa, N. (1974). *Jpn. J. Clin. Chem.* **3,** 229.
Zsigmondy, R., and Carius, C. (1927). *Chem. Ber.* **60B,** 1047.

10

Design, Operation, and Maintenance of A 5-mgd Wastewater Reclamation Reverse Osmosis Plant

I. NUSBAUM
Consulting Engineer
San Diego, California

DAVID G. ARGO
Orange County Water District
Fountain Valley, California

ISBN 0-12-085480-5

I. Introduction

The Orange County Water District (OCWD) has constructed Water Factory 21 (WF-21) and a series of injection wells near the Pacific coast in Fountain Valley, Calif., to provide a hydraulic barrier against ocean water incursion and to supplement the domestic water supply. The source of water used for recharge is biologically treated municipal wastewater, which is subjected to advanced waste treatment (AWT). Because the injected water is combined with the potable groundwater supply, the design of WF-21 incorporated the best available technology, including demineralization, to insure that only highest-quality water would be injected.

Water Factory-21 is a 0.66 m^3/sec (15 mgd) AWT plant that includes the following processes:

(1) lime treatment for suspended solids, heavy metals and organic removal,
(2) air stripping for ammonia and volatile organics removal,
(3) recarbonation–chlorination for pH adjustment and algae control,
(4) mixed media filtration for additional suspended solids removal,
(5) activated carbon absorption for removal of dissolved organics and heavy metals,
(6) reverse osmosis (RO) demineralization for removal of inorganics and organics, and
(7) final chlorination for disinfection and partial ammonia removal.

The drought in California, the pending reduction of Colorado River water availability to the state, the growing population and demands on the water resources of California have emphasized the immediate need to extend present water supplies to meet future demands. Because of the high quality of water reclaimed by WF-21, this facility provides an excellent opportunity to study the feasibility of augmenting domestic water supplies with reclaimed wastewater. But there is limited information on AWT's ability to remove inorganic, organic, and biological contaminants; questions have been raised by regulatory agencies responsible for protecting public health. It is essential that information be developed that will verify the effectiveness and reliability of these treatment technologies for removing all contaminants of health concern from municipal wastewater effluent. There is also a lack of information regarding the cost and the long-term effectiveness of many of these AWT technologies, particularly wastewater demineralization. Because of these concerns, a study was undertaken to provide necessary research data. The specific objectives of the project were

(1) to produce cost data so large-scale wastewater RO membrane projects can be evaluated for application feasibility in AWT facilities,

(2) to evaluate various pretreatment systems for wastewater demineralization: Ultrafiltration (UF), ozonation, lime clarification, and filtration,

(3) to provide detailed information on the long-term operation of large-scale RO systems for evaluating the feasibility of demineralization for wastewater reclamation,

(4) to provide water-quality data that demonstrate the reliability of AWT and RO for producing a potable-quality water from municipal wastewater, and

(5) to provide detailed information on the effectiveness of AWT and RO for organics and virus removal.

II. Design and Operation of a 5-mgd Reverse Osmosis Plant for Water Reclamation

This chapter is devoted to the planning, development, design, and operation of a 5-mgd RO plant operating on an AWT effluent. The demineralized water is required to reduce the total dissolved solids (TDS) in the AWT effluent to less than 500 mg/liter and to meet all the established requirements for groundwater injection. The injected reclaimed water is used to create a fresh water barrier to seawater intrusion of a major groundwater basin in southern California. The water in the basin is used for municipal and industrial purposes.

This chapter will cover some background information on the Orange County Water District and the AWT as a prelude to the discussion of the planning, design, and operation of the RO demineralization facility. Data are included on operating problems, mechanical and membrane, and on costs. Accompanying the operation of the 5-mgd system is a continuing RO pilot study involving pretreatment and new RO membrane elements. A comparison of the results obtained by the pilot studies in parallel with the large plant is made. Current information on the operating costs of the AWT and the RO processes is included.

III. Orange County Water District Responsibilities and Background Leading to Present Facilities

The OCWD was formed by an act of the California legislature in 1933 to protect and manage Orange County's massive groundwater basin, which had been overdrafted by excessive pumping and diversions of the Santa Ana River by upstream users. The act specifically provides for

(1) groundwater basin management,

(2) groundwater basin conservation, including quality and quantity of water, and

(3) protection of Orange County's water rights and the natural flow of the Santa Ana River.

The District encompasses more than 202,000 acres generally overlying the coastal basin of the Santa Ana River within the county. This represents about 40% of the county land area containing nearly 1.4 million people, almost 90% of the county's total.

The District's primary responsibility is to protect and preserve Orange County's groundwater basin, which provides approximately 65% of all water used within the district. Groundwater recharge has been an ongoing project since formation of the district in 1933. Orange County's tremendous growth was made possible by its natural groundwater reserve, a reliable source of water. As competition increased for the available supply, however, groundwater levels dropped. A report prepared in 1925 at the request of Orange County concluded that serious degradation of fresh groundwater supplies would result from seawater intrusion if measures were not taken to halt the drop in the water table. Fortunately, for the next 20-yr period, the county received above-average rainfall, and seawater encroachment was minimal. Continued development of the area and transition from an agricultural to urban economy, however, caused an accumulated overdraft of the groundwater basin of 500,000 acre feet* by 1956, permitting seawater to migrate approximately 3.5 mi inland through the underground aquifers. To prevent further destruction of the basin groundwater supply and to sustain an adequate amount of groundwater to meet increasing demands, the OCWD began purchasing surplus Colorado River water in 1949, and Northern California water from the State Water Project in 1973 for groundwater recharge.

The OCWD in 1962 began investigating methods of establishing an

* One acre foot is 325,900 gal.

effective barrier to seawater intrusion. It ultimately determined that the most practical method would consist of a pumping trough near the coast and a series of injection wells inland of the intrusion zone. It was calculated that 20,000–30,000 acre feet of injection water was needed each year for a reliable facility.

The shortage of local water, growing difficulties of increasing import supply, improving techniques in water treatment, and rising costs of power and construction were the principal factors influencing the District's decision to construct facilities to reclaim wastewater and participate with the federal Office of Saline Water in a seawater desalter to produce the injection-barrier supply.

The base supply of water for the Orange County area was Colorado River water, with an average quality of 750 ppm TDS. As this water passed through the community, it accumulated an organic and dissolved inorganic load. The organic and suspended materials were readily removed through conventional treatment processes, but the dissolved mineral load remained virtually unchanged. Consequently, the product water of the conventional AWT processes (tertiary) ranged from 1100 to 1400 mg/liter (ppm) TDS, a brackish supply that had to be diluted with a higher-quality water before it was suitable for injection into the county's fresh water reserve.

To develop a water source low in dissolved solids to blend with the reclaimed water, the district obtained the support of the then Office of Saline Water (OSW) of the U.S. Department of the Interior for a 3-mgd seawater desalting plant. The OCWD proceeded to construct a 15-mgd AWT plant and the support facilities for the seawater distillation plant.

Prior to completion and testing of the seawater distillation plant, funding for the study was terminated. The OCWD had authorized the preparation of a feasibility report of alternative methods for the operation of WF-21.

The report concluded there were advantages to a wastewater desalting program at WF-21 as follows:

(1) Increase dependable water supply to injection barrier.
(2) Increase operating capacity of wastewater reclamation plant.
(3) Decrease dependence on deep wells.
(4) Develop an alternative to desalting seawater.
(5) Demonstrate the desalting process and cost on highly treated wastewater. This information will also be valuable for the district's proposed Anaheim Forebay wastewater reclamation and desalting plant.
(6) Utilize the existing pretreatment facilities, substantially reducing desalting costs.

The District qualified electrodialysis (ED) and RO as the demineralization methods that could be utilized. Based on the proposals received, a decision was made to use RO. Bids had been requested on a first-stage 1-mgd plant and 5-mgd facility. Economic considerations favored the 5-mgd system. The plant selected and utilized 8-in.-diameter spiral-wound RO elements using cellulose acetate membranes.

IV. Water Factory 21 Process Description

Water Factory 21's advanced wastewater reclamation facilities were designed to treat 0.66 m^3/sec (15 mgd) of municipal activated sludge effluent by the processes indicated in Fig. 1. These processes include lime clarification with sludge recalcining, ammonia stripping, recarbonation, breakpoint chlorination, mixed-media filtration, activated-carbon adsorption with carbon regeneration, postchlorination, and RO demineralization. Also indicated in Fig. 1 are the sampling locations, designated as Q1, Q2, etc. The Orange County facility provides for evaluation of the initial AWT processes as well as the newer RO demineralization technology. All unit processes are designed with dual or parallel units to assure operation of any given unit process at near-design capacity during the total study period. Major design criteria for each process are listed in detail in Argo (1979b), McCarthy *et al.* (1980), and Trussell *et al.* (1979).

A. Chemical Clarification

Chemical clarification is accomplished in separate rapid mix, flocculation, and sedimentation basins. Lime is used as the primary coagulant and is added in slurry form to the rapid-mix basin. Lime feed is automatically controlled to achieve an optimum pH of 11.0, a dose of 350 to 400 mg/liter as calcium oxide is sufficient to maintain this pH level. The three-stage flocculation basins are operated with G values of 100, 25, and 20 sec^{-1} in the first, second, and third basins, respectively. Detention time is approximately 10 min in each compartment. An anionic polymer is used as a settling aid in the third-stage basin, usually added in a dose of approximately 0.1 mg/liter to improve clarification. The water flows from the bottom of the third flocculation basin to the settling basin, which is

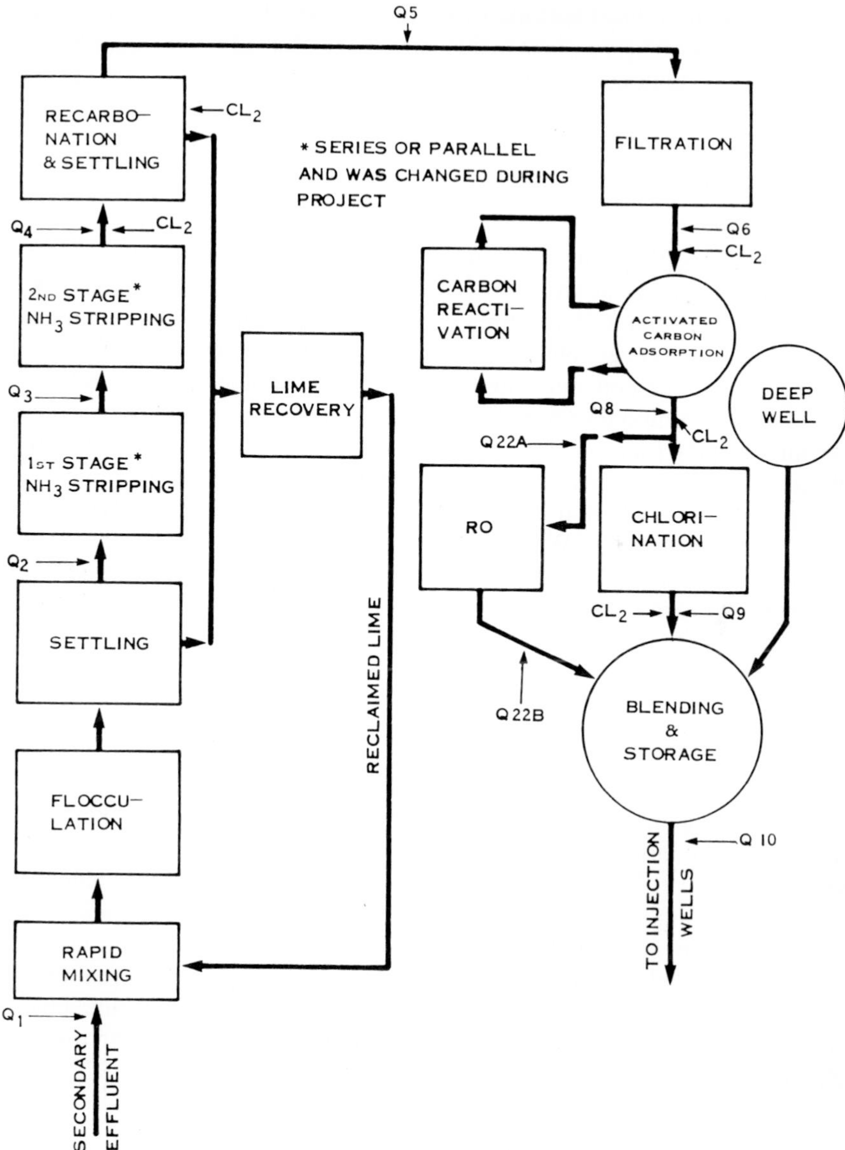

Fig. 1 Wastewater reclamation process diagram.

equipped with inclined settling tubes to improve clarification. Results of lime clarification have shown this process to be effective in reducing turbidity, phosphates, organics, and suspended solids.

B. AIR STRIPPING

Following settling, air stripping is accomplished in a countercurrent, induced draft tower at an air-to-water ratio of 300 m^3/m^3. Originally, the tower was operated to strip ammonia nitrogen from the secondary municipal effluent. However, changes in the secondary treatment system have reduced ammonia concentrations in the WF-21 influent to levels below 5 mg/liter. The ammonia tower fans have, therefore, been shut down, and the air stripping process is now used for removal of volatile organics only. Water from the chemical clarifier is pumped to the top of the ammonia stripping towers, where it is allowed to cascade over 7.6 m of polypropylene splash bar packing. With the tower fans not operating, the only draft is caused by natural ventilation. Although little ammonia removal is achieved through the towers, this process has been very effective in removing a wide range of volatile organic compounds.

C. RECARBONATION

Following air stripping, the pH of the treated wastewater is adjusted in the recarbonation basin. Carbon dioxide is added to lower the pH to approximately 7.5. This basin also serves as a chlorine contact chamber; generally 10 mg/liter of chlorine is added, primarily as a disinfectant. During certain periods the basin is used as a breakpoint chlorination chamber and chlorine dosages are increased to provide for breakpoint chlorination of ammonia nitrogen. During this phase of operation a chlorine-to-ammonia nitrogen weight ratio of 9 or greater is required to reduce ammonia nitrogen levels to less than 1 mg/liter.

D. MIXED MEDIA FILTRATION

The recarbonated effluent passes through open, gravity, mixed-media filter beds designed for a hydraulic loading rate of 0.2 m^3/m^2 min. The filter media, 0.76-m deep, consists of layers of coarse coal, silica sand, and garnet, supported by a layer of silica and garnet gravel with a Leopold underdrain. Alum and polymer are occasionally added to improve clarification.

E. Activated Carbon Adsorption

The water is then pumped to the top of one of 17 downflow granular activated carbon (GAC) contactors which contain Calgon Filtrasorb 300 carbon. The contactors operate in parallel, each having an empty bed contact time of 34 min. The hydraulic loading rate for each column is 0.2 m^3/m^2 min. Following GAC adsorption, the flow from the AWT plant is divided. Two-thirds goes to the final chlorination basin for postchlorination, followed by 30 min of contact time at design flow. The other one-third of the activated carbon effluent is diverted to the 0.22 m^3/sec RO plant, which removes the dissolved solids from the reclaimed wastewater.

V. 5-mgd Reverse Osmosis Performance

The primary purpose of RO is to remove dissolved inorganics. Wastewater demineralization is required at WF-21 because the waters being treated contain approximately 1000 mg/liter TDS. In southern California, all wastewaters have this high TDS concentration because a major water source is the Colorado River, which has a TDS of about 750 mg/liter. Because the reclaimed wastewater to be injected must meet stringent mineral quality requirements, it is necessary to demineralize part of the water treated at WF-21.

The RO water-treatment process utilizes a semipermeable membrane to remove dissolved solids from water solutions. The most widely used general purpose semipermeable membrane is a 100 μm (0.004 in.) modified cellulose acetate (CA) film. It is asymmetric, with one surface consisting of a relatively dense layer approximately 2000 Å (0.2 μm) thick, and the remaining film a relatively porous, spongy mass that is approximately two-thirds water. The membrane is prepared by casting a solution (15–25% by weight) of CA in a solvent system of acetone and formamide. After casting, the membrane is immersed in water where it gels. It is placed in another water bath at 60 to 90°C for annealing to increase the salt rejection characteristics of the membrane. The basic theory of the process is that water under high pressure will pass through the CA membrane more rapidly than dissolved solids. The product water flow through a semipermeable membrane is expressed as

$$F_w = A(\Delta p - \Delta \pi), \tag{1}$$

where F_w denotes water flux (g/cm² sec), A the water permeability constant (g/cm² sec atm), Δp the pressure differential applied across the membrane (atm), and $\Delta \pi$ the osmotic pressure differential across the membrane (atm). The salt flux through the membrane can be expressed as

$$F_s = B(C_1 - C_2), \tag{2}$$

where F_s denotes salt flux (g/cm² sec), B the salt permeability constant (cm/sec), and $C_1 - C_2$ the concentration gradient across the membrane (g/cm³).

The water permeability and salt permeability constants are characteristic of the particular membrane used and the processing it has received.

An examination of Eqs. (1) and (2) shows that the water flux is dependent on the applied pressure, whereas the salt flux is not. As the pressure of the feedwater is increased, the flow of water through the membrane should increase while the flow of salt remains essentially constant. It follows that both the quantity and the quality of the purified product should increase with increased driving pressure.

The water flux clearly increases as the available pressure differential increases. It is also evident that the water flux decreases as the salinity of the feed increases, because the osmotic pressure contribution increases with increasing salinity. Further, as more water passes through the membrane when the feed passes through the unit, the salinity of the feedwater becomes higher (concentrated). The osmotic pressure contribution of the concentrate becomes higher, and this results in a lower water flux with increasing water recovery percentages.

Finally, because the salinity of the feed concentrate increases with increasing product water removal, the quality of the product water decreases with increasing water recovery.

Reverse osmosis system designs have evolved into two distinctly different approaches. They are the hollow fine fiber and the spiral wound. The design selected for the RO facility at WF-21 utilizes the spiral-wound concept.

Figure 2 shows how the spiral-wound module is constructed. Basically, it consists of two membrane sheets separated by a porous support material. The 8-in. 8150HR elements used in the Orange County system contain 12 leaves. The support material is a polyester tricot knit fabric commonly referred to as tricot. The material supports the membrane

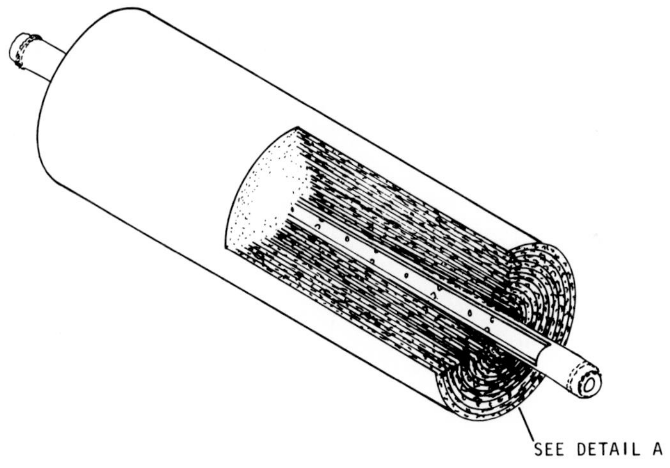

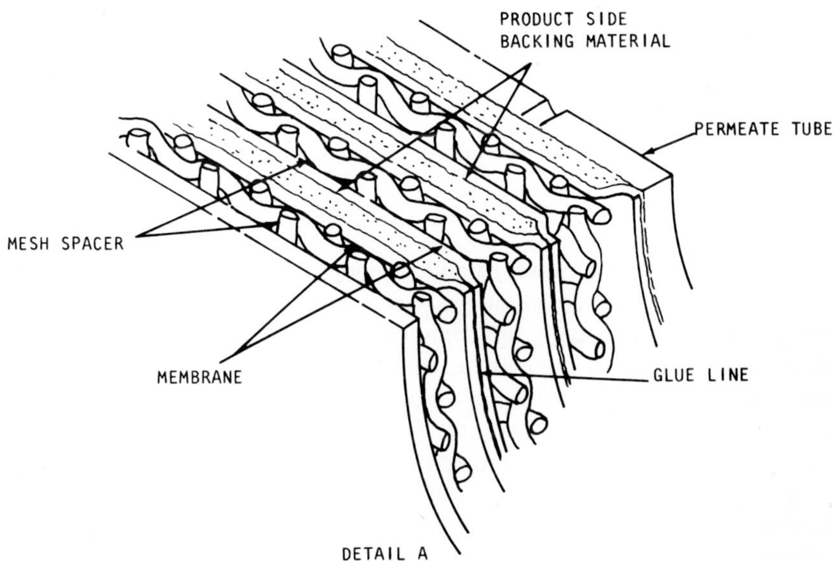

Fig. 2 Details of spiral-wound module construction.

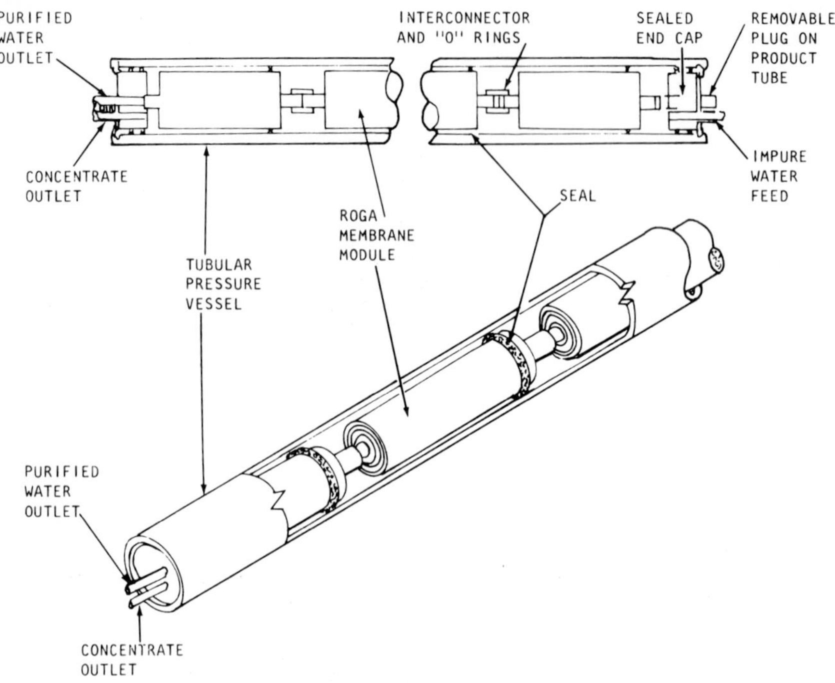

Fig. 3 Module pack assembly.

against the operating pressure and provides a flow path for the excess product water. This "envelope" is sealed around three sides with a waterproof cement to prevent contamination of the product water by the feed or brine. The fourth edge is sealed to a hollow plastic tube that has perforations inside the edge seal area so product water can be removed from the porous support material. This long envelope is rolled up about the central tube in a spiral, along with a mesh spacer that separates the facing membrane surfaces to form a brine feed channel, and promotes turbulence of the feedwater as it passes through the module. A polypropylene Vexar* spacer is currently used in most spiral modules. Several such modules are connected in series, equipped with a peripheral brine seal, and slipped into commercially available steel or fiberglass reinforced plastic (FRP) pipes that act as the pressure vessel (Fig. 3).

During operation, the feedwater enters the upstream end of the mod-

* DuPont.

ule and flows axially through the module. Some water flows through the membrane surface, where it travels down the porous backing material to the central tube. The brine, which is more concentrated than the feed because of the purified water removal, exits from the module and enters as the feed into the second module in series.

The peripheral brine seal ensures that all of the flow passes through the module. The feed then passes successively through sequential modules connected in series and exits from the pressure vessel at the end cap through a stainless steel U bend. The product from each module is collected in the central tube (product water tube) and flows through the central tube of each module in series and out of the system, through the end cap, in a separate PVC product water line.

The spiral module offers several advantages, including high packing density. Packing densities greater than 200 ft^2/ft^3 have been achieved, higher density systems are presently under development. In addition to having a high packing density to minimize pressure-vessel volume, the spiral design allows use of low-cost carbon steel or FRP pressure vessels.

VI. Detailed Process Description

The RO plant design at WF-21 required 90% rejection of all salts, while achieving an overall product water recovery of 85%. The specifications required operation over a wide range of temperatures and included limits on chemical and power consumption. The detailed plant-design criteria are given in Table I and a flow schematic is in Fig. 4.

The RO system consists of three subsections. The first includes pretreatment, which is required to adjust the RO feedwater stream to a condition that assures maximum performance efficiency. The second portion of the process consists of the core of the RO system, wherein the pretreated feedwater is pressurized, acidified, and then pumped through the membrane elements in the pressure vessel assemblies. It is during this process, under a nominal operating pressure of 400 psig, that the water is demineralized by passage through a semipermeable membrane. The last portion of the RO process is posttreatment, which consists of adjusting the final product water to conditions compatible with achieving potable water. A more detailed explanation of each of these subsections is provided subsequently.

TABLE I

Major Design Criteria OCWD 5-mgd RO Plant

- General performance requirements[a]
  - Minimum permeate flow rate: 5.00 mgd
  - Maximum concentrate flow rate: 0.88 mgd
  - Feed flow rate: 5.88 mgd
  - Design feed water temperature: 65°F
  - Annual throughput requirement: 1,679 Mgal
  - Minimum permeate water recovery: 85%
  - Minimum salt rejection: 90%
  - Concentrate pH: 5.0–8.0
  - Permeate pH: 6.5–8.0
  - Contract completion time: 670 calendar days
  - Maximum noise level outside RO building: 55 db_a
  - Pretreatment
    - Feed water source
      - Normal: activated carbon adsorption effluent
      - Optional: mixed-media filtration effluent
    - Filter feed pumps
      - Type: vertical turbine, single stage
      - Number: 3 (including 1 standby)
      - Capacity: 2,045 gpm (each)
      - TDH: 58 ft
      - Power: 50 hp each
    - Scale inhibitor feeder
      - Number: 1
      - Design rate: 4.1 lb/hr (2 mg/liter)
      - Maximum capacity: 21.5 lb/hr (10.5 mg/liter)
      - Inhibitor: Sodium hexametaphosphate
    - Chlorinators
      - Number: 1
      - Capacity: 500 lb/day (10.2 mg/liter)
      - Note: Backup to this unit from WF21 chlorinators
    - Feed clearwell
      - Number: 1
      - Total capacity: 15,000 gal
      - Average detention time at 5.88 mgd: 3.67 min
  - RO flow rates
    - Feed flow
      - Per section: 681 gpm
      - Per unit: 2043 gpm
      - Total plant: 4086 gpm
    - Permeate flow
      - Per section: 579 gpm
      - Per unit: 1737 gpm
      - Total plant: 3474 gpm

TABLE I (*continued*)

Concentrate flow
Per section: 102 gpm
Per unit: 306 gpm
Total plant: 612 gpm
Posttreatment
Decarbonators
Number: 2 (both normally in operation)
Type: countercurrent packed bed
Air flow rate: 3 f^3/m gpm permeate
Hydraulic loading: 25 gpm/ft^2
Permeate clearwell
Number: 1
Total capacity: 8,850 gal
Average detention time @ 5.0 mgd: 2.55 min
Permeate pumps
Type: Vertical turbine, single stage
Number: 3 (includes 1 standby)
Capacity: 1740 gpm each
TDH: 25 ft
Power: 20 hp each
Electrical energy requirements

	Voltage	Total installed (hp)
	2300	2700
	480	333
Total	2780	2033

Cartridge filters
Number: 4 (includes 1 standby)
Elements: 240-10 in. or 120-20 in. polypropylene cartridges per filter
Rating: 25 μm
RO feed pumps clearwell
Number: 1
Total capacity: 33,700 gal
Average detention time at 5.88 mgd: 8.25 min
RO feed pumps
Type: Vertical turbine, 17 stages
Number: 3 (includes 1 standby)
Capacity: 2045 gpm each
TDH: 1386 ft maximum
924 ft normal
Power: 900 hp each

(*continues*)

TABLE I *(continued)*

Acid feeders
- Acid: Concentrated sulfuric acid (93% or 66°Be′)
- pH: RO feed adjusted to pH 5.5
- Type: Positive displacement
- Number: 3 (includes 1 standby)
- Capacity: 0.175 gpm each, design
0.35 gpm each, maximum

Reverse osmosis
- RO membranes, sections, units
  - Number of units: 2 (both normally in operation)
  - Number of sections: 3 per unit
  - Number of pressure vessels: 35 per section
  - Number of membranes: 6 per pressure vessel
  - Total number
    - Units: 2
    - Sections: 6
    - Pressure vessels: 210
    - Membranes: 1,260
  - Pressure vessel array per section: 20–10–5
  - Pressure vessel length: 21 ft 10 in.
  - Nominal pressure vessel diameter: 8 in.
  - Membrane diameter: 8 in.
  - Membrane length: 40 in.
  - Membrane type: Spiral wound, cellulose acetate

Support systems
- Air compressors
  - Number: 2 (includes 1 standby)
  - Capacity: 22 cfm each
  - TDH: 125 psig
  - Power: 7.5 hp each
- Cleaning system
  - Tanks: 2 at 1500 gal each
  - Pumps: 2 at 880 gpm each, 93 ft TDH, 30 hp each
- Flushing system
  - Tanks: 2 at 6500 gal each
  - Pumps: 2 at 300 gpm each, 100 ft TDH, 15 hp each

[a] Conversion factors are 1 mgd = 3785 m^3/day, 1 mil gal = 3785 m^3, 1 gpm = 0.0631 sec^{-1}, 1 hp = 0.7457 kW, 1 lb = 0.454 kg, 1 gal = 3.785 liter, 1 in. = 2.54 cm, 1 ft = 0.3048, 1 f^3min/gpm = 0.4437 m^3/min per sec^{-1}, 1 gpm/ft^2 = 40.7 1/min m^2, 1 psig = 0.0703 kg/cm^2.

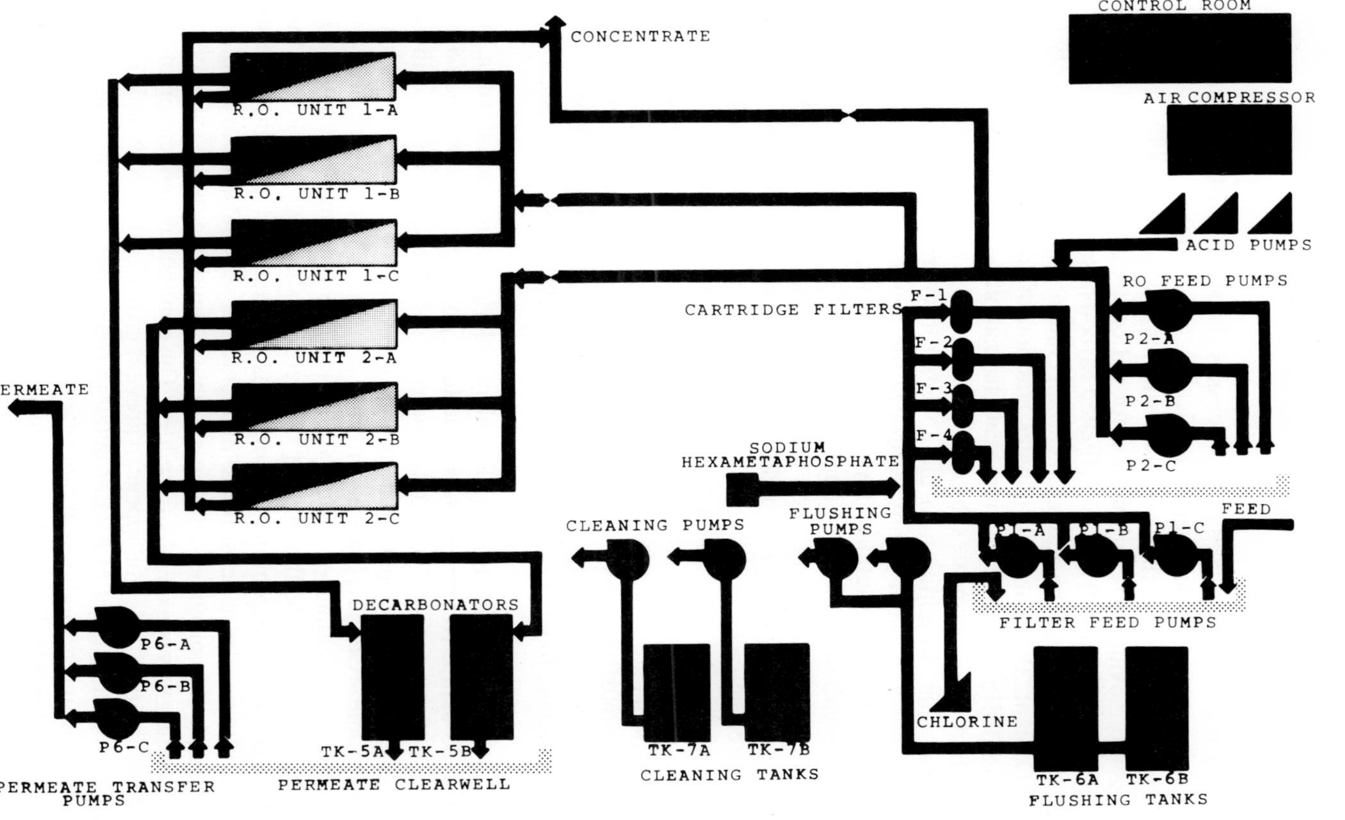

Fig. 4 5-mgd RO flow schematic.

A. PRETREATMENT

The wastewater reclamation processes ahead of the RO pretreatment section consist of: (1) chemical clarification for phosphate, trace metals, and suspended solids removal; (2) air stripping for organics and ammonia nitrogen removal; (3) recarbonation for pH control; (4) filtration for suspended solids removal; and (5) activated carbon adsorption for removal of dissolved organics. The activated carbon effluent then flows by gravity to the RO plant. Pretreatment includes addition of sodium hexametaphosphate (SHMP) to inhibit calcium scale formation.

When water permeates the membrane, almost all the salt remains on the brine side. In addition to the concentration that takes place in the brine because of water removal through the membrane, a concentration effect occurs in the boundary layer because of the limited circulation and exchange with the bulk of the fluid. As the water is demineralized, a point is reached where the saturation limits are exceeded and precipitation of calcium sulfate or other sparingly soluble salts occurs. Because calcium carbonate and calcium sulfate are the most commonly occurring salts in natural water, the most common scale control method that has been used is threshold treatment with SHMP. This precipitation inhibitor represses both calcium carbonate and calcium sulfate scale formation. Dry SHMP is fed continuously to the RO feed sump to provide a dose of approximately 2 mg/liter. The observation that some inorganic precipitation was occurring, however, has led to an increase in the SHMP dose to about 5 mg/liter.

Chlorine solution from the RO plant chlorinator is also added in the filter feed sump. Reverse osmosis modules provide a large surface area for the attachment and growth of bacterial slimes and molds, and these organisms may cause membrane fouling or even module plugging. There is also some evidence that the enzyme system of some organisms will degrade CA. A chlorine residual of 0.5 mg/liter is maintained automatically to help mitigate the growth of bacteria, slime, and molds on the surface of the RO membrane. Caution must be exercised, though, because the membrane is also sensitive to high chlorine residuals. To avoid damage, the plant is provided with an automatic shutdown in the event chlorine residual levels exceed 1.0 mg/liter. The continuous application of 0.5 mg/liter chlorine does not appear to prevent growth sufficiently at WF-21. This will be discussed in greater detail in the results section.

Following the addition of chlorine and SHMP to the filter feed sump, the water is pumped through three cartridge filters that remove suspended solids that could damage or foul the RO membranes. The 25 polypropylene cartridge filters are replaceable.

B. REVERSE OSMOSIS TREATMENT

The water from the cartridge filters discharges into the high-pressure RO feed sump. The conditioned feedwater is then pumped to the RO membranes by two 900-hp vertical turbine high-pressure feed pumps, each capable of providing 3 mgd at an operating pressure of up to 540 psi. To increase system reliability, a third high-pressure feed pump is provided as a standby.

Water Factory 21 uses 8150HR CA membranes, manufactured by Fluid Systems Division of UOP, Inc. Each of the elements is an 8-in. diameter spiral-wound type, approximately 40-in. long. These modules are rated at about 4000 gpd. There are 6 modules connected end to end within each of the single pressure vessels that are approximately 20-ft long. The 8-in. pressure vessels are constructed of FRP, which is corrosion resistant.

Sulfuric acid is added downstream of the feed pumps before the water enters the RO modules to prevent membrane hydrolysis and to convert carbonate alkalinity to carbon dioxide for scale control.

An important limiting factor in the life of CA membranes is the rate of membrane hydrolysis. Cellulose acetate will hydrolyze to cellulose and acetic acid, depending upon the temperature and pH. The rate of hydrolysis is at a minimum at about pH 4.7. Acid is also added because at pH 5.5 approximately 75% of the total alkalinity has been converted to dissolved carbon dioxide, thereby reducing the potential of calcium carbonate scale. Thus, the RO plant is operated at pH 5.5 to bring the membrane into a reasonable part of the hydrolysis curve and reduce the possibility of carbonate scaling. It was decided to inject the acid on the downstream side of the high-pressure feed pumps. This design permitted the economical and time saving use of standard construction materials for the high-pressure feed pumps.

There are two RO units, each consisting of three sections, with operation of all units and sections being essentially identical. Each section, as originally designed, contains 35 pressure vessels arranged in a 20, 10, 5 array. The pressurized feedwater arrives at each section manifold and provides equal supply to the first 20 vessels. The feedwater is separated into two streams during passage through the semipermeable membrane elements.

After passage the product water is collected and retained in a separate stream while the concentrate from the first 20 vessels becomes the feedwater for the second assembly of 10 vessels. This process is continued as the concentrate from these 10 vessels becomes the feedwater for the final 5 vessels. Thus the system is designed to recover 85% of the initial feedwater with 15% brine discharged from the third pass directed to waste. The system is operated at constant brine flow. As the membranes become fouled, the feed flow control valves provide the additional pressure necessary to maintain the system at a constant 85% recovery.

The system as designed, with 35 pressure vessels in each section, and operated at pressures from 540 to 600 psi, has not met rated capacity during the first 12,000 hr of operation. (Problems will be described in detail in Section VII.) The system is being modified to provide more membrane and allow for lower-pressure operation at 400 psi. The modified units will contain 42 pressure vessels arranged in a 24, 12, 6 array.

C. POSTTREATMENT

The product water obtained from the RO units is supplied to the posttreatment section for final adjustment before discharge and use. The product water flow from each RO unit is directed to the top of one of two decarbonators where it is distributed over polypropylene packing. As water flows down, small droplets are formed to achieve high surface area for aeration provided by the decarbonator blower. In this manner dissolved gases, primarily carbon dioxide, are removed before the product water is collected in the reservoir provided below each tank. Following decarbonation, the product water is pumped by transfer pumps to the blending reservoir where it is mixed with other water produced at WF-21 before injection.

VII. Performance

A. DESIGN ASSUMPTIONS

One RO design problem that has never been completely resolved is that of predicting membrane capacity. This must be predicted to maintain flux and rejection performance within design specifications over an extended period of time. The performance level specified for WF-21 is 5 mgd, with 90% salt rejection. Among brackish water sources, groundwater normally provides a supply of uniform quality and temperature; performance is predictable within narrow limits. However, this is not the case with municipal wastewater. Demineralization of municipal wastewater provides the ultimate challenge for RO treatment technology, as the quantity, quality, and temperature of the feedwater continually fluctuates. Although conventional and advanced wastewater treatments tend to moderate the extremes, it is still difficult to predict the quantity of membrane demineralization equipment and level of performance required.

Prior to the installation of WF-21, RO experience on municipal waste streams was limited to relatively small installations. Fluid Systems had classically used the 4-in.–diameter 4101 element for these installations. This spiral element is characterized by a brine spacer that is approximately 25% thicker than the spacer used in 8150HR elements supplied for WF-21. This additional opening in the brine channel makes the 4101 more tolerant of suspended solids and the ensuing pressure drops that can occur when this is a problem. However, because the AWT plant at WF-21 included virtually every process for producing a suitable RO feedwater, Fluid Systems decided prior to plant construction the ROGA 8150HR element could be used, thereby optimizing membrane packing density. This decision has been substantiated because there have been no significant problems associated with excessive ΔP across the various RO units at WF-21.

The model 4101 element also differs from the 8150HR with respect to water flux. Design flow for the single 4101 at 400 psi net is 11.1 gfd. Under the same conditions, the 8150HR has a flux of 13.3 gfd. In view of the extensive pretreatment available, this difference in water flux was not considered to be significant because it represented only a minor extension of the "state of the art."

Based on the preceding, Fluid Systems selected the 8150HR element for WF-21. Time has shown that this selection was probably warranted. Hindsight has shown, however, that a number of poor assumptions and/or decisions were made that were contributing factors in the systems' failure to perform to specifications. These decisions were further confounded by early start-up fouling (carbon), which will be discussed subsequently. The reasons for these decisions were apparently economics, an overly optimistic expectation of operating temperatures, and the assumption that no serious fouling would occur. It is worthy to note that Fluid Systems addressed the problems and is correcting the deficiencies on a mutually agreed-on schedule.

Briefly, the decisions made that contributed to the eventual failures were the following:

(1) Available membrane area was reduced by 17%, translating to higher operating pressures.
(2) Optimism with respect to operating temperatures, and possible oversight of the evaporative cooling effect resulting from the ammonia stripping columns.
(3) Miscalculation of the effects of particulate carbon in the initial feedwater, and its effect on water flux.
(4) Overly optimistic expectations of low-temperature effects on the strength of materials used to fabricate the spiral elements.

There were other problems, discussed later, that occurred at WF-21 also affecting performance. The preceding four items were major contributors. Initial design, however, could have provided a functional system, had the operating temperature remained above 65°F and had the system not been started on a highly fouling type of water. The decisions were made, however, and as a result, the contractor had to install additional membranes as well as replace the initial load of elements. In spite of these problems and deficiencies, the plant has produced nominally 80% of the rated capacity and provided valuable information and a means to establish sound design criteria for water reclamation from municipal effluents. A more detailed discussion of the preceding factors and other considerations follows.

B. START-UP PERFORMANCE

A summary of the operating results achieved during the RO start-up and acceptance test is shown in Tables II and III. The data in these tables are inconsistent, because of differences in flow measurements taken from different flow meters. Table II is based on the flow readings that measured total system output, whereas Table III gives a summation of the individual flows measured for each of the six subsections of the RO facility. During start-up, individual subunits (1A–2C) were operated at pressures varying from 490 to 560 psig. The average pressure shown in both tables is the average for the subunits on that day. The average temperature is based on grab samples measured every 4 hr each day.

Table II indicates the plant achieved the design requirements. The average flow was calculated to be 5.2 mgd, and an average system operating pressure of 538 psig was required to maintain this production level.

TABLE II

Start-up Performance

Date	Average pressure (psig)	Average feed water temp °F	Total product flow (mgd)	Rejection (%)	Recovery (%)	Guaranteed[a] salt removal (ton/day)				Electrical[a] consumption (kW hr/ton) salt removed	SHMP[a] consumption (lb/ton) salt removed	Sulfuric acid[a] consumption (lb/ton) salt removed
						TDS[b]	Na[c]	Cl[d]	SO_4[e]			
5/9/77	539	54	5.3	95	85.7	34.7	7.3	8.5	11.5	1074	3.5	121
5/10/77	543	54	5.2	95	85.5	34.0	7.2	8.5	11.4	1166	3.2	149
5/11/77	505	57	5.0	94	85.0	30.9	7.0	8.2	11.0	1191	3.9	147
5/12/77	541	54	5.3	96	85.6	35.9	7.3	8.5	11.4	1023	2.9	201
5/13/77	549	56	5.4	94	85.9	34.9	7.4	8.4	11.6	1049	3.4	163
5/14/77	551	56	5.0	94	85.3	35.8	6.9	8.1	11.0	1027	3.3	165
Average	538	55.2	5.2	95	85.5	33.6	7.3	8.3	11.3	1106	3.6	155
Guaranteed	—	—	5.0	90	85	31.4	6.7	7.8	10.1	1108	7.8	243

[a] Adjusted to design feedwater conditions.
[b] TDS = 1400 mg/liter.
[c] Na = 300 mg/liter.
[d] Cl = 350 mg/liter.
[e] SO_4 = 450 mg/liter.

TABLE III

Start-up Performance

Date	Average pressure (psig)	Product flow (gpm)						Total product flow (mgd)	Rejection (%)
		1A	1B	1C	2A	2B	2C		
5/9/77	539	480	535	550	565	550	545	4.64	95
5/10/77	543	480	535	540	550	545	520	4.56	95
5/11/77	505	470	510	520	510	500	490	4.32	94
5/12/77	541	480	510	500	510	505	495	4.32	96
5/13/77	549	480	515	530	535	560	510	4.51	94
5/14/77	551	465	500	520	520	560	500	4.41	94

Required removal levels of TDS, Na, Cl, and SO_4 were also achieved. The daily average tons of these constituents removed and the power and chemical consumptions, adjusted to reflect design feedwater conditions, are compared with the guaranteed values. During start-up, the plant had no trouble meeting the 90% salt-rejection requirement. Using the flow data presented in Table II, when normalized for pressure and temperature, the rate of flux decline was calculated to be −0.02 during start-up. This is a typical value characteristic of cellulosic membrane and was an indication that the plant did operate satisfactorily during start-up.

A later, more detailed analysis of start-up and early operating data indicates that the plant probably did not meet the design conditions for flow. The sum of the flows produced by the individual subsections is less than what was measured by the total product flow meter. It is almost impossible to determine after the fact which flow measurements were more accurate, but mass-balance calculations and calibration checks of the RO flow instruments indicate that the individual flow measurements are generally more accurate than the total flow meter. A review of these data was made and the results are in Table III. This table indicates the plant did not produce the nominal rate of 5 mgd during the start-up and acceptance test.

While the plant was operated at high pressure (538 psig), the average daily output during start-up was calculated to be only 4.46 mgd by totaling the individual subsection flows. It is believed that these results were probably more accurate, and this conclusion seems to be confirmed by the data collected after start-up as well as the data now being collected. Thus it appears the overly optimistic flux design criterion of 13.2 gal/ft^2 day was not justified. This will be explained in greater detail later in the section. Although Table III shows the plant did not produce the required amount of water, it had little trouble meeting the 90% salt-removal requirement. Using the flow data given in Table II and calculating the flux decline during start-up, the result is an M value of -0.10. This value is considerably higher than the -0.02 initially calculated using the total system output. Thus it would appear there was a rapid flux decline during start-up that was initially overlooked because of a flow measurement error in the total system flow meter. The causes and impact of this rapid flux loss are discussed in greater detail later in this chapter.

C. Results

During start-up and the last year of operation, the performance of the RO plant has been monitored for several water quality parameters. The summary of removal efficiency for a few key water-quality parameters after 100 hr of operation is shown in Table IV. A gradual decline in salt rejection is normal with time. Table V summarizes the average rejection achieved for various parameters during the first 12,000 hr of operation. By comparing the removals given in each table, one can measure the decline in salt rejection of the RO system with time (12,000 hr). It was hoped that salt rejection would remain above 90% during the first 12,000 hr of operation, but this did not occur and the causes of the decline in rejection performance are discussed later in this section.

The removal of dissolved minerals decreased to about 85–88% and average COD removal dropped to 89%. Although RO has been considered primarily a process for removing dissolved inorganics, it has also proved to be highly effective for COD removal. Another study of OCWD for EPA

TABLE IV

RO Plant Efficiency Average Removal[a] After 100 Hr

Parameter	RO influent	RO effluent	Percent removal
EC	1460	70	95
Sodium	196	11	94
Chloride	280	16	94
Sulfate	220	0.8	99
COD	25	1.5	94
Ammonia	5.6	0.5	91
Calcium	118.0	2.0	98

[a] In milligrams per liter.

TABLE V

Removal of Contaminants by Reverse Osmosis Demineralization

Contaminant	Unit	Influent concentration Q-22A	Percent removal
EC	μmho	1367.0	87
pH		5.6	
Na	mg/liter	160.0	85
Cl	mg/liter	195.0	85
SO_4	mg/liter	223.0	98
TH	mg/liter	223.0	94
Ca	mg/liter	85.0	97
NH_3 nitrogen	mg/liter	2.9	85
NO_3 nitrogen	mg/liter	11.0	49
TKN	mg/liter	3.6	85
COD	mg/liter	13.15	89
TOC	mg/liter	7.0	63
Ag	μg/liter	0.8	—
As	μg/liter	<5.0	—
Ba	μg/liter	14.8	55
Cd	μg/liter	10.0	96
Cr	μg/liter	7.6	65
Cu	μg/liter	34.0	73
Fe	μg/liter	23.8	74
Hg	μg/liter	0.5	—
Mn	μg/liter	3.9	70
Pb	μg/liter	2.4	50
Se	μg/liter	<5.0	—
Zn	μg/liter	<100.0	—

indicated that although RO is extremely effective in removing the high molecular weight (MW) organics (COD/TOC), which represent the majority of organics in secondary effluents, it is ineffective in removing most trace organic material with MW less than 200. The higher MW materials are generally referred to as humic and fulvic acids. Many are hydrophilic materials found in secondary effluents that are not always removed well by GAC. Thus for wastewater reclamation, RO is a good complementary process to GAC, lime treatment, and air stripping for organics removal.

The average rejection of nitrate by the cellulose membrane was 49%. Also shown in Table V is the rejection of many heavy metals. Generally, RO is effective in removing these trace elements.

Figure 5 plots rejection and product water flow of the total plant as a

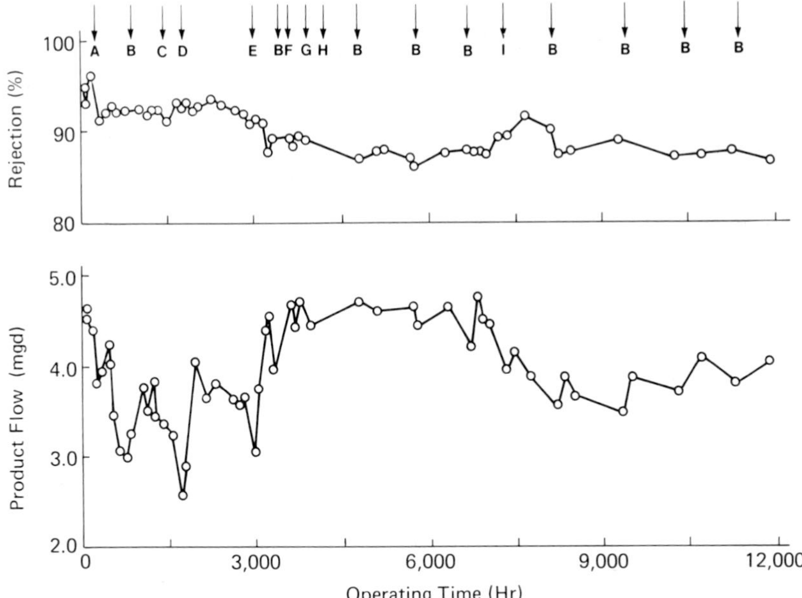

Fig. 5 Total plant rejection and product flow versus time. (A) Carbon columns converted from upflow to downflow configuration; (B) unit cleaned; (C) pressure vessel failure and acid leak to system resulting in low pH feed; (D) ammonia stripping, tower fan speed reduced from high to low and unit cleaned; (E) ammonia stripping tower fans shut off; (F) plant shutdown for 5 days because of influent valve control problems. Possible acid leak into system; (G) plant shutdown for 4 days because of acid day tank leak. Acid leak into system; (H) plant shutdown for 22 days because of chlorine system repairs; (I) unit cleaned. Cleaning followed by *in situ* sizing of membrane.

function of operating time, Fig. 6 shows feedwater pressure and temperature as a function of operating time. The total plant product water flow is the total of the product water flow from each of the subunits.

The flows shown are actual flows; i.e., they have not been normalized to a specified temperature and pressure. The legend that accompanies Figs. 5–13 describes key events that occurred during the evaluation period. Figure 5 graphically shows that during the time period evaluation, the plant did not achieve its nominal rated capacity of 5 mgd. The maximum product flow was 4.77 mgd and the minimum product flow was 2.69 mgd. The mean flow was 3.94 mgd with a standard deviation of 0.54 mgd. The feed pressure varied from subunit to subunit, and the pressure shown in Fig. 6 is the average pressure for the six subunits at a particular time. The maximum average pressure was 593 psig, and the minimum average pressure was 416 psig, with a mean pressure of 538 psig and a standard

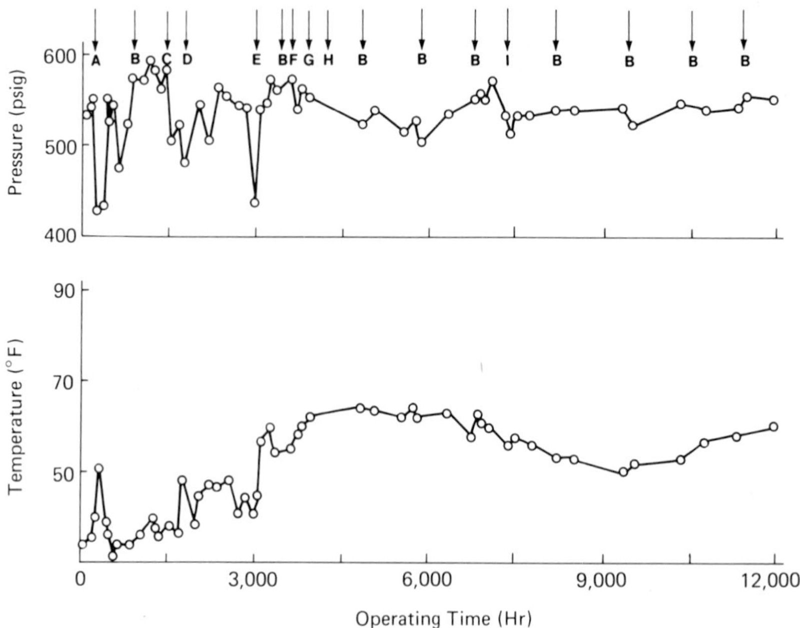

Fig. 6 Total plant feed pressure and temperature versus time. (A)–(I) as in Fig. 5.

deviation of 35 psig. The feed temperature varied from 84 to 54°F during the evaluation period.

Feedwater temperature, as shown in Fig. 6, varies seasonally as a result of the method of operation in the AWT plant (ammonia stripping towers). At the beginning of the evaluation period, the fans in the ammonia stripping towers were at maximum speed and this, coupled with low relative humidity in southern California, resulted in low feedwater temperatures (54–60°F). After about 1800 hr the fan speed was reduced with an increase in RO feedwater temperature as the result. After 3000 hr the fans were turned off and the RO feedwater temperature significantly increased. Because product water flow is proportional to feedwater temperature, there was a significant increase in production as a result of shutting down the tower fans (Fig. 5).

Figure 5 shows the 5-mgd RO plant performance for rejection during the 12,000-hr evaluation period. The plant rejection is based on conductivity of the total plant feedwater and product water. The maximum conductivity rejection based on feedwater was 96.7% at 190 hr, and the minimum rejection was 86.5% at 11,980 hr. Figure 7 plots rejection of sodium and chloride during the evaluation period. The maximum chloride rejection

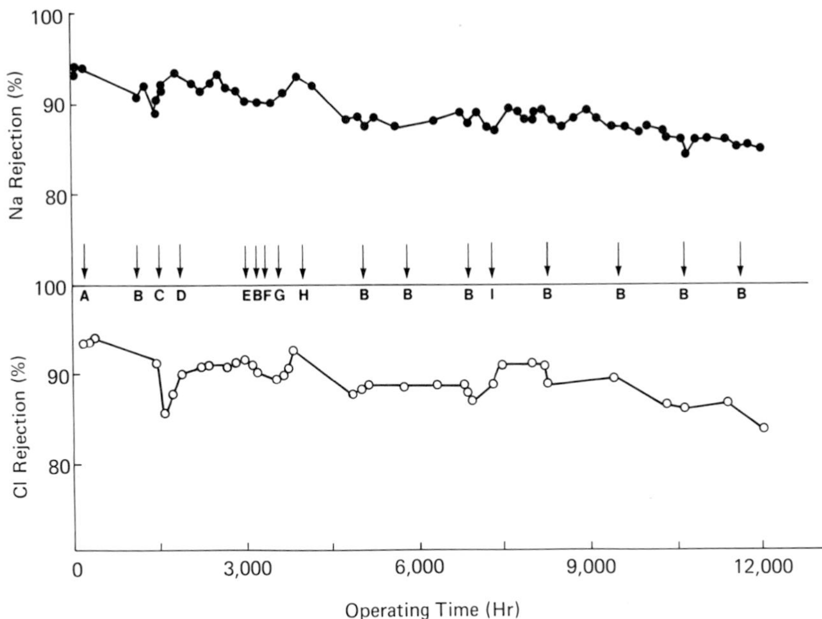

Fig. 7 Total plant sodium and chloride rejection versus time. (A)–(I) as in Fig. 5.

was 94.1% at 80 hr, and the minimum rejection was 83.6% at 12,000 hr. The maximum sodium rejection was 94.2% at 82 hr, and the minimum rejection was 84% at 10,650 hr. Comparison of the rejection plots for conductivity, chloride, and sodium shows the same characteristic pattern. A significant decline of 1 to 3% occurred in the first 1000–1500 hr of operation. There was a gradual decline with another sudden decrease occurring around the 4000-hr mark. Following 4000 hr, the rejection performance seemed to stabilize, except at the 7500-hr mark when treatment of the membrane temporarily increased removal. The causes for and the detailed explanation of the observed rejection performance are presented in Section VIII.

Because there were wide variations in feedwater temperature and pressure during the evaluation period, product flow data for all 6 subunits were normalized to 77°F and 540 psig. The results for subunits 1–6 are shown in Figs. 8–13, in which the normalized product water flow is plotted as a function of operating time. A comparison of these figures indicates all of the six subunits have similar operating results and, consequently, further remarks will be confined to the performance of subunit 1A as shown in Fig. 8.

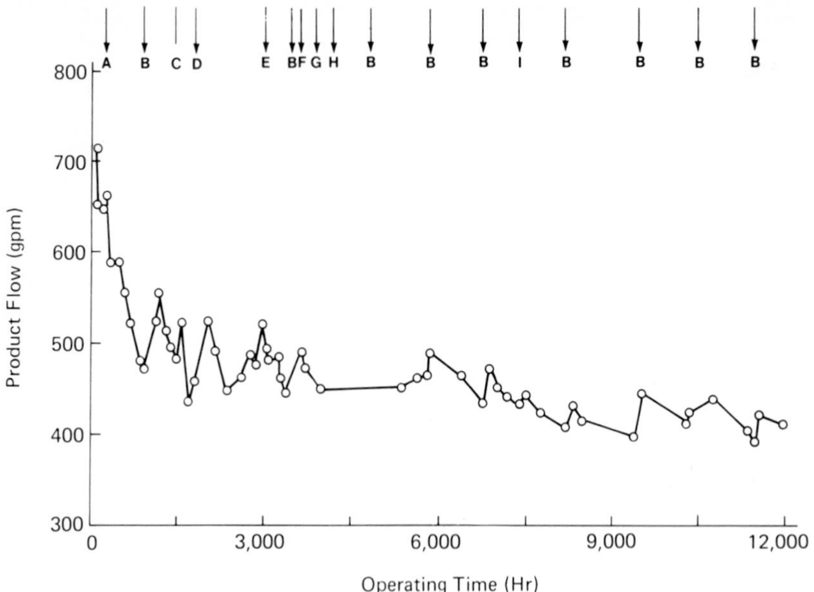

Fig. 8 Subunit 1A flow versus time. Flow normalized to 77°F and 540 psig. (A)–(I) as in Fig. 5.

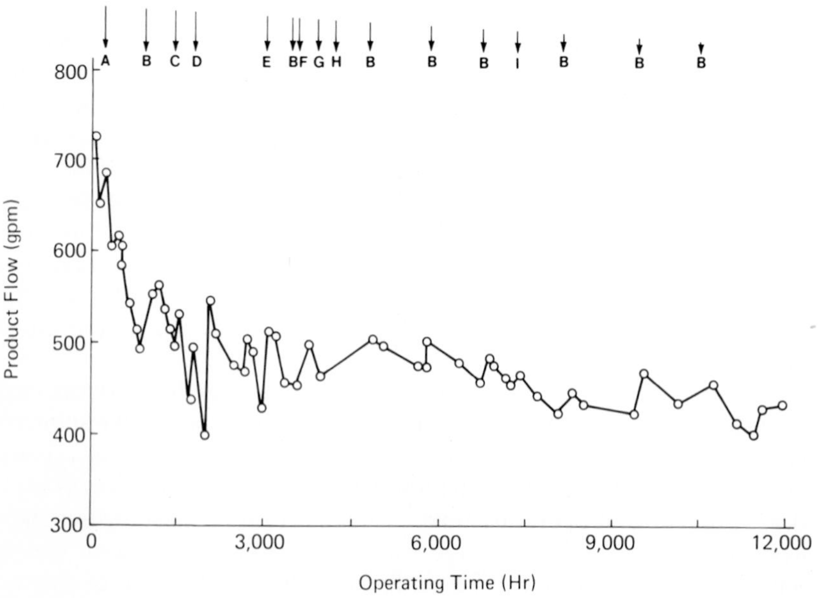

Fig. 9 Subunit 1B flow versus time. Flow normalized to 77°F and 540 psig. (A)–(I) as in Fig. 5.

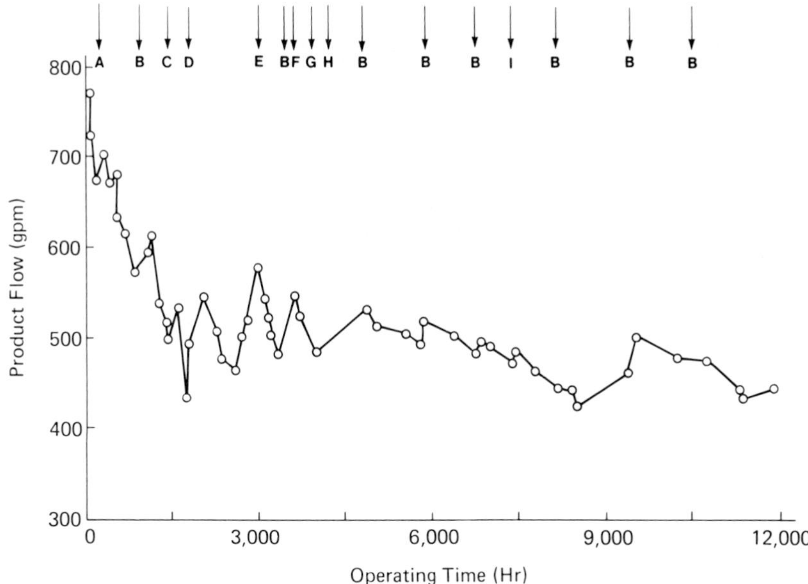

Fig. 10 Subunit 1C flow versus time. Flow normalized to 77°F and 540 psig. (A)–(I) as in Fig. 5.

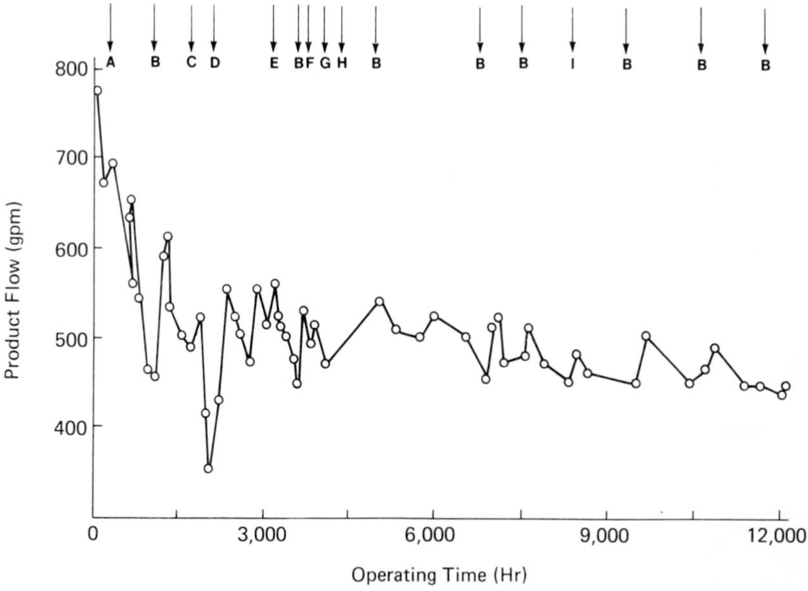

Fig. 11 Subunit 2A flow versus time. Flow normalized to 77°F and 540 psig. (A)–(I) as in Fig. 5.

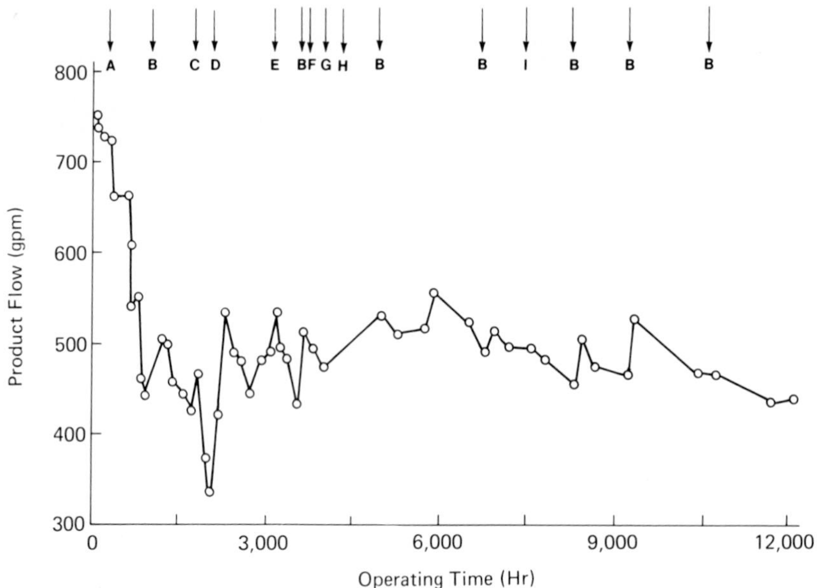

Fig. 12 Subunit 2B flow versus time. Flow normalized to 77°F and 540 psig. (A)–(I) as in Fig. 5.

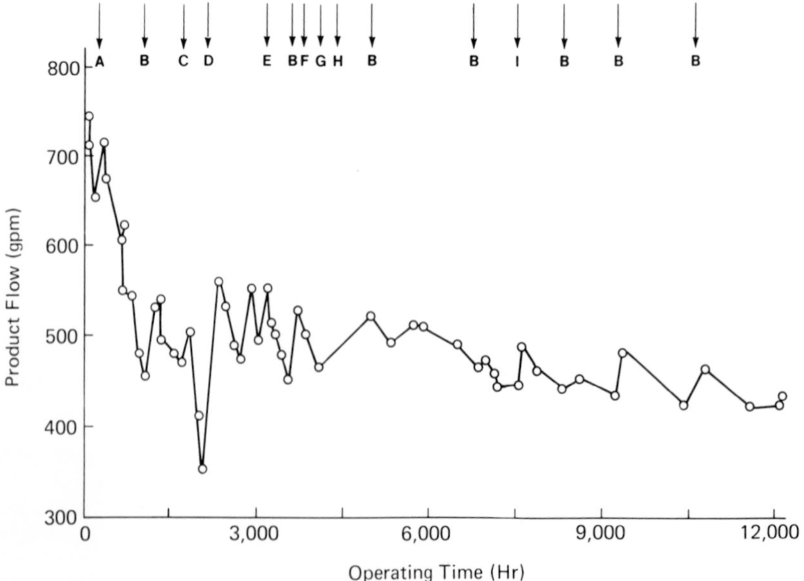

Fig. 13 Subunit 2C flow versus time. Flow normalized to 77°F and 540 psig. (A)–(I) as in Fig. 5.

VIII. Discussion

The results of the first 12,000 hr of operation indicate that the capacity had decreased by approximately 20%, operating at a level of 80% the design condition. Salt removal had also decreased by almost 10% (to 88%), which is below the 90% design requirement. Based on actual operating experience with CA membrane, a decline in performance would be expected. The actual observed performance at Orange County, however, had been worse than anticipated. Table VI lists a number of causes the changes in observed performance at WF-21 can be attributed to and their effect on both flux and salt passage. Also shown is the importance of each condition and, based on the conclusions made during this study, its impact on the changes observed at WF-21. These causes have been classified as having either a negligible, minor, possibly significant, or significant effect. Gross leaks in elements because of construction were not included in the

TABLE VI

Summary of Operating Conditions That Affect Membrane Performance of Water Factory 21 5-mgd System and Their Significance

	Effect		
Condition	Flux	Salt passage	Significance
Membrane hydrolysis			
Normal	Increase	Increase	Negligible
Acid accelerated	Increase	Increase	Significant
Base accelerated	Increase	Increase	Possibly sig.
Loss of colloid (size)	Increase	Increase	Minor
Membrane fouling[a]			
By activated carbon	Decrease	Decrease/increase	Minor
Organics	Decrease	Decrease/increase	Possibly sig.
Microbiological	Decrease	Decrease/increase	Possibly sig.
Inorganic	Decrease	Decrease/increase	Significant
Effect of high pressure, high temperature	—	—	Significant
Membrane compaction			
Effect of high pressure, high temperature	Decrease	—	Significant
Physical effects			
Intrusion	Decrease	Increase	Significant
Membrane embossing	Increase	Increase	Significant
Biological degradation	Increase	Increase	Negligible

[a] A small amount of fouling may decrease salt passage until the degree of fouling increases to a point that will interfere with boundary layer conditions.

table, as these effects would be immediately obvious and the elements replaced. It is important to note that no elements were replaced because of gross leaks. It was, however, common to correct leaks that occurred at the product water tube O-ring seals following installation of new elements, before the system went into service. Almost all effects noted in Table VI developed to some degree in the first year of continuous operation of the 5-mgd RO facility, making any analysis extremely complex.

This section, however, discusses the effects of all conditions noted in the table and concludes with a summary of corrective measures taken to eliminate or minimize these effects.

It should be recognized that many innovative concepts were introduced in the design, construction, and operation of the Orange County RO system for water reclamation. It should also be noted that RO demineralization has been demonstrated to be effective in removing many contaminants besides inorganics from reclaimed wastewater. Although performance declined over the first year, the RO system operated with good on-line availability and produced a considerable quantity of demineralized water for blending and injection into the Orange County Coastal Project. The mechanical reliability was good (greater than 90% on-line operation), and these types of problems were limited to three basic areas.

The first was the failure of the 8-in. RFP pressure vessels. The initial design used a collar-type end cap connection held in place by an epoxy adhesive. After approximately 300–400 hr of operation, two failed when the collar separated from the pressure vessel. Fluid Systems indicated the failure was a result of improper epoxy application during manufacture. Because it was impossible to predict how many vessels were defective (approximately seven more vessels either failed or were replaced), a structural steel restraining barrier was constructed to prevent the collars from separating away from the end of the vessels. This prevented any catastrophic failure mode and provided personnel protection. The failed vessels were replaced with a redesigned integrally wound RFP vessel that eliminated the collar connection.

Another mechanical problem involved the high-pressure acid-injection system. Placement of the acid-injection point downstream of the high-pressure feed pumps reduced the time of delivery and cost of the pumps. But the system was not without its problems. Originally, the high-pressure acid feed and dilution pumps and mixing piping was made of 316 stainless steel that included some threaded connections. This material, unable to withstand the corrosive conditions, quickly developed leaks, particularly at the threaded connections. To prevent the leaks, which pose an obvious safety hazard, the piping was replaced with all flanged and welded Carpenter 20 alloy. This material significantly reduced the frequency of leaks and operated satisfactorily.

The other problem area was instrumentation reliability. As discussed in the section on start-up, measurement of flows, pressure, conductivity, etc., the instrumentation required continuous cleaning, calibration, and adjustment. The instrumentation used on the project was state of the art, but the instruments were frequently inaccurate. Operation of the Orange County system illustrated the need for frequent (no less than once per month) calibration and cleaning of all instrumentation by skilled instrument technicians.

A. MEMBRANE HYDROLYSIS

Normal membrane hydrolysis can be expected to occur with all CA RO membranes. The minimum rate of hydrolysis occurs at pH 4.5 and rapidly increases with changes of ± 3 pH units. Hydrolysis is also accelerated with increasing temperature. One major reason for operating cellulose acetate membranes within the pH range of 5.5 to 6 is to achieve conditions that minimize hydrolysis and increase membrane life. In general, except as discussed later, WF-21 operated within this range and the effects of normal membrane hydrolysis during the first year of operation were negligible.

1. *Acid-Accelerated Hydrolysis*

A review of Figs. 5 and 7, which show total plant rejection based on conductivity, chloride, and sodium as a function of operating time, reveals that shortly after start-up the rejection decreased from about 95 to 93%. This level was maintained until 2300 hr of operation, when the rejection began to deteriorate rapidly. From 2300 hr until about 4800 hr, the rejection steadily decreased to about 87%. This level was maintained during the remainder of the reporting period with one exception. Rejection was temporarily restored to about 90% after 7200 hr when the plant was cleaned and the membrane sized. This is discussed later. During the course of operation, the plant was periodically shut down for modifications and/or repairs. Following one of these shutdowns at about 1500 hr to modify pressure vessels, after restarting the system it was observed that acid had leaked into the feed piping. This resulted in the subsequent exposure of a number of lead elements in each subunit to low pH water. A few of the most severely damaged elements were replaced, but many of the elements that had undergone moderate acid hydrolysis were kept in service. As seen in Fig. 5, the rejection dropped from 93 to 91% shortly thereafter, presumably as a result of acid damage.

After 3600 hr the plant was shut down for 5 days as a result of influent valve control problems. No acid leaks in the feedwater piping were observed at this time, but a possibility exists such an incident did occur and was undetected. The plant was again shut down at about 3900 hr for 5 days to repair a leak in the acid day tank. After restarting, it was determined that there was an acid leak into the feedwater piping. Again, the rather sharp, sudden decrease in salt removal would indicate some acid damage. After 4200 hr the plant was inoperative for 22 days to repair the chlorine-injection system. During this period, there was another opportunity for an acid leak into the feedwater piping, although a later investigation indicated probably none took place. A review of Figs. 5 and 7 shows there was a steady decline in the rejection performance from 2300 to 4800 hr that apparently can be attributed to acid leakage and acid-accelerated membrane hydrolysis.

After the series of shutdowns during the 2300- to 4800-hr period, it was discovered that acid from the acid day tank leaked back through the high-pressure acid feed pumps and/or acid-dilution water pumps because of the static head of the acid level maintained in the day tank. The heavier sulfuric acid drained through the pumps into the high-pressure feed manifold, displacing some of the water. On restarting the plant, concentrated sulfuric acid was pumped into the lead first-half elements. This effect is confirmed by data given later in Table XI. It indicates all lead elements tested individually appeared to have been exposed to acid damage. To prevent a recurrence of this problem, new operating procedures were developed. Manually operated block valves on the discharge of the acid day tank were always closed when the RO facility was shut down because the check valves installed on the acid and acid-dilution water pumps were not adequate to prevent backflow through the pumps into the feed manifold. This procedure was developed after the 4800-hr period and results indicate it corrected the problem.

In summary, it was concluded that acid-accelerated hydrolysis, which occurred on at least two separate occasions, was significant in increasing salt passage in the Orange County RO system.

2. *Base-Accelerated Hydrolysis*

The 5-mgd RO system was cleaned 11 times; this may have resulted in some membrane deterioration caused by base-accelerated membrane hydrolysis. The cleaning solution used in all cleanings was designated as cleaning solution B; its composition is as shown in the accompanying tabulation.

Chemical	Amount
Sodium tripolyphosphate	250 lb
Versene–100	100 lb
Sodium carboxymethylcellulose	80 lb
Triton X–100	6 liters
Water	1500 gal.

This solution is very alkaline and is normally adjusted to a lower pH with sulfuric acid. The maximum recommended cleaning temperature is 120°F. After the second cleaning, formaldehyde (1%) was added for bacterial disinfection. The cleaning procedure at WF-21 involved pumping the pH-adjusted (8.5) cleaning solution for 10 minutes through the first pass. The same solution was then pumped through the second pass for 10 min. Checking the pH of the cleaning solution indicated that, because of dilution with the lower pH water in the elements, the pH had been reduced to 7.5. The cleaning solution was then reheated to 120°F while pumping through the third pass. Usually, this required 30–45 min before 120°F was reached. The cleaning solution was maintained at 120°F and pumped through the third pass, the second pass, and the first pass, each for a 1-hr period.

Following cleaning, the unit was flushed to waste until all cleaning solution was displaced. Initially, Fluid Systems recommended an adjustment of the cleaning solution to a pH of 8.5 in anticipation of better cleaning performance. Although the pH of the cleaning solution was diluted by water within the modules, it is possible some of the elements were exposed to a pH of greater than 7.5 and a temperature of up to 120°F for short periods of time during each of the cleaning cycles.

A review of Fig. 5 shows that cleaning was effective in flux restoration in the early stages of operation, but they were much less effective as operation continued. The graph indicates an increase in flux of approximately 5–10% following cleaning. The effects of cleaning are more graphically illustrated by Figs. 8–13, which show flux restoration cleaning for individual subunits, and Fig. 14, which plots the cleaning effect for a typical first-pass 8150HR element.

The cleanings had little effect on rejection restoration. Only slight improvements were observed after the early cleanings. This was later followed by declines in rejection following each cleaning cycle (see Figs. 5 and 7). Following cleaning number 11, in which only subunits 1A and 2A were cleaned, the results were sufficiently negative (no flow increase but increased salt passage) that the remaining subunits were not cleaned.

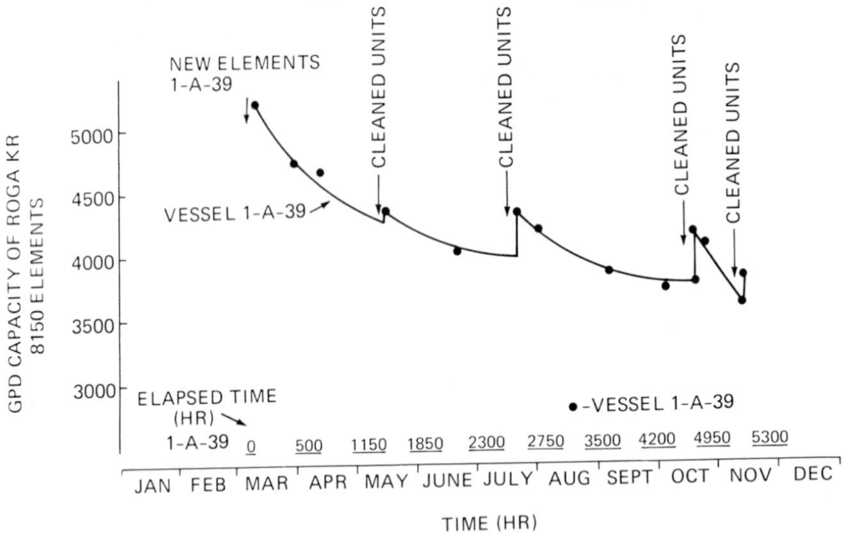

Fig. 14 The effects of cleaning on restoring flux of a typical first pass vessel; 5-mgd RO plant, Orange County, California.

It was concluded that base-accelerated hydrolysis during cleaning was possibly significant, but did not play a major role in the observed decline in the Orange County system. Although cleaning probably had little effect on hydrolysis, it probably caused a loss of colloid.

3. *Loss of Colloid*

The CA membrane used in the 8150HR spiral-wound elements was treated with a supplemental polymer following fabrication of the element. This polymer is known as Colloid 189 and its application to cellulose-acetate membranes was developed under OWRT contract. Colloid 189 is commonly used in the fabric industry to size fabrics, the membrane elements treated with the polymer are commonly referred to as having been "sized." The sizing material is applied to the completed membrane element in a water solution at ambient temperatures and 400 psig. The result is a decrease in salt transport (increase in rejection) and a decrease in product water flux. The beneficial effects can be maintained for extended periods of time under proper pH control. It is known, however, that the sizing material can be removed from the membrane by increasing the pH of the feedwater.

Following the seventh cleaning, Colloid 189 was applied to the elements *in situ*. It is seen in Fig. 5 that the rejection based on conductivity increased from 87 to 90%, not as great an increase as had been expected. Because effective field application of the sizing compound is dependent on a thoroughly efficient cleaning prior to application, it is hypothesized that the seventh cleaning was not as thorough as desired. Thus fouling on the surface of the membrane interfered with the application of the size. It can be seen in Fig. 7 that following sizing, the rejection increased from 87 to 89% for sodium and 88 to 91% for chloride. It can also be seen in Fig. 5 that rejection dropped from 90 to 88% immediately following the eighth cleaning. This suggests that most of the colloid retained by sizing was removed during the next cleaning.

At this time it is impossible to quantify the effectiveness of the cleaning technique or the contribution to membrane deterioration caused by the cleaning or loss of colloid. There is enough evidence to state a possibly significant amount of colloid loss (size) caused by the cleanings at WF-21. Additional work is planned with the objective of developing more effective and less-harmful cleaning solutions and techniques for use at WF-21.

B. MEMBRANE FOULING

1. *Carbon*

To meet groundwater recharge requirements, WF-21's AWT processes included GAC treatment. A number of studies using GAC indicate that adsorption of organics from wastewater was more cost effective in upflow columns, as they provide better contact time and more efficient utilization of the carbon bed depth. Upflow columns have been used successfully in other conventional AWT applications. Although the carbon columns were designed to minimize fluidization at a design rate of 5 gpm/ft^2, some fluidization of the carbon columns did occur. The interparticle movement within the carbon bed resulted in a continuous stream of less than 15-μm–sized carbon fines bleeding from the activated carbon columns that had previously gone undetected. With cartridge filters and RO membranes downstream, the presence of these fines was quickly detected and recognized as a potential fouling problem.

During the pilot study, an effort was made to minimize bleed-off of carbon fines by packing the columns tightly with GAC to restrict bed expansion. This, however, was unsuccessful and it was finally decided to operate the carbon system in downflow, thus eliminating fluidization of carbon particles within the carbon bed. A decision was made to convert each of the carbon columns to downflow to provide a higher quality of feedwater for the RO plant. Because the conversion would take 6–8 weeks, and because the construction of the 5-mgd RO plant was approaching completion, it was decided to run the start-up and acceptance test on the available upflow feedwater containing carbon fines. Based on the data obtained during the pilot-plant operation, it was believed that the amount of carbon fines that would enter the RO modules during this short initial operation period of about 120 hr would have little, if any, effect. Thus the start-up and acceptance test (May 9–14, 1977) was performed on water treated by GAC contactors operated in upflow mode. After this initial test period, which accounted for approximately 240 hr of actual plant operation, the RO facility was shut down and the carbon columns were converted to downflow operation.

It is difficult to determine whether the carbon fines were responsible for major changes in membrane performance. Although it was first believed the carbon fines that passed through the cartridge filters were a major foulant, later investigations indicated that the small amount of carbon particulates found on the surface of the membrane elements could be washed off with relative ease. Although there is some earlier evidence to indicate a performance loss because of carbon fines when comparing some original elements exposed to carbon with other elements installed after conversion of the carbon contactors to downflow, it is difficult to make an exact comparison. In fact, the new elements installed after converting to downflow were also exposed to carbon fines because of the cleaning procedures that followed. The carbon introduced into the entire system during the first 2400 hr of operation was subsequently introduced into all elements during the following cleaning cycles. During cleaning, water would be pumped through all elements at a given flow rate for approximately 1 hr, and the carbon fines washed from the original element were reintroduced into the elements installed after the column modifications. Thus it is impossible to compare the performance of original tubes with those that had new membranes installed after the carbon-column conversion.

Because carbon fines are discrete particles, if carbon fouling were significant, one would expect to measure an increase in pressure drop (Δp) across the system. Discrete particle fouling generally increases Δp as the brine-side water passages become fouled. Because no significant increase in Δp occurred and the carbon particles were not bound to the membrane surface, it is believed carbon fouling was minor.

In conclusion, it appears that the problem caused by carbon fines was eliminated by converting the columns to downflow. After this modification, very little carbon was picked up on the cartridge filters, and monitoring experiments showed virtually no carbon passing through. It was suggested that the carbon fines may have enhanced adsorption of organics onto the membrane surface. This possibility will continue to be investigated. It would appear, however, that the discrete carbon fines have little impact on the performance decline of the 5-mgd system. Probably the most significant impact of the carbon fine fouling was the obscuring of other possible causes of the performance decline during initial system operation. This resulted in overlooking other causes that, if detected earlier, could have allowed corrective action before irreversible damage to the membrane occurred. In any case, because the impact of carbon fines remained something of a question, it was recommended that only downflow carbon columns precede RO systems.

2. *Organic*

A surface coating of fouling material has been observed on some of the dissected elements that can seriously impede water permeation. Analysis of this coating, removed from elements that had been in service after approximately 3000 hr, indicated approximately 40% of the material was organic. The results of these early investigations suggest that part of this material could be complex aromatic carboxylic acids known as humic and fulvic acids. These have previously been observed as foulants in RO systems, particularly when associated with metals. If these organics are, in fact, fouling the membrane, high operating pressures of 500–600 psi such as have occurred at WF-21 may increase the severity of any organic problem. This would be consistent with the data, which indicate cleaning was more successful during the initial phases of operation and that later cleanings had diminishing results.

It is significant and contradictory that pilot tests conducted on lime-coagulated and clarified secondary effluent (the same feed as the 5-mgd plant, less mixed-media filtration and GAC treatment) operated successfully for 6 months at 400 psi without serious fouling problems. The unit was not cleaned during the entire 6-month test. It should be noted that the organic carbon concentration of the RO feedwater to the lime–RO pilot plant was 3 to 5 times greater than GAC treated water, which typically has a COD concentration of 10–15 mg/liter. These facts suggest that the higher organic concentrations in the RO feedwater are not necessarily significant, and raise some additional questions about the source of the organic fouling that may be taking place.

Another analysis of the fouling material from elements after 12,000 hr of operation has revealed a slightly different pattern than originally observed. Analysis of the surface coating from the removed elements reveals a distinct pattern of fouling. In these elements, organics contribute up to 90% of the material removed from the surface of the membrane from first-pass elements. The amounts of organic fouling decreased through the system until the last elements in the third pass showed only 50% organic material. The procedure used to analyze the surface coating removed from the dissected elements was: The elements were cut open and membrane leaves exposed. Then the surface of the membrane was scraped while wet with a flat metal spatula until 100 g of the material was collected. A mixture of the fouling material was placed in a 103°F oven for drying. To determine the percentage of organic material present, the dried sample was placed in a muffle furnace at 1700–1800°F and the loss of weight on ignition determined. It was found that the lead elements exhibited considerable weight loss on ignition, whereas downstream second- and third-pass elements showed a lesser weight loss, indicating a smaller percentage of organics in the foulant material. Table VII shows the percent volatile solids for five different elements removed from the first, second, and third passes, six elements in each pass. The modules are numbered individually, and the positions of the elements are designated by numbers 1–6, with the lead element being No. 1 and the last No. 6.

In summary, it is thought that organic fouling (humic and fulvic substances) of the 5-mgd system may be significant, but the performance of the lime–RO pilot system tends to be contradictory. This is an area for considerable work, and a full investigation is planned. An attempt will be made to determine the composition of this fouling material on the surface

TABLE VII

Percentage Volatile Solids In Membrane Deposits

Module# position	Volatile organic	Inorganic
V3487 First pass #6 element	89.5	10.5
Z0316 Second pass #1 element	89.5	10.5
Z0322 Second pass #6 element	63.4	36.6
V3586 Third pass #1 element	76.0	23.0
Y5392 Third pass #6 element	46.2	53.8

of the used membranes, and if it is found to be organic, cleaning procedures can be developed for its removal.

3. *Microbiological*

Another possible source of organic material that may be contributing to the performance decline of the RO system is microbiological growth on the membrane surface. An investigation of some of the used membranes from the 5-mgd system indicated that possibly significant biological growth was taking place on the membrane surface. The material removed from the membrane surface was slimy, with total plate counts of approximately 600,000/ml. Analysis of the RO product water also showed an average TPC of 100/ml and average total coliform concentration of more than 3/100 ml. In fact, if further study shows most of the organic fouling found on the membrane surface is a result of biological growth, it would offer one possible explanation for the difference in performance observed between the 5-mgd system and the lime–RO pilot plant.

The lime-clarified feedwater is low in bacterial count, due to disinfection resulting from a high pH (11.0). Therefore, few, if any, microorganisms are in the feedwater to the RO system. The GAC columns, however, have been observed to be extremely favorable media for microorganisms culturing, and the water discharged from these columns has a high plate count. The low level of chlorination before the 5-mgd system is inadequate to control these organisms. Thus it may be possible, because the lime treated water is essentially free of microorganisms, that the improved performance observed is a result of decreased biological activity. If so, this would explain why the lime–RO water, which contains considerably higher levels of dissolved organics as measured by COD or TOC than the activated-carbon–treated water, resulted in less system fouling. This may be one reason why the lime–RO system outperformed the 5-mgd system. More details of the lime–RO pilot-plant study are presented in the section on Continuing Studies.

Because no rapid decline in salt rejection had been observed, it is believed that any effect of biological activity on the membrane performance in the Orange County system had been limited to a decline in flux only. If any actual membrane degradation due to enzymatic activity was taking place, a rapid deterioration in salt rejection would be observed. Studies are planned, however, to identify any bacteria growing within the system. Pure cultures of these organisms will be established, then innoculated on fresh CA membrane. After varying periods of time, the membrane will be tested in a membrane test cell to measure changes, if any, in membrane performance.

In conclusion, it is felt that biological fouling is possibly playing a significant role in the decline of the system's performance. Studies are planned to substantiate the role of biological fouling, identify and quantify the organisms growing within the system, and explore alternative operations and disinfection procedures to minimize their effects. The need for better disinfection procedures to control biological growth within RO systems is an area in need of further research.

4. *Inorganic*

An analysis of membranes removed from the Orange County system revealed a substantial portion of the material on the membrane surface as inorganic. Vessels in the second and third passes appeared to be more heavily coated with inorganics than the first-pass vessels. While much of the material appeared to be some type of organic, as much as 50–60% of the material removed from some of the third-pass vessels consisted of inorganic precipitates. The material that was removed from five different vessels in the first, second, and third passes was dried, ashed at 1700°F to remove volatiles, and dissolved in acid to be analyzed for calcium, iron, chloride, sulfate, silica, aluminum, phosphorus, and other inorganics. Results of the analysis, shown in Table VIII, indicate that the inorganic fraction of the material removed from the surface of the membrane increased through the system as the feedwater became more concentrated. The material contained in the second-pass elements appeared to contain a larger percentage of iron (3–4% as Fe) and, on ashing, the residue had the characteristic reddish-brown color of ferric oxide. The second-pass vessels also contained considerable calcium and phosphorus. Dried ash from the third-pass elements had a characteristic white color and contained even more calcium than the second-pass elements, with 30–40% calcium oxide and 17% phosphorus. The data indicate that calcium phosphate precipitation occurs within the elements on concentration. This suggests that the adjustment to pH 5.5 varies and is perhaps higher at times than indicated, because calcium phosphate is normally soluble at lower pH. Another possibility is the precipitation of an organic calcium phosphate complex that is insoluble, even at pH 5.5. A review of feedwater quality indicates phosphorus concentrations average 0.4 mg/liter. This concentration appears to be too low to explain the presence of such a high percentage of phosphate in the sample. It is possible, therefore, that the precipitating phosphorus comes from the addition of the SHMP. Normally,

TABLE VIII

Inorganic Materials Found on Used Elements From WF-21 5-mgd RO Plant

	Module no. and position				
Constituent[a]	V3487 1st pass #6	Z0316 2nd pass #1	Z0322 2nd pass #6	V3586 3rd pass #1	Y5392 3rd pass #6
Volatile	89.5	89.5	63.4	76	46.2
Nonvolatile	10.5	10.5	36.6	24	53.8
CaO	18.8	20.0	27.0	37.4	27.0
Mg	0.1	0.6	0.2	0.3	0.2
K	0.7	0.7	0.2	0.3	0.2
Al	3.1	3.1	6.8	6.4	8.9
Fe	2.8	13.0	4.8	3.2	3.5
Cr	1.0	1.2	0.4	0.7	0.25
Cu	0.1	0.1	0.03	0.05	0.01
SO_4	4.0	4.0	0.2	0.5	0.06
Cl	16.0	8.7	3.4	5.6	3.3
PO_4 as P	6.0	7.0	17.0	16.0	17.0
NO_3	0.1	0.1	0.04	0.06	0.04
SiO_2	1.8	0.4	0.1	<0.1	<0.1
Insoluble (HCl)	37.0	27.0	3.5	7.0	1.3

[a] Percentage by weight.
[b] Dissolves in HCl.

SHMP would not hydrolize within the detention time available in the treatment system, but based on observed results, this possibility cannot be overlooked. The dried material removed from the membranes also contained 3–4% aluminum, apparently from the use of alum in the pretreatment system. Aluminum is amphoteric and is least soluble around pH 5.5.

Thus even though SHMP was used as a scale inhibitor and the pH of the RO system was adjusted to 5.5, it appears that inorganic precipitation of sparingly soluble salts, particularly in the second and third passes, was playing a major role in contributing to fouling and flux decline. In fact, the addition of SHMP may have created part of the problem. More detailed analyses to identify and quantify these inorganic precipitates will be made in other studies. Once the inorganic portion of the fouling is better characterized, an attempt will be made to develop more effective cleaning solutions or modify the present mode of operation to prevent or reduce the amount of inorganic precipitation that occurs within the system by adjusting pH or using different inhibitors.

C. Membrane Compaction

The largest flux decline was experienced within the first 1000 hr of operation. Subsequent cleanings were effective in partial restoration of flux, but each cleaning seemed to be less effective than the preceding ones. To evaluate the effect of high-pressure–high-temperature operation, the compaction slopes with the various phases of operation were determined for subunit 1A. In a clean system, on dilute sodium chloride feedwater, membrane compaction at a specific pressure was determined by flux measurement as a function of time. When these data are plotted on a log–log plot, results are usually linear, and the slope of the plot is expressed as the *compaction slope*. In the actual RO plant, the decrease in flux as a function of time results from membrane compaction and the following:

(1) Membrane intrusion into the product water layer of material.
(2) Deposition of suspended solids on the membrane.
(3) Deposition of dissolved solids precipitated on concentration in the process.

Thus throughout this report, the definition of flux decline or compaction includes all these factors.

The compaction slope for unit 1A was determined by linear regression. The results are shown in Table IX.

Cellulose acetate membrane compaction increases with increase of feedwater pressure and/or temperature. The data summarized in Table IX illustrate that during the first 1000 hr, compaction slope exceeded 0.10 because of the high–feed-pressure operation (521 psig). This extremely high rate of compaction occurred even though during the same period of

TABLE IX

Compaction Slopes-Subunit 1A

Data points (hr)			
From	To	Significant event	Compaction slope
70	190	Start-up acceptance test	−0.100
70	1,070	—	−0.133
70	1,798	Fans high to low	−0.129
70	3,070	Fans off	−0.113
70	11,980	End of period	−0.97

time, the feedwater temperature was 57°F. Thus it appears that an excessive rate of compaction can occur even when feedwater temperature is reduced if the system is operated at elevated pressures between 500 and 600 psig. The effect of the high-pressure operation on compaction is illustrated by comparing the results in Table VIII with those in Table X.

During June 1979, subunit 1A was loaded with new HR8150 CA membranes and new membrane elements. Originally unit 1A contained a 20, 10, 5 configuration and was designed for a 13.2 gfd/ft^2 flux rate. During June, unit 1A was modified and seven additional pressure vessels added to change it to a 24, 12, 6 array. This modification increased the amount of membrane within the unit by 20%, and this additional membrane capacity reduced the flux design to about 10 gfd/ft^2. The results of the initial operation of subunit 1A with the additional membrane elements, given in Table X show this unit was able to produce a greater capacity of 580 gpm at 350 psig. The results of the initial operation have been normalized to 77°F and 400 psig.

The compaction slope of the normalized product flow as a function of time as shown in Table IX for the modified is 0.004. This value was obtained at significantly higher temperatures than experienced by subunit 1A during the original stages of operation in the reporting period. Following modification, unit 1A was operated at 350 psig and feedwater temperatures averaged 85°F. The difference between the original compaction slope of −0.13 for the first 1000 hr and −0.004 for the first 1000 hr of the modified unit clearly demonstrates the difference made by the lower-pres-

TABLE X

Subunit 1A Performance with New Elements

Time (hr)	Feedwater		Product flow (gpm)	
	Pressure (psig)	Temperature (°F)	Actual	Normalized
1	400	85	650	585
9	350	85	600	620
180	350	82	590	636
441	350	84	590	618
465	350	84	590	618
1,083	350	85	560	578
1,158	350	86	555	565
1,294	350	85	565	584
1,374	350	85	560	578
1,375	350	86	582	593

sure operation. These results prove that the original flux design criterion was overly optimistic and show that the major cause of flux decline in the Orange County system was the high-pressure operation necessitated by the lack of adequate membrane surface area installed during initial construction.

D. PHYSICAL EFFECTS

Besides increasing membrane compaction, high-pressure operation led to other problems. Visual inspection of the elements on dissection showed heavy intrusion and embossing of the membrane. Intrusion is the phenomenon whereby the membrane and the fabric carrier material are forced into the woven ribs of the product-water carrier material. The intrusion into the melamine-impregnated ribs decreases the area of flow for the product water in the product-water carrier material. The decrease in flow area results in an increase of parasitic differential pressure loss that decreases the applied pressure to the membrane. This in turn decreases the product water flow in the element. The term embossing is used when the weave pattern of the fabric material is detected in the CA membrane. Both intrusion and embossing are usually a result of high-pressure operation.

The product-water carrier material is a polyester with a tricot weave, and is commonly referred to as tricot. To impart compressive strength to this material, the cloth is impregnated with a melamine-formaldehyde resin cured at 200°C. Properly impregnated tricot will have a degree of stiffness, can be torn relatively easily, will not take up dye in a boiling solution of Rhodamine B, and will not lose significant weight in boiling water. During the visual inspection of elements removed from subunits 1A and 1B, it was noticed that some of the tricot in the elements did not possess the expected stiffness, nor could it be easily torn. Subsequently, it was determined that the tricot did not take up dye from the boiling solution of Rhodamine B. There was, however, significant weight loss on boiling in water. The loss of weight results from removal of improperly cured melamine resin by the boiling water. It is possible to remove improperly cured resin in the product water under long-term operation at ambient temperatures. In any event, there is sufficient evidence to warrant further investigation into the seriousness of this problem.

Fluid Systems fabricated a small spiral-wound element with suspect tricot from WF-21 dissected elements, and new tricot to determine if there was a flux decline attributable to deterioration of the product water carrier material. When tested, there appeared to be no significant difference in the element performance between used tricot and new tricot. The procedure used for this test was: The 8150HR elements that were suffering low flux and the tricot water carrier that was suspected of causing this effect were removed. The tricot removed from the OCWD element was used to roll two 7500 elements, and new tricot was used to roll two additional 7500 elements. The small test elements were wet tested using brackish water at 420 psig. The following results were obtained:

New tricot	125.4 gpd	93.7% rejection
New tricot	155.8 gpd	94.5% rejection
Used OCWD tricot	155.8 gpd	94.8% rejection
Used OCWD tricot	120.3 gpd	96.2% rejection

From these results, it appears there is no significant difference resulting from the change in physical properties of the used tricot from the Orange County elements. This test was not designed to prove that intrusion did not occur; in fact, it demonstrates the opposite. The actual loss of flux that occurred at Orange County was undoubtedly partially because of the intrusion of the membrane down into the tricot fabric. This was determined by physical examination of the elements. Because the used tricot performed comparably with the new tricot in the short run, low-pressure (400 psi) test indicates no intrusion by the membrane, and the transport properties of the used tricot are the same as new when no intrusion occurs. It is recommended that additional tests be conducted at high pressure over extended time to better duplicate conditions that occurred in the 5-mgd plant. It is also believed, on the basis of physical examination, that embossing of the membrane was very severe. The embossing and the intrusion observed is undoubtedly a result of the high-pressure operation. With the recent modification and addition of more membrane capacity in the subunits contained in unit 1, which provides for lower pressure operation, the effects of both intrusion and embossing will be minimized in future operation.

IX. Summary and Conclusions

The effects of hydrolysis, compaction, and fouling were demonstrated when single elements from pressure vessels 1-C-53, 1-C-60, and 1-C-67 were tested individually in a loop tester. These elements were tested in a single-element pressure vessel at WF-21 on plant feedwater at 540 psig. Brine flow was maintained at 25 gpm and the feedwater temperature varied from 84 to 86°F. The results of the element tests are in Table XI.

TABLE XI

Individual Element Test

Element			
Position	Number	Product flow (gpm)	Rejection (%)
Pressure vessel 1-C-53 first array			
1	H5-51156	2.5	60.6
2	Y-1845	2.3	79.5
3	K-1502	2.4	81.5
4	H5-51150	2.2	92.2
5	Y-1847	2.1	94.4
6	V-3487	2.3	94.2
Pressure vessel 1-C-60 second array			
1	Z-0316	2.8	94.5
2	Z-0325	2.8	93.4
3	Z-0351	2.4	93.3
4	Z-0342	2.4	93.7
5	Z-0354	2.0	93.8
6	Z-0322	1.8	92.8
Pressure vessel 1-C-67 third array			
1	V-3586	2.0	95.7
2	Y-5500	1.8	95.2
3	Y-5483	1.8	95.6
4	Y-5395	1.2	91.5
5	Y-5404	<1.0	88.5
6	Y-5392	<1.0	86.9

The first three elements in the first array pressure vessel demonstrate a low rejection, as would be expected resulting from accidental hydrolysis. Assuming an active membrane area of 300 ft^2, the first element (H5-51156) has a flux of 12 gfd/ft^2. This is considerably less than would be expected from a severely hydrolyzed membrane, and it may be attributable to a severely compacted membrane and intrusion. The second array elements show a decreasing flow rate with a slight decrease in rejection along the flow path. This would indicate fouling in a process stream that is being concentrated. Similarly, the third array elements show a significant decrease in flux as well as rejection along the flow path. This demonstrates fouling by precipitates. Elements from subunit 1B were dye tested and dissected. Visual observations confirm that elements in the first array were probably hydrolyzed and the elements from the second and third array were fouled, with the third array being most fouled.

In summary, the major problem with operation of the RO plant at WF-21 was that the plant never produced the rated capacity of 5 mgd. A minor problem was the plant was slightly below the goal of 90% salt removal from the feedwater after the first 3000 hr of operation. Results indicate the reason for the failure to achieve the production rate was because of insufficient membrane area being initially installed in the plant, and the reduced capacity because of membrane fouling. The lack of sufficient membrane area necessitated high feedwater-pressure operation that in turn caused excessive membrane compaction to further lower the already marginal capacity. The high-pressure operation also aggravated membrane fouling and made the fouling material more tenacious and difficult to remove.

Fluid Systems began the necessary corrective action to remedy the problem of insufficient membrane area. As previously mentioned, this was done by increasing the number of pressure vessels in subunits 1A, 1B, and 1C from 35 to 42. It is gratifying to note that by doing so, subunit 1A was able to meet its production requirement of 580 gpm with a feedwater pressure of 350 psig at a temperature of 86°F and a resulting flux decline slope of 0.004 for the first 1000 hr of operation. This low-pressure operation, combined with the future development and implementation of improved cleaning techniques, should provide considerable improvement in operation of WF-21.

A review of Table X shows fouling is most serious in the third pass of the 4-2-1 array. The potential causes of membrane fouling have been discussed in detail and procedures developed to determine what can be done to mitigate the effect of fouling on performance. Particular attention will be paid to monitoring the performance of the individual pressure vessels in each pass of each subunit. Cleaning procedures will be initiated when the monitoring of individual pressure vessels indicates it is required. In other words, future operations will attempt to prevent the occurrence of any irreversible fouling. Improved cleaning procedures are being developed by Fluid Systems for implementation at WF-21.

Although not a serious problem at WF-21, the loss of rejection appears to have been largely the result of accidental acid injection into the first-pass vessels while the plant was inoperative. Corrective action was taken, as previously described, to remove this deficiency in the plant design. No recurrence of this problem is expected.

X. Reverse Osmosis Demineralization Costs

Table XII reflects the 1980 operating and capital costs for RO demineralization at WF-21. A review of all the AWT plant costs are noted in Argo (1979a). The OCWD had a maintenance contract with the company that

TABLE XII

Water Factory 21 Operating and Capital Costs Reverse Osmosis Demineralization

Cost		$/Mgal
Capital		208.87
Operating		610.77
Operations labor	37.57	
Electricity	390.80	
H_2SO_4	18.21	
Cl_2	0.77	
SHMP	26.51	
Maintenance contract	136.32	
Gas	0.59	
Total		819.64

supplied the demineralization system. Membrane replacement and cleaning costs are included in the table under maintenance contract.

A summary of the capital and operating costs for each unit process within the AWT plant at WF-21 is provided in Table XIII. Capital cost for AWT is $444/Mgal and AWT operating and maintenance cost is $649/Mgal. Reverse osmosis capital cost is $209/Mgal, and operating and maintenance cost is $610/Mgal. The capital cost of the deep-well water blending

TABLE XIII

Water Factory 21 Operating and Capital Costs Summary

	$/Mgal		
Process	Capital	O&M	Total
Inf. pump and lime clarification	47	104	151
Solids handling	73	205	278
Ammonia stripping	123	32	155
Recarbonation	16	61	77
Filtration	37	21	58
Activated carbon adsorption	93	56	149
Activated carbon regeneration	34	92	126
Chlorination	21	78	99
AWT subtotal	444	649	1093
RO demineralization	209	610	819
Deep well water	59	147	206
Injection	46	21	67
Blended Water Cost	427	692	1119

system is $59/Mgal and the operating cost $147/Mg. The cost of injecting the waters produced by the AWT, RO and deep-well facilities is $46/Mgal capital and $21/Mg operating and maintenance. When AWT, RO, and deep well costs are combined and injections costs added, the cost of the blended water injected is $427/Mgal capital and $629/Mgal operating and maintenance, for a total cost of $1,119/Mgal.

How does this blended water cost compare with the cost of other water supplies available in Southern California? The OCWD, in previous reports (Argo, 1977), compared the cost of wastewater reclamation with the cost of imported waters being furnished Southern California through the Colorado River Aqueduct and the SWP. The cost of the conventional water supplies was adjusted to reflect the actual cost of providing this water to the community. It was shown that when tax and other subsidies, such as power rates, were removed, the cost of imported water would exceed the cost of wastewater reclamation. Estimates have also been made for furnishing additional water supplies to Southern California through longer, more energy-intensive aqueducts from sources as far away as the Columbia River. Preliminary estimates for these water supplies range from 1,500 to $3,000/Mgal.

A. ENERGY CONSIDERATIONS

Much of the water used to meet the needs of Orange County has been provided historically by large-scale importation of water from distant sources, primarily the SWP, which brings Northern California water from the Sacramento Delta to Southern California; and the Colorado River Aqueduct. At the time of their inception, development of these projects was well-within reasonable dollar and energy economic limits. Today, however, we are acutely aware that as a nation, we are not completely energy independent and that energy requirements will play a key role in determining the course of future water-supply projects.

As is the case with any water supply, energy is required to deliver the water to its point of use. Studies made by the OCWD (Argo, 1978) have indicated that when wastewater reclamation projects are evaluated on an energy-cost basis, they show even greater potential savings than when compared strictly on a dollar and cents basis. Because local water supplies are inadequate to meet the needs of Orange County this will continue to experience a very high energy cost for water. Current estimates for delivering Northern California water through the SWP to the Southern California area indicate an energy cost of approximately 10,000kW hr/Mgal.

The energy cost of furnishing Colorado River water to Southern California is 6,400 kW hr/Mgal.

Table XIV gives the primary energy requirements for wastewater reclamation at WF-21. This table shows the actual amount of energy used during one fiscal year to produce reclaimed water, which was injected back into the groundwater supply. When looking at the total energy cost of producing this reclaimed water and comparing it to the energy cost of supplying Northern California water, one can see that, even when demineralization is included, the energy cost of reclamation is less than the energy cost associated with pumping Northern California water to southern California. The Orange and Los Angeles counties Water Reuse Study has shown that for reuse applications where water quality requirements are less stringent than those imposed for groundwater recharge and injection at WF-21, even greater energy savings can be realized. Therefore, it would appear that wastewater reclamation will play a key role in meeting the future needs of Southern California.

With energy playing an increasingly important role in the economics of water supply, it is apparent that in many applications the least energy-intensive alternative will be reuse. Therefore, it is imperative that research on removing substances of health concern, such as organics and viruses, be expedited to provide the data necessary to address the questions being raised by regulatory officials regarding wastewater reuse for domestic purposes.

TABLE XIV

Water Factory 21 Energy Requirements For Wastewater Reclamation

Process	kW hr/yr	kW hr/Mgal
Influent pumping	456,253	191.6
Lime clarification	195,537	82.1
Solids handling/lime recalcination[a]	4,824,974	2,026.0
Air stripping	1,022,030	429.2
Recarbonation	462,074	194.0
Filtration	70,436	29.5
Activated carbon adsorption	1,293,406	543.2
Activated carbon regeneration[a]	1,111,634	466.8
Chlorination	232,816	227.0
AWT subtotal	9,669,160	4,060.0
RO demineralization	11,567,707	8,532.0
Deep wells	946,980	741.0
Injection	981,197	268.0
Blended water—total	23,165,044	6,331.0

[a] Includes energy cost for natural gas, assuming 1 kW hr = 10,500 BTU.

B. Continuing Studies

A major effort has continued on the evaluation of the effect of various pretreatment options on the performance of a spiral-wound RO system (Argo, 1980, in press). If some of the unit operations employed in the WF-21 AWT plant could be eliminated without adverse impact on the RO unit, substantial savings could result. These studies were undertaken using a pilot-plant scale unit and in no way interfere with the operation of the 5-mgd RO system.

The pilot plant consists of a 10,000 gpd (permeate rate) spiral-wound RO system including the basic pretreatment steps normally incorporated with the RO unit. The pilot plant was supplied by the manufacturer of the 5-mgd system. The system, as delivered, included pH control, addition of a threshold inhibitor, dual media filtration, cartridge filtration and the RO system with all the necessary instrumentation for monitoring the operation. Although small in scale, the pilot plant is essentially capable of reproducing, with considerable precision, what may be expected of a large scale operation. For all practical purposes, the RO portion of an RO demineralization system is subject to linear scale-up.

1. *Reverse Osmosis System*

The RO system consists of six stainless steel pressure vessels, each 10 ft long, rigidly attached to a steel frame and fed by a high-pressure centrifugal pump. The pump has a rated capacity of 13 gpm at 1000 psi.

The system employs two 4-in., two 3-in., and two 2-in., diameter vessels. Each of these contains three spiral-wound RO membrane elements. The six vessels are arranged in series so the concentrate from one becomes the feed for the next. The vessels are arranged in order of decreasing number (i.e., 4 in. to 3 in. to 2 in.) to maintain the necessary flow and pressure conditions to obtain an 85% recovery operation on such a small system.

The permeate outlet from each vessel is connected to a three-way valve so the permeate water from each vessel can be collected in a common header or sampled individually. Rotameters and sampling ports are arranged so individual or composite flows and conductivities can be measured conveniently. This arrangement of vessels, valving and instrumentation allows for the evaluation of cumulative system performance at 6 independent recovery levels, ranging from 22 to 85% by progressively measuring product quality and flow from vessels No. 1–6. The operating pressure and flow of the feed is controlled by adjusting the high-pressure feed valve and brine-discharge valve.

Other controls and instrumentation part of the RO system include: a pressure gauge for monitoring pump discharge and RO feed pressure; feed, concentrate, and permeate total-flow rotameters; conductivity monitors; and a low suction-pressure shut-off switch to protect the main feed pump. All high-pressure piping is stainless steel; all low-pressure piping is PVC.

Basically, two phases were completed. Phase one used CA spiral-wound elements of the same type used in the 5-mgd plant. Phase two utilized thin-film composite (TFC) polyamide membrane spiral-wound RO elements. A third phase under way uses a low-pressure (200–250 psi) TFC membrane element. All of the tests were run on the effluent from the AWT lime clarification process. Shortly after the first phase began, the dual media filter was eliminated. The lime clarifier effluent is pH controlled in two steps from a pH of 11 or higher to the operating range pH of 5.5 to 5.8. Single-step control proved to be unstable. The test demonstrated the feasibility of eliminating the ammonia stripping towers, recarbonation, mixed-media filtration, and carbon adsorption as process steps when RO is employed as one of the unit operations. Modifications to the AWT plant are planned that will make it possible to run a test on a full-plant scale.

Table XV shows typical results obtained with the CA membrane elements. The flux-decline slope values indicate membrane fouling during the test was not a problem. The following conclusions were drawn.

TABLE XV

Pilot Plant Operation
Typical Salt Rejections
CA Membrane

Constituent	Feed (mg/liter)	Permeate (mg/liter)	Percentage rejection
Calcium	68	nil	—
Magnesium	8	nil	—
Sodium	177	11	93.8
NH_4	10	1.1	89.0
HCO_3	1	6.1	—
SO_4	227	1.6	99.3
Cl	210	14	93.3
NO_3	20	5.8	71.0
SiO_2	13	1.5	88.5
TDS	749	42	94.4
COD	23	2	91.3

C. CONCLUSIONS

This pilot-plant program is considered a success. Several points have been demonstrated and several conclusions can be drawn:

(1) A spiral-wound RO system utilizing CA membranes can be operated satisfactorily on lime-clarified effluent.

(2) With the clarifier operating normally (i.e., without bulking of the sludge, etc.) the RO system could be operated without granular media filtration ahead of it. Nevertheless, it would be generally advisable to provide preconditioning and granular media filtration.

(3) Unless the lime-clarified effluent has a reasonable buffering capacity, significant care must be taken to prevent high pH excursions. It was concluded that a two-stage acid adjustment was needed to control pH.

(4) When a CA RO system is operated at 85% recovery on lime-clarified effluent, salt rejections and COD reductions in excess of 90% can be anticipated; and NH_3 nitrogen can be reduced from 5 to 10 ppm down to the indetectable level.

Table XVI shows typical results of the test run using the TFC membrane elements. In almost all respects, the performance was superior to that of the CA elements. An important factor in the operation of the TFC elements was the ability to clean this element with solutions having a pH as low as 1 or as high as 12. This permits the use of more vigorous cleaning techniques.

TABLE XVI

Pilot Plant Operation
Typical Salt Rejection
TFC Membrane

Constituent	Feed (mg/liter)	Permeate (mg/liter)	Percentage rejection
Sodium	159	1.4	99.1
NH_4	0.5	<0.1	>80.0
SO_4	320	1.0	99.7
Cl	186	7.5	96.0
NO_3	11	0.3	97.3
TDS	885	11.0	98.8
COD	20	<1.0	>95.0

The TFC elements demonstrated the capability of high nitrate rejection and improved organic rejection. Laboratory studies have already shown that TFC membrane is capable of rejecting a wider spectrum of organic substances than CA membrane can.

The following conclusions were drawn:

(1) Spiral-wound polyamide-type membranes operating on lime-clarified effluent can achieve salt rejections of 97% and COD rejections of 95% when operated at an 86% recovery rate. The NH_3 nitrogen was reduced from 2 mg/liter to the indetectable range.

(2) The polyamide-type membrane is successful in removing 97% of the nitrates from the feedwater stream. This is in sharp contrast to the cellulosic-type membranes where nitrate removals of 50% are common.

(3) Reasonable care must be taken to keep the pH within the recommended operating range of 5.5 to 5.8. In particular, operating at too high a pH may cause irreversible calcium carbonate scaling, necessitating membrane replacement.

(4) Membrane cleaning with 1% EDTA, 1% borax, and 1% trisodium phosphate solution restored the membranes to their original start-up flow rates. This indicates that flux decline was a result of membrane fouling and not compaction.

(5) Additional study is needed to determine the effect of repeated cleanings on the membranes. The cleaning described in conclusion (4) needs to be repeated for several cycles to determine its effect on the life of the membranes and degree to which membranes can be restored to their original flux rates.

In addition to these pilot studies, several other pilot units similar in design to those described were built. Spiral-wound RO elements containing composite membranes supplied by other manufacturers were tested at normal (~400 psig) and low operating pressures (200–300 psig). The low-pressure tests were particularly encouraging because a major reduction in power costs of about 30–40% could be anticipated. Membrane fouling continued to be a major operating problem. Although the fouling can be alleviated by regular and frequent cleaning, the results are erratic. A number of additional cleaning solutions recommended by the manufacturers of the elements were tried with varying success, but work on the causes of the membrane fouling, prevention of membrane fouling and development of effective cleaning solutions continues.

Studies using pilot plants were also conducted on electrodialysis-reversal (EDR) for demineralizing the same feed used in the pilot studies noted previously and on ultrafiltration (UF) as pretreatment for RO (Argo, 1980, 1981).

The attractions of EDR are expected to be lower operating costs, particularly in energy and chemical consumption; greater stability in flux and rejection; and, possibly, higher water recovery. The EDR studies did show good process stability with little decline in flux or change in performance. Additional work remains to be done. The high removals of organic matter attained with the RO equipment is not reached with EDR.

Several UF studies have been made as pretreatment for RO. In all cases a rapid flux decline is experienced and cannot always be counteracted. With some of the equipment used it was possible to maintain constant performance by frequent cleaning procedures. With the equipment now available it does not appear that UF will provide economic pretreatment for RO systems operating on treated wastewaters.

A number of interesting developments occurred in the operation and maintenance of the 5-mgd RO plant that used CA membrane elements. The work by the manufacturer in adding seven additional pressure tubes to each of the six sections of the RO system was completed. It was also discovered that the high-pressure pumps were operating at excessively high pressures with an increased power demand. By trimming the impellers in the multistage high-pressure RO pumps and limiting the maximum pressure, a substantial reduction in the RO energy requirements was experienced. In a published summary of energy and costs (McCarty, 1982), the energy requirement for RO was shown as 7570 kW hr/Mgal produced, as compared to the earlier value of 8532 kW hr/Mgal. The energy consumption was also affected by the increased number of elements now in service and some work done with low-pressure elements.

During the summer of 1981, a severe incident of biological fouling occurred that plugged the cartridge prefilters, requiring frequent replacement. It had been observed that a great increase in the number of microorganisms occurred in the GAC columns. The GAC has frequently been observed to provide a favorable media for bacterial proliferation. In fact, this problem was anticipated and a bypass constructed permitting the RO feed to avoid the GAC. Under the maintenance contract, bypassing the GAC required approval of the system manufacturer. The GAC was finally bypassed, reducing the problem substantially and probably resulting in a reduction of biological membrane fouling.

The membrane elements in several of the sections were replaced with CA elements using varying operating pressure and flux characteristics. The membranes were CA variants and included CA blend membranes with a higher acetyl content. The operating results reproduced in the referenced reports (Argo, 1980, 1981), show excellent promise of reducing energy costs while producing a water suitable for blending to inject into the groundwater basin.

References

Argo, D. G. (1977). "The Cost of Water Reclamation by Advanced Wastewater Treatment." Presented at the Water Pollution Control Federation Conference, Anaheim, California.

Argo, D. G. (1978). *Desalination,* **25,** 135–149.

Argo, D. G. (1979a). "Evaluation of Membrane Processes and their Role in Wastewater Reclamation." Prepared for Office of Water Research and Technology (U.S. Dept. of Interior) by the Orange County Water District.

Argo, D. G. (1979b). "Water Factory 21—Design and Operating Data." Office of Water Research and Technology (U.S. Dept. of Interior) (Contract No. NTIS PB-300 602).

Argo, D. G. (1980). "Evaluation of Membrane Processes and their Role in Wastewater Reclamation," Vol. 2. Prepared for the Office of Water Research and Technology (U.S. Dept. of Interior) (Contract No. 14-34-0001-8520).

Argo, D. G. (1981). "Evaluation of Membrane Processes and their Role in Wastewater Reclamation," Vol. 3. Prepared for the Office of Water Research and Technology (U.S. Dept. of Interior) (Contract No. 14-34-0001-8520).

Argo, D. G., and I. Nusbaum (1976). "Water Factory 21—An Alternative Approach." 4th Annual Conference, National Water Supply Improvement Association, July 11–15.

McCarty, P. L. *et al.* (1980). "Advanced Treatment for Wastewater Reclamation at Water Factory 21" (Technical Report No. 236). Department of Civil Engineering, Stanford University (Grant No. EPA-S-803873).

McCarty, P. L. *et al.* (1982). "Advanced Treatment for Wastewater Reclamation at Water Factory 21" (Technical Report No. 267). Department of Civil Engineering, Stanford University (Grant No. EPA-CS-806736-01-3).

Trussell, R. R. *et al.* (1979). "Water Factor 21—Virus Study." Office of Water Research and Technology (U.S. Dept. of Interior) (Contract 14-34-0001-8520).

11

Design and Operation of Desalting Systems Based on Membrane Processes

ALFRED N. ROGERS*
Bechtel Corporation
San Francisco, California

* Engineering consultant, Pleasanton, California.

ISBN 0-12-085480-5

Introduction

This chapter covers membrane plants of various types, the pretreatment of feed streams to the plants, and the posttreatment of the purified water.

I. The Membrane Plant—Reverse Osmosis

In reverse osmosis (RO) plants, pressure is applied to a feed stream to force a portion of its water content through a membrane. The membrane is comparatively impermeable to ions, dissolved materials of large molecular size, and colloids. As a result, the greater fraction of the rejected species remains behind and is discharged from the plant in the concentrate stream. Reverse osmosis is used for providing purified water from a saline feed stream, generally brackish water or seawater. In addition, it is finding increasing use in the concentration of industrial and sanitary wastes to minimize the volume requiring ultimate disposal while simultaneously providing a permeate stream suitable for use in agriculture or as process water in industry.

Reverse osmosis membranes are available in the form of hollow-fine fibers (HFF), spiral-wound units, tubes with the membrane either on the outside or inside of the supports, plate-and-frame, or leaf-type construction. Figure 1 is a schematic diagram showing the construction of the spiral-wound module.

In 1982, the spiral-wound and HFF plants dominated the water treatment market. Units of either type are enclosed in metal or reinforced plastic pressure vessels. A bank of pressure vessels is visible in the foreground of the power-plant water purification system in Fig. 2. The large

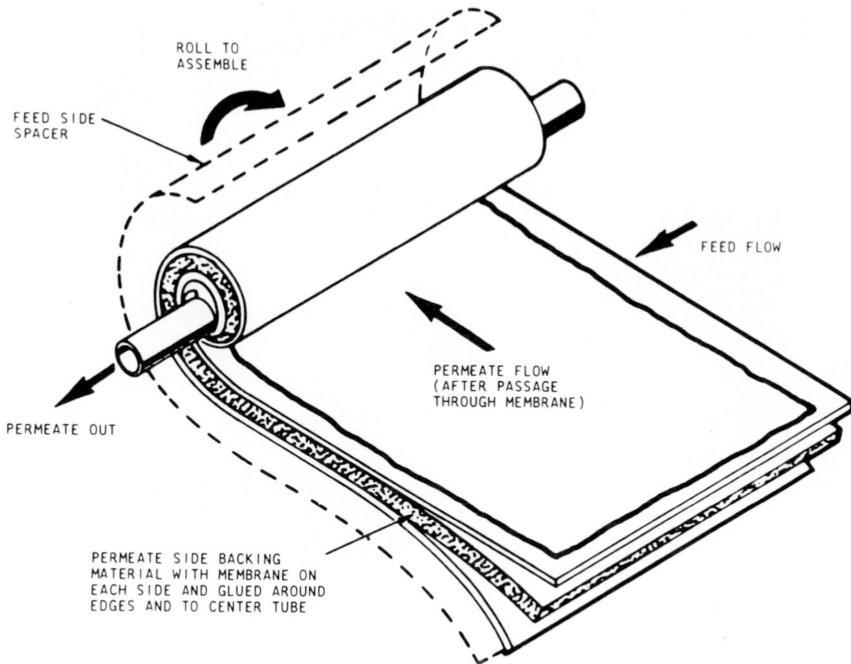

Fig. 1 Schematic diagram showing construction of a spiral-wound membrane module. (Photo courtesy of Fluid Systems Division of Universal Oil Products.)

vessel in the background contains the ion-exchange resin used for final "polishing" of the RO permeate to produce the high-purity water required for makeup to the high-pressure boilers (see Section V.D).

The tubular RO pilot plant, shown in Fig. 3, recovers purified water from a secondary sewage effluent at Firebaugh, California. The feed enters the interior of the RO tubes and the purified water that has passed through the membrane flows through each outer plastic pipe to a collection manifold at the end of each rack.

A. PURCHASING SPECIFICATIONS

The user of RO almost never designs or constructs the plant. Instead, it is bought as a complete package. The issuance of adequate specifications is mandatory. At the minimum, the supplier will expect a complete ionic analysis of the feed stream and, in addition, its pH, temperature

Fig. 2 A 216,000 gpd hollow-fine fiber RO plant at the Dayton Power and Light Company. (Photo courtesy of Illinois Water Treatment Company and DuPont.)

range, and suspended solids content. Also, the supplier will wish to know the required water recovery and the purity of the product. The supplier will comment on the pretreatment scheme and recommend changes that are considered essential for the protection of the membranes. Among other things, the supplier will recommend the acid and metaphosphate (or other additive) dosage required to prevent scaling at the target water recovery.

The plant supplier will guarantee membrane life. The guarantee will specify the production of permeate and the purity of the product during the membrane life, usually 3 yr. Under one type of guarantee, failure of the plant to meet the specified performance within the guarantee period results in the replacement of the defective membranes, generally on a prorata basis. If, under this type of guarantee, failure occurs after 2 yr of a 3-yr guarantee life period, the purchaser pays only two-thirds the cost of the replacement membranes. One supplier guarantees to replace defective membranes at no cost during the first year of plant operation, then a prorata replacement for the remainder of the warranty period.

All RO membranes suffer a flux decline during use. Some membranes suffer a smaller flux decline on standby even under the most favorable conditions of storage. It is possible to maintain the permeate production

Fig. 3 Tubular RO plant for water recovery from aqueous municipal waste. (Courtesy of the University of California.)

rate by increasing feed pressure. It is more common, however, to install additional modules during the later years of the plant's life if capacity drops below design. The user should include in the original purchase sufficient racks and connections to permit the installation of additional modules.

There are some details that the buyer can and should specify. In a competitive market, it is common to find suppliers who furnish high-pressure pumps that are minimally sufficient. The buyer would do well to include in the specifications the requirement for high-grade pumps. In the smaller sizes, a number of satisfactory pumps are available. For large plants, boiler feed pumps are highly desirable. The pump materials (casing, impellers, wear rings, and seals) must be compatible with the acidified feed to the RO plant. The purchaser should explore this carefully with *pump* manufacturers before writing the RO plant specifications.

A small but important component of the plant is the acid-metering system. Its failure will shut down the entire plant. Assuming that sulfuric acid is used, its point of entry into the system exposes nozzles and pipes to *dilute* acid, which is far more destructive to metals than is concentrated sulfuric acid. All components that may contact dilute acid during operation *and on shutdown* should be fabricated from Alloy 20.

The purchaser must assure that the space allotted to the RO plant provides ready accessibility. In particular, it will be necessary to remove defective modules from time to time. Adequate clearance for this operation must be provided.

The specifications should require readily accessible sampling connections to permit identifying defective modules during plant operation.

The manufacturer should be required to provide a complete system for the flushing and cleaning of membranes while in the pressure vessels. This system may not use the RO high-pressure pumps, which are totally unsuited for this purpose. Instead, a separate cleaning pump should be provided.

The RO plant may be erected outdoors provided the membrane modules are protected against temperature extremes, particularly against freezing and sunlight that is sufficiently intense to overheat the modules. Plastic piping installed outdoors must be resistant to sunlight. The control panels are preferably installed in a control room where they will be protected against weather, dust, and insects.

When a high on-stream factor is required, the purchaser should stock adequate spares. Most important are a complete set of impellers, wear rings, bearings, and seals for the high-pressure pump. If the feed pump is of the reciprocating type, a set of spare plungers, valves, valve seats, and springs should be stocked. In addition, spare metering pumps are highly desirable.

With an adequate set of purchase specifications and reasonable care in operating and maintaining the RO plant, many years of satisfactory and economical operation may be anticipated. The basics of RO design set forth in the following pages will serve as a valuable guide to the purchaser and user of RO equipment.

B. PROCESS DETAILS

1. *Water Recovery*

Reverse osmosis is generally directed toward the recovery of purified water from sea or brackish water or the reduction in volume of aqueous waste streams. In either case, it is desirable to remove the maximum possible fraction of water from the feed. The recovery (expressed as a percent of the feed) is limited by several considerations, most importantly the osmotic pressure of the reject, which is roughly proportional to its content of total dissolved solids (TDS). Because

$$\text{Effective pressure} = \text{Applied pressure} - \text{Osmotic pressure},$$

a high recovery, and hence a high TDS in the plant reject, decreases the effective pressure and thus the productivity (flux) of the plant.

The degree of concentration is also limited by the approach to saturation of scale formers such as silica or calcium sulfate in the reject brine. Except for these constraints, there is theoretically no limit to the fraction of water that can be recovered from the waste.

In practice, the size and complexity of the plant will increase with the fraction of water recovered. All suppliers of RO plants specify a minimum permissible reject flow rate per module to provide adequate turbulence within the module and avoid stagnant areas with high concentration of scale formers. To attain this minimum flow rate, high water recovery can be achieved only by staging, as shown in Fig. 4. The plant shown in Fig. 4 achieves a recovery of 87.5% of the water from a waste stream totaling 1.0 million gallons per day (mgd) and containing 2000 ppm of total dissolved solids. Each successive stage contains a smaller number of RO modules than the preceding stage, thus assuring a comparatively constant liquid velocity in the modules. Staging has the further advantage that the danger of scale formation is limited principally to the last stage, where it will affect fewer modules.

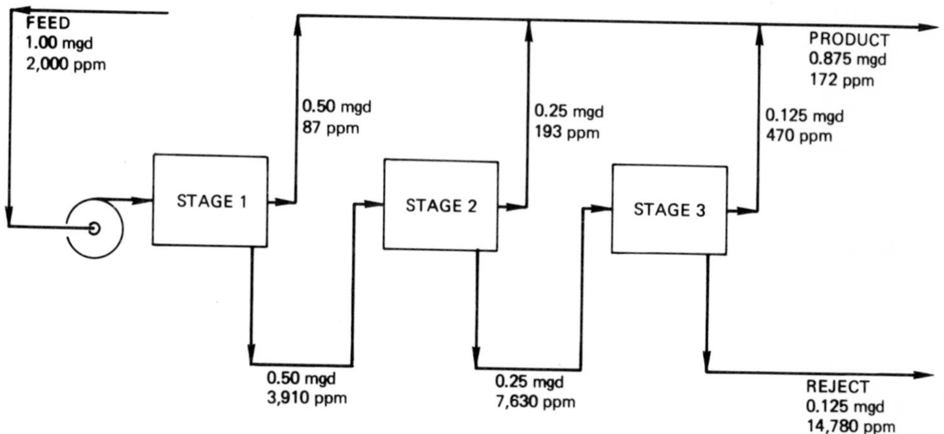

Fig. 4 High water recovery achieved by a three-stage RO plant.

2. Salt Rejection

The design of RO processes is influenced by salt-rejection requirements as well as by water recovery. At a feed pressure of 400 psi, membranes in common commercial use reject roughly 93% of the dissolved salts of the feed at 70–75% water recovery. The degree of rejection (or conversely, salt passage) is dependent on the particular ion under consideration. If the proposed use of the recovered water permits a higher salt content than the membrane plant delivers, the water can be blended with untreated (or partially pretreated) feed. In the example in Fig. 5, the output of usable water from the RO plant has been increased by 24% by blending the RO product stream with a portion of the incoming feedwater to achieve a final concentration of 500 ppm of TDS. From the standpoint of the disposal of plant wastes, the volume of the reject from the 2.336 mgd of incoming feed water has been reduced 14% by blending.

A reduction in product purity can, on the other hand, be achieved by lowering the pressure of feed to the RO plant (see Section B.4 below). This is attractive for two reasons:

(1) A lower feed pressure results in a reduction in pumping power,
(2) the rate of decline in water flux through the membrane is much less at a reduced feed pressure.

The disadvantage of low-pressure operation is the resultant decrease in plant throughput, that is, the treatment of a given flow of feed will require an increase in plant investment.

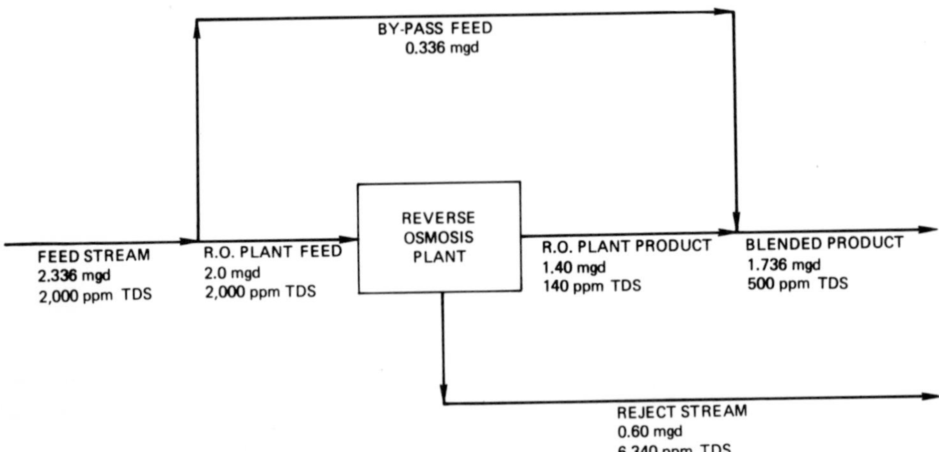

Fig. 5 Blending the RO permeate with a portion of the feed.

If, on the contrary, a higher purity is required, it can be attained by staging the product water as shown in Fig. 6. In the example shown, the purified water from stage 1 is further purified in stage 2, which recovers 80% of the water in its feed. The reject from stage 2 is purer than the initial waste stream and thus can be blended into the feed to stage 1. As a result of this reject recycle, the overall recovery of water from the waste stream reaches 65%. The 25 ppm purity of the reclaimed water is comparable with that of an excellent grade of distilled water. The two-stage RO treatment has the further advantage of greatly diminishing the probability of bacteria passage into the product, which is consequently acceptable for pharmaceutical use.

For some purposes such as boiler feed water or water for use in the manufacture of electronic equipment, the product of a two-stage RO plant can be further polished by passage through a mixed-bed ion exchanger. With a feed containing only 25 ppm of TDS infrequent regeneration of the ion-exchange system is required and little regenerant chemicals. In general, however, it is more economical to install a single-stage RO system followed by an ion-exchange unit.

3. *Effect of Feed Temperatures*

An important operating variable is the temperature of the feed to the RO plant. An approximate rule of thumb states that a 1°C (1.8°F) rise in temperature results in a 3% increase in water flux. Thus it may be advisable to heat a cold feed stream. The fuel consumption can be held to a

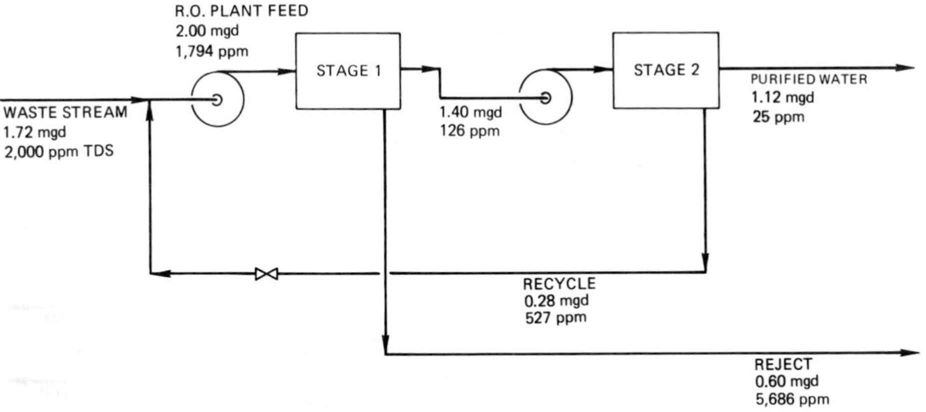

Fig. 6 Staging the RO permeate to increase product purity.

minimum by cooling the purified water product regeneratively against the feed (see Fig. 7). At high recoveries, it may not be profitable to attempt to recover the additional heat in the *reject* stream. The advantages of high-temperature operation are restricted by the 94–104°F upper limit imposed by the membrane manufacturers. This restriction results from the increase in hydrolysis experienced by cellulose acetate (CA) at elevated temperatures and, in the case of polyamide membranes, the deterioration in membrane properties when the glass transition temperature is exceeded. In addition, the potting compound currently used as a tube sheet for hollow-fine fiber modules is subject to compression set at higher temperatures. It is likely that this temperature limit may be increased as newer membranes and improved materials of construction reach the market. In 1977, a composite membrane was developed and is reported to withstand temperatures as high as 180°F (Channabasappa, 1977, private communication).

Sometimes the feed to the membrane plant may be a warm waste stream from an industrial process. If, however, heat must be supplied from an outside source, the process economics must be carefully analyzed before investing in the heat exchangers shown in Fig. 7 to ascertain that the improvement in RO throughput at higher temperatures justifies the added equipment costs.

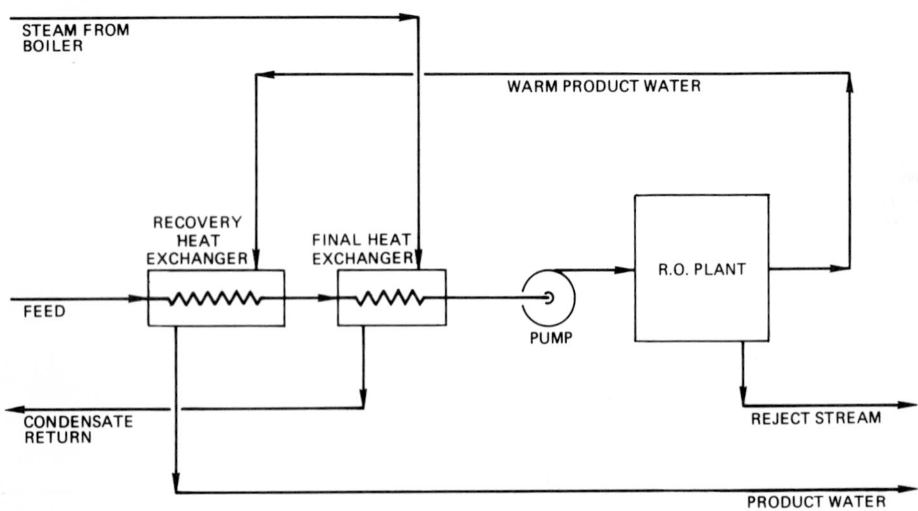

Fig. 7 Heating the RO feed to increase membrane flux.

4. *Effect of Operating Pressure*

Another important variable in the operation of an RO plant is the feed pressure. The passage of water through the membrane is opposed by the osmotic pressure of the *reject* stream. The osmotic pressure is roughly 0.01 times the dissolved salt content of the reject stream, expressed as parts per million. For example, when the reject contains 10,000 ppm of TDS, its osmotic pressure will be approximately 100 psi. Applied pressure in excess of 100 psi will cause some product to flow through the membrane. The higher the pressure, the greater the flux. Because salt passage is independent of pressure, the increased flux attained at higher pressures will result in a purer product. Conversely, a low operating pressure entails a lower consumption of pumping power. In addition, low-pressure operation will increase membrane life because membrane compaction is thereby decreased. Reverse osmosis plants for brackish waters and many industrial wastes generally operate at about 400 psi. A membrane that appeared on the market in 1977 is claimed to yield a high flux at feed pressures as low as 250 psi with a brackish water feed. Where a lower product purity is acceptable and, consequently, a reduction in feed pressure is permitted, the purchaser of the plant should so specify. The plant supplier will then furnish a "loose" membrane yielding a high flux but giving reduced permeate purity.

When the feed is of high salinity, the osmotic pressure of the reject becomes substantial. For example, if the RO plant recovers 40% of the water content from a normal seawater feed, the rejected brine will exert an osmotic pressure of roughly 600 psi. At the membrane surface, polarization effects will result in an even higher concentration and, consequently, a higher osmotic pressure. Thus for seawater RO, the applied pressure must be 800–1000 psi.

C. CLEANING OF THE MEMBRANES

With adequate pretreatment, trouble-free operation of an RO plant may be anticipated over long periods. Occasional membrane cleaning may be required. A failure in the pretreatment train, for example, may result in alkaline scaling. This may be removed by flushing the system with an acid solution. Some membrane manufacturers recommend the use of a 2 wt% solution of citric acid adjusted to pH 4 with ammonia. Others advise flushing with water acidified to pH 2.5 with hydrochloric acid. For a badly scaled membrane, rinsing with strong acid solutions (10 vol% HCl) has been used. If the membrane becomes coated with slime, cleaning is performed by circulating a detergent solution through the modules. Whatever the problem, it is essential that the plant operator follow the membrane supplier's cleaning procedures to avoid irreversible membrane damage.

D. PLANT MAINTENANCE

The RO modules themselves require practically no maintenance except for the occasional replacement of gaskets or O-rings. Serious maloperation, however, may damage membranes irreparably. Experience indicates that membranes can be expected to perform well for at least five years if the feed is pretreated adequately, and some manufacturers now offer a pro-rata guarantee for five years. Several suppliers furnish a polymer emulsion that, when pumped through an RO module with excessive salt passage, improves the purity of the RO permeate, but with some reduction in water flux. The acid system, when constructed of suitable materials, will give little trouble except possibly for replacement of the diffuser and acid-mixing section of the feed line every few years.

The principal source of maintenance problems is the high-pressure pump. Although a pump may operate successfully for four or five years without attention, cases have been reported in which the impellers of high-pressure centrifugal pumps required replacement every year. Reciprocating pumps need new valves, valve seats, and springs at periodic intervals.

II. The Membrane Plant—Electrodialysis

Electrodialysis (ED) removes dissolved matter from water in contrast to RO, which removes the water and leaves the solutes behind. An applied voltage drives the ions through ion-selective membranes, as described in Section III.B, leaving purified water behind. As a result, ED does not remove silica, suspended matter, many organic compounds, or any material that is not ionized. A second consequence is that the electrolyzing power consumption is directly proportional to the ion content of the feed. This contrasts with RO, in which the separative work is generally a small fraction of the power consumed and thus is less strongly influenced by feed concentration. Consequently, ED operating costs are favorable for low feed concentrations and become less so as concentration increases.

One exception to this is the use of ED for the concentration of seawater, prior to crystallization, in the manufacture of salt. A purified water product is *not* delivered; instead, the diluted seawater is rejected to the sea. At present, this process is used only in Japan.

The selective removal of ionized material is advantageous in the purification of some industrial products from accompanying salts. For example, ED has been used successfully for many years to remove dissolved salts from cheese whey.

A typical ED installation is shown in Fig. 8. The cartridge filters (see Section IV.H) are housed in the cylindrical vessel on the extreme left. Following the control panel in the center foreground are six ED stacks arranged in two parallel trains, three stacks in series in each train. A plant of this size can deliver from 100,000 to 264,000 gpd of purified water, the exact quantity depending on feed concentration, temperature, and required water purity.

Fig. 8 A six-stack ED plant. (Courtesy Ionics, Inc.)

A. PURCHASING SPECIFICATIONS

As in the case of RO plants, the purchaser of an ED system almost always buys a complete package from the supplier. The package includes the recirculating pumps and control panel (frequently mounted on a skid), membrane stack(s), and components of the clean-in-place system. The control system generally comprises

(1) a level controller in the treated water storage tank to start and stop the ED unit;

(2) pressure sensors for the feed water and concentrate streams to shut down the entire system in response to a high- or low-pressure signal and, if desired, actuate an alarm; and

(3) a conductivity monitor to divert off-specification treated water to drains.

Suppliers of ED equipment normally send a field engineer to the purchaser's plant to supervise the erection and start-up of the unit. The cost of this service is included in the purchase price of the plant.

The bid invitation will include the source and ionic analysis of the feed stream, its pH, temperature range, and concentration of suspended solids. Product purity and water recovery must be specified. The plant supplier should be consulted as to the adequacy of the proposed feed pretreatment system.

In the ED system, the feed and recirculation pump are not as troublesome as in RO because of the lower pressures and higher feed pH in ED plants. Standard water pumps are frequently supplied. The purchaser would do well, however, to discuss pump materials with the ED plant supplier before placing the order for the plant. Special consideration should be given to the brine-pump construction material, which pressurizes a highly concentrated stream. In small ED plants, this brine pump is sometimes used for cleaning in place and must also be resistant to cleaning chemicals. If acid is to be injected into the feed to adjust its pH, all components of the injection system that may contact dilute acid should be constructed of Alloy 20.

The specifications should require the provision of readily accessible connections for injecting cleaning chemicals into the ED stack(s) and for draining the spent solutions from the system.

The purchaser must bear in mind that stack disassembly may occasionally be necessary. Sufficient space must be provided.

The plant may be erected outdoors if precautions are taken against temperature extremes. In such cases, plastic piping must be protected against exposure to sunlight. When large plants are located outdoors, it is preferable to locate the instrumentation in a control room.

In the construction of the ED stacks themselves, each supplier has his or her own design. The purchaser has no control of such details. Acquaintance with the basis for the various systems, as explained in this chapter, will be of value in selecting a supplier.

B. PLANT DETAILS

Figure 9 presents a schematic section of an ED stack. In the figure, the feed stream is distributed to each of the flow channels. As the direct current flows between the electrodes, the cations (positive ions) flow through the cation-selective membranes (C) but are stopped by the anion-

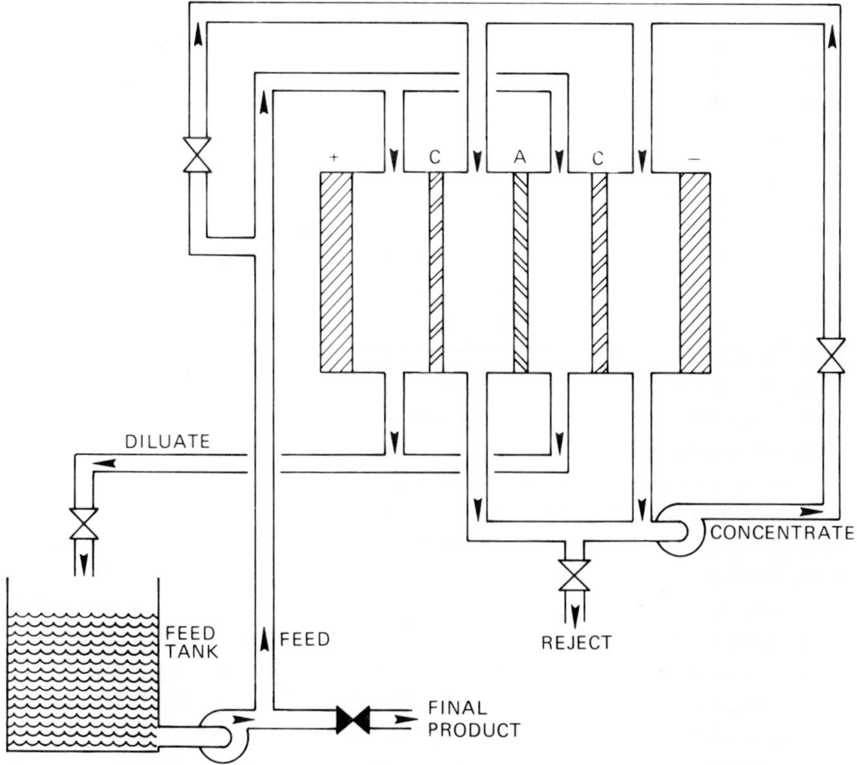

Fig. 9 Schematic diagram of an ED stage (batchwise operation).

selective membranes (A). For the anions (negative ions) of the feed stream, the converse is true. As a result, the dissolved salts accumulate in every other flow channel to form the concentrate. The channels that have been depleted in salt concentration are said to carry the diluate.

The rate of transfer of ions is proportional to the current in conformance with Faraday's law. As the current flow increases, the applied voltage rises almost linearly. The current transfers not only ions of dissolved salts, but also the hydrogen and hydroxyl ions contributed by the water. As the applied voltage increases, a point is reached at which the H^+ and OH^- ions of the water begin to enter into undesirable reactions at the membrane surfaces, among them the deposition of scale. The point at which the voltage rises steeply with a small increase in current is called the polarization point. To avoid the undesirable side reactions and the local overheating of membranes that occurs at high current densities, it is a desirable engineering practice to restrict stack current to 70 to 80% of the polarization value.

Each cell pair in the stack consists of a cation membrane, a spacer, an anion membrane, and a second spacer. Spacers are of two types: Sheet flow and tortuous path. In the former, the feed enters at one edge, flows across the membrane as a sheet of liquid, and leaves the stack from the opposite edge. The spacer is provided with integrally molded turbulence promoters. Despite the turbulence promoters, the flow is nonuniform across the membrane. The tortuous path spacer, by contrast, confines the flow to S-shaped channels with molded-in-place turbulators thus assuring a definite flow path. The resulting improvement in flow pattern decreases the possibility of bypassing some regions of the membrane surface and also serves to sweep salt-depleted brine away from the membranes, thereby decreasing stack resistance. The disadvantage of the tortuous-path spacer is the increase in pressure drop and the decrease in utilization of the membrane surface because some of the area is obstructed by the channel boundaries.

Between membranes and spacers, gaskets are required to confine and direct fluid flow. Some ED suppliers cast the gaskets integrally with the spacers, other supply separate gaskets. The former are easier to handle in disassembling and reassembling the stack. When gaskets are not an integral part of the spacers, poor gasket alignment will result in excessive liquid leakage from the stack. For this reason, at least one manufacturer of stacks having separate gaskets insists that stacks requiring disassembly be returned to the factory. This may involve costly shipping charges.

It is common practice to operate an ED plant with partial blowdown of concentrate, recycling a portion of the concentrate stream in order to decrease the volume of reject brine and improve the overall conductivity of the cell. The resulting decrease in voltage drop reduces power consumption. The total salinity of the recycle stream may be allowed to rise to as much as 0.4 *N*. There are several reasons it seldom exceeds 0.2 *N*.

(1) The rate of back diffusion of salts from the concentrate into the diluate increases as the concentration of the concentrate stream increases.
(2) At high brine concentrations, the solubility of scale-forming compounds such as $CaSO_4$ and $CaCO_3$ may be exceeded.
(3) Internal short circuits of the current by highly concentrated brine will begin to introduce operational difficulties.
(4) Osmotic pressure drives an appreciable percentage of purified water back into the brine at high brine concentrations (Valcour, 1977).

The flow rates of concentrates and diluate are generally maintained approximately equal to insure comparable pressures in both compartments and to minimize cross leakage and membrane distortion. Because the diluate usually is not purified sufficiently in one pass through an ED stage, either a batch system or a continuous system with a number of stages may be employed.

In Fig. 9, the diluate is returned to the feed tank and recirculated to the ED stack. Operation is continued until the contents of the feed tank attain the required purity. This is known as batch-type operation. It may be operated by manual control, thus requiring almost no instrumentation. If desired, it may be automated by modulating the voltage across the stack as the salinity changes. In addition, the change in conductivity of the liquid may actuate a switch to start the pump, transferring the product to the storage tank when the required purity is attained; immediately thereafter, a fresh batch is admitted and the cycle restarted. The batch system requires a minimum of membrane area. Its disadvantages are that

(1) the supply tank must be large enough to hold a complete batch of feed,
(2) the prolonged holdup and exposure of the feed encourage bacterial growth, and
(3) current efficiencies are low.

The batch system is normally used when the salinity is greater than 7000 ppm.

The simplest continuous flow sheet contains a single stage of ED. For greater purity of the product (diluate), a multistage plant is used.

It should be noted that the purpose of staging is not the extraction of greater water fractions from the feed and, thereby, reducing the volume of the plant reject. Instead, each stage upgrades the purity of the diluate from the preceding stage. A single ED stage may be expected to remove 45–55% of the dissolved salts in the feed, that is, its purified effluent will have a salt content of 55 to 45% of the original feed salinity. Figure 10 presents as an example the production of water containing 410 ppm of dissolved solids from a 4500 ppm feed by means of a three-stage ED plant. One feed pump is shown in the feed stream and one in the concentrate recycle. The 50 psi pressure applied by the pump is sufficient to supply as many as 6 stages in series without intermediate feed pumps, but with some decrease in feed rate. Excessive feed pressure must be avoided to prevent excessive leakage. The flow rate in any stage must be sufficient to create adequate turbulence in the flow channels of the stack to keep the concentration polarization below the scaling limit.

To conserve floor space, several stages may be incorporated in a single stack with intermediate additional electrodes and suitable internal manifolding. The cell pairs between any pair of electrodes constitute one electrical stage. Within a single electrical stage, manifolding may be provided to induce the streams to flow back and forth between the electrodes several times. Each transit of the streams is referred to as a "hydraulic stage." Providing several hydraulic stages in each electrical stage decreases the number of electrodes required. It is apparent that as the feed proceeds from one hydraulic stage to the next in any electrical stage, the salinity of the feed decreases. To maintain uniform current efficiency, the number of cell pairs is decreased in each succeeding hydraulic stage in a given electrical stage. This arrangement is known as the "taper" of the stack.

Earlier in this chapter, it was noted that the splitting of water and the passage of hydroxyl ions at higher current densities may deposit alkaline scale on the anion-selective membrane. This condition is controlled by adding acid to the feed. The problem is particularly severe in the cathode compartment. The tendency for the latter to become alkaline is counter-

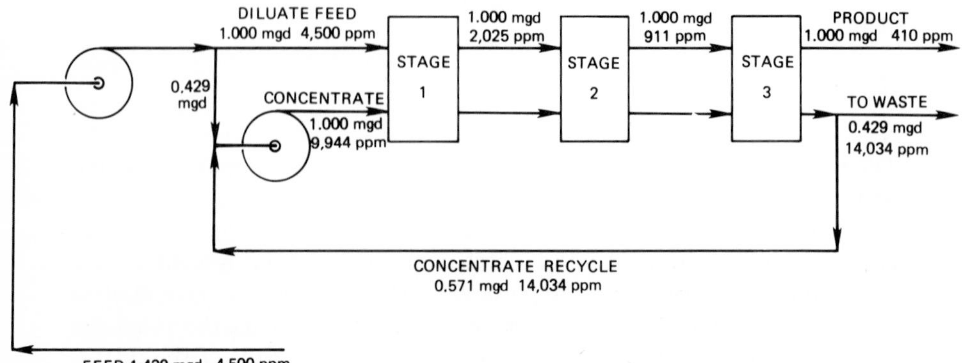

Fig. 10 Flowsheet of a three-stage ED plant.

acted by circulating through it the acidic effluent from the anode compartment along with some additional acid. A number of ED manufacturers combat these problems by periodically reversing the applied voltage while simultaneously rerouting the diluate and concentrate flows. Several manufacturers recommend reversal once a day, others every 15 min on an automatic cycle. As a result, electrode compartment scaling is avoided and the acid requirement of the feed is reduced or eliminated completely. Polarity reversal has the further advantage of purging the membrane surfaces of any deposited fouling materials. The only apparent disadvantage of this system is the loss of productivity while sweeping the fluid from the cells following each current reversal. This loss is reported to amount to about 15% of the plant output for plants on a 15-min reversal cycle. The loss of off-specification water produced during the voltage reversal can be minimized by routing it to the recirculating brine stream or by returning it to the feedwater tank.

One of the most important operating variables is the feed temperature. As operating temperature increases, the resistance of aqueous solutions decrease, resulting in a reduction in power consumption. A further benefit resulting from operation at elevated temperatures is the improvement in salt transport through the membrane. Each 1°F (0.55°C) increase in temperature improves the "cut" (salt removal per stage) by approximately 1% (Katz, 1971). One of the commercially available ED plants is claimed to operate successfully up to 110°F on a continuous basis and up to 120°F for short period of time.

C. Cleaning of the Membranes

Cleaning of an ED plant is an infrequent procedure provided the pretreatment of the feed is adequate, 5- or 10- μm filters are used ahead of the stack, and conservative operating conditions are maintained. Generally, a decrease in purity of the product water indicates the need for cleaning. For in-place cleaning, a 5 vol% solution of commercial hydrochloric acid is circulated through the stack for 30 to 60 min to remove scale and inorganic slime. This is followed by a water rinse. Organic contaminants are removed by a subsequent flushing with a 5 wt% solution of NaCl adjusted to pH 13 with Na OH. The operation is simple if flushing connections are properly located on the stack. Cleaning in place may be required at intervals of 1 to 4 weeks, depending on the quality of the feedwater. The waste should be stored in a tank or pond and adjusted to a neutral pH prior to disposal.

If the membranes have become severely fouled, it may be necessary to resort to the more time consuming method of disassembling the stack and cleaning the membranes individually. Manufacturers recommend sponging, brushing, or even scouring the membrane surfaces with steel wool. Highly skilled labor is not required for these operations. The frequency of stack disassembly and manual cleaning will depend on feed quality and operating conditions, varying from a monthly to a semiannual operation.

D. PLANT MAINTENANCE

As noted previously, little maintenance is required by the feed and circulating pumps if suitable construction materials were used. If it is necessary to inject additives into the feed stream, experience indicates the likelihood of replacing valves, seats, springs, and diaphragms on the metering pumps at least every 6 months.

Approximately 10% of the membranes will require yearly replacement. The replacement rate may be less with a well-water feed stream, somewhat greater with surface waters or aqueous water streams. The membranes most likely to fail are those nearest the electrodes, where more severe temperature and chemical environments are encountered. When the stack is disassembled, it may be necessary to replace spacers and gaskets (when separate gaskets are used) if they have distorted in service or have been damaged as a result of stack disassembly.

A word of caution here: If pressure is maintained on the assembled stack by tie bolts rather than a hydraulic ram, the manufacturer's directions must be carefully followed as to the torque and the sequence of bolt tightening during reassembly of the stack. Failure to do so may result in excessive liquid leakage from the assembled stack or irreparable damage to the stack components.

III. The Membrane Plant—Ultrafiltration

Ultrafiltration (UF) is used to concentrate suspended or colloidal matter in an aqeous stream and deliver water essentially free from such contaminants. It is also used to remove dissolved macromolecules. It has found a market in

(1) the concentration of pharmaceuticals,
(2) the fractionation and recovery of marketable products from cheese whey,
(3) the clarification of fruit juices,
(4) the recovery of dyes and particles of water-based paint from rinse streams,
(5) the concentration of the polyvinyl alcohol sizing material in textile rinse water,
(6) the concentration of highly stable oil emulsions from waste and rinse waters from metal cutting and rolling operations,
(7) as a prefilter ahead of a RO plant, and
(8) as a final filter for deionized water intended for use in electronic component manufacture.

For all of these purposes, the aqueous feed stream is forced under pressure through a membrane with pore diameter selected to remove particles in the desired size range. As an example, one commercial membrane with pores ranging from 15 to 50 Å in diameter has a cut-off at molecular weight (MW) 16,000, above which rejection is 95–100% (Schell, 1977, private communication). This same membrane has zero rejection when the MW is as low as 5000. Other membranes of 480-Å pore size reject solutes of MW over 300,000 (Michaels, 1968). The reader must not infer, however, that UF can be generally used for molecular fractionation, separating 5000-MW molecules from those of 300,000. The interaction between the large and small molecules will prevent successful fractionation and result in the retention of *both* sizes (Eykamp, 1976).

A. PURCHASING SPECIFICATIONS

To insure satisfactory performance of a UF plant, the purchaser must give the supplier a complete description of the anticipated feed, in particular:

(1) its chemical composition and pH,
(2) the concentration and particle size distribution of all suspended solids and colloidal material in the feed,
(3) the particle size of suspended and colloidal matter that must be removed and the desired completeness of removal,
(4) feed temperature,
(5) anticipated feed flow rate and plant turndown requirement,
(6) percent water recovery desired, and
(7) maximum pressure drop permissible.

Before the supplier will guarantee the equipment, he or she will usually perform a pilot-plant test on the actual feed, unless the particular application has been successfully demonstrated in the past.

The supplier will generally recommend that adequate connections and valving be provided to permit the flushing of clogged or slimed modules. The slight added expense involved is justified by the increased ease of membrane cleaning. As an added recommendation, the equipment should be easily disassembled for manual cleaning when on-line flushing is ineffective.

For UF, as for other membrane plants, sufficient space must be allocated in the plant for disassembly. Outdoor installation is permissible if the membrane equipment is not subjected to freezing or overheating. Instrumentation, however, is best located in a control room.

The purchaser has a choice between hollow fibers, flat-sheet membranes wound in spiral modules, tubular units, flat-sheet units in plate-and-frame configuration, and flat-sheet units of the leaf-type construction. The hollow-fiber and spiral-wound units have the advantage of large area per unit volume. In contrast, tubular and flat-sheet modules have a comparatively small area for permeation per unit module volume. The open configuration ultrafilters, however, can be used for feeds containing fibrous materials that cannot be processed readily in small channel ultrafilters. Of the open configuration units, the tubular type has the further advantage of permitting cleaning by brushing or by forcing sponge balls through the tubes, a procedure that may be necessary when treating highly contaminated liquids such as secondary sewage effluent.

B. Plant Details

1. *General Considerations*

The feed to a UF plant is normally the effluent from another process. A storage tank is recommended to serve as buffer ahead of the UF plant to assure a more uniform feed stream. Some pH adjustment may be required. Because the ion content of the feed is not altered by passage through the membrane, as will be discussed, there is no need for a softening step or additive injection ahead of the membrane modules. Also, prefiltration (except for the screening of large objects) is not required because UF is, in effect, a filtration process. Some pretreatment steps, however, are frequently beneficial. For example, if a feed is undergoing flocculation, the throughput of the ultrafilter can be greatly improved by allowing adequate time for the floc to form and separating it by decantation, centrifuging, or filtration prior to UF. The precipitation of any material from the feed while passing through the ultrafilter is to be avoided. As another example, the oxidation of ferrous iron by the atmosphere to insoluble ferric compounds in the UF unit is highly detrimental. Pretreatment is particularly important for feeds to hollow-fiber ultrafilters, which are more likely to be clogged than are those of more open configuration.

2. *Limitations on Flux*

Flux is defined as the rate of flow of liquid *through* the membrane per unit membrane area. The flux of pure water through a membrane is roughly proportional to the applied pressure. For feed streams other than pure water, the permeate, which is the liquid flowing through the membrane, leaves behind its charge of suspended matter or macromolecules. The suspended matter is swept away by the liquid flowing through the system, the feed stream flowing *across* the membrane.

At low applied pressures, generally under one atmosphere, the flow through the membrane is proportional to pressure even with suspended solids or macromolecules in the feed stream. At higher pressures, however, the proportionality no longer exists. The buildup rate of particles or macromolecules reaches a point at which they form a gel layer, a thixotropic or solid layer that imposes an obstacle to further flow through the membrane (Fenton-May and Hill, 1971). At this point, the flux decreases to a small fraction of the value predicted from laboratory tests with pure water at the same pressure. Flux then becomes dependent only on the nature of the feed stream, the extent that the membrane is fouled, and the Reynolds number (Re) to the nth power, as predicted by the Sherwood correlation (Sherwood, 1937, p. 39). The Reynolds number is that of the stream flowing *past* (not through) the membrane. For high values of Re, n will vary from 0.8 for true solutions to 2.0 for colloids. For low values of Re such as those encountered in hollow fibers, n will be as low as 0.5.

Because the gel layer consists of the material that does not pass through the membrane, it is not surprising that the limiting flux is almost inversely proportional to the concentration of suspended solids or macromolecules in the feed. At increasing feed stream velocities there is a tendency to sweep away the particles concentrating at the membrane surface. The speed that the material, rejected at the membrane surface, can be brought back into the bulk stream is controlling. Thus the limiting flux increases almost linearly with feed flow rate (Fenton-May and Hill,

1971). A higher feed flow rate can be attained by recirculating part of the feed stream. The resulting improvement in membrane performance will increase plant output, but at the price of an increase in pumping power.

If the flow of feed past the membrane is decreased, the limiting flux through the membrane, that is, the flux where the gel layer is formed, returns to its previous value, i.e., the formation and destruction of the gel layer are reversible phenomena. This is in contrast to fouling, which is not reversible. Once a fouling layer is formed, flux can no longer be returned to its original value by merely decreasing the feed flow rate past the membrane surface (see the subsequent discussion of cleaning).

3. *Effect of Temperature*

One other variable is under the control of the plant operator, namely, the temperature of the liquid. A higher temperature is beneficial because it lowers the viscosity of water and, consequently, increases its permeation rate and its Reynolds number, as discussed in the previous section. In a test on the UF of cheese whey, an increase in temperature from 15 to 45°C tripled the limiting flux (Fenton-May and Hill, 1971; Schell, 1977, private communication). Therefore, if the fluid being processed can withstand an elevated temperature, a large increase in productivity can be attained by heating the fluid to the highest temperature that can be tolerated by the UF system. One manufacturer claims its membranes can function continuously at temperatures up to 99°C (Young, 1977, private communication). Unfortunately, it is not possible to take advantage of higher temperature operation in the case of feeds containing solids that are coagulated or denatured on heating.

C. Cleaning of the Membranes

Any drop-off in water flux when the plant is operated under design conditions is to be regarded as a possible sign of fouling, and cleaning should be initiated promptly. Membrane fouling may be the result of many causes, among them bacteria or fungal slimes, iron or manganese hydroxide, calcium phosphate, fats, or biological degradation products. The selection of the cleaning method is dependent on both the membrane material and the nature of the foulant. The simplest technique is a water rinse or a backflush with ultrafiltered water, either from an on-stream UF plant or from a storage tank. Backflushing can be used only if the membrane is an integral part of its support structure and, therefore, will not be lifted off the support by the reverse water flow. If the membrane is not of the polyamide type, 50–100 ppm of sodium hypochlorite are added to the flush water to control bacterial or fungal growths. A convenient rule of thumb for the addition of chlorine to flush water is $(Cl)T < 300$, where (CL) is the equivalent chlorine content in parts per million and T the rinse time in minutes per day. For example, a rinse stream containing 10 ppm of chlorine can be applied daily for 30 min. If, by contrast, rinsing occurs every 5 days, the free chlorine content may be increased to 50 ppm and still have the same effect on membrane longevity. As an effective and widely used alternative cleaning method, detergent and enzyme mixtures are circulated through UF systems, particularly those fouled by food products such as cheese whey, which deposit difficult-to-remove fats, calcium phosphate, and organic decomposition products. For example, a solution containing small amounts of a calcium sequestrant has been found to gradually restore the flux to almost its initial value (Eykamp, 1976). Under some circumstances it is necessary to use a solution containing a detergent, a builder, and a proteolytic enzyme (Eykamp, 1977, private communication). Proteins, lipids, lipoproteins, and some organic decomposition products are very difficult to remove. Before initiating any cleaning process, the plant operator should check with the UF system supplier to make certain the membrane can withstand the treatment.

Tubular UF units, which are used for slime-forming feeds such as secondary sewage effluent, are sometimes cleaned by brushing or by sending sponge balls through the tubes under hydraulic pressure. A system is available for circulating sponge balls through the tubes continually while the UF system is operating.*

D. PLANT MAINTENANCE

Because pumps for UF systems operate at comparatively low pressures, maintenance is not a serious problem provided that pump materials are compatable with the process fluids.

At modest UF operating pressures, membrane compaction is slow. If, however, the membrane is operated in a plasticizing environment, that is, at increased pressures and elevated temperatures or in the presence of certain solvents, compaction may become appreciable. But in general any flux reduction will be almost entirely attributable to fouling.

Should the cleaning methods described in the previous section fail to

* The sponge balls and circulating system are available in the United States from Amertap, the American affiliate of Taprogge in Hamburg, Germany.

restore fouled membranes to a satisfactory flux level, the faulty membranes must be replaced. If an adequate supply of membrane modules, gaskets, O rings, and spacers is maintained, membrane replacement should be a simple matter and result in little plant down-time.

If acid is injected into the feed to adjust its pH, the usual problems with an acid injection system may be anticipated, i.e., failure of the valves, seats, springs, and diaphragms of the acid-metering pump. Replacement of the defective components is generally a simple matter.

IV. Pretreatment

The success of any membrane process depends on adequate feed pretreatment. For a thorough discussion of water treatment, the reader is referred to a number of books on the subject (Betz, 1976; Nordell, 1961; Powell, 1954). In this chapter, only sufficient details are included to provide a clear understanding of the problems involved and methods available for coping with them.

The deleterious constituents in various feedwaters may be ionic, colloidal, suspended, or bacterial. Pretreatment steps available for dealing with each of these are discussed here. Some of these treatment procedures will remove more than one type of impurity from the feed.

A. SOFTENING BY ION EXCHANGE

For all feedwaters, whether of natural or industrial origin, the presence of hardness may require a softening step, as discussed under the individual membrane sections of this chapter. Several commercial techniques are available, one of which is ion exchange.

As the water passes through a bed of ion-exchange resin, the alkaline earth ions, calcium, magnesium, barium, and strontium that constitute hardness are exchanged for ions on the resin:

$$Na_2R + Ca^{2+} \longrightarrow CaR + 2Na^+ \tag{1}$$

where R symbolizes the resin molecule. A similar equation applies to magnesium, barium, or strontium. After the bed capacity is depleted, the spent resin is backwashed to loosen the bed and carry away any suspended impurities that may have been deposited. Then it is regenerated by contact with a comparatively concentrated regenerant solution:

$$CaR + 2NaCl \longrightarrow Na_2R + CaCl_2 \tag{2}$$

The replacement ion, shown here as Na^+, may be of some other alkali metal, ammonium ion, or hydrogen ion, the choice generally dictated by economics. Where a low pH is desirable in the treated water, hydrochloric or sulfuric acid may be used as regenerant. With any regenerant, an excess of about 10% over the dosage theoretically calculated must be used to attain an acceptable degree of resin regeneration. A novel approach involves the use of sodium sulfate as regenerant, leading to the reaction

$$CaR + Na_2SO_4 \longrightarrow Na_2R + CaSO_4 \tag{3}$$

The low solubility of the calcium sulfate produced results in its precipitation from solution and sharply reduces the calcium ion content of the spent regenerant solution, thereby driving the reaction to the right. As a result, regeneration of the resin is more complete at any specified excess of regenerant ion. Sodium sulfate has the merit of low cost and ready availability. Where a substantial fraction of the hardness exists in the form of magnesium, this technique has little merit because the magnesium sulfate formed is extremely soluble and remains in the regenerant stream, thereby diminishing the extraction of magnesium ion from the resin.

A feed stream to the membrane system high in alkali ions (such as sodium or potassium) may be softened by the preceeding ion-exchange process with a minimal requirement for added chemicals if the membrane process recovers a high fraction of the water from the feed. In this case, the reject stream from the membrane plant serves as regenerant as shown schematically in Fig. 11. If water recovery by the membrane plant is approximately 67% or greater, that is, if the total dissolved salts are concentrated to more than three times their original value, it may be possible to operate without purchasing any regenerant chemicals.

With the exception of this method, all ion-exchange processes have the disadvantage of increasing the dissolved solids requiring ultimate disposal. Each equivalent of hardness ion removed from the water entails adding to the system approximately 1.1 equivalents of a replacement cation and 1.1 equivalents of associated anion. The increased solids load adds to the cost of disposal whether it be by ponding, deep-well disposal, or transportation to the ocean or other environmentally acceptable disposal site. These disposal costs are discussed in detail in Chapter 13.

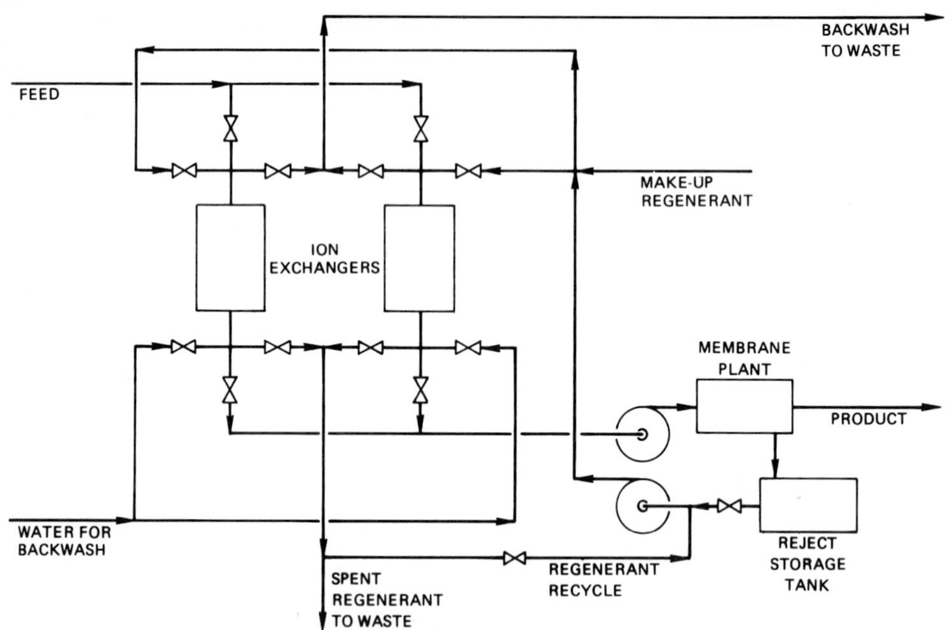

Fig. 11 Ion exchangers regenerated by the reject stream from the membrane plant.

B. LIME AND LIME–SODA SOFTENING

Perhaps the oldest and most common method for the softening of water is lime treatment, with the addition of soda ash or caustic soda if required by the water chemistry. The reactions with lime are

$$Ca(OH)_2 + Ca(HCO_3)_2 \longrightarrow 2CaCO_3 + 2H_2O \tag{4}$$

$$Ca(OH)_2 + Mg^{2+} \longrightarrow Mg(OH)_2 + Ca^{2+} \tag{5}$$

Equation (4) indicates that each pound of calcium ion stoichiometrically requires 1.85 lb of hydrated lime for precipitation. Equation (5) states that each pound of Mg ion requires 3.05 lb of hydrated lime. Equation (4), however, assumes the concentration of HCO_3^- generally expressed as bicarbonate alkalinity is adequate in the feed to complete the reaction to the degree of softening required. The Ca precipitated in Eq. (4) includes the ion originally present in the feed and the Ca added as lime reagent for Eq. (5). Any carbonate deficit requires its addition in the form of soda ash or carbon dioxide gas. The soda ash reaction,

$$Na_2CO_3 + Ca^{2+} \longrightarrow CaCO_3 + 2Na^+ \tag{6}$$

requires 2.65 lb of 100% pure sodium carbonate to precipitate each pound of calcium ion. In computing the dosage, the operator must take into account that commercial hydrated lime contains 93% $Ca(OH)_2$ and soda ash is approximately 98% pure Na_2CO_3. In addition, an excess of either reagent will be required to drive the respective softening reactions to the desired degree of completion. The reader will find a detailed discussion of this subject in books on water chemistry (Betz, 1976; Nordell, 1961; Powell, 1954).

The actual reagent in Eqs. (4) and (5) is the hydroxyl ion, which can be provided equally well by caustic soda, NaOH. To replace each pound of $Ca(OH)_2$ as softening agent, 1.08 pounds of 100% pure NaOH are required.

The $CaCO_3$ and $Mg(OH)_2$ precipitated during the softening are removed in the form of a bulky sludge. From time to time it has been suggested that this waste material be marketed for the beneficiation of agricultural land or as a soil stabilizer in road construction. Neither method, however, has attained widespread acceptance. As an alternative, consideration has been given to disposal by sale to lime manufacturers. Lime manufacturers have expressed grave reservations concerning possible problems in the calcining kiln arising from the formation of clinkers by impurities such as silica or iron in the sludge. Consequently, there is a strong likelihood that the operator of a small- to medium-sized membrane plant will have to incur the expense of disposing of the sludge in a landfill.

To reduce the amount of softening sludge requiring disposal, the owner of a large membrane plant will be justified in considering the installation of sludge dewatering equipment and a kiln. The kiln reactions are

$$CaCO_3 + \Delta \longrightarrow CaO + CO_2 \qquad (7)$$

$$Mg(OH)_2 + \Delta \longrightarrow MgO + H_2O \qquad (8)$$

It is apparent from these equations that the lime inventory is being increased constantly by the ions precipitated from the feed. Thus the disposal problem still persists for a fraction of the sludge precipitated. The preferred procedure is to divide the sludge into two portions either by a two-stage softening process or by selective sludge centrifugation. When properly performed, one portion of the sludge is comparatively pure and suitable for reuse whereas the other portion, high in impurities, is discarded as landfill.

Calcining a portion of the sludge entails an investment in sludge dewatering equipment and a kiln and also operating costs from the associated labor, power consumption, and the requirement for large quantities of fuel. On the credit side of the ledger, the calcined product furnishes the raw material for the softening process of Eq. (4) and (5), and the carbon dioxide can be delivered to the softener to provide the carbonate ion for Eq. (6).

As an added plus, a modest build-up of magnesia in the calcined sludge is beneficial in reducing the silica content of the wastewater during the softening process. Membrane fouling by silica is commonly prevented by adding "active" magnesium oxide to the softener. Freshly precipitated magnesium hydroxide is considered to be even more effective for this purpose. The silica removal effectiveness of the MgO from the calcined sludge plus that precipitated from the feed during softening must be determined by tests on the actual feed.

C. REMOVAL OF IRON AND MANGANESE

Among the more troublesome impurities found in natural waters are dissolved iron and manganese compounds. In the reduced state, these ions are very soluble and cause no difficulties, particularly at low pH values. In the absence of air, ferrous ion concentrations up to 3 ppm can be tolerated in the feed without harm to the membrane system. After exposure to atmospheric oxygen, however, they are oxidized to the higher valence state in which they readily undergo hydrolysis:

$$Fe^{2+} \longrightarrow Fe^{3+} \tag{9}$$

$$Fe^{3+} + 3H_2O \longrightarrow Fe(OH)_3 + 3H^+ \tag{10}$$

The manganese ion undergoes an analogous series of reactions. The hydroxides (or hydrated oxides) thus generated, form a slimy film on membranes so as to decrease the flux and selectivity of ED membranes. In UF, these slimes may cause rapid and irreversible degradation of membrane performance. Preventive measures are comparatively simple. To facilitate their removal, iron and manganese are oxidized to the higher valence state by contact with atmospheric oxygen in an aerator. The simplest but least effective version is the trickling aerator, in which water is allowed to fall over a series of slats that natural air currents flow across. More effective is the packed column, in which water descends through a packed bed up which a stream of air is forced by a small blower. The precipitated hydroxides may be removed by filtration.

When softening is used, precipitation of the oxidized iron and manganese proceeds to completion at the elevated pH in the softener. The precipitates, together with suspended and colloidal matter, settle in the softener. Formation of easily settled flocs is promoted by using additives such as ferric sulfate and organic coagulants. The formation of large flocs that settle comparatively fast is promoted by recycling a fraction of the sludge so as to mix it with the incoming feed. With proper settler design, the supernate is comparatively clear as it overflows.

An alternate removal method is to pass the feedwater through a bed of manganese greensand, which oxidizes the iron and manganese ions and filters out the resulting precipitate. Great care must be exercized so that no leakage of the potassium permanganate solution occurs, which is used as regenerant, because permanganate will deposit a slimy film on the membrane surface.

D. FILTRATION

The residual solids content of the settled liquid, consisting of suspended matter originally in the feed and particles of floc carried over, is removed by passage through a filter, either of the gravity or the pressure type. Sand is frequently used in a single-medium filter. In multimedia filters, a layer of sand is deposited on a layer of crushed anthracite, sometimes with an additional layer of garnet or ilmenite. The particle sizes of the bed components are selected so the media will settle naturally in the liquid to yield a bed with the desired gradation in particle size. For example, a satisfactory dualmedia bed consists of No. $1\frac{1}{2}$ anthracite at an 18-in. depth, and No. 16 crushed silica at a 12-in. depth. For a multimedia filter, the following is recommended: No. $1\frac{1}{2}$ anthracite at a 17-in. depth, No. 30 crushed silica at a 9-in. depth, and No. 60–80 crushed garnet at a 4-in. depth. The base should be No. 16 crushed garnet, gravel and pebbles. The number designations are the nominal mesh sizes of the respective media.

The effectiveness of the filter in removing ultrafine particles may be improved by adding commercially available coagulants to the feed.

The particles that are filtered from the liquid penetrate the bed for a depth of several inches and are deposited in the interstices between the granules of the bed. Liquid flow rate is generally of the order of 2–5 gpm per square foot of bed in a gravity-type filter, but may run as high as 10–15 gpm per square foot in a well-designed pressure filter. To maintain this flow, the pressure applied in a pressurized filter will increase with time. Conversely, if a gravity filter or a pressure filter with constant head is used, the flow of liquid will decrease as the bed becomes loaded with precipitates. When the pressure becomes excessive in the former, or the flow rate too low in the latter, operation is interrupted for cleaning. A strong stream of clean, filtered water is introduced through perforated pipes located near the bottom of the bed. The upward flow expands the bed and breaks up clusters of the fill material.

In an improved version, the perforated pipes rotate in a horizontal plane through the bed, breaking loose any clumps cemented together by the precipitate in the bed. The impurities float off from the top of the bed. For stubborn deposits that cement the filter media particles, a preliminary air scouring is beneficial. At the conclusion of the backwash cycle, a properly designed bed settles back to its initial configuration. It is apparent that the clearwell, which serves as a buffer tank between the filters and the membrane plant, must have adequate volume to supply the required backwash. In addition, a separate pump is required. As a rough rule of thumb, backwashing requires a flow of 2.5 to 4.0 gpm per square foot of bed and, if the filter has not been permitted to clog excessively between washings, can be completed in about 10 min.

The clarification and filtration steps just outlined are excellent for large desalination plants. For small plants, the required investment would be excessive. In such cases, UF provides an excellent method for removing small particles that might clog RO membranes or deposit on ED membranes. Ultrafiltration is particularly useful for the pretreatment of wastewaters containing colloidal impurities.

E. CHLORINATION AND CHLORINE REMOVAL

The equipment and lines used in Sections IV.A–IV.D are subject to sliming by bacterial and fungus growths. Some of the bacteria may be carried through the system to the membrane units where they will impair the functioning of an ED stack or destroy CA, RO, or UF membranes. In addition, pieces of slime will clog small passages in the membrane plant. It is common practice, therefore, to chlorinate the feed prior to softening. As a biocide, chlorine is frequently applied in "shock" treatment. For example, the dosage may be adjusted to give 1–2 ppm chlorine residual for a period of 1 hr out of 24. During treatment, the chlorine not only destroys organic matter, but oxidizes ferrous and manganous ions in a manner similar to the action of oxygen in Eq. (9).

For feedwaters containing hydrogen sulfide, chlorination can serve an additional function, namely, the removal of hydrogen sulfide:

$$Cl_2 + H_2O = HClO + HCl \tag{11}$$

$$H_2S + HClO = HCl + H_2O + S \tag{12}$$

Unfortunately, the sulfur separates as a fine precipitate that deposits in the lines. Furthermore, in plants where softening and filtration are not required, sulfur particles tend to coat and clog subsequent process equipment. Under such conditions, an overdose of chlorine will eliminate the deposition of sulfur

$$H_2S + 4HClO = 4HCl + H_2SO_4 \tag{13}$$

The requirement for chlorine for complete oxidation of H_2S is quadruple that for sulfur precipitation, a consequential cost item for feedwaters having a high sulfide content.

With few exceptions, membranes are sensitive to the presence of chlorine in water. Manufacturers of the new polysulfone membranes claim that several parts per million of chlorine are permissible in their feed stream. Cellulose acetate membranes and the ion-selective membranes used in ED plants, however, are limited in exposure to a few tenths of 1 ppm of chlorine. For polyamide membranes, the chlorine tolerance is 0.10 ppm if the feed pH does not exceed 8.0, 0.25 ppm at higher pH values. For dechlorination, the feed stream is passed through an activated carbon filter. Commercial equipment is available in the form of pressure vessels packed with a bed of activated carbon 6–8 ft in depth and supported on a base of crushed granite. The vessel is sized to permit a flow rate of 1.0 to 1.5 gpm per square foot of cross section.

Although the behavior of the activated carbon is sometimes described as adsorption, the chlorine is actually removed by reaction with the carbon, gradually reducing the size of the carbon granules. The carbon fines thus produced, together with suspended matter in the feed (for example, the sulfur precipitated from sulfide-containing waters), will eventually increase the pressure drop across the bed to the point where a shutdown is required for backwashing.

The cost of carbon replacement, consumption of power to overcome pressure drop, and capital charges of the activated carbon process encourage the use of an alternative such as the reduction of residual chlorine by bisulfite:

$$Na_2SO_3 + HClO = Na_2SO_4 + HCl \tag{14}$$

Each part per million of residual chlorine requires for its destruction 3.6 ppm of sodium sulfite or 2.7 ppm of sodium metabisulfite. Even for CA membranes, a constant *low* level of chlorine in the feed to the membrane modules is undesirable because cellulose-destroying bacteria appear to thrive in a mildly oxidizing environment. Consequently, it is advisable to maintain a 5 ppm residual of sodium sulfite in the feed after the chlorination step to insure a reducing environment.

F. pH CONTROL

A comparatively soft feed may be sent directly to the membrane plant without a presoftening step. For an unsoftened water, or even when feed softening is used, enough calcium ion may remain to form calcium carbonate scale in the system. Consequently, pH control is generally a part of the pretreatment chain. The tendency for the formation of calcium carbonate scale in membrane plants is expressed in terms of the Langelier index (Langelier, 1936; 1946). Because the dissolved ions are concentrated in the reject stream of either an RO or ED plant, it is essential to maintain the Langelier index of the *reject brine* at or slightly below zero at the membrane surface. The reduction in the index is achieved by acidifying the feed stream to the membrane plant:

$$Ca(HCO_3)_2 + H_2SO_4 = CaSO_4 + 2CO_2(dg) + 2H_2O. \tag{15}$$

The (dg) indicates that the liberated CO_2 is retained in solution as a result of the pressure applied by the feed pump, thereby reducing the pH to the extent required to attain the target Langelier index. It is apparent that the acid consumption of a partially softened feed is considerably less than in the absence of softening. The sulfuric acid, commercially available as a 93% solution (66° Baume), is stored in a steel tank protected by an acid resistant coating such as a baked phenolic. The acid is metered directly into the feed line ahead of the cartridge filter and high-pressure pump. A pH meter downstream from the point of injection transmits a signal to the metering pump, either varying the pump stroke or the pump revolutions per minute so as to maintain the required pH of the feed water. An alternate device consists of a jet through which the feed stream flows on its way to the membrane plant. The acid is aspirated into the feed by the vacuum created in the jet. The acid dosage is controlled by a valve in the acid line regulated by a pH meter downstream of the acid injection point. If the composition of the feed is relatively constant, the automatic valve may be replaced by a valve manually adjusted as dictated by the pH of "grab samples," resulting in a simple and inexpensive system.

The operator wishing to avoid the hazards of storing and handling strong acid may substitute a solution of "solid acid." Crystalline sodium bisulfate is dissolved in a feed tank and metered into the aqueous feed stream in the same manner as sulfuric acid. This form of acid is costly and, consequently, not used in large membrane plants.

Aside from the acid cost, this type of pH control has the disadvantage of increasing the sulfate ion and, consequently, decreasing the solubility of calcium in the water (see the following discussion of calcium sulfate). Hydrochloric instead of sulfuric acid may be used, but at a substantial cost penalty. As an alternative, acidification to the required Langelier index can be achieved by injecting carbon dioxide into the feedwater. Where the feed is softened and the softener sludge calcined, "no cost" carbon dioxide is available in the off gases from the kiln to substitute for the costly acid of Eq. (15). The gaseous carbon dioxide is converted to the CO_2(dg) of the right side of Eq. (15) by injecting the gas into the feedwater through a diffuser nozzle. The injection pressure must be slightly in excess of the feed pressure (40–60 psi for ED, 320–450 psi for RO, 800–1000 psi for seawater RO). In addition, the hot kiln exhaust must be cooled prior to entering the compressor. It thus becomes apparent that the CO_2 required for acidification is by no means free.

The material requirements of acidification are somewhat diminished in an ED plant operating on the polarity reversal principle. Feeds for either RO or constant polarity ED plants customarily are adjusted to 5.5 to 6.0 pH values. For polarity reversal plants, pH values in the neighborhood of 7.0 have been reported (Geishecker, 1977, private communication).

G. Calcium Sulfate Control

Avoiding calcium carbonate and magnesium hydroxide scale through proper pretreatment and feed acidification still leaves the problem of calcium sulfate scaling. The solubility of calcium sulfate in water is expressed by

$$K_{sp} = [Ca^{2+}][SO_4^{2-}], \tag{16}$$

where K_{sp} is the solubility product and the quantities in the brackets are the molar concentrations of the respective ions. When the product of the molarities of the calcium and sulfate ions exceeds the solubility product, calcium sulfate scale will deposit. The same considerations apply to barium and strontium sulfates. A number of commercially available additives retard the deposition of scale or alter its crystalline structure so as to prevent its adherence to the surface of membranes and other plant components. Favorable experience in scale control has been reported for additives containing acrylic polymers, aminoethlene phosphonic acid derivatives, and sodium hexametaphosphate (SHMP) (Vetter, 1972). A solution of the selected additive is injected into the feed by a metering pump.

The danger of scaling increases with the concentration of the offending compound *at the membrane surface*. When the water recovery represents a high percentage of the feed, the reject stream will contain a corresponding high concentration of the scale formers and, therefore, require a higher concentration of scale-control additive. It must be emphasized that scale deposition is a kinetic phenomenon. If the reject stream leaves the module at a sufficiently high velocity and if stagnant areas in the equipment are avoided, it is possible to exceed the solubility limits of scale formers and still not inject additives. The need for additives and their concentration are best determined by tests under conditions simulating actual production-plant operation. These tests must be of sufficient duration (at least 2 weeks at the proposed additive concentration) in a pilot plant supplied with the *actual* feedwater, not a synthetic feed.

The equipment necessary to inject the additive is very simple. Two feed tanks, each with a propeller stirrer, are used to dilute the additive to the required concentration. The solution is prepared in one tank while the second supplies the pump. If SHMP is the additive selected, the batch size should not exceed a 24-hr supply because of the tendency of SHMP to decompose during long standing-water contact. As in the acid system, the solution is delivered to the feedstream by a metering pump. For the scale preventative, however, it is customary to rely on manual adjustment of the stroke or the revolutions per minute of the pump rather than automatic control.

For the owner of a small plant who wishes to avoid the cost and complexity of mixing, storage tanks, and metering pumps, a metaphosphate scale-control additive is available as the slightly soluble calcium–sodium compound, sold in the form of solid blocks. The feed passes over a block of the additive retained in a perforated holder, from which the metaphosphate is slowly leached. This form of metaphosphate is comparatively costly and not used in large water-treatment plants.

H. CARTRIDGE FILTERS

Even with the most elaborate pretreatment system it is essential to remove ultrafine particulates as a last step before entering the high-pressure pump, which delivers the feed to the membrane unit. Most manufacturers recommend using a 10-micron (or even 5-micron) filter. One of the most common types contains a number of string-wound cartridges in a pressure vessel. If the pretreated feed is very low in suspended soils, a set of filters will serve for as long as 6 months before an increase in pressure drop (generally 10 psi for clean filters) dictates the need for replacement. Under unfavorable conditions, however, the life of a set of filter cartridges may be reduced to as little as 2 weeks, contributing appreciably to the cost of the product water. As an alternative, there is a filter that is claimed to be regenerable by backwashing.

V. Posttreatment

Purified water, recovered by a membrane process, may require additional treatment before it is suitable for reuse. Some of the more common post-treatment steps are discussed here.

A. DECARBONATION

A high CO_2 concentration is undesirable in the purified water from a membrane plant for two reasons:

(1) It increases corrosive attack on components of the water-distribution system, and
(2) it consumes large quantities of costly chemicals used to render the pH of the product water suitable for human consumption or industrial use.

If acid is not injected into the feedstream to the membrane, an objectionably high CO_2 content almost never occurs in the product water. By contrast, acidification of the feed converts the bicarbonate ion into dissolved CO_2, which migrates readily through an osmotic membrane, establishing a dissolved CO_2 content in the permeate almost identical with that in the acidified feed.

Decarbonation is readily accomplished by exposing a large surface of liquid to the atmosphere. The product can be allowed to trickle across a bed of rocks or crushed stone. A much more effective and compact design permits the water to trickle down over wooden slats, similar in construction to a cooling tower.

The common wooden-slat tower, with natural air circulation across the falling film of water, can be expected to yield an effluent containing at best about 30 ppm of CO_2 by weight. The most effective, although most expensive system, consists of a vertical column with 4 to 8 ft of packing depth. Plastic pall rings or various patented designs of plastic fill are preferred over ceramic packing because of their light weight, low pressure drop, and large effective surface. The CO_2 is stripped from the water by a slow air current blown up through the bed. Under favorable operating conditions, a well-designed stripping column can reduce the CO_2 content of the water to as low as 7 ppm.

B. SILICA REMOVAL

The posttreatment for ED will, in general, involve silica removal because ED in contrast to RO, does not reduce the silica content of water. Silica is conveniently removed from the ED product by an ion-exchange treatment in which the water is passed through a cation resin in the hydrogen form, converting the dissolved silica to silicic acid. The latter is readily removed by passage of the water through a bed of strongly basic anion exchanger, which can reduce the silica content to 0.05 to 0.2 ppm. Silica removal by this system requires that the water be decarbonated between the cation and anion exchange steps.

C. pH ADJUSTMENT

If the product water is required to have a neutral pH, the controlled addition of alkali may be necessary. For this purpose, a soda ash or sodium hydroxide solution is injected into the product water by a metering pump of which the stroke or the revolutions per minute is controlled by a pH meter in the water line sufficiently far downstream from the injection point to insure thorough mixing. An alternate injection method is to use an aspirator, as discussed in Section IF. If the composition of the product steam and its flow rate are sufficiently steady, a pump delivering constant flow of injected alkali is permissible with only occasional manual adjustment of a valve in the alkali line, as dictated by the pH of periodic "grab samples."

D. DEIONIZATION

The product water from commercial membrane plants will usually contain from 7 to 10% of the TDS concentration of the feedstream. If this product water is intended as make-up to high-pressure boilers, the residual dissolved solids content may be unacceptable since boiler feed purity of about 1–2 ppm TDS is generally specified. Such purity can be attained by passing the water through a mixed-bed ion exchanger. A mixed-bed ion-exchange posttreatment is applied also to water used in the manufacture of electronic components. The frequency of regeneration and the consumption of regenerant chemicals will be minimal because the bulk of the ion removal will have been accomplished by the membrane process.

E. FILTRATION

Ultrapure waters used in the electronics industry must be free not only from dissolved salts but from suspended matter because any residual particles remaining after the final water rinse may result in a malfunction of the electronic component. After the membrane and mixed-bed ion-exchange steps, a thorough filtration is highly desirable for water intended for such use. Most of the particulate matter is, of course, removed from the feedwater by the cartridge filters used ahead of RO and ED plants. In addition, passage through the RO membranes can be expected to yield a very clean product. The deionization step, however, may introduce resin particles and rust flakes from steel equipment and lines. As an added precaution, it is advisable to subject the water product to a UF treatment prior to its use in the manufacture of electronic components. The final UF step can reduce total suspended solids to less than 1 ppm, and the turbidity to as low as 0.21 Jackson Turbidity Units (Breslau *et al.*, 1975).

F. COLOR AND ODOR CONTROL

If the water produced by a membrane process is intended for human consumption, its flavor, appearance, and odor must be considered. Minute amounts of undesirable constituents may remain in the water even after all the previous steps. The recommended process for their removal consists of passing the water through a bed of activated carbon. In this type of service, a long bed life may be anticipated. Bed channeling is minimized by operating the bed in an upflow mode. Bed washing, which is required infrequently, consists of passing *filtered* water up through the bed at a velocity sufficient to suspend the carbon granules. Except in cases of gross maloperation, it is not necessary to replace the original charge of activated carbon. Small additions of carbon are occasionally needed to replace losses.

G. Chlorination

Microorganisms and viruses are not removed from the feed by ED. Although they cannot pass through RO and most UF membranes, there is always the possibility of a flaw in one of the membranes or mechanical seals. Consequently, chlorination of the product water from any membrane system is required if the water is intended for human consumption. The chlorination step is highly desirable for the prevention of slimes and algae growth in the product water system unless the presence of chlorine is prohibited in any process for which the water is intended.

Although the owner of a treatment plant can readily assemble a chlorine-injection system, it is generally preferable to buy the system complete from any one of a number of reputable suppliers.

VI. Conclusion

This chapter is intended to give the reader a general understanding of the principles involved in the purification of aqueous streams by membrane processes. Sufficient detail has been included to alert the operator to problem areas. As a word of caution, if the feedstream differs widely from those successfully processed by industry in the past, of if the requirements of the purified water or waste concentrate are unusual, it is recommended that the choice of process and equipment be preceded by adequate pilot-plant testing.

References

Betz (1976). ''Handbook of Industrial Water Conditioning.'' Betz, Trevose, Pennsylvania.

Breslau, B. R., Agranat, E. A., Testa, A. J., Messinger, S., and Cross, R. A. (1975). *Chem. Eng. Prog.* **71,** (12), 74–80.

Eykamp, W. (1976). Paper presented at the 69th Annual Meeting of AIChE, Chicago, Illinois.

Fenton-May, R. I., and Hill, C. G. (1971). *J. Food Sci.* **36,** 14.

Katz, W. E. (1971). *Ind. Water Eng.* **8**(6), 29–31.

Langelier, W. F. (1936). *J. Am. Water Works Assoc.* **28,** 1500.

Langelier, W. F. *J. Am. Water Works Assoc.* **38,** 169.

Messinger, S. (1974). Twelfth Liberty Bell Corrosion Course, Philadelphia, Pennsylvania.

Sponsored by the Philadelphia Section of the National Association of Corrosion Engineers.
Michaels, A. S. (1968). *Chem. Eng. Prog.* **64**(12), 31.
Nordell, E. (1961). "Water Treatment for Industrial and Other Uses." Reinhold, New York.
Powell, S. T. (1954). "Water Conditioning for Industry." McGraw-Hill, New York.
Sherwood, T. K. (1937). "Absorption and Extraction." McGraw-Hill, New York.
Valcour, H. C. (1977). National Water Supply Improvement Association, National Meeting, San Diego, California.
Vetter, O. J. (1972). *J. Pet. Technol.* **24,** 997–1006.

12

Economics of the Application of Membrane Processes

PART 1: Desalting Brackish and Seawaters

PINHAS GLUECKSTERN

NATHAN ARAD

Mekoroth Water Co. Ltd.
Tel Aviv, Israel

I. Introduction*

The cost of desalted water is determined by a large number of factors, such as site, process technology, plant capacity, and economic parame-

* Because of cost escalation and the changing inflation rate resulting from the economic down turn of the early 1980s, estimation of the current costs for membrane treatment remains an elusive target. To account for escalation, the reader may wish to use one of the published escalation tables or figures for the change of equipment cost indices, such as those in the quarterly cost roundups published in *Engineering News Record,* McGraw Hill, New York.

ISBN 0-12-085480-5

ters. Similar to other production facilities, unit water costs are generally composed of capital costs, costs related to energy and materials consumption, and other operating and maintenance costs.

For membrane plants with extensive feedwater pretreatment, the capital costs are usually divided into two major cost components: desalting costs and pretreatment costs.

The two major economic characteristics of a desalting plant are the specific capital cost, expressed as dollars per gallon day (\$/gd), and the unit water cost, expressed as dollars per thousand gallons (\$/kgal) or dollars per cubic meter (\$/m^3). For plants with extensive pretreatment requirements these costs are usually expressed for the desalting and pretreatment functions separately.

For a given process type, the specific capital cost and the unit water costs are most affected by plant capacity, site conditions (especially feedwater type) and the economic ground rules, such as interest rate, plant operating factor, and unit costs of energy, materials, and labor.

Obviously, generalized cost data can only be indicative for any specific application while more accurate cost data must be evaluated for every specific case based on the applicable site conditions and economic ground rules.

Most of the cost data presented in the following sections is based on generalized site conditions defining the feedwater type only by its general ion composition, total salinity, and temperature. It is, therefore, emphasized that these costs can be applied for indicative purposes only. To make it possible to apply these costs for different economic conditions, the effect of the major economic parameters on the water cost, as a percentage of the base cost, is presented.

II. Desalting Brackish Waters

A. General

In contrast to distillation processes, the energy consumption of membrane processes is strongly affected by the salinity of the treated saline water. At prevailing high energy prices, the desalting of brackish water using membrane process has, therefore, a strong economic edge above seawater desalting, which was, until the late 1970s, almost exclusively performed by thermal desalting plants.

According to published data (El-Ramly and Congdon, 1981) 1,239 plants for brackish water desalting using membrane processes of 25,000 gpd (gallons per day) or larger were in operation or construction throughout the world as of June 30, 1980. These plants are capable of producing about 463 mgd (million gallons per day) of fresh water for municipal and industrial uses. Membrane processes account for 24% of total plant capacity from all processes (1922 mgd) and 56% of the total number of plants. It is, however, interesting to note that during the period 1977–1980 (since the previous inventory report was prepared), worldwide sales of membrane plants of 25,000 gpd or larger accounted for about 75% of all plants sold during this period (549 plants with a combined capacity of 270 mgd), but their combined capacity was only about 27%. This trend of increasing the portion of membrane plants is currently even stronger because for most cases where brackish water is available, the cost of desalting by membrane processes is only about one-third of the cost of seawater desalting using thermal plants.

B. ESTIMATION OF INVESTMENTS

1. *Capital Cost Components*

(1) Desalting equipment. Special process equipment such as the electrodialysis (ED) stacks or the reverse osmosis (RO) module assemblies, and auxiliary process equipment such as the pumps and drives, valves and piping, instrumentation and accessory electrical equipment.

(2) Membranes and ED gaskets and spacers. Reverse Osmosis membrane modules or ED membranes and other replaceable items such as the gaskets and spacers.

(3) Site development. Land cost, site preparation, structures, intake and discharge, and brine disposal.

(4) Pretreatment and posttreatment. This item is sensitive to the type of feedwater. For some cases the required pretreatment includes only simple sand and micron filtration and chemicals dosing; for other cases, removal of organic contents and inorganic contaminants such as iron and manganese might require a complex and costly pretreatment system.

The posttreatment usually consists of a decarbonator and/or dosing of neutralizing agent to adjust the acidity (pH) of the product water to the required value.

(5) Indirect costs. This category includes the costs of engineering design and supervision, interest during construction, and the owner's cost for training and start-up. These costs are in most cases proportional to the total direct capital costs. Engineering design and supervision ranges between 5 and 10%, interest during constructions depends on the plant capacity and interest rate but is usually also in the range 5–10%, and owners cost for training and start-up can be assumed as 2–5% of the total direct costs. The total indirect costs will therefore be, in most cases, between 15 and 25% of the total direct costs.

(6) Other costs. These costs include shipment and erection of the factory-assembled desalting equipment (which is usually quoted FOB from the manufacturer), and an allowance for contingency. The first cost is usually estimated as 10% of the equipment cost and the second as 10% of the total capital cost.

2. *Generalized Capital Costs*

Generalized capital costs of ED and RO plants were compiled by Hittman Association Inc., (Curraw *et al.*, 1976) in a study made for the U.S. Office of Water Resources and Technology (OWRT), for two typical brackish waters, defined as "Sodium Chloride" (type 1) and "Calcium Sulfate–Bicarbonate" (type 2).

The relative ion composition of these two representing water types are shown in Table I. These data were originally prepared by the Ameri-

TABLE I

Relative Ion Composition of the Brackish Water Types Used in the Economic Evaluation

	Cations			Anions		
Ion type	Na^+	Ca^{2+}	Mg^{2+}	Cl^-	SO_4^{2-}	HCO_3^-
Sodium chloride water (type 1) (% of total)	30.0	4.5	0.5	41.5	6.1	17.4
Calcium sulfate–bicarbonate water (type 2) (% of total)	7.0	13.1	6.5	12.2	33.0	28.2

can Water Works Association Research Foundation (AWWARF, 1973) for a wide range of plant capacities and feedwater salinities, using 1972 prices. According to the authors of the Hittman study, a reasonable approximation of current prices can be made by applying the escalation factors represented by the Engineering News Record Building Cost Index (ENRBCI), published regularly by McGraw–Hill (Engineering News Record, 1972–1982).

Selected data for the generalized capital cost, adjusted to mid-1977 prices are shown in Figs. 1 and 2, for type 1 and 2 waters, respectively. Unpublished cost data indicate that the prevailing cost in 1981 and early 1982 was in fair agreement with the capital cost shown in Figs. 1 and 2 after applying the ENRBCI escalation factors.

C. ESTIMATION OF ANNUAL AND UNIT WATER COSTS

1. *Economic Ground Rules*

To evaluate the annual and the corresponding unit water costs, a set of economic ground rules must be used. At least the following must be defined: Annual interest rate, annual expenses for insurances and taxes, plant operating factor, plant life, membrane life, required operating and maintenance (O and M) staff, averaged annual expenses for O and M staff ($/man year), unit energy cost ($/kW hr), cost of supplies and materials (usually an allowance proportional to the direct or total capital costs is applied), and unit costs of applied chemicals.

2. *Generalized Unit Water Costs*

Based on the capital cost estimates presented in Figs. 1 and 2, water costs were calculated by applying the ground rules defined in Table II.

The resulting generalized unit water costs are presented in Figs. 3 and 4 for type 1 and 2 feedwaters, respectively.

A cost breakdown and a sensitivity analysis indicating the effect of the major economic parameters on the unit water costs of 1 and 5-mgd plant capacities are shown in Tables III and IV for type 1 and 2 feedwaters, respectively.

Reported cost evaluations (Glueckstern, 1979) indicate that the unit

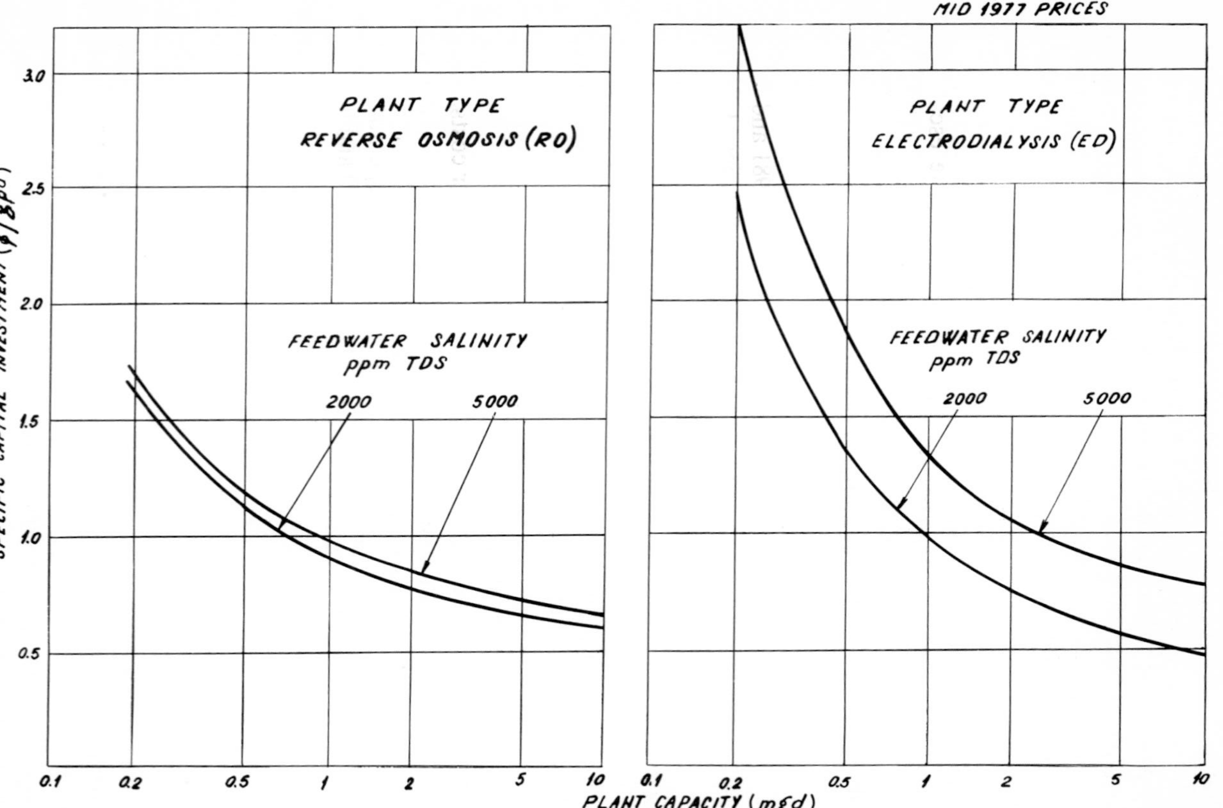

Fig. 1 Specific capital investments of ED and RO plants desalting brackish waters of the sodium chloride type versus plant capacity.

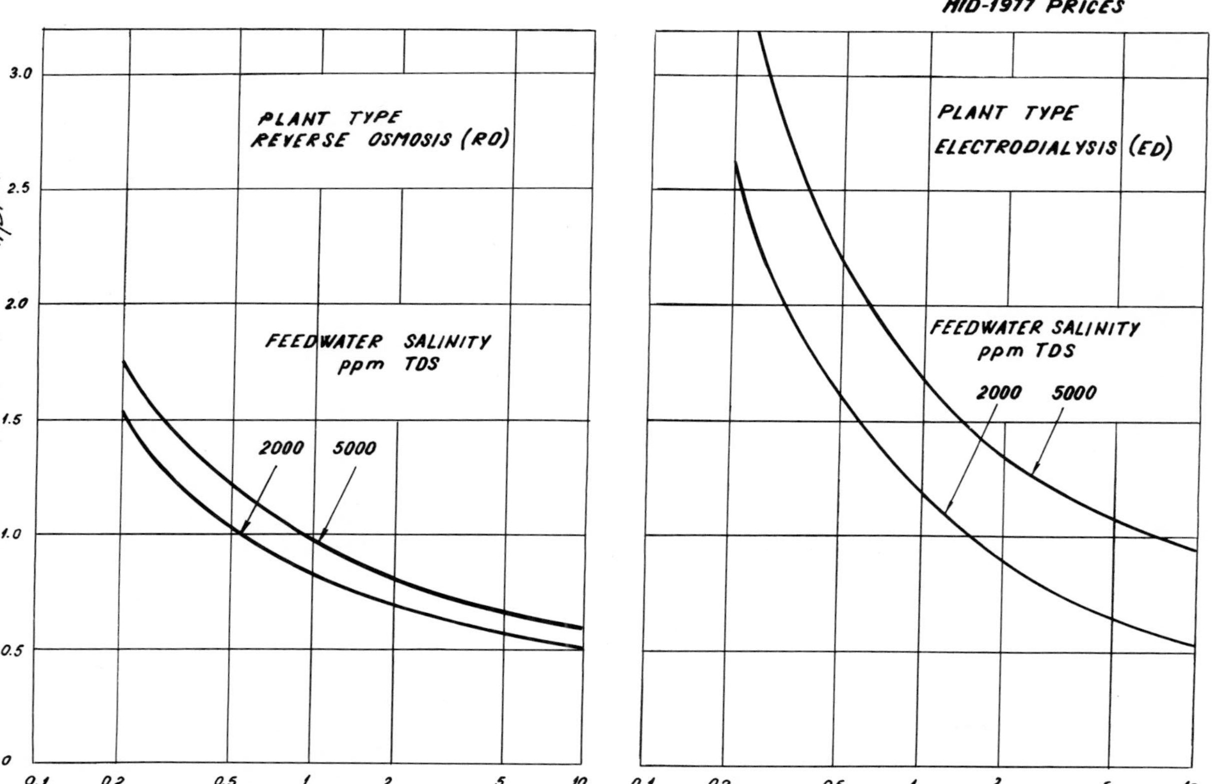

Fig. 2 Specific capital investments of ED and RO plants desalting brackish waters of the calcium sulfate–bicarbonate type versus plant capacity.

TABLE II

Economic Ground Rules and Costing Factors Assumed for the Evaluation of Unit Water Cost

Interest rate (%)	8.0
Insurances and taxes (%)	1.0
Plant operating factor (%)	90.0
Plant life (yr)	20.0
RO membrane life (yr)	3.0
ED membrane life (yr)	5.0
Operation and maintenance staff for 1-mgd plants[a]	5.0
Average staff cost ($/man yr)	18,000.0
Power cost ($/kW hr)	0.04
Annual costs for supplies and materials (% of total capital cost)	1.0
Unit prices of chemicals	
Sulfuric acid ($/t)	40.0
Sodium hexamethaphosphate ($/kg)	.60
Other chemicals ($/m^3)	.02

[a] For capacities other than 1 mgd a scale factor of 0.5 was applied.

water cost shown in Figs. 3 and 4 (in mid-1977 dollars) are representative of plants with favorable site conditions and operation with a high plant factor. In other cases, the more accurate cost evaluation would show higher costs of up to 20% or more.

The unit water costs shown, as well as these reported in the previously mentioned reference, do not take into account the possible reduction in unit water cost by blending the product water with raw water. For desalting low-salinity brackish water by the RO process, this factor may reduce the unit water cost after blending to 500 ppm TDS by a value of, in some cases, 20%.

It should also be mentioned that from 1980 to 1982, the remarkable progress in RO technology caused substantial cost reduction.

Glueckstern *et al.*, 1981, as well as unpublished cost evaluations, indicated a 20–30% unit water cost reduction from RO plants in comparison with the cost prevailing in the late 1970s. This cost reduction (in real prices) was caused by improvements in RO membrane and power recovery technology, affecting membrane replacement and power cost. The reduction in these two cost components compensate the price inflation in equipment cost. The unit water costs from RO plants shown in Figs. 3 and 4 are, therefore, in relative agreement with the prevailing costs in 1981 and early 1982.

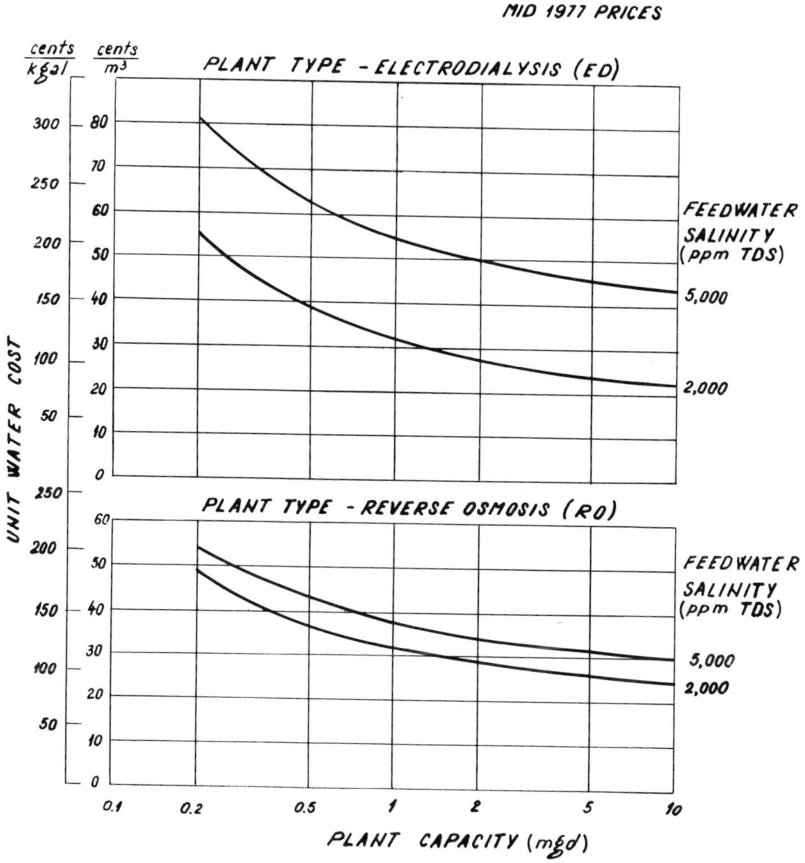

Fig. 3 Unit water costs of ED and RO plants desalting brackish waters of the sodium chloride type versus plant capacity.

D. TYPICAL CAPITAL AND UNIT WATER COST OF OPERATING PLANTS

Table V summarizes typical capital and unit water costs for operating ED and RO plants, which were evaluated by DSS Engineers, Inc., for OWRT (5). The capital charges of these plants were based on a 20-yr plant life, a 6% annual interest rate, and the actual operating-plant factors experienced during the corresponding operating period when the costing was made. It should, however, be noted that the rather low operating-plant factors for some of the plants are not necessarily because of plant inoperability, but are because of other factors, especially low water demand.

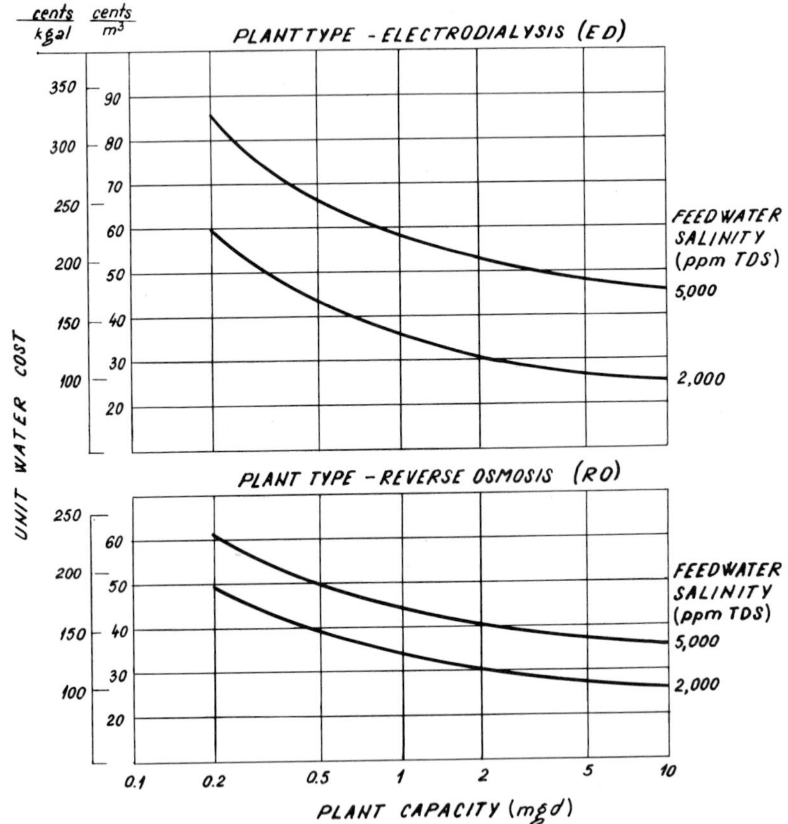

Fig. 4 Unit water costs of ED and RO plants desalting brackish waters of the calcium sulfate–bicarbonate type versus plant capacity.

III. Desalting Seawater

A. GENERAL

Although membrane processes for seawater desalting have been demonstrated at several locations, and commercial RO plants with capacities of up to 12,000 m^3/day are in operation (Hickman *et al.*, 1979; Boesh, 1981), some of the more important operating factors affecting costs cannot yet be based on established operating and cost data.

TABLE III

Generalized Unit Water Costs of ED and RO Plants Desalting Brackish Water of the Sodium Chloride Type

Costs	ED				RO			
	Plant capacity (mgd)				Plant capacity (mgd)			
	1[a]	5[a]	1[b]	5[b]	1[a]	5[a]	1[b]	5[b]
Total capital cost[c] ($/gpd)	.97	.58	1.32	.86	.89	.65	.96	.71
Unit water cost[a] (¢/m^3)								
Capital cost	8.4	4.9	11.3	7.2	7.7	5.5	8.2	5.9
Power cost	11.6	11.6	28.3	28.3	8.9	8.9	10.9	10.9
Membrane replacement	2.6	2.6	4.3	4.3	5.0	5.0	6.1	6.1
Operation and maintenance[d]	9.9	5.7	11.3	6.8	10.6	6.4	13.0	8.8
Total	32.5	24.8	55.2	46.6	32.2	25.8	38.2	31.7
Effect of economic narameters (percent of base								
25% change in capital cost or fixed charges	6.5	4.9	5.1	3.9	6.0	5.3	5.4	4.7
1 ¢/kW hr change in power cost	8.9	11.7	12.8	15.1	6.9	8.6	7.2	8.6
50% change in membrane replacement cost	4.0	5.2	3.9	4.6	7.8	9.7	8.0	9.6
50% change in operation and maintenance cost	15.2	11.5	10.2	7.3	16.5	12.4	17.0	13.9

[a] Feedwater salinity 2000 ppm TDS.
[b] Feedwater salinity 5000 ppm TDS.
[c] Excluding brine disposal.
[d] Operating and maintenance labor, supplies and materials including chemicals.

The current ED technology applied for desalination of brackish waters is not economically attractive for seawater conversion, mainly because of its relatively high specific-energy consumption. In the early 1980s, however, programs to develop high-temperature operation, thin membranes and improved membrane spacers can be expected to improve the economics of this process considerably.

The application of RO technology for seawater conversion is much more advanced than ED technology, but the costing for this process is strongly site dependent and is still changing rapidly with time. This is because of several operational uncertainties that strongly affect the economics of the process; namely, membrane life and the required seawater pretreatment. These factors are affected primarily by the quality of the raw seawater and the developmental status of membrane technology.

Most of the published data do not include site development or own-

TABLE IV

Generalized Unit Water Costs of ED and RO Plants Desalting Brackish Waters of the Calcium Sulfate—Bicarbonate Type[a]

	ED				RO			
	Plant capacity (mgd)				Plant capacity (mgd)			
Cost	1[b]	5[b]	1[c]	5[c]	1[b]	5[b]	1[c]	5[c]
Total capital cost[d] (\$/gpd)	1.20	.65	1.69	1.08	.84	.58	.98	.66
Unit water cost[d] (¢/m³)								
Capital cost	10.5	5.6	14.6	9.2	7.3	4.9	8.5	5.6
Power cost	11.6	11.6	28.3	28.3	8.9	8.9	15.9	15.9
Membrane replacement	2.6	2.6	4.3	4.3	5.0	5.0	5.6	5.6
Operation and maintenance[e]	11.4	6.9	12.0	7.5	12.9	8.7	14.4	10.1
Total	36.1	26.7	59.2	49.3	34.1	27.5	44.4	37.2
Effect of economic parameters (percent of base cost)								
25% change in capital cost or fixed charges	7.3	5.2	6.2	4.7	5.4	4.5	4.7	3.8
1 ¢/kW hr change in power cost	8.0	10.9	12.1	14.6	6.5	10.3	9.0	10.7
50% change in menbrane replacement cost	3.6	4.9	3.7	4.5	7.4	9.1	6.3	7.5
50% change in operation and maintenance cost	15.8	12.9	10.1	7.6	19.0	15.8	16.2	13.6

[a] 1977 cost study.
[b] Feedwater salinity is 2000 ppm TDS.
[c] Feedwater salinity is 5000 ppm TDS.
[d] Excluding brine disposal.
[e] Operating and maintenance labor, supplies and materials including chemicals.

er's cost and are based on upscaling of current or projected technology. In the first part of this analysis (Section III.B), published data for small plants are presented without adjustments for unaccounted cost elements (such as site costs). In the second part of this analysis (Section III.C), which is for plants in the 1–10 mgd range, some of the data are also presented without adjustments regarding cost components, updated prices, and consistency of economic ground rules, but an attempt has also been made to analyze the economics based on an updated set of ground rules, equipment, and material prices (mid-1977 prices).

A short discussion of the effect of RO membrane improvements and power recovery technology as of 1981 on RO seawater desalting cost is also presented.

TABLE V

Typical Cost Data of Operating ED and RO Plants[a]

Plant location	Year	Type	Feed TDS (ppm)	Capacity (1000 gpd)	Capital cost		Plant factor (%)	Unit water cost ($/m^3)			Operating period[b]
					$1000	$/gpd		Capital charges	Operating cost	Total	
Siesta Key, Florida	1969	ED	1307	1830	1004	0.55	60.7	0.055	0.082	0.137	1/74–9/74 (9 mo)
Gillete, Wyoming	1972	ED	1840	1500	666	0.44	21.6	0.195	0.134	0.329	7/73–6/74 (12 mo)
Sanibel Island, Florida	1973	ED	2930	1200	456	0.38	62.1	0.107	0.150	0.267	1/74–6/74 (6 mo)
Sorrento Shores, Florida	1973	ED	2786	70	88	1.26	87.0	0.089	0.235	0.324	5/74–2/75 (10 mo)
Ocean Reef, Florida	1971	RO	6300	930	560	0.60	75.0	0.065	0.171	0.236	2/74–6/74 (5 mo)
Rotonda West, Florida	1972	RO	7000	500	386	0.77	47.5	0.107	0.225	0.332	7/73–6/74 (12 mo)

[a] Based on American Water Works Association Research Foundation (1973).
[b] For which unit water costs were evaluated.

The reported cost data of the commercial RO plants just beginning operation are not detailed enough and are too specific regarding site conditions for comparison with the reported cost studies.

B. INITIAL COST ESTIMATES OF SMALL PLANTS

1. *Reverse Osmosis*

Commercial RO membranes, capable of desalting seawater in a single pass, were introduced by the DuPont Company in 1973. In 1976, the DuPont and Dow Chemical companies reported (Macgowan *et al.*, 1976; Rosenblatt *et al.*, 1976) estimated desalting costs for single-pass seawater desalting based on the performance of their membrane modules developed under sponsorship of OWRT.

The cost breakdown presented by the Dow Company for a 50,000 gpd (190 m^3/day) plant consisting of 20 8-in. cellulose triacetate hollow-fiber membrane modules, based on the actual construction and operation of a half-size (25,000 gpd) pilot plant was the more detailed and was, therefore, selected for the present analysis.

The construction cost of the plant, excluding the site development, buildings, seawater intake and pretreatment system, was estimated at approximately \$177,000 (mid-1975 prices). This cost included \$36,000 for replaceable membranes. The cost of the pretreatment system was not reported directly, but from the stated depreciation cost it can be concluded that its cost was about 7% higher than the RO plant, excludng replaceable membranes. The total cost was, therefore, about \$300,000, or \$6/gpd.

The specific power consumption of the plant, which was based on an 800 psig (56 kg/cm^2) operating pressure, a 30% product recovery, and a 60% pumping efficiency, was approximately 13 kW hr/m^3.

The operation and maintenance cost, including operating and maintenance labor and materials but not membrane replacement, was estimated as \$.18/m^3.

By assuming an electrical power cost of \$.04/kW hr, a capital fixed charge rate of 10% per year, and a 300 day/yr operation at full capacity, the following unit water costs were obtained:

Capital cost (neglecting site costs):	\$ .53/m^3
Power cost:	.52/m^3
Operation and maintenance cost:	.18/m^3
Membrane replacement cost (1 yr/ 2 yr/ 3 yr life):	.63/.32/.21/m^3
Total unit water cost (1 yr/ 2 yr/ 3 yr life):	\$1.86/1.55/1.44/m^3

2. *Electrodialysis*

The first ED seawater desalting plant in the world began operation in September 1974 in Japan. The capital and operating costs of this small capacity (120 m^3/day) plant have been reported (Seto *et al.*, 1976).

The construction cost of the plant, excluding site development, buildings and seawater intake, was 47.7 million Japanese yen. At an exchange rate of approximately 250 yen/$U.S., this amounts to approximately $190,000 or about $6/gpd (1974 prices).

The specific power consumption of the plant was reported to be 16.2 kW hr/m^3, and the reported operation and maintenance cost, including membrane replacement, based on a 10-yr membrane life, was 80 Japanese yen. By assuming an electrical power cost of $.04/kW hr, a capital fixed charge rate of 10% per year, a 300 day per year operation at full capacity, and an exchange rate of 250 Japanese yen per U.S. dollar, the following unit water costs were obtained:

Capital cost (neglecting site costs)	$ .53/m^3
Power cost	.65/m^3
Operation and maintenance cost	.32/m^3
Total	$1.50/$m^3$

It should be noted that this plant operates automatically and has, therefore, a relatively low labor cost (25 yen/m^3).

C. COSTING OF LARGE PLANTS BASED ON CURRENT AND PROJECTED TECHNOLOGY

1. *Reverse Osmosis*

a. 1975 DuPont Cost Study Applying OWRT Ground Rules. As of mid-1977, only small-scale RO units were in operation and no commercial cost data for large plants were available. For such plants only a few cost studies, based on an incompletely tested membrane technology, have been reported.

One of the most comprehensive cost studies was made by the DuPont Company for OWRT, and was published in November 1975. This study evaluated the capital investment and operating cost for 1- and 5-mgd RO plants based on various noncellulosic membrane modules and various feedwater pretreatment systems (Hornburg *et al.*, 1975).

According to the results of the just-mentioned study, the investment cost, excluding the feedwater pretreatment system, ranged between \$2.4 and \$3.8/gpd for 1-mgd plants, and between \$1.9 and \$3.1/gpd for 5-mgd plants. The additional investment cost for the pretreatment systems ranged between \$.49 and \$.82/gpd for the 1-mgd plants, and between \$.34 and \$.59/gpd for the 5-mgd plants.

The corresponding unit water costs were evaluated for the following main economic ground rules: annual interest rate of 5.25%, annual fixed charge rate (excluding membranes) of 7.95%, plant factor of 90%, electric power cost of 0.7 ¢/kW hr, and a membrane life between 1 and 3 yr. Based on these ground rules the following unit water costs were obtained:

	Plant capacity (mgd)			
Plant capacity, mgd	1		5	
Unit water cost	\$/kgal	\$/m^3	\$kgal	\$/m^3
For 1-yr membrane life	3.75–5.30	.99–1.40	2.35–3.52	.62–.93
For 3-yr membrane life	3.18–4.73	.84–1.25	1.85–2.91	.49–.77

The lower cost figures were obtained for 8-in. diameter hollow-fiber modules and the higher cost figures for 4-in. diameter spiral-wound modules.

b. 1977 Cost Study with Modified Ground Rules. The basic cost data of RO membranes plant equipment used in the DuPont study, updated to mid-1977 prices, we used for estimating large-scale plants in the 1- to 10-mgd range. The site costs and the indirect cost were based on a study of large RO and other desalting plant types operating in conjunction with small nuclear reactors (Glueckstern and Kantor, 1977).

The unit water costs were evaluated with the same economic ground rules as defined in Table II to estimate the cost of membrane plants desalting brackish water.

The main technical parameters and costing bases of these cost estimates are summarized in Table VI.

The resulting capital investments and unit water costs of 1- , 5- , and 10-mgd plants, operating with standard seawater salinity (35,000 ppm TDS) at a temperature of 25°C, along with a sensitivity analysis of the main economic parameters, are summarized in Table VII.

The effect of feedwater salinity and temperature on capital investment and unit water costs was evaluated with the aid of a computer program developed at ORNL (Glueckstern *et al.*, 1976). The results of this evaluation are illustrated in Figs. 5 and 6.

TABLE VI

Bases and Assumptions for the Evaluation of Large RO Seawater Plants[a]

Technical parameters	Data
Membrane type	B-10 Aromatic Polyamide Hollow Fiber
Nominal module diameter (in.)	8
Initial productivity (gpd)	5000[b]
Salt rejection (%)	99
Log flux decline rate (m)	−0.05
Membrane life (yr)	3
Operating pressure (kg/cm^2)	56 (800 psig)
Overall product recovery (%)	30
Seawater concentration (ppm TDS)	35,000
Seawater temperature (°C)	25
Pretreatment	Alum, sand filtration, diatomaceous earth filtration, and UV sterilization
Overall efficiency of high-pressure pumps	76%
Overall efficiency of power recovery turbine	65%
Costing bases	
Basic RO equipment and pretreatment	Based on Glueckstern and Kantor (1977)
Site development and indirect costs	Based on Glueckstern (1979)
RO membrane replacement cost	$2000/membrane module
Economic ground rules	See Table III

[a] 1–10 mgd.
[b] For 35,000 ppm TDS feedwater at 25°C, 800 psig pressure and 30% recovery.

c. Effect of Recent Developments on the Economics of Reverse Osmosis Seawater Desalting. Developments in RO membrane technology, such as the increase in allowable operating pressure for hollow-fiber membranes (B-10 DuPont), longer guaranteed membrane life, and the reported progress in Japan (Kurihara *et al.*, 1981; Matsui *et al.*, 1981), significantly reduce the cost compared with the technology developed in the 1970s.

Improved membrane technology made it possible to reduce capital investments and even more significantly, membrane replacement cost.

Another development affecting energy consumption is related to the progress in the technology of power recovery (Woodcock and White, 1981). More efficient power-recovery turbines, and higher product recovery rates made possible by higher values of membrane salt rejection resulted in significantly lower specific energy consumption.

For large plants (25 mgd or larger), specific capital investment of 2 to 2.5 $/gpd and specific energy requirements of 4 to 5 kW hr/m^3 were considered applicable in the early 1980s (Glueckstern *et al.*, 1981). The result-

TABLE VII

Capital Investments and Unit Water Costs of Large RO Seawater Plants[a]

Costs and investment	Plant capacity (mgd)								
	Mid-1970s						Projected[d] with power recovery		
	Without power recovery			With power recovery					
	1	5	10	1	5	10	1	5	10
Specific investment ($/gpd)	4.49	3.22	2.90	4.66	3.28	2.94	3.94	2.66	2.33
Unit water cost, (¢/m^3)									
Capital cost	39.2	27.7	24.9	40.7	28.3	25.3	34.6	23.1	20.1
Power cost	29.5	29.5	29.5	21.3	21.3	21.3	22.7	22.7	22.7
Membrane replacement	16.8	16.8	16.8	16.8	16.8	16.8	11.8	11.8	11.8
Operation and maintenance	16.8	11.8	10.6	16.8	11.8	10.6	15.6	10.5	9.3
Total	102.3	85.8	81.8	95.6	78.2	74.0	84.7	68.1	63.9
Effect of economic parameters (percent of base cost)									
25% change in capital cost or fixed charges[b]	9.6	8.1	7.4	10.6	9.0	8.4	10.2	8.5	7.9
1 ¢/kW hr change in power cost[c]	7.2	8.6	9.0	5.6	6.8	7.1	6.7	8.3	8.9
50% change in membrane replacement	8.2	9.8	10.3	8.8	10.7	11.2	7.0	8.7	9.2
50% change in operation and maintenance cost	8.2	6.9	6.5	8.8	7.5	7.1	9.2	7.7	7.3

[a] 1977 cost study.

[b] Reference fixed charges of 11.185%.

[c] Reference power cost of 4 ¢/kW hr.

[d] 12 in. modules operating at 1000 psig (70 kg/cm^2) and 35% recovery; membrane cost is 70% of 1977 Cost.

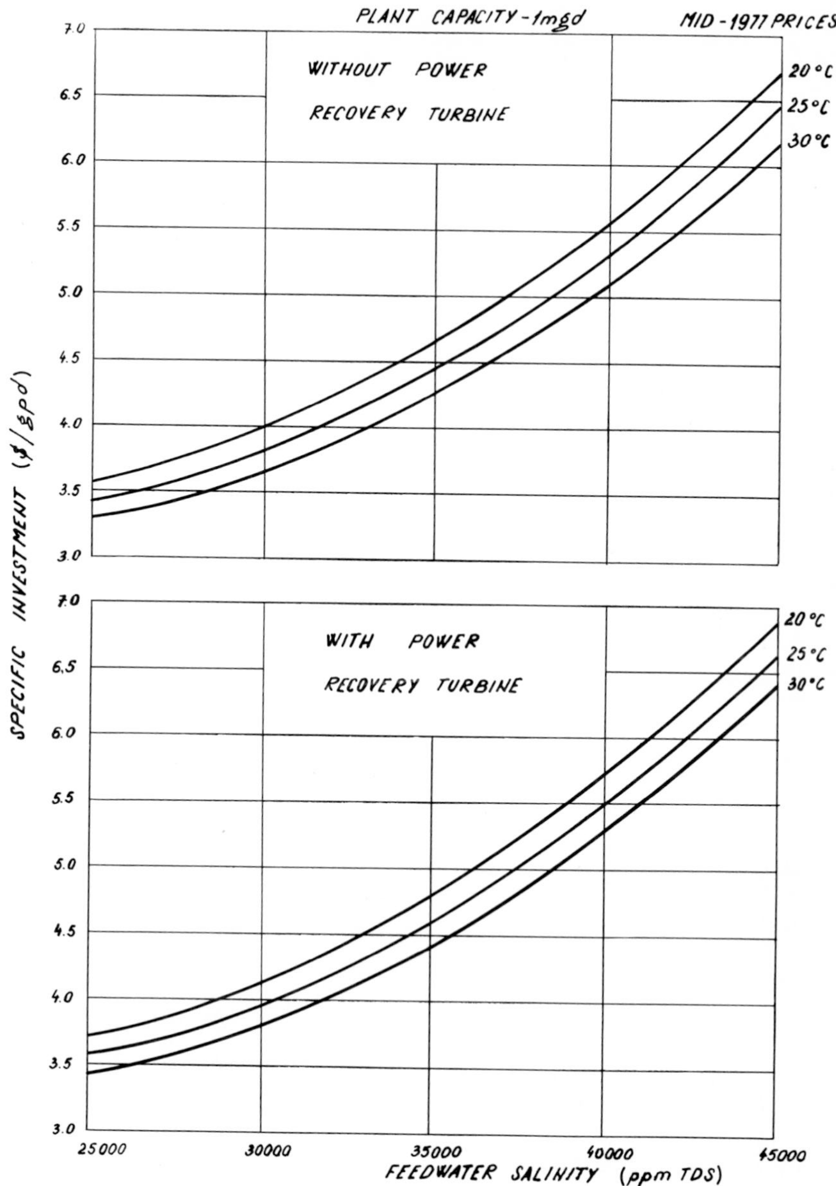

Fig. 5 Specific investments of RO seawater plants versus feedwater salinity and temperature.

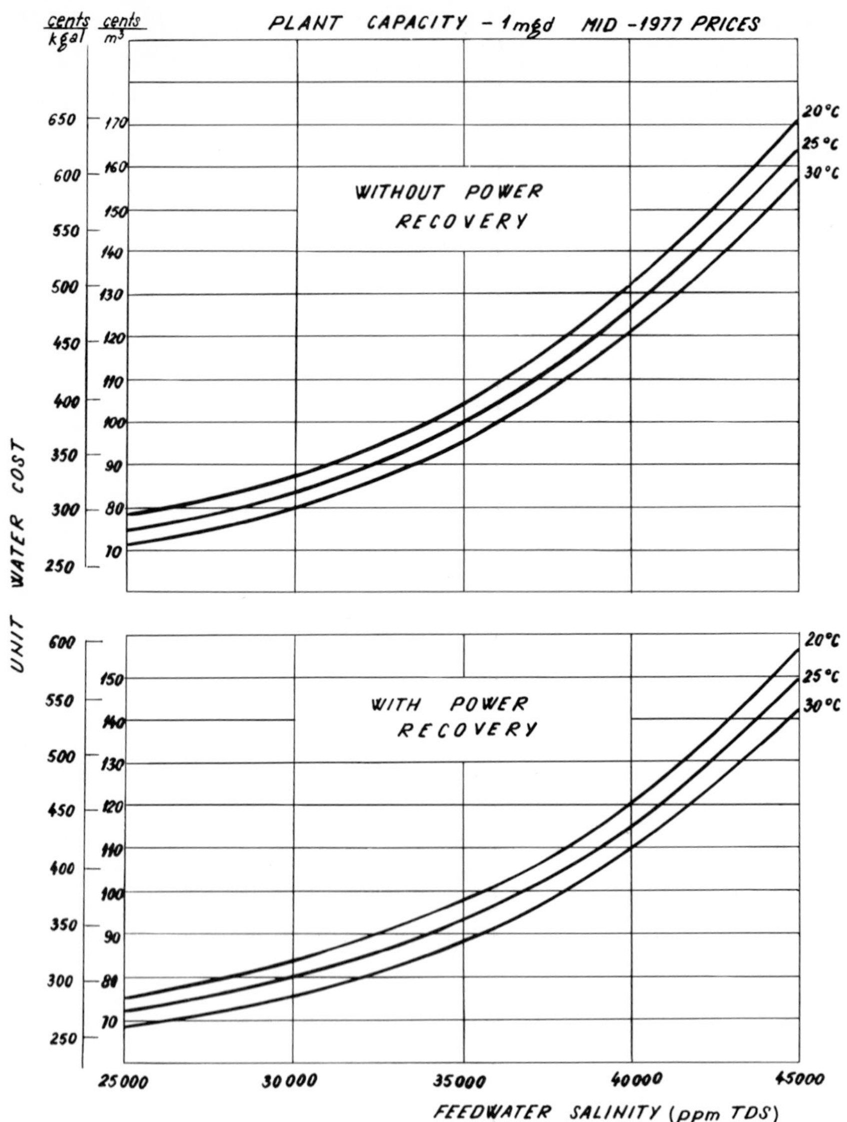

Fig. 6 Unit water costs of RO seawater plants versus feedwater salinity and temperature.

ing unit water cost, based on these figures and economic ground rules defined in Table II, range between 0.60 and 0.70 \$/m^3. These cost figures are similar or lower than those estimated in the 1977 cost study (see Table VII), extrapolated to larger plant sizes.

From the preceding, it can be concluded that the developments in RO technology have compensated or more than compensated the equipment and materials price inflation during mid-1977.

The range of cost figures based on mid-1970s technology and those projected for improved technology, evaluated for mid-1977 prices (Table VII), may, therefore, present a fair approximation for the costs of RO seawater desalting prevailing in the early 1980s.

2. *High-Temperature Electrodialysis*

As already discussed, conventional ED is a relatively high-energy consumer when used to convert high-saline waters. Therefore, it cannot compete with the RO process to desalt seawater. The current development work relating to high-temperature ED may, however, make this process much more attractive, primarily because of the potential of a significant reduction in energy consumption.

An economic evaluation of this process was made by the U.S. Bureau of Reclamation by assuming a reasonable range of membrane costs and lifetimes. Based on this data an estimate of ED equipment and operating cost of a 165-m^3/hr (1.05 mgd) plant was published at the 5th International Symposium on Fresh Water from the Sea (Leitz, 1976).

Based on the previously mentioned basic data, an estimate of the total capital investment and the corresponding unit water cost, in accordance with the economic ground rules summarized in Table II, was made. The reported ED equipment cost was escalated to mid-1977 prices and all other site-related costs were estimated with the relevant costing factors used for RO plants. The resulting capital investment and unit water cost, along with a sensitivity analysis of the main economic parameters, are summarized in Table VIII.

Other cost studies (Parsi *et al.*, 1980) reported an estimated cost of \$4.69 per gpd for a 1-mgd plant. The total cost, including indirect cost, was estimated at \$6.18 per gpd, with the costs being based on the second quarter of 1980. The resulting unit water cost, evaluated for a 2.5 cent/kW hr power cost and a capital recovery factor of 0.1174, was \$1.24 per cubic meter. After adjustment to the higher power cost and somewhat lower capital recovery factor, as per Table II, the resulting unit water cost amounts to approximately \$1.4 per m^3. This cost is close to the one obtained by escalating the cost from the 1977 study to 1980 prices.

TABLE VIII

Capital Investment and Unit Water Cost of High-Temperature ED Seawater Desalting Plant[a]

General design data[b]	
Plant capacity	165 m^3/hr (1.05 mgd)
Current density	350 A/m^2
Operating temperature	65°C
Power consumption	11 kW hr/m^3
Capital cost ($$10^6$)	
Desalting equipment (excluding membranes)[b]	1.97
Membranes at $40/$m^2$	0.68
Site development, buildings, seawater intake, and pretreatment system	1.10
Total direct cost	3.75
Indirect cost (30%)	1.13
Total capital cost	4.88
Unit water cost[c] (¢/m^3)	
Capital cost	42.6
Power cost	44.0
Membrane replacement (5-yr membrane life)	10.4
Operation and maintenance including chemicals	14.4
Total	111.4
Effect of economic parameters (percent of base cost)	
25% change in capital cost or fixed charges	9.6
1 ¢/kW hr change in power cost	10.1
50% change in membrane replacement cost	4.8
50% in operation and maintenance cost	6.6

[a] 1977 cost study.

[b] Based on U.S. Bureau of Reclamation (Leitz, 1976) cost data adjusted to mid-1977 prices.

[c] Economic ground rules—same as for RO seawater plants, except membrane life.

D. Comparative Economics and Overall Sensitivity Analysis of RO Technology

1. *Effect of Energy Cost and Cost Comparison with Distillation*

An economic comparison of RO seawater desalting with the most frequently used distillation technology, multistage flash evaporation (MSF), is illustrated in Figs. 7 and 8 for 1- and 10-mgd plants, respectively. The comparison was made for technologies prevailing in the mid-

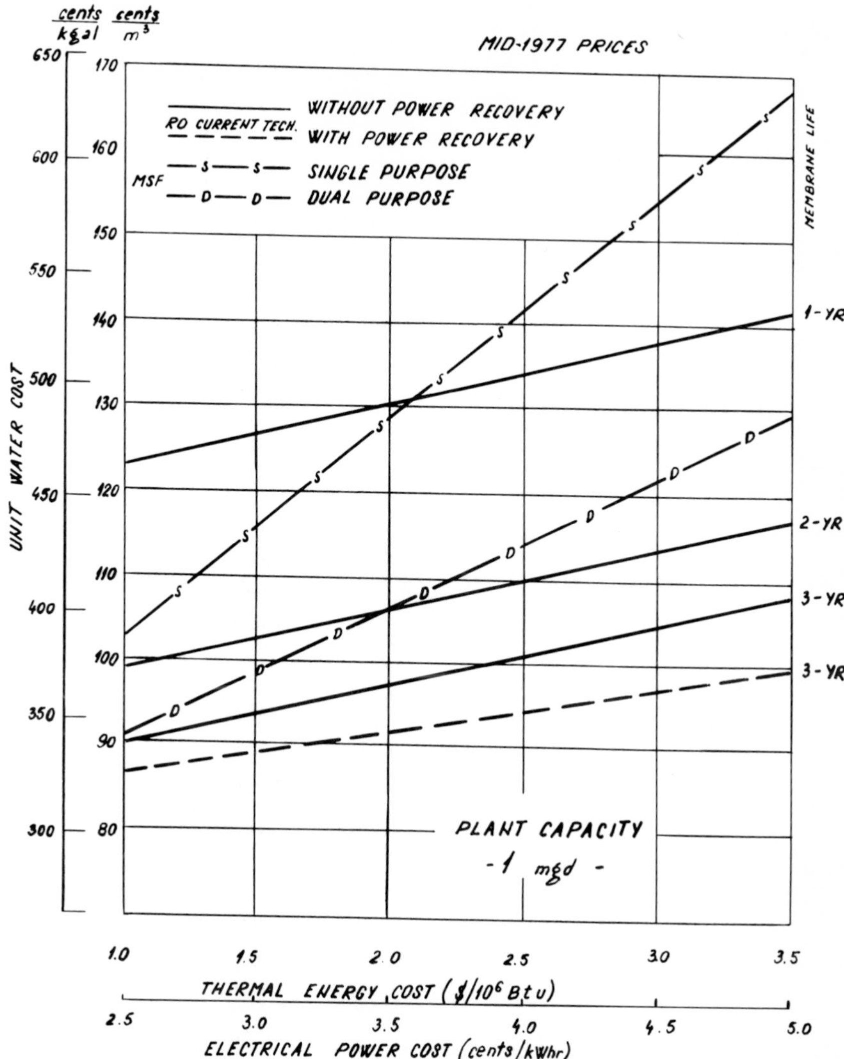

Fig. 7 Unit water costs of 1-mgd RO seawater and MSF single and dual-purpose plants versus energy costs.

1970s. The cost data for the single- and dual-purpose MSF plants compared were taken from a study made at Oak Ridge National Laboratory (Glueckstern and Reed, 1976), and reevaluated to mid-1977 prices and the economic ground rules applied in the RO cost study.

From the results shown in Fig. 7, it is evident that for plant capacities in the 1-mgd range, RO technology was competitive with MSF plants,

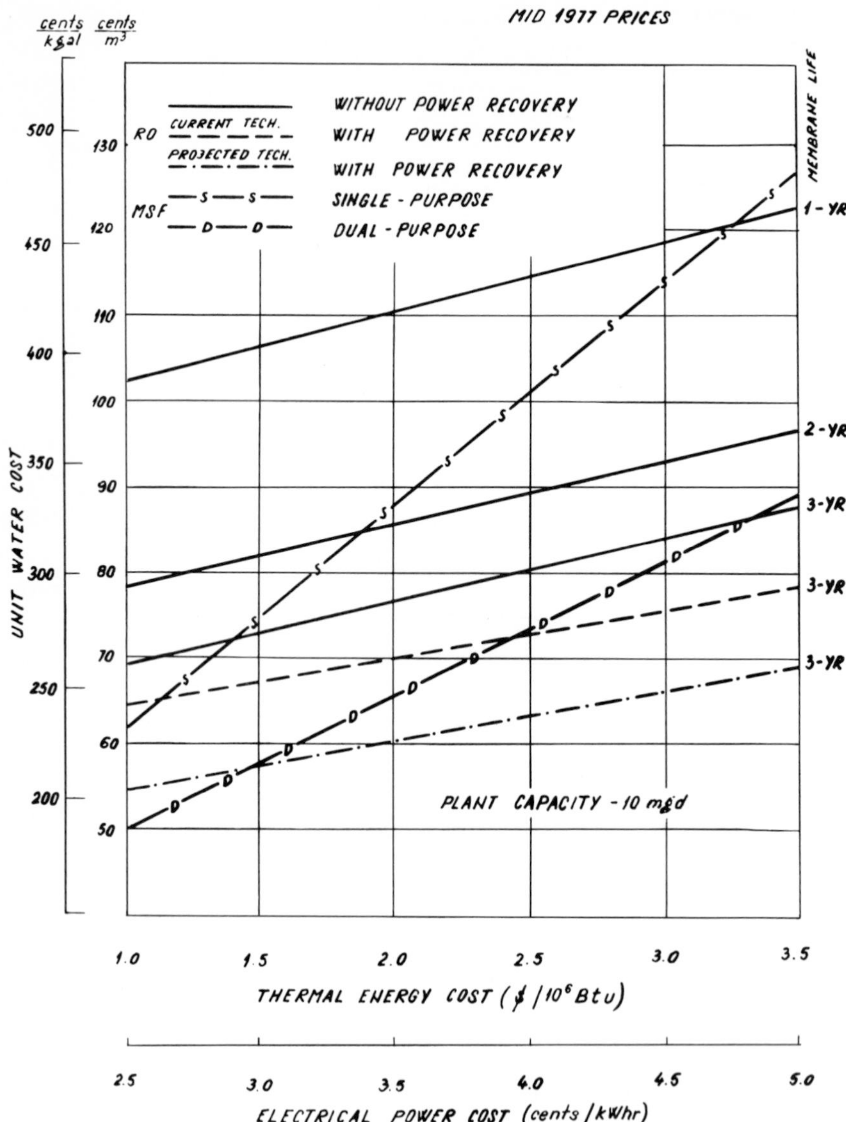

Fig. 8 Unit water costs of 10-mgd RO seawater and MSF single and dual-purpose plants versus energy costs.

even for prevailing low energy costs, providing a reasonably long membrane life could be realized. Obviously, at high energy cost levels the economics of RO technology was much more competitive. In the 10-mgd range (Fig. 8), competition of RO technology with dual-purpose MSF plants was possible only at higher energy cost levels, while realizing a membrane life of not less than 3 yr. For projected RO technology it was concluded in 1977 that RO would be able to compete with thermal dual-purpose plants even for locations with prevailing low energy costs (see Fig. 8). Other cost studies (Glueckstern, 1979) confirmed this trend of increasing competition of the RO process.

2. *Overall Sensitivity Analysis of Reverse Osmosis Technology*

The present evaluation was based on a set of technological and cost parameters that are still changing rapidly with the developmental status of membrane technology. The more important factors affecting economics, which are not yet fully established, are membrane initial and replacement costs, membrane lifetime and flux decline, and the strongly site-dependent raw seawater pretreatment requirements. Several additional technological parameters, however, such as the efficiencies of high-pressure pumps and power-recovery turbines, and the cost of the seawater intake system may also considerably effect the economics of the RO process. Therefore, a brief sensitivity analysis of these and several other factors was made to determine their effect on the unit water cost. As reference case for this investigation, the 1-mgd plant of the 1977 cost study was selected. The effect of these parameters varied singularly, while all others were kept at their reference design values. They are summarized in Table IX. The combined effect of varying several parameters simultaneously, representing overall sets, optimistic and pessimistic assumptions were also evaluated and are shown in Table IX.

The results of the sensitivity analysis emphasized again the strong effect of membrane life on the unit water cost. The other factors characterizing the membrane, however, such as log flux decline and, of course, the unit membrane cost, will also have a major influence on the potential competitiveness of the RO process for seawater desalting. The cost of the pretreatment and the seawater systems, as well as the other "nonmembrane" factors, such as the pump efficiencies and the power-recovery turbine, have only a minor effect alone.

The combination of several of these factors may, however, increase or decrease the unit water cost substantially.

TABLE IX

Effect of Technological and Cost Parameters on the Unit Water Cost[a] of a 1-mgd RO Seawater Plant

	Cost increase		Cost saving	
Parameters varied	(¢/m³)	(%)	(¢/m³)	(%)
Unit membrane initial and replacement cost				
33.3% higher	8.3	8.8		
33.3% lower			8.3	8.8
Membrane life (yr)				
Lower (2)	8.7	9.2		
Higher (5)			7.0	7.4
Log flux decline (m)				
Higher −0.08	5.1	5.4		
Lower −0.03			2.9	3.1
Operating pressure				
Lower 750 psig (52.7 kg/cm²)	6.2	6.6		
Higher 900 psig (63.3 kg/cm²)			3.2	3.4
Overall efficiency of high-pressure pumps				
Lower (68%)	3.3	3.5		
Higher (84%)			2.6	2.7
Overall efficiency of power recovery turbine				
Lower (60%)	0.6	0.6		
Higher (75%)			1.2	1.3
Capital cost of pretreatment system				
Higher (+50%)	6.3	6.7		
Lower (−50%)			6.3	6.7
Capital cost seawater intake				
Higher (+50%)	1.7	1.8		
Lower (−50%)			1.7	1.8
Combined effect of all cost increasing factors	51.8	54.8		
Combined effect of all cost savings			26.9	28.4

[a] Unit water cost from 1977 cost study = 94.6 ¢/m³.

3. *Comparative Economics of Projected RO Seawater Technology with Distillation*

The comparative economics of membrane processes are still dependent on several technological and economical factors that are not fully established. These factors can be analyzed by comparing RO technology with the most economic alternative of using a low-temperature thermal process coupled to a power station.

The capital cost of these two alternatives for large-scale desalting plants, in both cases utilizing the seawater intake of the power station, would be approximately the same, provided available seawater quality would not require an extensive pretreatment for the membrane process (Glueckstern *et al.*, 1981).

The operation and maintenance cost, including energy consumption and membrane replacement cost for the RO process would also be more or less the same.* For this case, the factors to be compared are the energy cost and the additional membrane replacement cost of the RO process.

The energy consumption of RO plants using efficient power recovery turbines while operating at 800 psi (56 kg/cm^3) and 35% recovery is approximately 4 kW hr/m^3. The low-temperature thermal process is approximately 6 kW hr/m^3 (Adar, 1977). Studies of RO technology predict a specific membrane cost (excluding vessel) of less than \$0.5/gpd (Matsui *et al.*, 1981). This cost could probably be realized for large-scale production, but may also be lower if further improvement in membrane technology can be achieved.

For this hypothesis, the factors involved in comparing the RO process for seawater desalting with the most economic thermal process narrow down to the following: (1) unit energy costs (cents/kW hr), (2) specific membrane cost (\$/gpd), and (3) membrane life.

The lower energy cost of the RO process (approximately 2 kW hr/m^3 less) would range between 6 and 8 ¢/m^3 (assuming a low cost of 3 ¢/kW hr for a large nuclear power station and 4 ¢/kW hr for a conventional power station), and could, therefore, fully compensate the unit membrane replacement cost if a 5-yr membrane life could be realized or, alternatively, by a shorter membrane life if further reduction in specific membrane cost will be achieved.

The preceding comparative analysis has not taken into account several factors that are very difficult to define. Some of them such as the comparative plant-operating factor can be economically analyzed, but others such as long-range prediction of energy cost, especially for the thermal process coupled to a power station, would be much more difficult to predict.

Concerning the plant-operating factor, the RO process has the advantage of not being dependent on a single power source, as well as probably having a higher desalting-plant availability because of much smaller module sizes and less complicated mechanical equipment.

* The chemical treatment cost would be higher for the RO process, but staff and maintenance of the mechanical equipment would be somewhat lower.

Thus, thermal processes coupled to power stations would have a plant factor of not more than (as assumed in Adar, 1977) 70%, compared with a 90% plant factor in the case of an RO plant. For a plant having a capital cost of $2/gpd and assuming a fixed charge rate of 10% per annum, the effect of the reduced thermal plant factor is to increase the capital cost component of the unit water cost by 4.5 ¢/m^3.

Concerning the problem of coupling a thermal process to an initially base-loaded power station, it is well known that in an expanding electricity grid the newer and usually more efficient power stations are base loaded, while older stations are gradually less utilized. Coupling of a thermal plant to a power station may, therefore, cause a potential economic burden that has to be taken into account, especially with the prevailing energy situation where shifting to alternative energy sources such as coal versus oil and nuclear versus conventional must be considered.

Other factors to be considered in favor of RO desalting compared to thermal desalting plants are (1) the possibility of installing capacity in several stages with a smaller economic penalty, and (2) potential advantage of utilizing more efficient membrane technologies every few years when membranes are replaced.

From the preceding discussion it can be concluded that RO seawater technology will probably be a serious contender for large-scale desalting.

IV. General Conclusions

Wherever brackish water is available, membrane processes are currently more economical than thermal desalting processes, even when coupled to a power station and utilizing low-grade heat. Moreover, in some cases capital investments of new membrane plants can be economically justified to replace existing distillation plants. This is because for distillation plants the operational costs alone (primarily energy costs) are significantly higher than the total unit water costs of membrane plants.

For seawater desalting, RO seawater plants are in many cases competitive with distillation processes, especially for small plants at locations with prevailing high energy costs. At the present state of technology and commercialization of RO seawater technology, no general conclusion can be made without a detailed investigation for any specific application.

It is generally accepted that membrane processes have the largest development potential for long-term desalting of all kinds of saline waters.

References

Adar, J. (1977). *Desalination* **20,** 143–154, Elsevier, Amsterdam.

American Water Works Association Research Foundation (1973). "Desalting Techniques for Water Supply Quality Improvement."

Boesh, W. W. (1981). *Desalination* **38,** 485–496, Elsevier, Amsterdam.

Curraw, H. M., Dykstra, D., and Denkeremath, D. (1976). "State-of-the-Art of Membrane and Ion Exchange Desalting Processes" (Final Report HIT–658 for OWRT). Hittman Associates,

El-Ramly, N. A., and Congdon, C. F. (1981). "Desalting Plants Inventory (Report No. 7. National Water Supply Association.

Engineering News Record. (1972–1982). McGraw-Hill, New York.

Glueckstern, P. (1979). *Desalination* **30,** 223–234. Elsevier, Amsterdam.

Glueckstern, P., and Kantor, Y. (1977). "Economic Evaluation of Using Small Sized Nuclear Reactors for Seawater Desalination." *In* IDEA International Congress on Desalination and Water Reuse, Tokyo, Vol. 1, pp. 101–110.

Glueckstern, P., and Reed, S. A. (1976). "A Comparative Investigation of the Economics of Seawater Desalting Based on Current and Advanced Distillation Concepts" (Report No. ORNL-TM-5229), Oak Ridge National Laboratory, Tennessee.

Glueckstern, P., Wilson, J. V., and Reed, S. A. (1976). "RO-75: A Fortran Code for Calculation and System Design Optimization of Reverse Osmosis Seawater Desalination Plants" (Report No. ORNL-TM-5231), Oak Ridge National Laboratory, Tennessee.

Glueckstern, P., Kantor, Y., and Wilf, M. (1981). "Application of Synthetic Membranes in Water Supply System in Israel." *In* Am. Chem. Soc. Symp. Ser., Synthetic Membranes, Vol. 1, 63–77.

Hickman, C. E., Jamjoom, I., Riedinger, A. B., and Seaton, R. E. (1979). *Desalination* **30,** 259–281.

Hornburg, C. D., Morin, O. J., and Hart, G. K. (1975). "Commercial Membrane Desalting Plants: Data and Analysis" (Final Rep. for the Office of Water Research and Technology). U.S. Department of the Interior, PB 253–490, Washington, D.C.

Kurihara, M., Harumiya, N., Kanamaru, N., Tonomura, T., and Nakasatomi, M. (1981). *Desalination* **38,** 449–460, Elsevier, Amsterdam.

Larson, T. J., and Leitner, G. (1979). *Desalination* **30,** 525–539, Elsevier, Amsterdam.

Leitz, F. B. (1976). "Desalination of Seawater by Electrodialysis." 5th International Symposium on Fresh Water from the Sea, Alghero, Vol. 3, pp. 105–114.

Macgowan, C. F., Ammons, D., Mahon, H., Wagener, E., and Davis, T. (1976). "Hollow Fiber Membrane Seawater Desalination." 5th International Symposium of Fresh Water from the Sea, Alghero, Vol. 4, pp. 385–396.

Matsui, H., Matsumoto, H., Kuzumoto, H., Sekino, M., Nimura, Y., and Ukai, T. (1981). *Desalination* **38,** 441–448, Elsevier, Amsterdam.

Parsi, E. J., Prato, T. A., and Susa, T. J. (1980). "High Temperature Electrodialysis for Seawater Desalting." Proceedings 7th International Symposium of Fresh Water from the Sea, Vol. 2, pp. 449–458.

Rosenblatt, N. W., Agrawal, J. P., Brandt, D. C., and McGinnis, P. R. (1976). "Live Seawater Experience with Hollow Fiber Permeators." 5th International Symposium of Fresh Water from the Sea, Alghero, Vol. 4, pp. 397–408.

Seto, T., Ehara, L., Komoni, R., Yamaguchi, A., and Miwa, T. (1976). "Seawater Desalination by Electrodialysis." 5th International Symposium of Fresh Water from the Sea, Alghero, Vol. 3, pp. 131–138.

Woodcock, D. J., and White, M. I. (1981). *Desalination* **39,** 447–458, Elsevier, Amsterdam.

13

Economics of the Application of Membrane Processes

Part 2: Wastewater Treatment

A. N. ROGERS*
Bechtel Corp.
San Francisco

* Engineering Consultant, Pleasanton, California.

ISBN 0-12-085480-5

I. Basis for the Economics of Waste Processing*

A. FACTORS INFLUENCING THE COST CALCULATIONS

Many techniques have been considered in the past for the reduction in the volume of aqueous waste streams. The ultimate choice of process will be influenced largely by economics. The ideal guide for the potential user would be a published chart of the investment and operating cost of each process. It is immediately apparent, however, that a chart or graph of this type would have little value. Not only do costs vary with plant size, but they are strongly influenced by the nature of the waste to be processed and the constraints placed on the streams of waste concentrate and purified water issuing from the waste-treatment plant. A more useful approach is to evaluate each of the cost components entering into the various process steps. The reader then has merely to select those which apply to his particular situation and, by adding the cost items, can rapidly arrive at a realistic estimate of the total plant investment and operating cost.

With the exception of several applications of ultrafiltration (UF) (to follow), the economics of membrane processes is influenced more strongly by the thoroughness of pretreatment than by the source of the feed. A feed that has not been properly pretreated, no matter what its source, will necessitate frequent shutdowns of the membrane plant for cleaning or, even worse, may destroy the membranes in a short time. If, by contrast, pretreatment is adequate, the membrane plant will perform equally well on aqueous wastes as on brackish water or seawater.

The determining factor in membrane-plant cost is the concentration of total dissolved solids (TDS) in the feed stream. In Chapter 11, the influence of feed concentration on plant design was discussed in some detail. It was noted that a high concentration in the feed may require a number of stages to achieve the required purity in the purified water. Alternatively, if the objective is to reduce the waste volume to the greatest possible extent, the *reject* stream of a stage must be subjected to further water extraction in subsequent stages. In batch-type electrodialysis (ED), the increase in circulation that is required for a more concen-

* Because of cost escalation and the changing inflation rate resulting from the economic down turn of the early 1980s, estimation of the current costs for membrane treatment remains an elusive target. To account for escalation, the reader may wish to use one of the published escalation tables or figures for the change of equipment cost indices, such as those in the quarterly cost roundups published in *Engineering News Record,* McGraw Hill, New York.

trated feed will reduce the output per unit membrane area. In any type of membrane operation, a higher feed TDS will require a more complex and costly plant than will a more dilute feed stream.

These remarks apply to UF as well as to reverse osmosis (RO) and ED but for different reasons (see Chapter 11). Consequently, two sets of costs are presented for each type of membrane plant, one for a high-concentration and one for a low-concentration feed. High- and low-concentration streams cannot be defined exactly, because they depend on many variables, among them the nature of the solute (or colloidal matter) and the type of membrane. At intermediate concentrations, costs will lie between the two extremes shown in the tables and figures. In general, however, a straight-line interpolation cannot be expected to yield an accurate cost figure.

One other variable has been considered, namely, plant size. A small plant is frequently sold as a skid-mounted package complete with pumps, cartridge filters, acid and polyphosphate feeding equipment, and instrumentation. In larger sizes, pumps and instruments are still supplied by most manufacturers of membrane systems, but in separate packages. The membrane plant is assembled in modular form. A large plant consists of a number of small racks (for RO or UF) or stacks (for ED), frequently served by a single control panel and a single set of pumps.

Because of the modular construction, it is apparent that the purchase price of the plant per gallon per day of feed levels off for large plants, roughly around 3–5 million gallons per day (mgd). Similarly for operating costs, the economy of size vanishes above a plant size of about 5 mgd. Each additional block of this capacity will require an operating and maintenance crew the same size as the first large block. Also, unit chemical costs are the same for a 100-mgd plant as for one of 5-mgd capacity, and quantity discounts on replacement membranes reach a constant value for the large plants.

Some variables cannot be factored into an overall cost picture. As an example, acid and scale-control-chemical costs will depend on the feed composition. For such variables, curves are presented in this chapter relating chemical consumption to the composition of the feed stream.

An important consideration in calculating costs is the need for standby equipment. Scheduled maintenance can be made to coincide with the maintenance cycle of the process plant that produces the waste requiring treatment. *Unscheduled* shutdown of the membrane plant can be costly in terms of down time of the associated process plant. Several alternatives are available for dealing with this problem. The most reliable, as well as the most costly, is the installation of three completely self-contained membrane waste-treatment plants, each of 50% capacity. Because there is little probability that two of the three membrane plants will be down at any one time, this design has high reliability. Investment in the membrane plant may be reduced either by eliminating one of the 50% capacity units or by decreasing the capacity of each unit in the three-membrane plant. In either event, sufficient storage capacity must be provided to accept all the unprocessed aqueous waste discharged by the associated process plant during an anticipated partial or total membrane unit shutdown. Such storage, in the form of lined ponds, would be costly. The cost of storage varies from less than .10 per kilogallon to more than $1.00 (Parker, 1975). For the RO and ED economics, an average value of $.65 per kgal is used here for the cost of feed storage. For UF, it is assumed that the wastes usually processed will require storage in steel tanks, costing $80/kgal (Richardson, 1977–8). The cost of storage is added to purchase price and installation expenses to yield the total installed plant cost. (Where UF is used merely for the reduction of solids in municipal water, the plant investment shown in this section errs on the high side.)

For simplicity, each of the plants discussed here is calculated as a single unit of 100% capacity. A buffer is provided by including the cost of storage for four days' feed, which should be adequate to allow for the most common interruptions encountered in the membrane plant operation.

If the feed is stored during a membrane plant outage, the plant must have sufficient excess capacity to "catch up" during normal operation by gradually processing the stored liquid. Experience indicates that with few exceptions, it is reasonable to assume a 90% stream factor. That is, the membrane plant should be capable of processing 111% of the normal anticipated feed rate. It is assumed in this chapter that the published purchase price of a membrane plant covers this 11% excess capacity.

Another interesting possibility should be mentioned here. As an alternative to the user purchasing a plant and its operation, one manufacturer of membrane systems has installed a plant at his own cost, levying a use charge. After five years of operation, the plant will be transferred to the user at no cost. (Tidball, 1976, private communication).

B. BASIC ASSUMPTIONS

In the desalination of brackish water or seawater, when the objective is producing purified water, plant investment is given as a function of the plant's water output, and operating costs are expressed in cents per kilogallon of product water. In contrast, the treatment of waste streams is generally intended to reduce the volume of aqueous waste requiring storage or disposal. Consequently, the costs in this section are expressed in terms of gallons of *feed* to the plant.

The equipment and operating costs herein are based insofar as possible on published prices or vendors' quotations. Costs are effective as of October 1977. To bring costs up to date, a number of indices are available to the reader, among them the ENR Building Cost Index, CE Plant Cost Index, and the M&S Equipment Cost Index. (See details under References.) Remember that escalation by any of the available indices during a period of more than 2 or 3 yr will yield good results when comparing various process steps but must not be regarded as an exact cost prediction.

The installed cost of the membrane plants is based on their purchase price as shown in Table I. It is occasionally assumed that the installation of a small plant costs "practically nothing." It is frequently located in a basement or empty corner of the plant which is "there anyway." To keep costs in true perspective, the project should be charged with the prorated

TABLE I

Basis for Installed Costs

	% Outdoor installation[a]	% Indoor installation[b]
Plant equipment	63	55
Site development Grading, roads, fences, service buildings	2	2
Building	4	12
Electric utilities Main transformer, wiring, switchgear	3	3
Civil/structural/misc. equipment Foundations, cranes, air compressors, acid tanks, fork lift trucks	(Included in building cost)	
Piping, valves, and instruments (Other than supplied with units) Shutoff valves, manifolds, and metering.	11	11
Engineering services Includes general contractor's overhead, supervision, and fee	17	17
Total construction cost	100	100

[a] Some portions of an outdoor plant may be protected by a roof.
[b] The cost curves in this section are based on an indoor plant.

value of the space occupied and the services provided (air, water, power) to the membrane plant.

It is also assumed that published prices of membrane-plant sales have a built-in allowance for stream factor based on operating experience. Consequently, no allowance has been included here for stream factor either in plant investment or in fixed charges. Chemicals and power will be consumed only when the membrane plant is in operation. As for labor, it is common practice to assign the operating staff to maintenance duties when the plant is down. Stream factor corrections, therefore, are not included in the cost calculations in this section.

For all operations and maintenance (O&M) cost curves presented here, the basis for the fixed charges is

(1) 20-yr plant life,
(2) zero salvage value at the end of plant life,
(3) 10% annual interest rate on capital, and
(4) 2% of capital for annual taxes and insurance charges.

The operating cost per kilogallon of feed is the sum of the fixed charges and variable costs. The variable costs in this section include

(1) labor,
(2) maintenance materials, and
(3) power costs.

They do *not* include

(1) cost of acid,
(2) cost of hexametaphosphate,
(3) price of replacement membranes, and
(4) cost of chlorine.

Expenditures for these items vary with the nature of the feed and operating conditions. Curves are included to permit the user to determine cost as applied to his particular needs. Such costs are to be added to the total basic cost obtained from the appropriate graph.

The components of the variable costs were calculated on the following basis.

Labor It is difficult to generalize as to labor costs. Pay scales for the skills required in a membrane plant vary by a factor of three throughout the United States and even more widely in other countries. For the cost curves in this section, a base rate of $5.50 per hour was used to represent the average pay of the operators and maintenance crew. In addition to this, 10% was applied to cover supervision. Fringe benefits are included at

30% and labor burden at 70% of the sum of labor and supervision costs, bringing the total to $12.10 per hour of direct labor.

Maintenance Materials Materials in this category include paint, general maintenance supplies, replacement cartridges for filters, pump parts, and pump and valve replacement. In addition, an allowance is included for cleaning chemicals, which may include citric acid, detergents, enzymes, and chelating agents. The frequency of cleaning will range from semimonthly to semiannually. The difficulty and frequency of cleaning (and, thus, the consumption of chemicals) will depend on the nature and concentration of the feed. A lump sum figure of 5% per annum of plant investment was included to cover all these costs.

Power A value of 2.5 cents per kWh was used in deriving the following Variable Cost Curves.

II. Reverse Osmosis

A. VARIABLES AFFECTING PROCESS ECONOMICS

For purposes of this cost analysis, two types of feeds are considered. Those containing up to 3000 ppm of TDS, which are approximately the concentration of the softened Wellton-Mohawk drains delivered to the Bureau of Reclamation's Yuma desalination plant (Taylor and Haugseth, 1977), are classed here as low concentration. Many single-stage RO plants successfully concentrate streams containing more than 3000 ppm; however, to promote a high flux at moderately applied pressure and to minimize scale formation, it is common practice to provide several stages of water purification as the concentration reaches or exceeds this value. Feeds containing 18,000 or more ppm of TDS are considered here as high-concentration feeds.

In the high range of feed concentrations, a multistage plant is essential in providing the required purity of product water. It is also possible to attain a low salt concentration in the purified water by using seawater membranes, which reject 99% of the feed ions in a *single* stage. That is, a 35,000 ppm seawater feed will result in a purified water stream containing about 350 ppm of total dissolved solids. Plants employing seawater membranes, however, fall into the high cost category:

(1) Seawater type membranes are more costly than membranes designed for lower concentrations.

(2) A single-stage membrane is subjected to the full 800–1000 psi pressure difference applied to high TDS feeds, thereby subjecting it to more rapid deterioration from membrane compaction.
(3) The warranty on seawater type membranes extends to only three years' operation as opposed to the five-year warranty on the low-concentration membranes.

The high pressure required by concentrated feeds and the lower water recovery per stage adds to the operating cost because of the increase in pumping power.

For unit process equipment in many industries, the investment per unit of output decreases as the plant becomes larger. This is true for RO, where a large purchaser not only receives a quantity discount on the membrane units, but also reduces the investment in peripheral equipment by installing a single bank of cartridge filters and pumps to serve many blocks of membrane vessels. For large plants, however, greater flexibility is possible if each block can be operated separately. For plant turndown or for cleaning or repairing a small group of membranes, only a portion of the entire plant need be shut down. The balance of the plant can continue

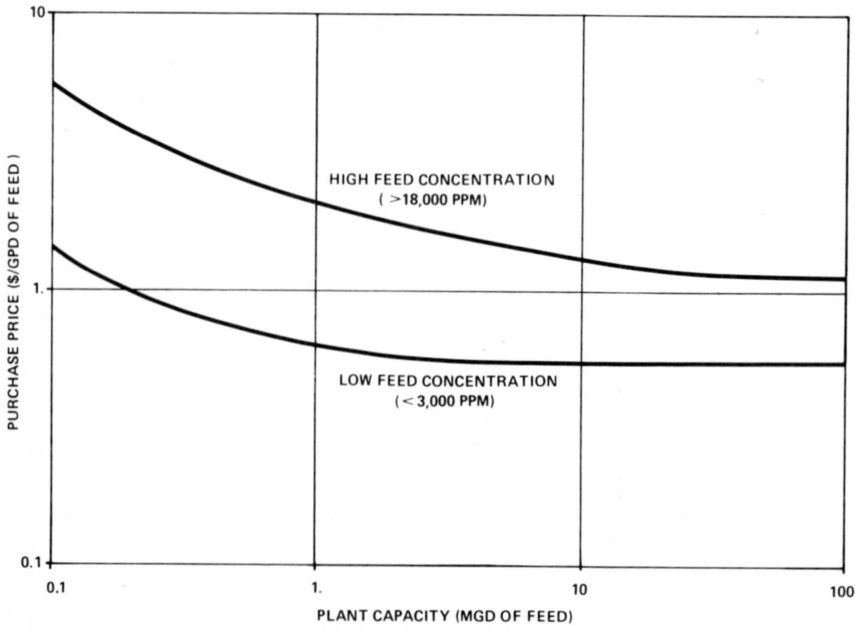

Fig. 1 Unit purchase price of RO plants versus plant capacity.

to operate. When the plant is composed of a number of identical blocks, each block can have its own pumps. Alternatively, all the blocks can have a common source of filtered, pressurized feed, with each block capable of isolation from the feed, product, and reject headers by suitable valving.

B. PLANT INVESTMENT

Figure 1 plots the unit purchase price (in dollars per gpd of feed) of RO plants for the processing of aqueous wastes as a function of plant capacity. Two curves are shown, one for a high-concentration and one for a low-concentration feed. On the basis of these prices, the total plant investment is shown in Fig. 2, as calculated by the application of the add-ons of Table 1. The total plant investment includes the estimated cost of four days of feed storage, as discussed in the previous introduction.

C. OPERATING AND MAINTENANCE COSTS

Each of the variable costs components is discussed here.

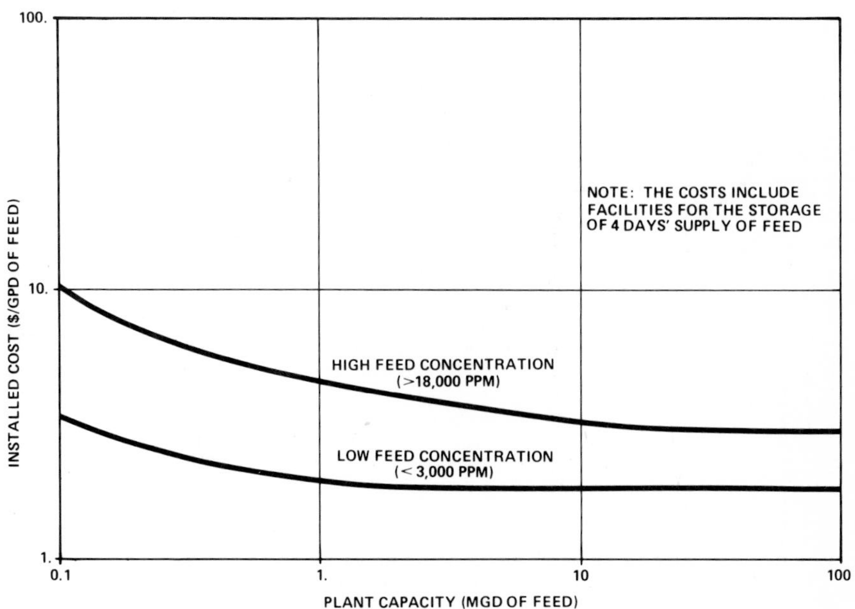

Fig. 2 Total RO-plant investment versus plant capacity.

1. Power Consumption

Reverse osmosis requires energy in the form of hydrostatic pressure. The applied pressure is determined partly by the osmotic pressure *of the reject stream* from the membrane plant.

In treating aqueous wastes containing 3000 ppm or less, the feed is delivered at 250–400 psi pressure. For higher feed concentrations, pressures as high as 1000 psi are used with high-concentration or seawater membranes; or a number of stages at lower pressure can be used with more conventional membranes. In either event, the power requirement increases with feed concentration. The energy costs shown in Fig. 3 are based on published values of power consumption for feeds of high and low concentration, respectively. (Dupee and Andrews, 1977). The cost of power was assumed to be 2.5 cents per kWh for the purpose of this calculation. The effect of size was factored in by using centrifugal-pump efficiencies ranging from 35% for small pumps to 88% for those over 200 gpm. For the smaller plants, the bottom curves indicate cost of pumping power if piston pumps, at an efficiency of 75–88%, are substituted for centrifugal pumps. Piston pumps are not commercially available in large capacities.

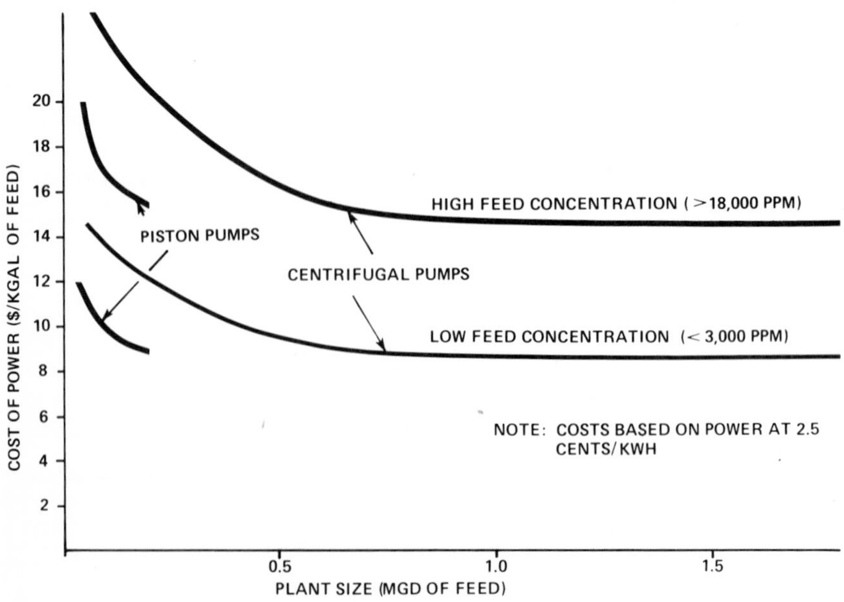

Fig. 3 Cost of power as a function of RO-plant size.

2. *Labor*

The target of the membrane-plant suppliers is to offer an automatic system requiring no operator attention. This goal is difficult to achieve, but in general, labor costs are moderate. Smaller plants that are not a critical part of a waste-treatment system can be left unattended for hours. Periodically filling the tanks with additives (acid and/or metaphosphate) is the only requirement.

The plant operator devotes a portion of his time to maintenance. For plants of intermediate size, say 200,000–1,000,000 gpd, the operator of the pretreatment and posttreatment systems occasionally checks the membrane plant, and maintenance is supplied by workers from other portions of the plant. Discussion with a number of plant owners indicates the assignment of roughly 40% of one person's time for maintenance per shift on plants in this size range. No sophisticated tasks are performed by the maintenance operator because parts are sent out to a shop when machining is required.

For capacities of 1 mgd or more, it is common practice to assign at least one full-time operator per shift to the membrane plant, whose responsibilities include cleaning and replacing the membranes. The duties of the operators will also include analyzing the feed and product streams. If the product water is destined for human consumption or is to be discharged to surface water, its analysis is usually mandated by law.

Large plants, because they cover a sizeable area, are more efficiently operated by assigning a complete crew to each block of, say, 20 mgd capacity. The block operating crew will consist of no less than one full-time operator in the control room and two on the floor, plus operators assigned to the pretreatment and posttreatment units. Also, at least two maintenance persons per shift will be required per 20 mgd block.

The above considerations provide the basis for the labor cost curve in Fig. 4.

It must be emphasized that the above estimates of labor cost are not absolute. The owner of the water treatment plant may conclude that they can get along with fewer operators than indicated here. Inadequate monitoring of plant operations, however, will be reflected in higher maintenance cost and more-frequent membrane replacement, which could possibly result in an *increase* in the total cost of water treatment.

3. *Maintenance Materials*

This component of variable costs is discussed in the introduction to this section. Particular emphasis is placed here on pump problems. Expe-

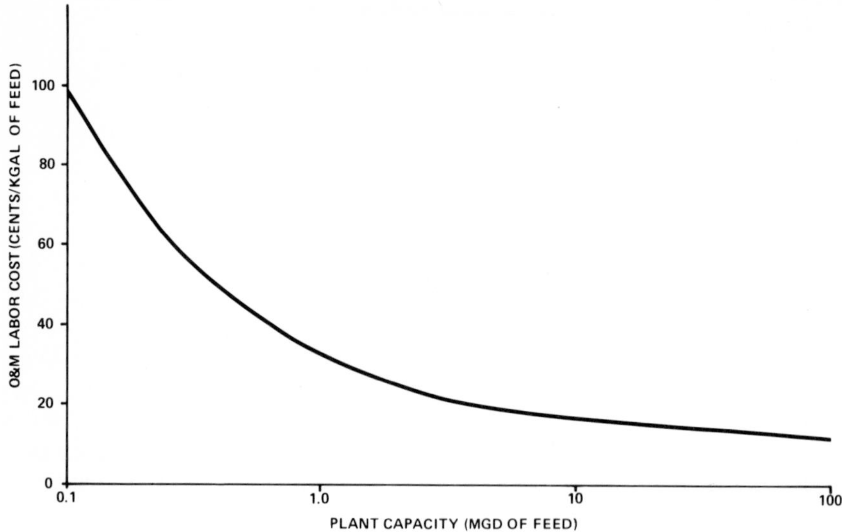

Fig. 4 Effect of RO-plant capacity on O&M labor cost.

rience indicates that the high-pressure RO pumps require continual attention and frequent maintenance. If they are properly serviced, a 20-year life expectancy is reasonable, although at the end of that period, few of the original pump parts will remain. Reciprocating pumps, which are occasionally used in small RO units, require replacement of valves, seats, gaskets, and plungers. For centrifugal pumps, which dominate the RO field, instances have been reported in which the impellers required replacement after one to two years' operation, and pump casings after four. In addition, seals and bearings require frequent replacement.

Two additional major components of maintenance material costs are cleaning chemicals and replacement elements for the cartridge filters.

These cost items are covered here by an annual charge of 5% of plant investment, an average value substantiated by actual field experience. The results are plotted in Fig. 5.

The following cost items are *not* included in the variable cost curves.

4. *Membrane Replacement*

A cost item that is strongly influenced by the nature of the feed stream is membrane replacement. For brackish-water RO, a number of installations have reported a membrane service life of 4–5 years before

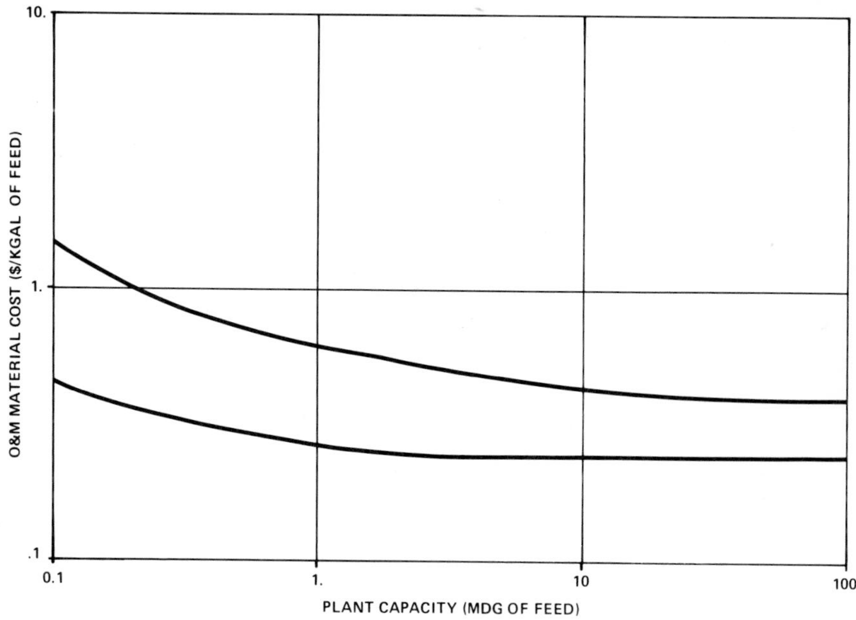

Fig. 5 Effect of RO-plant capacity on O&M material cost.

replacement is required. For wastewater service, on the other hand, there is little experience to indicate that a membrane life of more than two years can be anticipated, and many suppliers limit their warranty to one year for wastewater feeds. Figure 6 plots the contribution of membrane replacement to operating cost on the basis of a two-year operating life. For comparison, a five-year service-life curve has been included.

Membrane failures are attributable to the greater probability of membrane attack or scale formation in the RO modules by a wastewater feed. It is impossible to attach a cost to such failures because the likelihood of their occurrence depends on many variables, among them the thoroughness of feed pretreatment, the degree and quality of instrumentation, and the skill and motivation of the plant operators. The user should be aware that failures because of maloperation or lapses in pretreatment may void the membrane warranty and add substantially to wastewater processing costs.

This discussion and the curves in Fig. 6 apply to wastes of a wide range of concentrations. When the TDS of the feed to the membrane plant is high, it is possible to use special membranes. Earlier in this chapter, it was remarked that "seawater" membranes are available for the process-

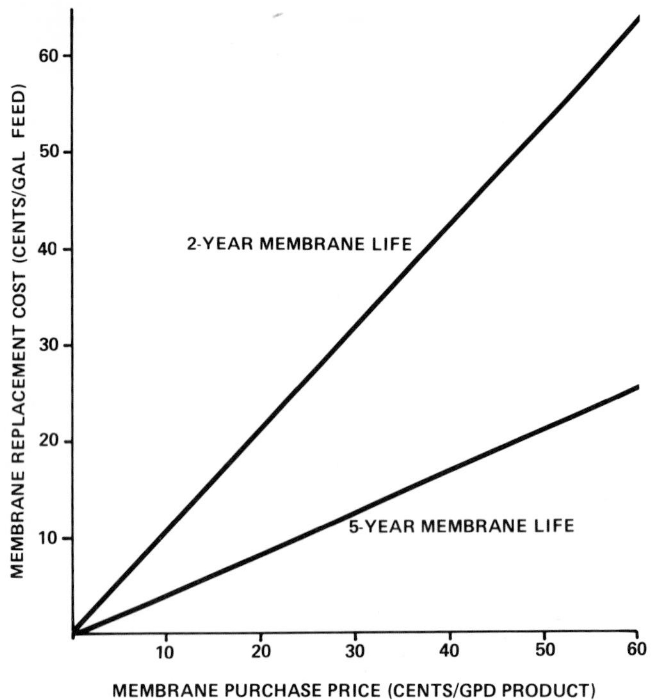

Fig. 6 RO-membrane-replacement costs for two assumed service lives.

ing of high-concentration aqueous wastes in a single stage, usually with limited water recovery. For aqueous wastes of high TDS using such membranes at a 35% water recovery, the membrane replacement cost ranges from \$1.32 to \$1.70 per kilogallon of feed, depending on the quantity of membranes purchased. This is based on a one-year membrane warranty in conformance with commercial practice for such membranes. If the emphasis is on the reduction in waste volume, the 35% water recovery corresponds to only small volume reduction, and a second stage of concentration may be required. (For an extremely concentrated waste stream, it is likely that volume reduction can be accomplished more economically by distillation.)

5. *Cost of pH Control*

Chemicals injected into the feed to prevent scale formation in the membrane plant may add substantially to the operating cost. The chemi-

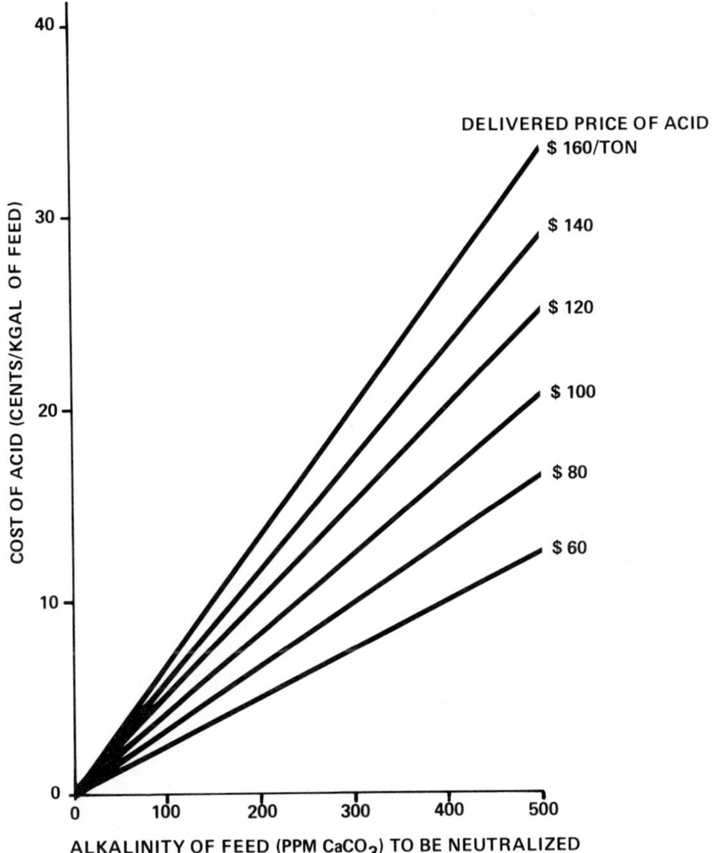

Fig. 7 Cost of pH adjustment versus alkalinity of the feed.

cal requirement depends on the chemistry of the feed stream. To prevent the precipitation of calcium carbonate when the dissolved solids are concentrated in an RO plant, acid is added to the feed to provide a Langelier index of zero or a slightly negative value *in the reject stream*. (Langelier, 1936, 1946). Sulfuric acid is used because of its price advantage over hydrochloric acid. The curves in Fig. 7 relate the cost of pH adjustment to the alkalinity to be neutralized. October 1977 prices of sulfuric acid range from \$50 to \$55 per ton in tank car lots fob works (basis 100% H_2SO_4). In small lots and out-of-the-way locations, however, the delivered price is as high as \$160 per ton. Therefore, the cost curves of Fig. 7 have been extended to include the top price. From these curves, it is apparent that the complete neutralization of even a moderate alkalinity of, for example, 150 ppm will add \$.05 to the cost of processing a kilogallon of feed at a delivered price of \$80.00 per ton of acid, which is common in many parts of the United States.

6. *Inhibition of Scale*

After the softening process in the pretreatment train (see Chapter 11), the residual calcium ion in the feed stream to the membrane plant may still be sufficient to precipitate calcium sulfate scale in the membrane modules if the sulfate ion is high. The scaling threshold is expressed by the product of the molar concentrations of calcium and sulfate ion in the *reject* stream from the membrane plant. This distinction is important because the concentration of offending ions in the reject stream may be several times as great as that in the feed stream. The control of calcium sulfate scaling can be achieved with the aid of a number of threshold scale inhibitors, among them sodium hexametaphosphate (SHMP). Threshold inhibitors function by interfering with the formation of crystals of the offending species. Thus, the required inhibitor concentration is not stoichiometrically related to the concentration of the scale former. Experience has shown, however, that the dosage of additive depends on the $CaSO_4$ in solution in a manner predicted approximately by Fig. 8, which indicates the cost of

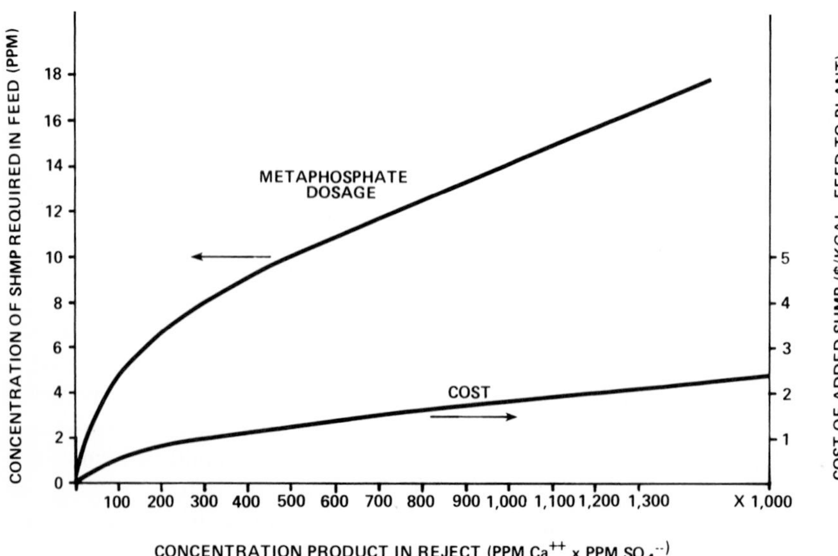

Fig. 8 Cost contribution of scale inhibitor.

the required additive. The abscissa is presented in terms of ppm of the ions rather than molarity for ease of calculation. Figure 8 errs on the side of safety because the solubility of calcium sulfate is frequently increased by the presence of certain other salts in solution.

A number of other scale-control agents are available, some of them more effective than SHMP against other impurities in the wastewater feed (see Chapter 11). For example, additives based on aminomethylene phosphonic acid (AMP) prevent not only $CaSO_4$ scaling but are effective in controlling $BaSO_4$ and $SrSO_4$, which are occasionally troublesome in wastewaters (Vetter, 1972). The newer compounds are more expensive than SHMP, ranging from \$.65 to \$1.10 per pound, and the required dosage is roughly the same as that of SHMP.

Total Operating Cost

The sum of the fixed cost plus the individual O and M costs discussed above results in the total cost. These costs are plotted as a function of plant capacity in Figs. 9 and 10, the former for an aqueous waste feed of low TDS, the latter for high concentrations. As in many types of process

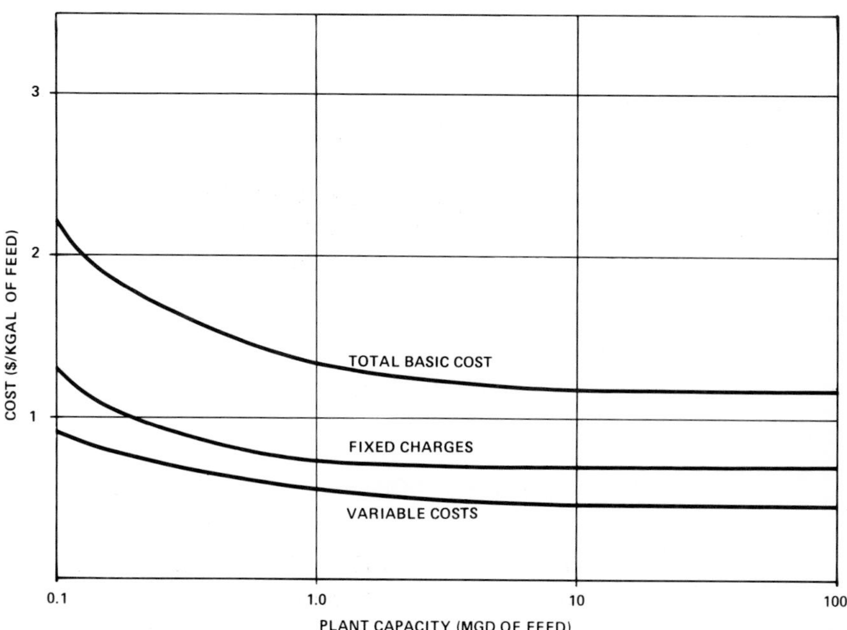

Fig. 9 Total RO basic cost for aqueous feeds of low concentration.

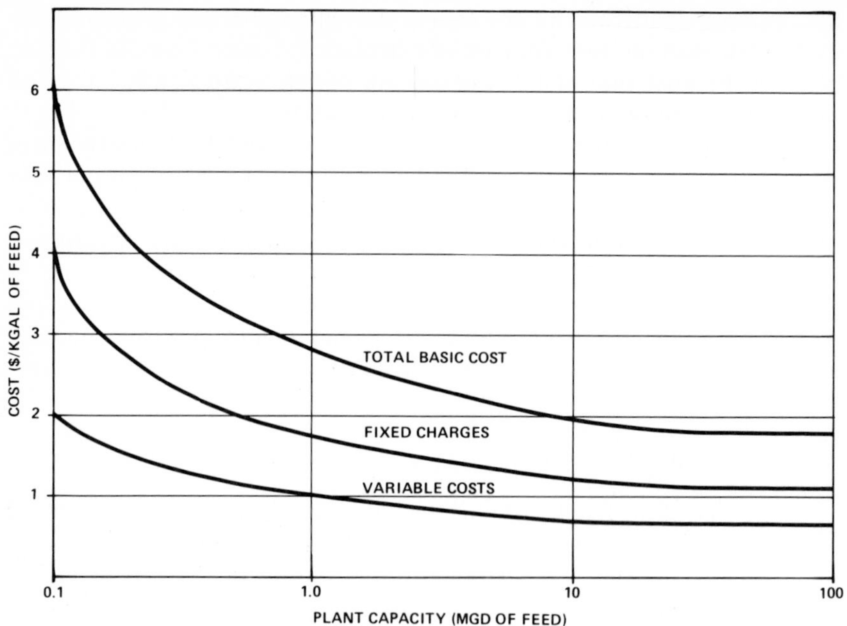

Fig. 10 Total RO basic cost for aqueous feeds of high concentration.

industries, the unit cost is high for small plants, decreasing sharply as plant size increases. The cost of treatment per kilogallon levels off for large plants.

As noted above, the *total* operating cost per kilogallon of aqueous feed is obtained by adding the respective total basic cost from Figs. 9 or 10 to the costs of membrane replacement, pH control, and scale control.

3. Electrodialysis

A. VARIABLES AFFECTING PROCESS ECONOMICS

The only commercial application of ED to high-salinity waters is the concentration of seawater to serve as feed to a salt crystallizer. A purified water stream is *not* produced in this process, and several of the transfer mechanisms and their resulting problems are entirely different from brackish water or wastewater processing. The concentration of seawater by ED falls outside the scope of this chapter.

For the treatment of aqueous wastes, the economics of ED are most favorable for dilute feeds. Because the salt rejection varies from about 45 to 55% per pass, a feed of high salinity must be subjected to a number of successive passes to produce a purified water (diluate) of dissolved solids content low enough for many uses. Where a fairly high TDS can be tolerated in the final diluate stream, feeds containing 4000 ppm or more of total dissolved solids can be processed by ED.

Two types of operation are considered in this cost section. In one, 70% of the water content of a 1500 ppm TDS feed is recovered for reuse. In the second, a 3000 ppm feed is concentrated to reduce its original volume by 83% and, in the process, to recover the purified water for reuse. The low-concentration application will require a two-stage ED plant. For the high-concentration feed, five stages are used to achieve the required volume reduction. The latter produces a reject stream containing more than 16,000 ppm of TDS.

In addition to the nature of the feed and degree of concentration desired, another important feature in any economic study is plant size. For larger plant capacities, a number of economies are possible.

(1) The number of stacks is decreased by using membranes of larger area.

(2) Several stages are incorporated in a single stack.

(3) A single pump feeds a number of stages in tandem.

(4) A single rectifier and control panel serves a number of stacks comprising a single "production block."

For extremely large plants, it is advantageous to provide a number of independent blocks, each furnished with its own pumps, rectifier, and control panel. This permits flexibility in plant turndown and isolation of banks of membranes for cleaning and replacement. Consequently, the plant cost per unit of throughput stabilizes as plant size increases.

B. Plant Investment

The unit purchase price (in dollars per gpd of feed) for ED plants to process aqueous wastes is plotted versus plant capacity in Fig. 11. As in the case of RO plants, the add-ons of Table 1 were applied to calculate the total plant investment plotted in Fig. 12.

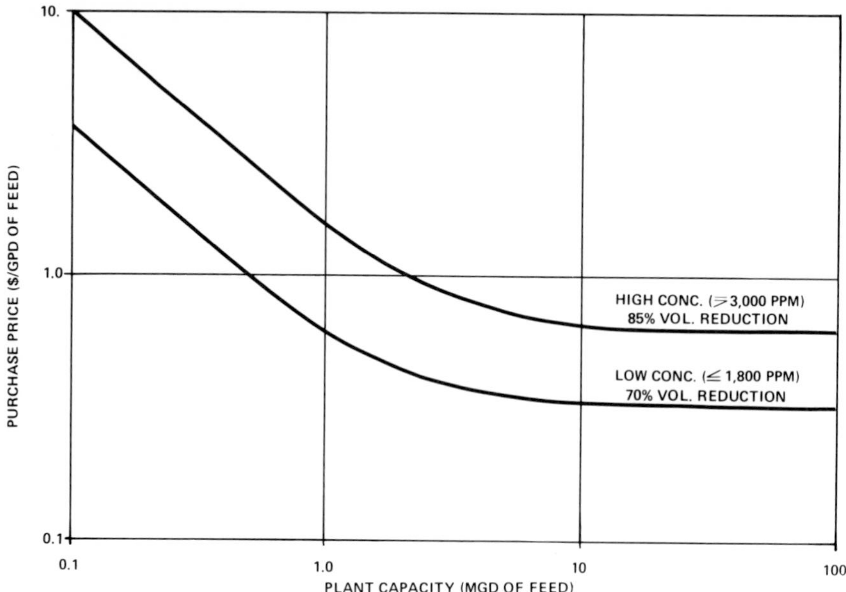

Fig. 11 Purchase price for ED plants as a function of plant capacity.

C. OPERATING AND MAINTENANCE COSTS

1. *Power Consumption*

In ED, the salts are removed from the water, whereas in RO or distillation, the water is removed from the salt. The separative work in ED is performed by direct current delivered by a rectifier that usually operates at extremely high efficiency. Because the salt ions in the liquid carry the greater portion of the electric current, power requirement is roughly proportional to salt removal.

More important than the energy loss in the power pack (transformer plus rectifier) are the inefficiencies of the ED process itself, as discussed in Chapter 11. As an example of power consumption, large units are warranted to require 4.6 kWh per 1000 gallons when reducing the total dissolved solids from 3000 ppm to 500 ppm. Manufacturers' warranties and actual plant experience form the basis for the curves in Fig. 13, where the power costs are plotted as a function of the ppm of TDS removed from the feed. The costs shown in Fig. 13 include a charge for the power consumed by the circulating pumps, which can represent as much as 15% of the total power requirement. The figure presents power costs based on

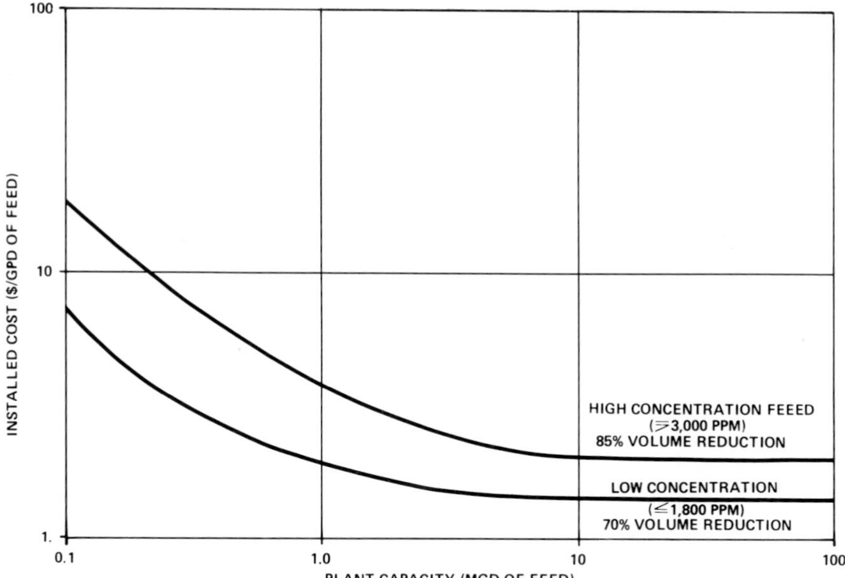

Fig. 12 Installed cost of ED plants.

rates of 2 and 4 cents per kWh, respectively. Intermediate values may be found by interpolation. For the variable costs calculation following, an intermediate rate of 2.5 cents per kWh was assumed.

A word of caution here: The warranted power consumption applies to newly installed plants. After 20,000 hours' operation (about $2\frac{1}{2}$ years), the actual power requirement may have increased by as much as 50%.

2. Labor

Small ED plants, like the RO plants discussed in Section 2 of this chapter, operate with a minimum of attention, usually only a periodic check of the meters and gauges. There are several items which do require the efforts of one or more operators:

(1) the preparation of acid and scale inhibitor solutions when required;

(2) the preparation of on-line cleaning solution and the monitoring of the on-line cleaning process;

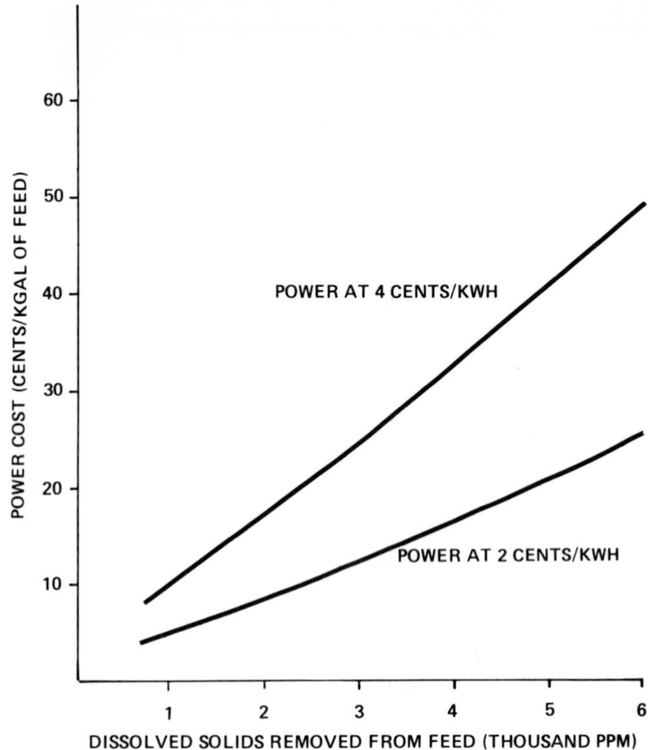

Fig. 13 Energy cost for water and waste processing by ED.

(3) detection and location of faulty cell pairs;

(4) stack disassembly for replacement of defective membranes, gaskets, and spacers;

(5) manual cleaning of the membranes when required;

(6) chemical analysis of the purified water.

Some manufacturers of ED plants specify that any stack disassembly will void the warranty and insist that defective stacks be returned to the factory. Thus, the user is spared the labor cost of disassembly and membrane cleaning. He must, however, incur the considerable expense of packaging the stack for shipment and the round-trip shipping costs.

In addition to the six items listed, labor is required for the repair or replacement of defective circulating and metering pumps.

Current experience indicates that a small plant processing 70,000 gpd of aqueous waste will require about three hours of O and M labor per day.

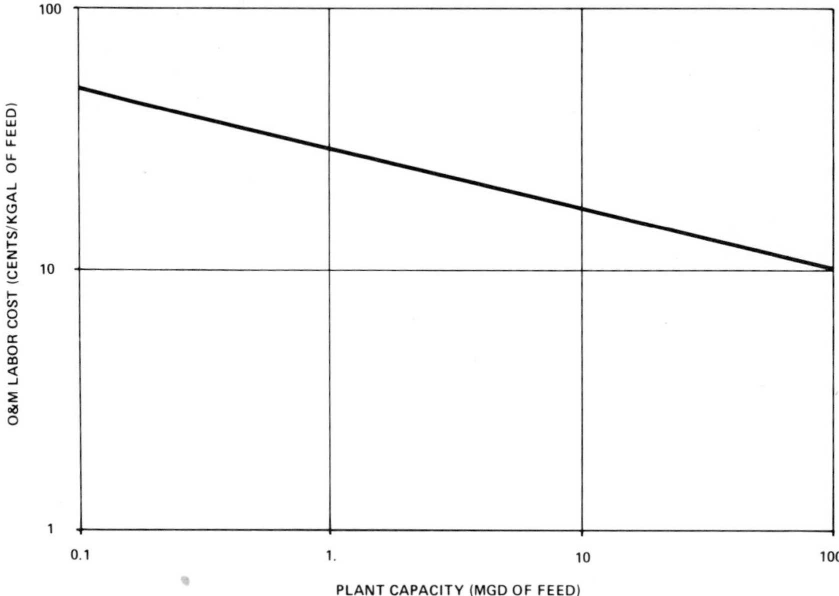

Fig. 14 O&M labor costs for ED plants.

A large plant, treating 4.5 mgd of aqueous waste, requires an average of 3 1/4 persons per shift on a three-shift basis. Figure 14 indicates a decreasing labor cost per kilogallon of feed with increasing plant size. Beyond a plant size of about 20 mgd, it is likely that large plants will be subdivided into operating blocks of 20 mgd or less, each block with its own operating crew.

It must be emphasized that these costs assume adequate pretreatment of the feed and conservative operation of the ED plant. Poor control of the pretreatment process, however, or incorrect operation of the ED plant may result in a drastic increase in the cost of maintenance labor.

3. *Cost of Maintenance Materials*

As in the case of RO plants, pump maintenance is an important item. The pressures delivered by the ED circulating and feed pumps is 90 psig or less; much lower than the pressures required by an RO plant. Although the service is not so severe, the ED feed stream may be just as corrosive. Periodic replacement is required for the impellers, seals, bearings, and casings of the feed and circulating pumps and for the components of the additive metering pumps.

A consequential item is the replacement cost of the cartridges in the cartridge filters. A definite figure cannot be given. The replacement frequency depends on the chemistry of the feed stream and the thoroughness of feed pretreatment. It can vary from once in two weeks to three or four times a year.

A third component is the cost of cleaning chemicals. A wastewater ED plant will require periodic in-place cleaning. Even with a good pretreatment system, in-place cleaning will be necessary every two to four weeks.

Because of the numerous variables involved, a precise estimate cannot be made for the cost of maintenance materials as defined. For the purposes of this chapter, Fig. 15 presents a summation of these cost items calculated as a percentage of the total installed plant cost. From the author's experience, 5% was selected as a representative value.

4. *Membrane Replacement*

Trade literature predicts membrane replacement at a rate of 10% per annum. With a difficult feed, industrial experience indicates 14% to be a

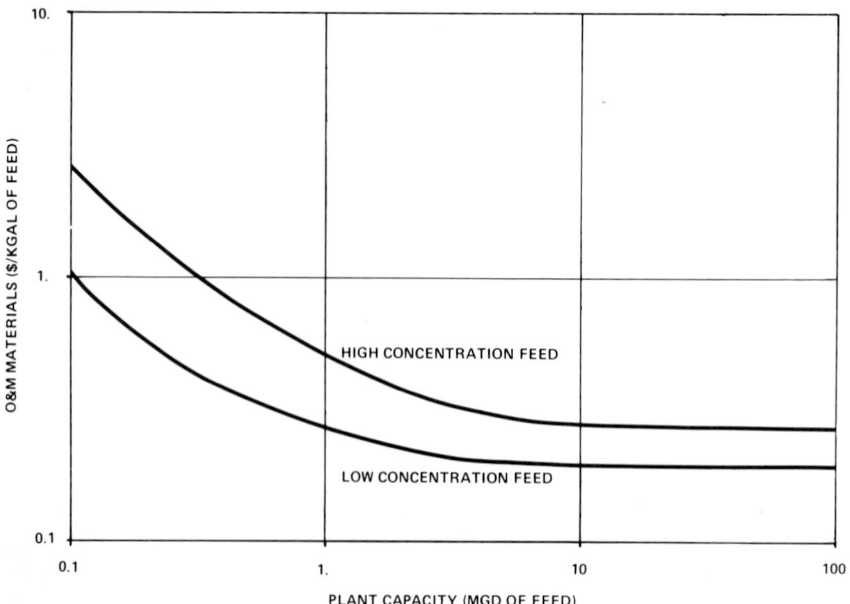

Fig. 15 Effect of ED plant capacity and feed concentration on O&M material cost.

safer estimate. Deterioration is not uniform. Most of the failures occur in the anion-selective membranes. In addition, membranes nearest the electrodes fail most frequently because of the more severe chemical and thermal environment there. Consequently, a shutdown to replace a few of the membranes may be anticipated as frequently as once in 8–12 months. During membrane replacement, defective spacers and gaskets will be replaced. Deterioration of gaskets and spacers is estimated at 5% per annum. In addition, the ED electrodes require recoating about once in five years. All of these costs are summed and presented as "membrane replacement" in Fig. 16.

5. *pH and Scale Control*

Figures 7 and 8 in Section II. (RO) are equally applicable to ED. One manufacturer states that incorporating polarity reversal in his ED units eliminates the need for scale inhibitors, pH adjustment of the feed stream, or acidification of the cathode compartment (Geishecker, 1977, private communication). Conservative operation, however, may dictate the need for small amounts of acid or inhibitor, even in a polarity-reversal plant. Because the requirement for such chemicals will vary with the feed and

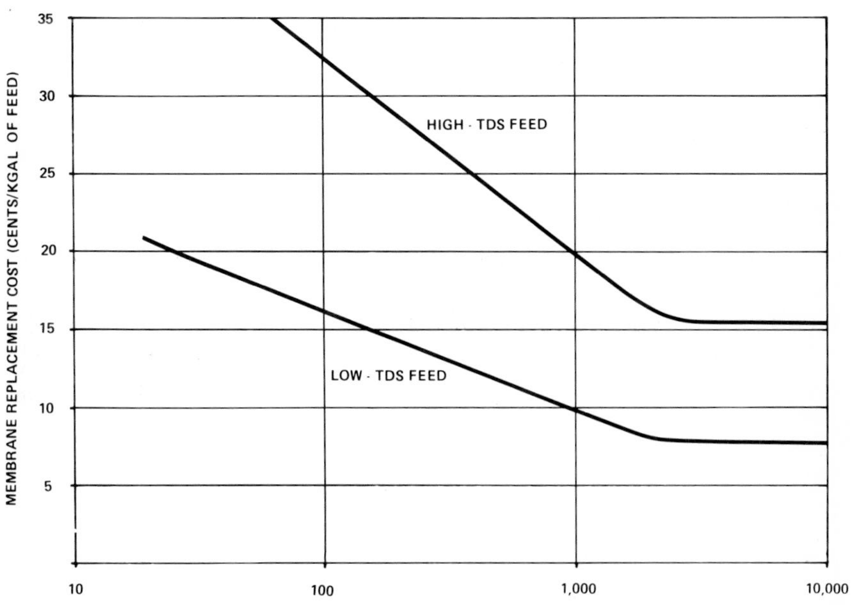

Fig. 16 ED membrane replacement costs.

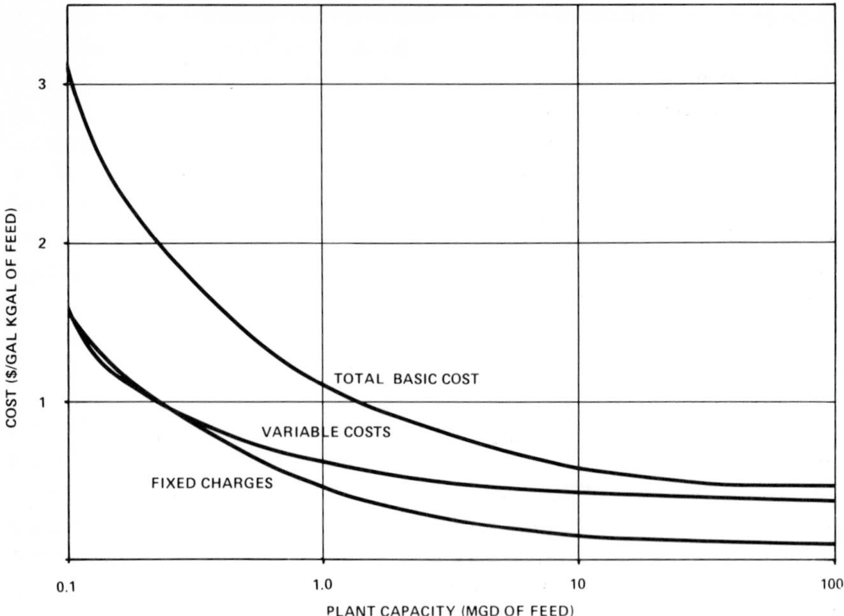

Fig. 17 Total basic ED cost for plants processing feeds of low concentration.

operating conditions, their costs are not included in the overall operating costs of the ED plant in Figs. 17 and 18.

6. *Total Operating Cost*

In Figs. 17 and 18, the fixed and variable charges are summed to yield the total basic cost curves. Figure 17 applies to feeds of low concentration, Fig. 18 to feeds of high concentration. To the total basic cost, the estimator must add the costs of membrane replacement and pH and scale control to obtain the total operating cost per kilogallon of feed.

IV. Ultrafiltration

A. Variables Affecting Process Economics

Ultrafiltration is of value in the concentration of wastes prior to disposal as well as for the concentration and recovery of valuable materials

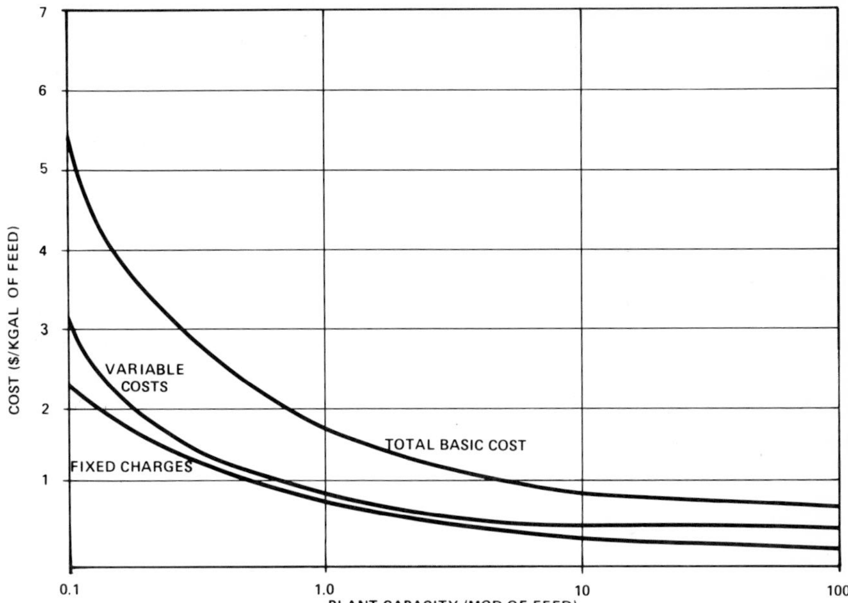

Fig. 18 Total basic ED cost for plants processing feeds of high concentration.

from dilute solutions. Unlike RO or ED, UF does not separate ions from aqueous wastes, but removes colloids, macromolecules, or fine suspended matter. As a result, costs are influenced not by TDS, but by the concentration of solids or macromolecules to be removed. As explained in Chapter 11, a high-solids loading in the feed will usually decrease the permissible flow rate of permeate through the membrane. That is, it will increase the membrane area required for a given feed rate to the plant. As an example of low-solids loading service, UF plants are used to remove traces of suspended solids from deionized water intended for use in the electronics industry. In the removal of suspended solids from municipal water supplies, a slightly higher solids loading is encountered in the feed, but still permits a high permeation rate. At the other end of the spectrum, the rinse water from a textile sizing plant will contain as much as 1 wt.% of polyvinyl alcohol. For such feeds, circulation rates *past* the membranes must be higher and permeation *through* the membranes slower, entailing an increased investment in membranes and pumps.

As in the case of RO and ED, there is a pronounced decrease in the installed cost as plant size increases. For large installations, however, the investment per unit of throughput levels off.

Ultrafiltration resembles RO in that its energy requirement is supplied as pumping power. The feed pressure for UF is only a fraction of that employed in RO, but the recirculation-to-feed ratio is generally great, resulting in a high power consumption.

B. PLANT INVESTMENT

Figure 19 presents the purchase price per gallon per day of feed for a UF plant for aqueous wastes, and Fig. 20 presents the total plant investment. In each figure, the lower curve applies to a feed with a low solids content and the upper to a feed of high solids content. Because generally the flux (flow rate *through* the membrane) decreases with increasing concentration of the feed stream, the reason for the high plant cost for wastes of high-solids loading is apparent.

C. OPERATING AND MAINTENANCE COSTS

1. *Fixed Power Consumption*

Chapter 11 emphasized the importance of minimizing the detrimental effects of the gel layer; that is, the stagnant, high-concentration layer

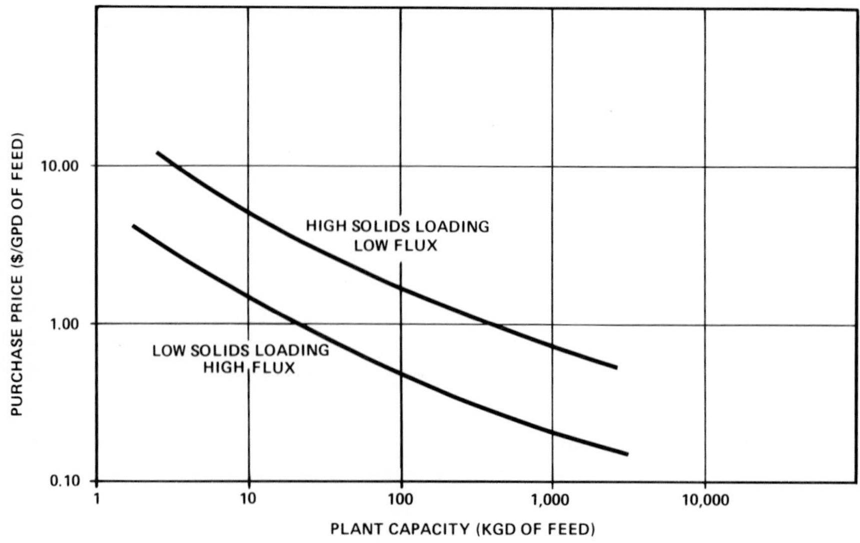

Fig. 19 Purchase price for UF plants as a function of plant capacity.

adjacent to the membrane surface. Control of this problem is achieved by rapid circulation of the feed stream *past* the membrane. The ratio of recycle to feed is quite high, reaching values of 50 : 1 for difficult feeds.

The dependence of pumping power cost on the recycle-to-feed ratio is shown in Fig. 21. Two curves are presented, one for a rate of 2 cents per kWh, the other for 4 cents per kWh. Values can be interpolated for intermediate rates. For the variable cost curves below, a rate of 2.5 cents per kWh was assumed.

2. *Labor*

Ultrafiltration plants are usually well automated. They require little operating labor. With the exception of units that remove suspended matter from comparatively pure water, however, UF systems require frequent cleaning, particularly when the feed consists of industrial waste streams. Consequently, experience indicates that small plants (less than 100,000 gpd of feed) will require the full time attention of one operator plus 40% of the time of a maintenance person on each shift. For large plants (more than 500,000 gpd), labor cost calculations are based on 40% of one operator's time plus 15% of the one maintenance person's time *per*

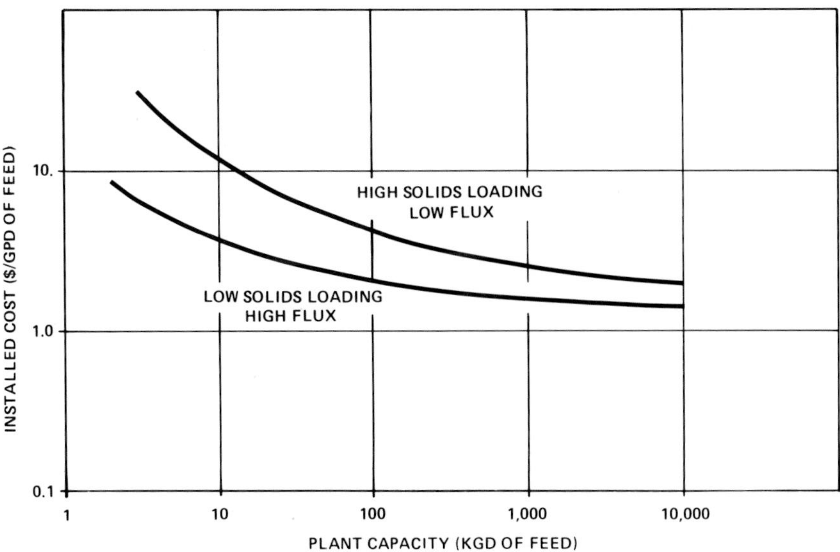

Fig. 20 Installed cost of UF plants versus plant capacity under various service conditions.

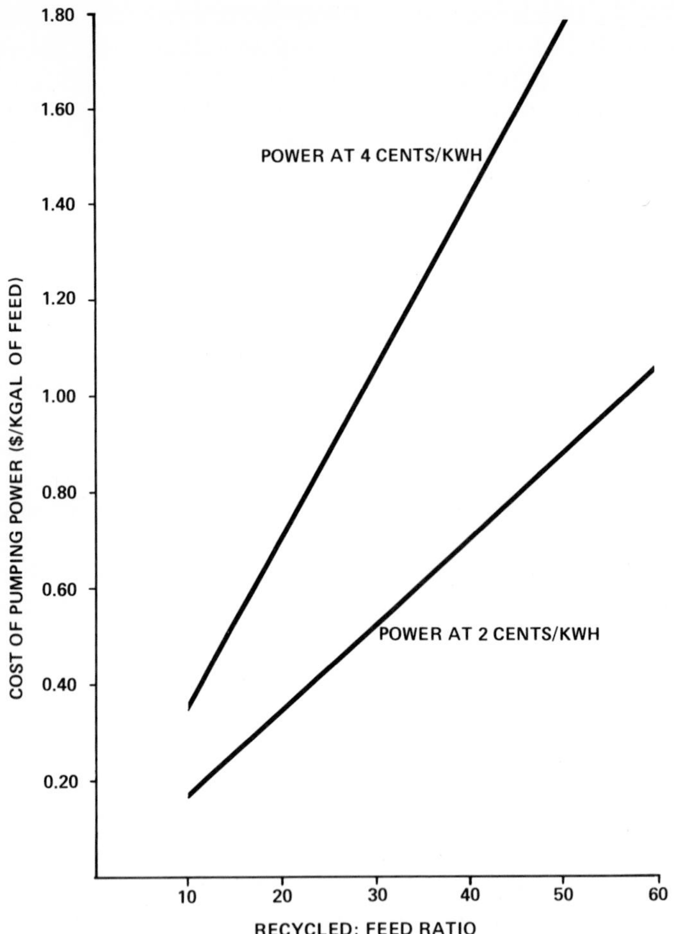

Fig. 21 Cost of power for UF processing as affected by operating conditions.

100,000 gpd of capacity per shift. The resulting labor costs per kilogallon of feed are plotted in Fig. 22.

3. Maintenance Materials

Ultrafiltration resembles the other membrane processes in its heavy reliance on pumps. The repair or replacement of pump components constitutes a good fraction of maintenance material costs.

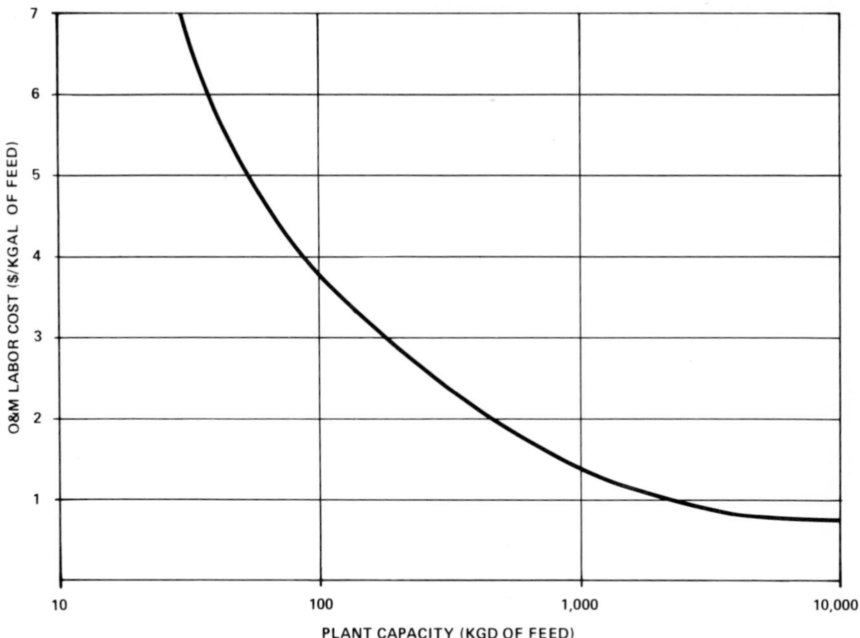

Fig. 22 O&M labor costs for UF plants.

A second large cost item is the cost of cleaning chemicals. Some feeds permit prolonged plant operation without membrane cleaning. On the other hand, some food products, of which cheese whey is the most notorious, cause serious fouling. Cleaning is accomplished by circulating enzymes and scale dispersants through the plant (Testa, 1977, private communication; Tidball, 1976, private communication; also see Chapter 11). Under severe conditions, cleaning may be required at least once a week. It would be difficult to cover all the possible cleaning techniques in this chapter.

Consequently, the curves in Fig. 23 were derived by applying an annual charge amounting to 5% of the plant investment.

4. *Membrane Replacement*

Aside from accidental damage resulting from scale formation or bacterial attack, the UF membranes should have a service life comparable to that of membranes in an RO plant. The UF membranes are subjected to a

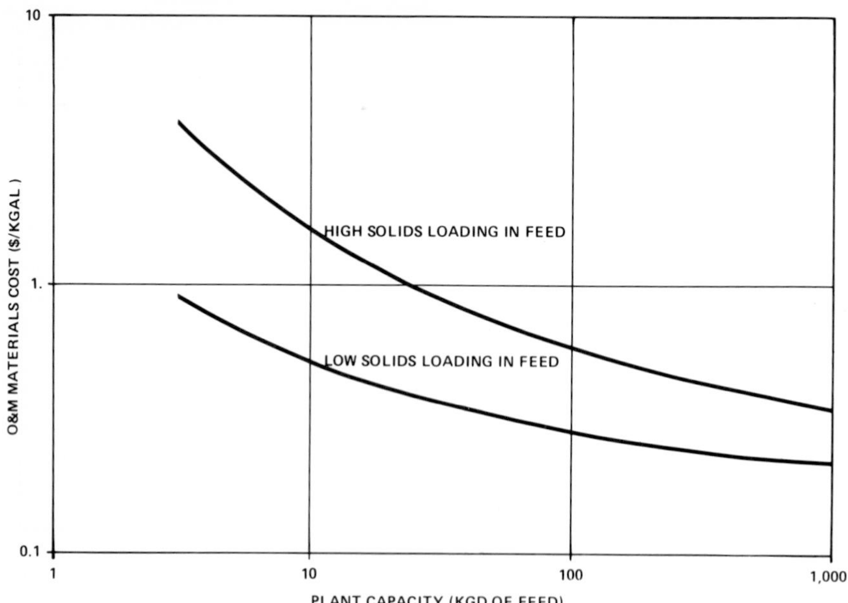

Fig. 23 O&M material costs for UF plants.

similar flux decay, commonly attributed to compaction. (Eykamp, 1977, private communication). For less-demanding applications, such as the removal of minute amounts of solids from water streams, performance is similar to that of RO membranes. For such conditions, the cost curves in Fig. 6 provide useful guidance.

For more severe applications, it is doubtful that manufacturers would offer a warranty on membranes without thorough preliminary pilot-plant testing. The membrane replacement cost can then be determined on the basis of the service warranty resulting from the pilot-plant tests.

5. *Total Operating Cost*

Figures 24 and 25 present the cost components and total basic cost for UF processing of feeds with low- and high-solids loading, respectively. To establish the total operating costs, a value must be added to the total basic cost to account for membrane replacement.

It is interesting to note that for UF, variable costs contribute much more to the total basic cost than do fixed charges. This results from the high-solids loading of waste streams generally delivered to the UF plant,

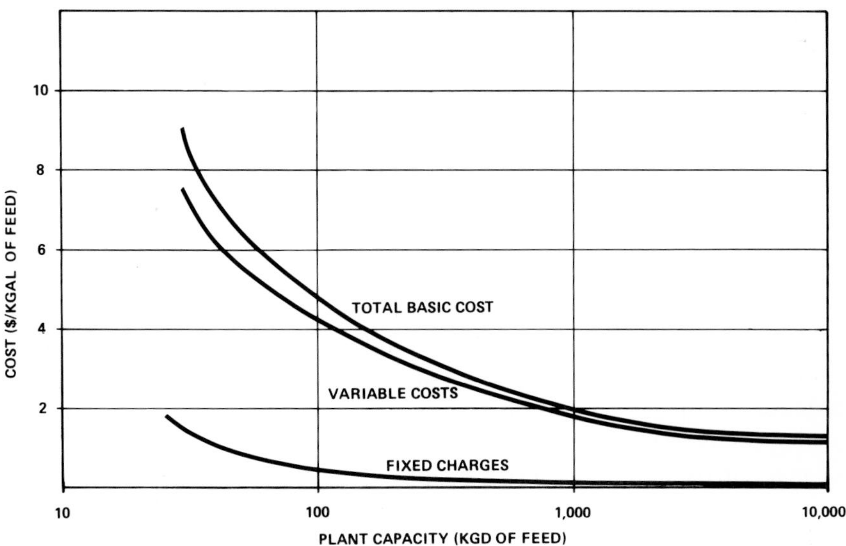

Fig. 24 Total basic UF cost for feeds with low solids loading.

the resulting increase in pumping power, and more frequent membrane cleaning.

V. Pretreatment

The pretreatment steps required for an aqueous waste stream prior to delivery to the membrane plant are similar to those used for brackish water or seawater feeds. These are primarily clarification, softening, removal of iron and manganese, filtration, chlorination, and the reduction or removal of residual chlorine. The principles involved are discussed in Chapter 11, and the economics are presented in the first section of Chapter 13. Two process steps require particular attention, namely, destruction of organic matter and removal of colloidal and suspended material. Such contaminants are particularly troublesome in wastes such as municipal sewage. After adequate digestion (anaerobic and/or aerobic), residual organic matter is removed by activated carbon. Colloidal and suspended matter are removed in settlers and clarifiers with the addition of coagulants if required. A final sand or mixed-media filtration step is necessary to insure freedom from fine particles that are carried over from the clarifier. Some wastes require special pretreatment steps. For example, cutting-oil wastes must be subjected to skimming and filtration before being concentrated in a membrane plant.

A. ACTIVATED CARBON

Granular activated carbon serves a double purpose: It acts as a filter to remove suspended and colloidal matter from the feed to the membrane plant. In addition, it provides a large surface for the growth of bacteria colonies that digest the organic matter in the feed. A bed of carbon, generally 5–10 feet in depth, is placed on a suitable support in the bottom of the vessel. The vessel is usually constructed of steel and protected against corrosion by an internal coating of plastic or rubber.

In downflow operation, the feed is passed down through the carbon bed until the pressure drop becomes excessive, at which time the bed is backwashed with clean water at an upflow rate adequate to fluidize the bed. Care must be taken to avoid excessive flow of wash water, which would result in loss of carbon. As an alternative, the bed may be operated in upflow mode, thereby eliminating the need for frequent backwashing.

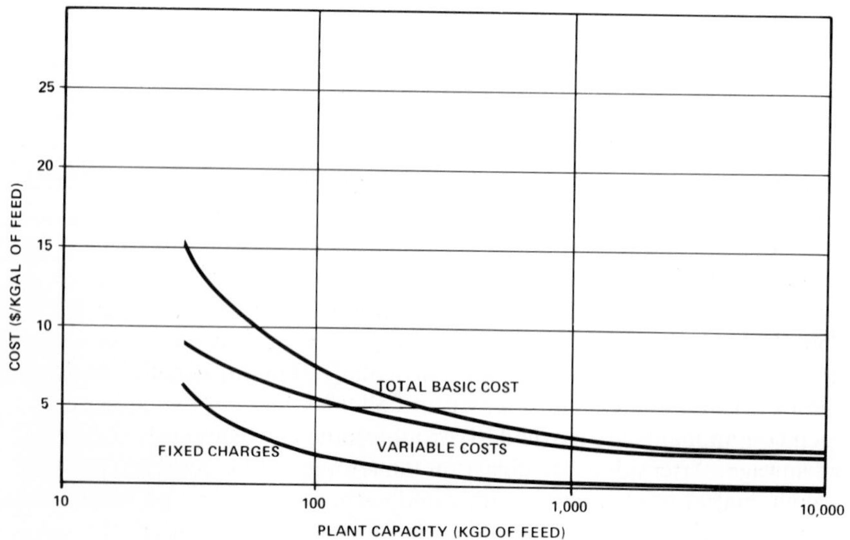

Fig. 25 Total basic UF cost for feeds with high-solids loading.

The upflow rate must be sufficient to just suspend the bed. When washing is required, the feed stream is replaced by clean water and the flow rate is increased until the bed is fully fluidized. Upflow operation requires careful operation to avoid hydraulic upsets and consequent loss of the carbon charge.

Figure 26 plots plant investment as a function of plant capacity. The curve represents the sum of the installed cost of the activated-carbon system, plus the initial carbon charge. Fixed charges, calculated on the same basis as those for the membrane plants, are plotted in Fig. 27. Power costs for this process are negligible. Also, labor costs can be neglected in a properly operated activated-carbon system. Maintenance costs are approximated by an annual charge of 5% of the equipment cost only (*not* including the carbon-charge cost). The largest contribution to operating costs results from carbon loss. Excessive flow rates in backwash will result in high loss rates. Even in a well-operated system, however, the size of the carbon granules is reduced by attrition to the point where the fines are ultimately washed out of the unit. The variable costs in Fig. 27 contain an allowance for carbon loss amounting to 1% of the total bed volume per month. The total O and M costs represent the sum of the fixed charges and variable costs.

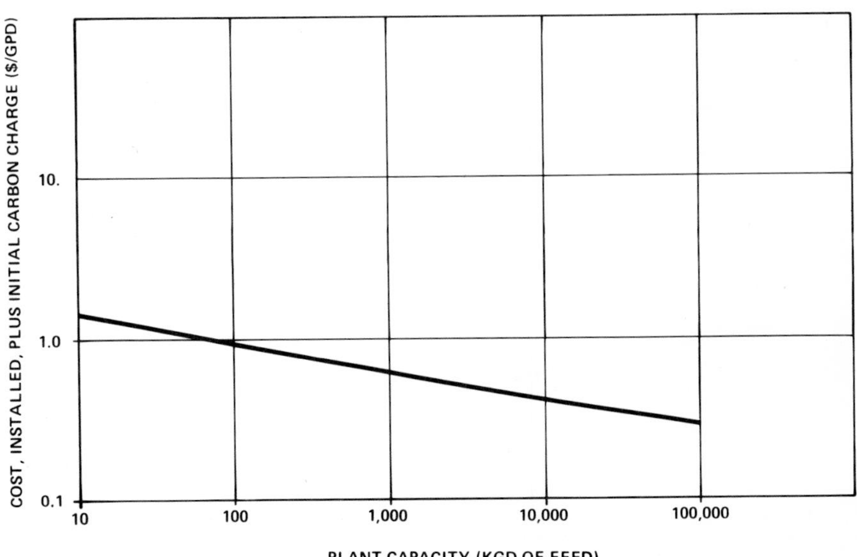

Fig. 26 Installed cost of granular activated-carbon system.

B. SOFTENING

The softening of an aqueous waste feed stream prior to membrane processing is similar to the softening required by a brackish feed. The economics of the latter are covered in the first section of this chapter. It is possible to substitute *powdered* activated carbon, added to the softener, for the granular-carbon pretreatment. This is a costly procedure, however, because the powdered carbon sells for double the price. In addition, the carbon leaving the softener in the lime sludge is discharged to waste.

C. SAND FILTRATION

The use of sand filtration to remove solids carryover after softening was discussed in the first section of this chapter. Even when no softening is required, however, filtration may be necessary to remove particulates from the aqueous waste.

A sand filter can be operated at rates of as high as 4 gpm per square foot. For the purpose of this analysis, the following assumptions have

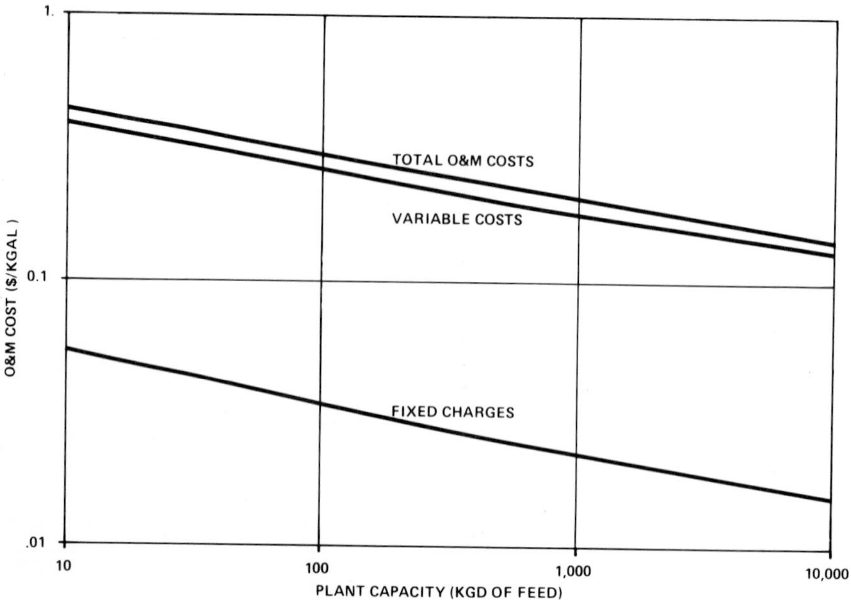

Fig. 27 Total O&M cost for activated-carbon treatment.

been made: a filtration rate of 3 gpm per sq. ft., mean pressure drop during filtration of 15 psi, backwash of 30 minutes every 24 hr and, backwash pressure drop of 25 psi.

Figure 28 summarizes the O and M costs for filtration. It is sometimes necessary to add a "conditioner" to the stream prior to sand filtration to agglomerate the particles of suspended matter (Clark, 1977, private communication). The price of the conditioner contributes little to the overall cost of filtration. The variable costs, comprising labor, power, and materials, constitute a small fraction of total operating costs. Fixed charges are responsible for the larger fraction. As in most of the processes involved in waste pretreatment, filtration costs are strongly influenced by plant size, and are quite moderate for plants in the mgd range.

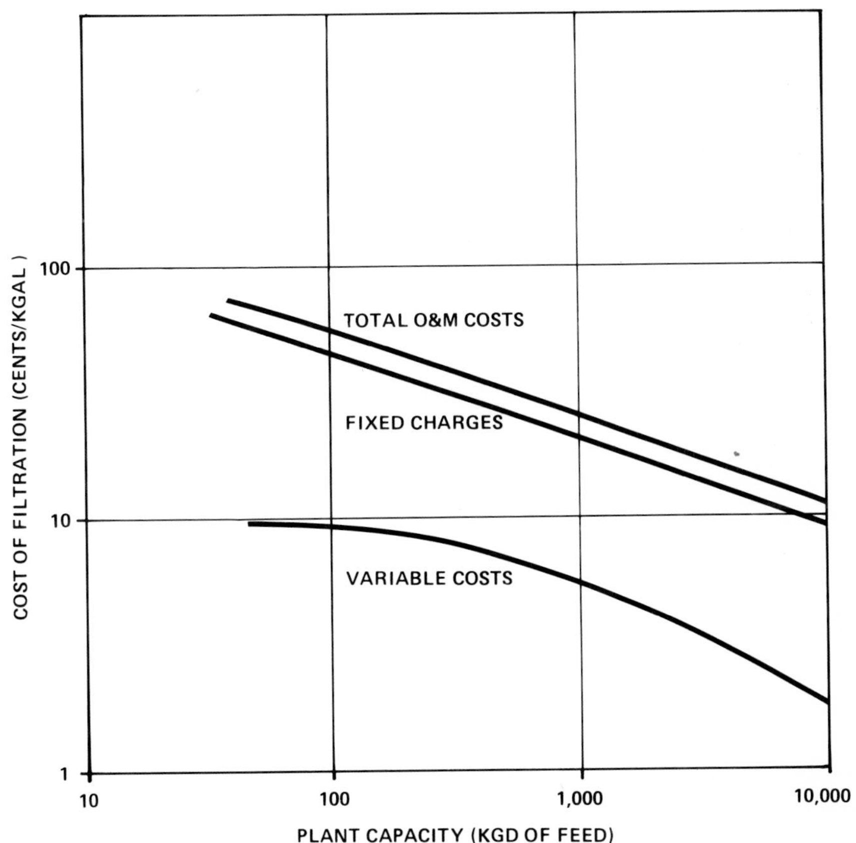

Fig. 28 O&M cost for sand filtration.

VI. Conclusion

The cost of treating an aqueous waste stream is the sum of pretreatment, membrane plant, and posttreatment costs. For each of these cost components, adjustments must be made to account for the feed composition, required concentrate concentration, and the target purity of the recovered water. Finally, an allowance must be included to cover the ultimate disposal of the concentrated waste stream if the concentrate is not recovered for sale or reuse.

References

The CE Plant Cost Index. *Chem. Eng.,* April 10, 1967, **74**(8), 197–198.

Dupee, T., and Andrews, W. (1977). Presented at the Fifth Annual Conference of the National Water Supply Improvement Association, San Diego, California, July, 1977.

ENR Building Cost Index appears regularly in the Engineering News Record.

Langelier, W. F. (1936). *J. Am. Water Works Assoc.* **28,** 1500.

Langelier, W. F. (1946). *J. Am. Water Works Assoc.* **38,** 169.

The Marshall and Swift Equipment Cost Index. *Chem. Eng.,* November 1947, **54,** 124–126 and also April 28, 1975, **182**(9).

Parker, C. L. (1975). *Pollut. Eng.* Nov. 32–37.

Richardson, (1977–1978). "Process Plant Construction Estimating Standards," Volume IV. Richardson Engineering Services, Solana Beach, CA.

Taylor, I. G., and Haugseth, L. A. (1977). *Ind. Water Eng.* **14**(2), 17.

Vetter, O. J. (1972). *J. Pet. Technol.* **24,** 997–1006.

Index

U

V

W